AF213495

ASTRONOMY AND ASTROPHYSICS LIBRARY

Series Editors:

G. Börner, Garching, Germany
A. Burkert, München, Germany
W. B. Burton, Charlottesville, VA, USA and
Leiden, The Netherlands
A. Coustenis, Meudon, France
M. A. Dopita, Canberra, Australia
B. Leibundgut, Garching, Germany
A. Maeder, Sauverny, Switzerland
P. Schneider, Bonn, Germany
V. Trimble, College Park, MD, and Irvine, CA, USA

For further volumes:
http://www.springer.com/series/848

Norbert S. Schulz

The Formation and Early Evolution of Stars

From Dust to Stars and Planets

Second Edition

Springer

Norbert S. Schulz
Massachusetts Inst. Technology
Kavli Institute for Astrophysics and Space Research
Cambridge, Massachusetts
USA

ISSN 0941-7834
ISBN 978-3-642-23925-0 ISBN 978-3-642-23926-7 (eBook)
DOI 10.1007/978-3-642-23926-7
Springer Heidelberg New York Dordrecht London

Library of Congress Control Number: 2012938880

© Springer-Verlag Berlin Heidelberg 2012, 2005
This work is subject to copyright. All rights are reserved by the Publisher, whether the whole or part of the material is concerned, specifically the rights of translation, reprinting, reuse of illustrations, recitation, broadcasting, reproduction on microfilms or in any other physical way, and transmission or information storage and retrieval, electronic adaptation, computer software, or by similar or dissimilar methodology now known or hereafter developed. Exempted from this legal reservation are brief excerpts in connection with reviews or scholarly analysis or material supplied specifically for the purpose of being entered and executed on a computer system, for exclusive use by the purchaser of the work. Duplication of this publication or parts thereof is permitted only under the provisions of the Copyright Law of the Publisher's location, in its current version, and permission for use must always be obtained from Springer. Permissions for use may be obtained through RightsLink at the Copyright Clearance Center. Violations are liable to prosecution under the respective Copyright Law.
The use of general descriptive names, registered names, trademarks, service marks, etc. in this publication does not imply, even in the absence of a specific statement, that such names are exempt from the relevant protective laws and regulations and therefore free for general use.
While the advice and information in this book are believed to be true and accurate at the date of publication, neither the authors nor the editors nor the publisher can accept any legal responsibility for any errors or omissions that may be made. The publisher makes no warranty, express or implied, with respect to the material contained herein.

Cover figure: The large-scale view of the giant hydrogen cloud and starforming region IC 1396 on the front cover was observed in the summer of 2004 from Fremont Peak State Park in California. The composite digital color image recorded emission from sulfur (red), hydrogen (green), and oxygen (blue). Observation and image processing by R. Crisp and reproduced here with his permission. From: http://www.narrowbandimaging.com.

Printed on acid-free paper

Springer is part of Springer Science+Business Media (www.springer.com)

To Erin, my family, and friends

Preface to the Second Edition

Half a decade after the first publication of *From Dust to Stars* there has been rapid and enormous progress in the field of stellar formation and evolution warranting a second edition. After a first review it became clear that the vast majority of the materials covered in the first edition remain very much valid; however, the need for many updates and extensions became obvious. In that respect the basic structure and content of the book has remained intact and existing chapters have received updates at various levels necessary. Three chapters have been added covering multiplicity in star formation (Chapter 7), massive star formation (Chapter 9), and a summary on the detection of exoplanets and basic planet formation (Chapter 13). Two chapters have received major additions, others just updates and minor revisions. The overall changes in the framework of the publication also suggested a switch in title to *The Formation and Early Evolution of Stars* with the subtitle *From Dust to Stars and Planets* to also acknowledge the extension towards planet formation.

Chapter 3 on interstellar matter now treats some basic physics about heating and cooling processes, cosmic rays, and radiative properties of dust. The issue of the mass functions also has a first discussion in that chapter, but it will re-surface at various occasions throughout the book as it now has become a central topic in the field. One may also wonder why there is a distinction of the initial mass function in general and a cluster mass function, which many studies claim to be the same, but after a review of all the material, the jury is still out on that issue. The chapter also received a much-needed extension to a discussion of star formation rates in the Milky Way, in the Local Group, and on cosmological scales.

Chapters 4, 5 and 6 have received only minor updates which might come to many as a surprise, given the enormous amount of material that has been provided through the five years of cold *Spitzer* operations, through the continuation of *HST* and *Chandra*, and now with the availability of *Herschel*. However, for these three chapters, the fundamentals have not really changed and here some updates and selected references to these new studies seemed to suffice. Much of this material has made its way into the book in later chapters.

The first new chapter (Chapter 7) has fixed a major shortcoming of the first edition which did not include much on multiplicity in star formation.

While theory yet has to explain binary formation on most observed scales, concepts of fragmentation which support multiplicity have been developed. Clearly the observational account is present and is currently the focus of many studies.

Chapter 8 again deals with basic physical phenomena in star formation and here only a few additions have been made to complement existing topics such as the *layered accretion* model recognizing that a lot of current research on accretion has moved toward the evolution of proto-planetary systems. The chapter has also remained relevant primarily for low-mass star formation as the following Chapter 9 has now added more topics related to massive stars. Massive star formation has been one of the most relevant topics in the field for the last decade and, at the time the first edition was published, too much was still in flow for a reliable summary. The second edition now fills this void with a review covering the formation of massive stars over a wide range of issues from a description of various observed early stages to massive star formation concepts, to ideas about the formation of the first stars.

Chapter 10 now describes the contribution of X-ray astronomy and here updates also have been extensive. The topic more than ever deserves its own chapter as this wavelength band is still widely ignored by many in star formation research. And yet the contributions, specifically with the big observatories *Chandra* and *XMM-Newton* still available to the community, are growing. What is also acknowledged more than ever is the fact that high-energy radiation plays a crucial role in the transformation of gas accretion to planet-forming disks. The accounts of active star forming regions are growing and so is the notion that it is never a single era event but it happens hierarchically where regions of collapsing stars are embedded in regions that are already actively forming planets. Here the *Chandra Carina Complex Project* proves to be an instructive study.

Extensive surveys by *Spitzer* such as *GLIMPSE* or the legacy program *Core to Disk* contribute vastly to our understanding of the evolution of protoplanetary nebulae and on several occasions there are references to these programs. In the first edition the transition from accretion towards planet formation has only been presented as an investigation of the Sun's origins and this is now Chapter 12. The discovery of so many exoplanets from the ground and now of even more from space with *Kepler* has led to increased efforts in the community, not only to search for planets but, to study the formation and early evolution of planetary systems. Chapter 13 thus has been added to provide a most actual account of exoplanetary research and provide some basic introduction into the formation of terrestrial and giant planets.

To reiterate what has been emphasized in the preface to the first edition for this second edition, the book includes the contributions of many scientists and the reference list grew to over 1,300 entries. And again many have helped me in the form of discussions, presentations, and reviews. My thanks for the first edition also extend to this second edition. Specific thanks go to M. Krumholz (Santa Cruz), J. H. Kastner (RIT), J. Winn, and P. Ford (both MIT) for scientific advice on the new extended material, and best of all for pointing out most recent developments I was not yet aware of. Much appreciation also goes to G. Allen and my fiancée Erin for very helpful proof-reading, N. Shen for her efforts on the bibliography, and to Springer for valuable comments to improve on the overall content. Final thanks go to Clive Horwood, whom I wish a happy life after retirement, and John Mason for getting this second edition going, and Ramon Khanna for managing its placement into Springer's Astronomy and Astrophysics library.

This revision also is very environmentally friendly owing to the use of modern cinema screen technology – most of the material and revised sections could be reviewed on screen and were pretty much paper- and toner-less. In fact, the only paper print of this manuscript came into existence for the copy-edit process at the very end of the process. All of the referred manuscripts were read and stored electronically saving many precious trees.

Massachusetts *Norbert S. Schulz*

Preface to the First Edition

The formation and the early evolution of stars is one of many intriguing aspects of astronomy and astrophysics. In the first half of the last century great strides were made in understanding the many aspects of stellar evolution, whereas answers to the question about the origins of stars usually remained much in the dark. The last decades, though, changed this predicament and produced a wealth of information. Star formation and early evolution is today a well established and integral part of astrophysics.

While approaching the writing of this book, I began to reflect on some experiences during my first research projects in the field, and immediately remembered the obvious lack of a reference guide on star formation issues. There exist numerous review articles, publications in various scientific journals, and conference proceedings on the subject. Thus today one has to read hundreds of papers to capture the essence of a single aspect. An almost unlimited resource for many years has been the *Protostars and Planets* series. About every seven years a large number of scientists from all over the world contribute numerous reviews on most research topics. Volume III published in 1993 and volume IV in 2,000 combined approximately 3,000 pages of review articles. For scientists who are seriously involved in star formation research they clearly are a 'must have' on the bookshelf – but then again, it is still 3,000 pages to read.

To date, there are not many monographs in print that highlight the physics of star formation. Noteworthy exception within the last ten years certainly is L. Hartmann's 2nd edition of *Accretion Processes in Star Formation* published in 2000. Also a few lecture notes appeared, including the *Physics of Star Formation in Galaxies* by F. Palla and H. Zinnecker, published in 2002, and the *Star Formation and Techniques in IR and mm-Wave Astronomy*, by T. Ray and S. V. W. Beckwith, published in 1994. These books are highly recommended resources. The scope, though, has to be expanded and should include many more modern aspects such as the properties of the interstellar medium, turbulence in star formation, high-energy emissions and properties of star forming regions, to name a few.

The formation and early evolution of stars today is a faster growing field of research than ever before and constitutes a frontier field of astronomy. It is not so much that the news of today will be out-dated tomorrow, but the appearance of more powerful and sensitive observatories as well as data-analysis techniques within the last 30 years provided not only a wider access to the electromagnetic spectrum, but is now constantly producing new and improved insights with astonishing details. For someone like me, for example, who, more than a decade ago, entered the field from the today still novel perspective of X-ray astronomy, it was somewhat difficult to see the true power of X-ray data with respect to star formation research. At the time there was no cohesive reference book available which summarized the current understanding of the field. The last ten years provided me with sufficient experience to generate such a summary for the astronomy community.

The information in this book contains observations, calculations, and results from over 900 papers and reviews, the majority of which where published within the last ten years. There are, of course, many more publications in the field and I had to make a biased selection. For any omissions in this respect I apologize. The information is very condensed and emphasis has been put on simple presentations. Specifically, most figures are illustrations rather than the original data plots. This was intended in order to have a consistent appearance throughout the book, but also to encourage the reader to look up the corresponding publication in the case of further interest. It is not a textbook for beginners, even though it contains enough explanatory diagrams and images that may encourage young students to become more engaged. The book also abstains from lengthy derivations and focuses more on the presentation of facts and definitions. In this respect it is very descriptive and a large amount of results are put into a common perspective. Though the amount of information may sometimes seem overwhelming, it is assured that this book will serve as a reference for a long time to come. I sincerely believe that a wide audience will find this book useful and attractive.

There are many friends, fellow scientists, and colleagues here at MIT and elsewhere who I am indebted to in many ways. My gratitude goes to Joel Kastner at the Rochester Institute of Technology, who from the beginning of the project reviewed my efforts and offered well-placed criticism and suggestions. Many others, within and outside the star formation research community, offered reviews and useful comments about the manuscript or parts of it. H. Zinnecker (who sacrificed his Christmas vacation) and R. Klessen (both AIP) and T. Preibisch (MPI for Radioastronomy) reviewed the content of the first manuscript. G. Allen, D. Dewey, P. Wojdowski, J. Davis, and D. P. Huenemoerder (all MIT) proofread all or parts of the book and provided many detailed and much needed comments. Thanks also goes to F. Palla (Firenze), D. Hollenbach (AMES), and H. Yorke (JPL) for suggestions and the provision of updated material. My special appreciation also goes to my longtime friend Annie Fléche who, over several months, weeded out most of the improper grammar and language. It seems though, that, except from the view of Oxford English purists, I am not that hopeless a case.

Many scientists within and close to the field of star formation were willing to review my original book proposal. These included, besides some of the contributors mentioned above: E. Feigelson (PSU), C. Clarke (Cambridge University),

T. Montmerle (Grenoble), J. H. M. M. Schmitt (Sternwarte Hamburg), and S. Rappaport (MIT). Their service is much appreciated.

The content of the book includes the contributions of a vast number of scientists and my own scientific inputs which, though I think they are great, are outweighed many times by all those who contributed. In this respect I am only the messenger. My thanks go to all who discussed the science with me on many meetings, conferences, and other occasions as well as to all the researchers contributing to the field. This includes all those who granted permission to include some of their published figures. It is needless to say that any remaining errors and misconceptions that may still be hidden somewhere (I hope not) are clearly my responsibility.

In this respect I also want to thank C. Horwood and his team at Springer-Praxis publishing for patiently making this book happen. Specifically John Mason provided many suggestions which improved the book in content and style.

Finally I want to thank all friends and colleagues who supported me during the project. This specifically includes C. Canizares (MIT), who, in the beginning, encouraged me to go through with the project. It includes also all members of the MIT Chandra and HETG teams for encouragements in many ways. My thanks also goes to the MIT Center for Space Research for letting me use many local resources – tons of recycling paper, printer toner and a brand new Mac G5 with cinema screen. Finally I need to thank the Chandra X-ray Center for tolerating my mental absence during many meetings and seminars.

One last remark: one manuscript reader and colleague asked me to emphasize somewhere in the book, how the modern view of stellar formation is no longer a boring story of the collapse of spheres, but includes exciting features such as accretion disks, outflows, magnetic fields, and jets. Well, I just did – and could not agree more.

Massachusetts *Norbert S. Schulz*

[illegible]

Contents

Chapter 1
About the Book

The study of the formation and early evolution of stars has been an ever growing part of astrophysical research. Traditionally, often in the shadow of its big brothers (i.e., the study of stellar structure, stellar atmospheres and the structure of galaxies), it has become evident that early stellar evolution research contributes essentially to these classical fields. There are a few new items on the list of traditional astrophysical studies, some of which are more related to features known from the extreme late stages of stellar evolution such as accretion, outflows, disks, and large-scale turbulence. In recent years the gap between theory and observations in massive star formation has begun to close as much attention has been focussed specifically on this topic. Newly emerging are protoplanetary systems, the formation and early evolution of planets, and related the detection of extrasolar planets. The field today requires the most powerful and sensitive instruments and telescopes mankind is able to provide. It utilizes the fastest computers and memory devices to simulate jets, evolving disks and outflows. A network of researchers all over the planet invest resources and time to contribute to the steadily growing knowledge about the origins of stars and planets and ultimately the birth of the Solar System.

The Formation and Early Evolution of Stars introduces the reader to a world of dense clouds and cores, stellar nurseries, the lives of young stellar objects and their interaction with the interstellar medium, and the formation and evolution of protoplanetary systems as we know it today. It attempts to provide a broad overview of the major topics in star formation and early stellar evolution. As such it now also extends its scope towards massive star formation and the development of towards planets. The book describes the complex physical processes involved both in the creation of stars and planets and developments during their young lives. It illustrates how these processes reveal themselves from radio wavelengths to high-energy X-rays and γ-rays, with special reference towards high-energy signatures. Several sections are also devoted to key analysis techniques which demonstrate how modern research in this field is pursued.

The sub-title of the book emphasizes the role of dust throughout star and planet formation and, as the reader will realize early on, each chapter demonstrates that dust is present at all phases of evolution. Of course, as catchy as the title sounds,

N.S. Schulz, *The Formation and Early Evolution of Stars*, Astronomy and Astrophysics Library, DOI 10.1007/978-3-642-23926-7_1, © Springer-Verlag Berlin Heidelberg 2012

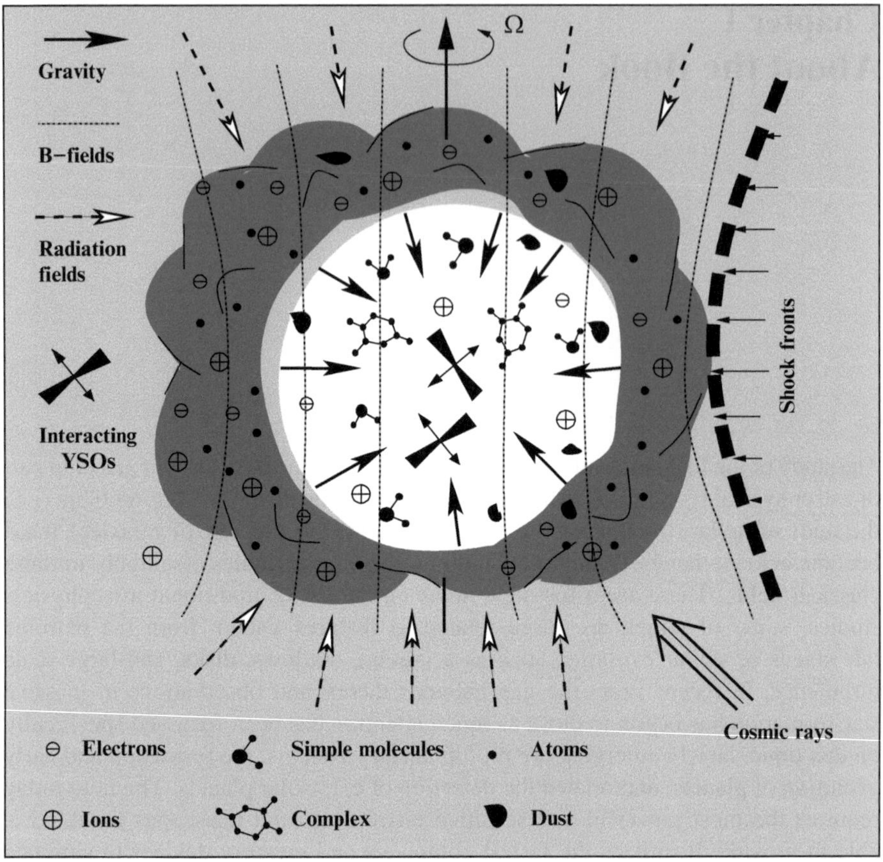

Fig. 1.1 Star-forming environments are a turbulent mix of gas clouds, molecules and dust which interact under the influence of ionized material, magnetic fields, turbulence, and foremost gravity. In addition, cosmic rays, external radiation fields, and traveling shock waves add to the complexity. Advanced stages of evolution have to deal with the interaction with radiation and outflows from various generations of newly born stars

there are many other ingredients required for stellar formation. In fact, so many items contribute, one has to concede that the riddle of star formation is far from being solved. These items include gases and molecules at various densities, force fields such as gravity and magnetic fields, rotation, shocks and turbulence, neutral and ionized matter, hot plasmas, cosmic rays, and various types of radiation fields, all playing in the symphony of stellar creation (see Fig. 1.1). The development of protoplanetary disks towards planets also involves coagulation and settlement of dust, the formation of planetesimals, a violent path towards planet formation and materials as are now found throughout the Solar System. In a sense this huge variety of physical entities provided a dilemma while writing this book. On the one hand, it is necessary to understand the underlying physics to be able to properly describe

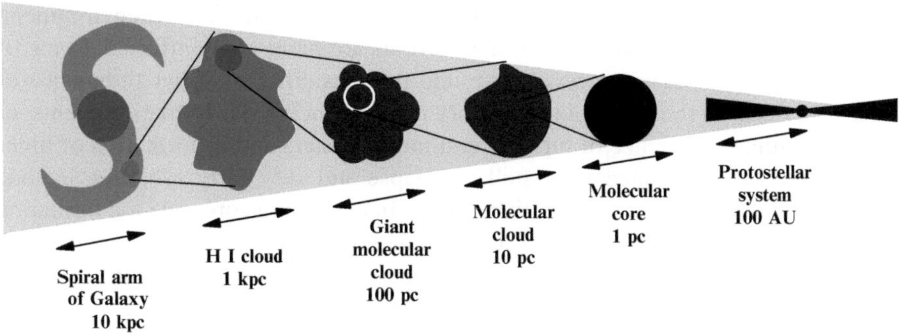

Fig. 1.2 Fragmentation is a multiscale process from the spiral arms of a galaxy down to protostars with characteristic lengths spanning from 10 kpc to 100 AU

the manifold of different processes involved. On the other hand there is a story to tell about the current views on star formation and the effort and resources required in the course of research. In this respect, the reader will find most of the needed basic physics and other information condensed into a few appendices with only a few practical equations and derivations in the main text.

The book has otherwise a very simple outline. In order to fully grasp the immense historic achievement that ultimately led to today's views, Chap. 2 offers an extensive historical perspective of research and developments spanning from ancient philosophies to the utilization of space-based observatories. This historical background specifically illustrates the necessity of the accumulation of basic physical laws and astrophysical facts over the course of centuries in order to be able to study the evolutionary aspects of stars. It also outlines the conquest of the electromagnetic spectrum that allows researchers today to look into all aspects of star-forming regions the sky offers to the observer.

Fragmentation is a quite fashionable concept in the current view of the structure of matter in the Galaxy and one that may well be valid throughout the entire universe. Matter in the Galaxy appears fragmented from the large scales of its spiral structure to the small scale of protostars, as is illustrated in Fig. 1.2. Chapter 3 deals with the distribution of matter in the Galaxy on large scales. This structure reveals itself as a huge recycling factory in which stars evolve and produce heavier elements in the main and last stages of their lives. During this entire life cycle matter is fed back into the medium to be further processed through the cycle. In progressing steps, starting with Chap. 3, the physical environments scale from the interstellar medium of the Galaxy to concepts of gravitational collapse in Chap. 5. Although there is yet a coherent picture of star formation to emerge, it seems that the main contributors and mechanisms have been identified. These will be described in some detail. Questions researchers face today relate to the definition of the circumstances under which these mechanisms regulate the star forming activity. As the reader will learn in Chap. 9 and Chap. 11, most stars do not form as isolated entities but in more or less large clusters. Although the basic physics of an assumed isolated stellar collapse will likely not notably change for a collapse in a cluster environment, mechanisms that eventually lead to these events are more likely affected. Some of today's leading discussions revolve around the feasibility of either turbulence or magnetic fields as the dominant regulatory mechanism for star formation events, or the predominant existence of binaries and multiple stars and/or formation of high-mass stars. Modern concepts of stellar collapse and very early evolution include observational facts about density distributions in collapsing cores and the formation of accretion disks, winds and other forms of mass outflow.

If there ever is a line to be drawn between the study of formation and early evolution of a low-mass protostar it likely has to be between Chap. 5 and Chap. 6. Such a division may arguably be artificial, but observational studies historically drew the line right there for a good reason. Even with the technologies available today it is still extremely difficult, if not impossible, to observe the creation of the protostar and its earliest period of growth. Phases between prestellar collapse and protostars remain obscured by impenetrable envelopes, which then become the objects of study. Consequently, very early protostellar evolution is the subject of theoretical concepts which then have to connect to the point of first visibility. These issues are mostly addressed in Chap. 6, which introduces the reader to various existing early evolution concepts, some of which are still highly controversial. Concepts include the birthline of stars, their class 0 to III classification based on their IR spectral energy distributions, and the ZAMS, as well as short descriptions of various young star phenomena. Figure 1.3 depicts a schematic view of the matter flow patterns around young stars. Angular momentum conservation and magnetic fields force inflowing matter into the formation of an accretion disk with the formation of jets and winds. The underlying physics of these phenomena is currently subject to intense study. Results from observations as well as computational studies are presented in Chap. 8, which goes on to emphasize star-disk magnetic configurations and the coronal activities of young stars. Much of the underlying physics for this chapter is presented in the first four Appendices.

Many young stars are now observed to be binary or multiple systems and accordingly the star formation process has to include the formation of multiple stars. To date little is known about how these mechanisms work other than in cases of extreme close and wide separations. Chapter 7 tries to provide some inputs on where research stands today and shows some examples of the variety of systems observed. In this context massive star formation theories might provide some further inputs in terms of how fragmentation shapes the final outcome of multiple systems and the formation of clusters through some more chaotic processes such as competitive accretion. At the end the jury is still out there and the issue of multiplicity is an open issue in star formation theory.

Observation and theory of massive star formation (Chap. 9) also has similar problems as, although progress has been made thoughout the first decade of the twenty-first century, there is no decisive common ground between the two. Monolithic core collapse and competitive accretion both have their merits; however, gravito-turbulent models as proposed for low-mass star formation still have to be considered as well. Here the observational account that started with *Spitzer* and

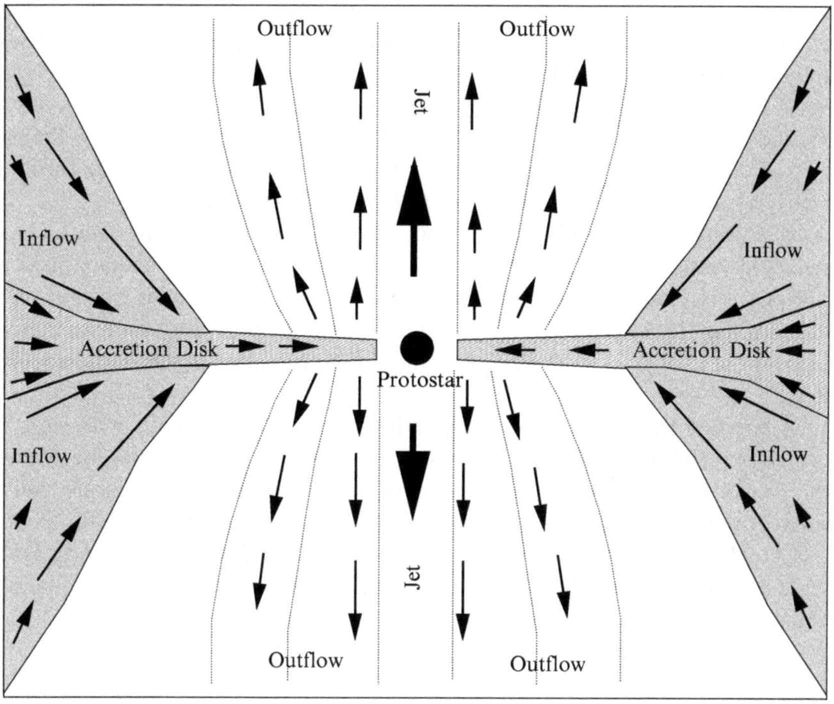

Fig. 1.3 Schematic view of matter flows expected in protostellar environments. The star still accretes from its primordial envelope, which feeds mass into an accretion disk with generally high accretion rates. How matter eventually reaches the protostar is relatively complex and unclear. In later phases the primordial envelope has gone and a star may still accrete matter out of the disk at very low rates. Specifically in early protostellar phases the system generates massive outflows, i.e., about 10% of the accreted material may by ejected through high-velocity winds. Some collimation may even be achieved as a result of acceleration in magnetic fields from the disk, resulting in jets

with now available potent ground capabilities combined with the ones in the near future should provide breakthroughs soon.

The treatment of stellar magnetic fields is necessary in anticipation of the review of high-energy signatures of young stars. Although throughout the entire book referrals to high-energy activity in stellar evolution are made, Chap. 10 specifically demonstrates how highly X-ray-active young stars are. The immediate environment of young stars is extremely hot and temperatures are orders of magnitude higher than in circumstellar envelopes of protostars and disks of young stars. X-ray astronomy has become a major part of the study of stellar evolution and in the light of new technologies it has evolved from the mere purpose of source detection to a detailed diagnostic tool to study young stellar objects.

The reader, up to this point, should now be familiar with many facts about the formation and evolution of young stars. The remaining Chaps., 11 and 12, then look back and reflect on two issues. Chap. 11 characterizes large star-forming regions containing not only the pre-stellar gaseous, dusty, and molecular clouds but also large ensembles and clusters of young stars at various young ages. Here properties of young embedded stellar clusters are presented, followed by an in-depth description of various types of star-forming regions, which include *Orion*, *ρ Ophiuchus*, *IC 1396*, and the large region of Carina. Second, an attempt is made to relate the early history of our Sun and the Solar System to the current knowledge of star formation and early evolution by investigating the *T Tauri* heritage of the Sun.

At this point, the ground work has been layed to investigate the fate of protoplanetary systems. While Chap. 12 emphasizes the fates of protoplanetary disks and a path possibly leading to the Solar Nebula from which our Solar System finally emerged, Chap. 13 adds the final piece to the puzzle by providing some insights on the account of exoplanetary systems and the actual formation of terrestrial and giant planets.

Last but not least, a series of appendices provide the reader with essential information. The first three appendices are devoted to important background physics covering gas dynamics, aspects of magnetohydrodynamics, as well as radiative transfer. Appendix D describes several examples of modern spectral-analysis techniques used in star formation research. The last three appendices then provide descriptions of abbreviations, instrumentation, and the physical quantities used throughout the book.

Chapter 2
Historical Background

It is a common perception that astronomy is one of the oldest occupations in the history of mankind. While this is probably true, ancient views contain very little about the origins of stars. Their everlasting presence in the night sky made stars widely used benchmarks for navigation. Though it always was and still is a spectacular event once a new light, a nova, a new star appears in the sky. Such new lights are either illuminated moving bodies within our Solar System, or a supernova and thus the death of a star, or some other phases in the late evolution of stars. Never is a normal star really born in these cases. The birth of a star always happens in the darkness of cosmic dust and is therefore not visible to human eyes (see Figure 2.1). In fact, when a newborn star finally becomes visible, it is already at the stage of kindergarten in terms of human growth. It takes the most modern of observational techniques and the entire accessible bandwidth of the electromagnetic spectrum to peek into the hatcheries of stars.

2.1 And There Was Light?

A historical introduction to stellar formation is strictly limited to the most recent time periods. Modern science does not recognize too many beliefs from ancient periods as facts. For example, timescales are specifically important for the physical mechanisms of the formation and early evolution of stars. Biblical records leave no doubt that the world was created by God in six days and the formation of the Sun and stars was a hard day's job. Allegorically speaking there is nothing wrong with that unless the attempt is made to match these biblical timescales with physically observed time spans. Then timescales from thousands to millions of years are relevant, whereas days and weeks hardly appear in this context. Today it is known that it takes about 100,000 years for a molecular cloud to collapse and more than many million years for most stars to contract enough to start hydrogen fusion, not to speak of creating solar systems and planets. One also realizes that planets take even longer to become habitable. In the case of the Earth it took billions of years.

N.S. Schulz, *The Formation and Early Evolution of Stars*, Astronomy and Astrophysics Library, DOI 10.1007/978-3-642-23926-7_2, © Springer-Verlag Berlin Heidelberg 2012

Fig. 2.1 This breathtaking image shows a newborn proto-Sun associated with the Herbig–Haro outflow object *HH 46* (see the bright loop-like strings pointing away from the star). The inset in the lower left corner shows the optical image of the same region. The proto-Sun with its outflow is hidden in a dense cloud. Credit: NASA/JPL/Caltec/A.Noriega-Crespo(SSC/Caltech)

The following few sections are an attempt to briefly summarize the road from ancient views to the point when the understanding of physical processes in connection with observations of stellar properties reached a level that permitted the first physical treatment of stellar formation and evolution. Though the presented material also introduces the reader to some basic facts of modern physics and astronomy, the emphasis of this chapter is not a substitute for introductory textbooks on physics and astronomy. The chapter resembles more an investigation of the developments and scientific milestones that led to the pursuit of modern star formation studies.

2.1.1 From Ptolemy to Newton

It took humankind until the dawn of the New Age to put basic pieces together and to accept proof over belief and superstition. The geocentric concept of Claudius Ptolemaeus, or Ptolemy, dominated the views of the world from ancient times to the 16th century. He walked the Earth approximately between the years 175 and 100 BC and, although he lived in the Egyptian city Alexandria, he was more a scientist of hellenistic origin and many of his views are based on the cosmological concepts of Aristotle (384 BC), a student of Plato. In his work 'Hypotheses of the

Planets' Ptolemy describes a system of the worlds where Earth as a sphere reigns at the very center of a concentric system of eight spheres containing the Moon, Mercury, Venus, the Sun, Mars, Jupiter, and Saturn. The eighth and last sphere belonged to the stars. They all had the same distance from earth and were fixed to the sphere. Constellations, as well as their size, consistency and color were eternal. The question about the structure of planets and stars was not pursued and the mystic element called 'ether' was introduced instead to fill the space within and between celestial entities – a concept that lasted until the 20th century. Sometimes stars were also referred to as 'crystalline'. For over 1,500 years Ptolemy's work was the main astronomical resource in Europe and the Orient.

After the medieval period, Earth resided uncontested at the center of the universe and the stars were still either lights fixed to the celestial sphere or little holes in a sphere surrounded by heaven's fire. How much the Ptolemaic scheme was imprinted into the most fundamental beliefs is shown in a picture from a bible print in the 16th century depicting the traditional Judeo-Christian view of the *Genesis*, the Bible's version of the creation in which God makes the Earth and the Cosmos in six days (see Fig. 2.2). Even after Nicolaus Copernicus published his famous book series *De Revolutionibus orbium coelestium libri sex* in 1543, which featured today's heliocentric system, Ptolemy's model prevailed for quite some time. The first indication that something could be missing in the Ptolemaic system came with the observation of a supernova in the constellation Cassiopeia by Tycho Brahe in 1572. Following Brahe's legacy it was at last Johannes Kepler with his publication *De Harmonice Mundi* in 1619 and Galileo Galilei's 'Il Saggiatore' from 1623 that not only placed the Sun as the center of the solar system but also established observations as a powerful means to oppose the clerical dogma.

This was not only a triumph of science, it had specific relevance from the standpoint of stellar evolution as it was realized that the Sun and the planets are one system. When Isaac Newton published his *Naturalis philosophiae principia mathematica* in 1687 the formal groundwork of celestial mechanics was laid.

2.1.2 Stars Far – Parallax

A remaining issue with Ptolemy's picture which posed quite a severe problem for the Copernican system was the fact that Ptolemy postulated stationary stars pinned to the celestial sphere at equal distance. If, however, Earth moves around the Sun one should be able to observe an apparent motion in the star's positions on the sky.

The only way out of the problem was to postulate that stars are so far away that the expected yearly displacement is too small to measure. In fact, the angle between two observations at two fixed positions should give the distance to the stars. Such an angle is called *parallax*. For quite a long time it seemed that Ptolemy's postulate would prevail as all attempts to find this angle were unsuccessful. It was a rocky road from E. Halley's discovery in 1718 that stars do have proper motions to the first successful measurement of the parallax of 61 Cygni by F. W. Bessel at

Fig. 2.2 Genesis view from Martin Luther's *Biblia*, published by Hans Lufft at Wittenberg in 1534. The impression (by Lucas Cranach) shows the Earth at the center and the Sun and stars in the *waters of the firmament* positioned at the inner edge of an outer sphere. Credit: from *Gestaltung religiöser Kunst im Unterricht*, Leipzig, Germany

a distance of 11.1 light years. The angle measured was only a fraction of an arc second (0.31") and represented the first high-precision parallax measurement. Most recently the astrometry satellite *Hipparcos*, launched in 1989, determined parallaxes

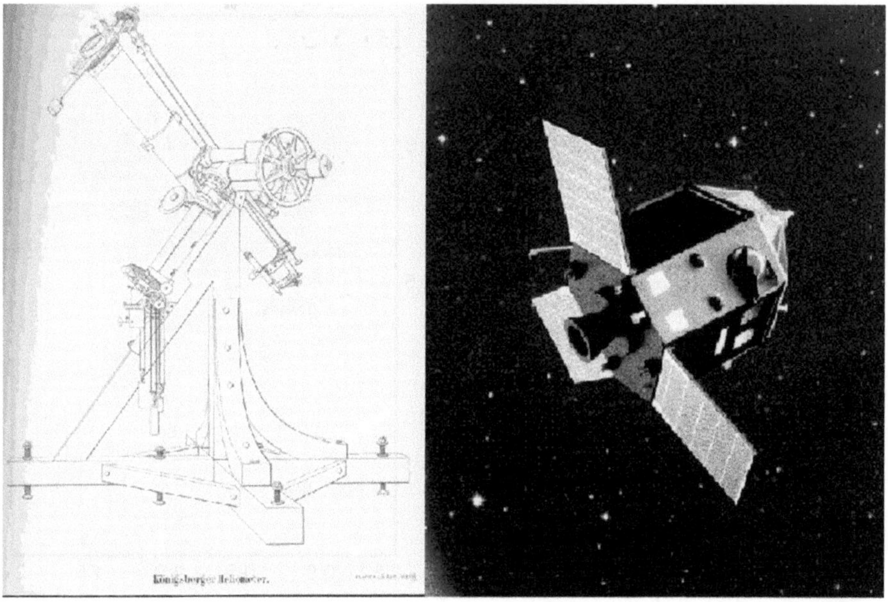

Fig. 2.3 (*left*) The Königsberger Heliometer Bessel installed in 1829 to perform parallax measurements with a resolution of 0.05 arc seconds. Credit: The Dudley Observatory, Drawing from the 1830s, Lith. Anst. v. J.G. Bach, Leipzig [5]. (*right*) An artist's impression of the astrometry satellite *Hipparcos* launched by ESA in 1989. The satellite allowed parallax measurements with 0.00097 arc seconds resolution. Credit: ESA/ESOC

of over 120,000 stars with a precision of 0.001 arc seconds. Data from satellites like *Hipparcos* are essential for today's astronomical research (see Fig. 2.3).

2.1.3 Stars Bright – Photometry

All astronomy preceeding the 20th century was related to the perception of the human eye. The 19th century marked a strong rise in the field of stellar photometry. About a hundred years before first attempts were made to define a scale for the brightness of stars, P. Bouguer published some of the earliest photometric measurements in 1729. He believed that the human eye was quite a good indicator of whether two objects have the same brightness and tested this by comparing the apparent brightness of the Moon to that of a standard candle flame. A more quantitative definition was introduced by N. Pogson in 1850. He defined a brightness logarithm on the basis of a decrease in brightness S by the relation

$$\frac{S_1}{S_2} = 10^{0.4(m_1 - m_2)} \tag{2.1}$$

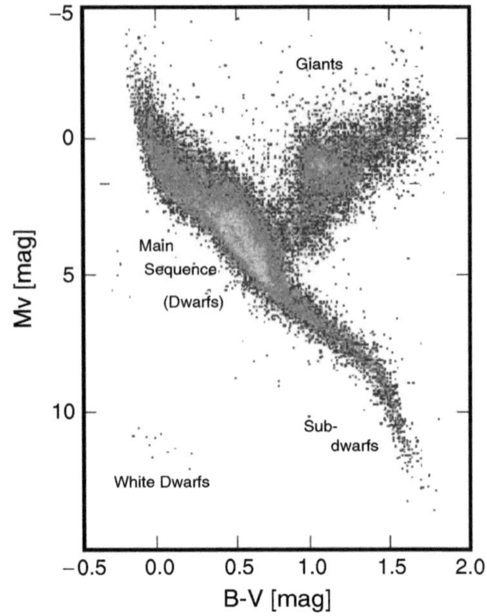

Fig. 2.4 Color-magnitude diagram for 41,704 single stars from the Hipparcos Catalogue with a color error of $\sigma_{B-V} \leq 0.05$ mag. The grayscale indicates the number of stars in a cell of 0.01 mag in color index $B - V$ and 0.05 mag in absolute magnitude M_V. The magnitude was limited to < -5 in the sample. Data from: ESA, 1997, The Hipparcos and Tycho Catalogues, ESA SP-1200

for each step $\Delta m = m_1 - m_2$, which he called 'magnitudo'. One of the greatest astronomical achievements in the 19th century was the publication of various photometric catalogs by F. Argelander, E. Schönfeld and E. C. Pickering containing a total of over 500,000 stars.

It was K. Schwarzschild [1] who opened the door into the 20th century with the creation of the first photographic catalog containing color indices, i.e., photographic minus visual brightness, of unprecedented quality. The key element was the recognition that the color index is a good indicator of the color and thus the temperature of a star. The use of photoelectric devices to perform photometry was first pursued in the early 20th century [2, 16, 17]. The UBV-band system developed in 1951 [18] determines magnitudes in three color bands, the ultraviolet band (U, ~ 3,500 Å), the blue band (B, ~ 4,000 Å), and the visual band (V, ~ 5,500 Å). Today photomultipliers are used to determine magnitudes with an accuracy of less than 0.01 mag and effective temperature measurements of stars to better than 1 percent [19]. Figure 2.4 shows a color-magnitude diagram of over 40,000 stars from data obtained by *Hipparcos*.

2.1.4 Star Light – Spectroscopy

The 19th century also marked, parallel to the development in photometry, the beginning of stellar spectroscopy. Newton had already studied the refraction of light using optical prisms and found out that white sunlight can be dispersed

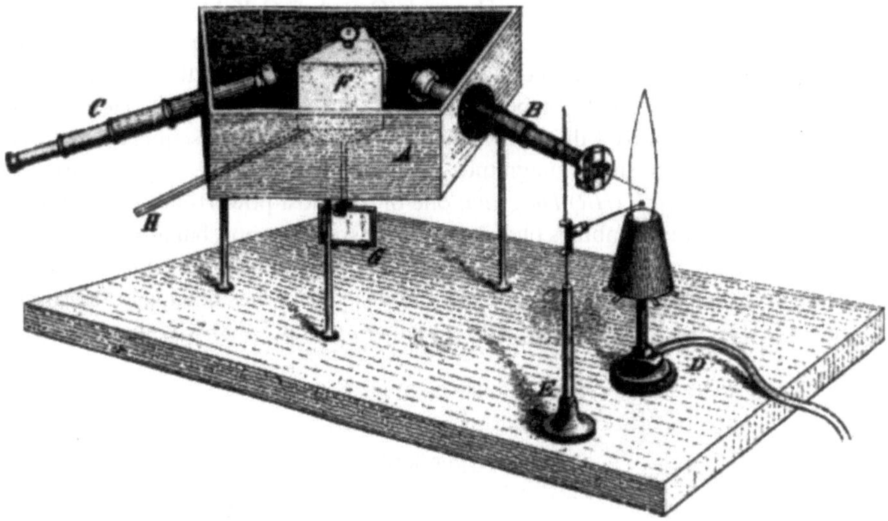

Fig. 2.5 Kirchhoff's and Bunsen's apparatus used for the observation of spectra. The gas flame was incinerated with different chemical elements. A rotating crystal (F) at the center diffracted the light from the flame, the diffracted light was then observed with a telescope (C). Credit: Kirchhoff and Bunsen [20]

into its colors from blue to red. However it was J. Fraunhofer in 1814 whose detection of dark lines in the spectrum of the Sun and similar lines in stars in 1823, represented a first step towards astrophysics. Not only was it remarkable that the Sun and stars showed similar spectra, but also that the strong lines in the solar spectrum indicated a chemical relation. After publishing the *Chemische Analyse durch Spektralbeobachtungen* in 1860, G. Kirchhoff and R. Bunsen established spectral analysis as an astrophysical tool (see Fig. 2.5).

This discovery sparked a range of activities which ultimately led to an understanding of radiative laws and the classification of stellar spectra against the one from the Sun. What was needed were laboratory measurements to identify elements with observed wavelengths, stellar spectra, and magnitudes and a theory of radiation. When pursuing his spectral analysis in the 1860s Kirchhoff realized that there must be a relation between the absorption and the emission of light. He also noted that various colors in the spectral band correspond to different temperatures – the basis of modern UBV photometry. This led him to the formulation of what is known today as Kirchhoff's law (see Appendix C) which now is one of the most fundamental laws in radiative theory. Kirchhoff assumed 'Hohlraum Strahlung', radiation from a hollow body in thermodynamical equilibrium, which had a spectral shape of what is today simply referred to as a blackbody spectrum (see Fig. A.2).

Vigorous studies were pursued at the Harvard Observatory under E. C. Pickering and A. J. Cannon at the turn of the century and ultimately led to the *Henry Draper Catalogue* released between 1918 and 1924 which contained over 225,000 spectra [21]. This study led to the Harvard (O-B-A-F-G-K-M) classification (or Draper classification) used today. At the time there were some speculations that this classification sequence reflects an evolutionary sequence, though the aspect that stars could evolve had not been exploited yet (see below). E. Hertzsprung and H. N. Russell realized that these classes not only form a sequence, but also correlate with absolute stellar magnitude. This correlation is the very definition of the *Hertzsprung-Russell (HR) Diagram*, one of the most powerful tools in modern astronomy. It not only combines photometry with spectroscopy, but also empirically allows us to make predictions of the size of stars. The Harvard classification shows that stars vary by orders of magnitude in their basic properties like mass, radius and luminosity. The Sun, for example, with a temperature of 5,700 K is then classified

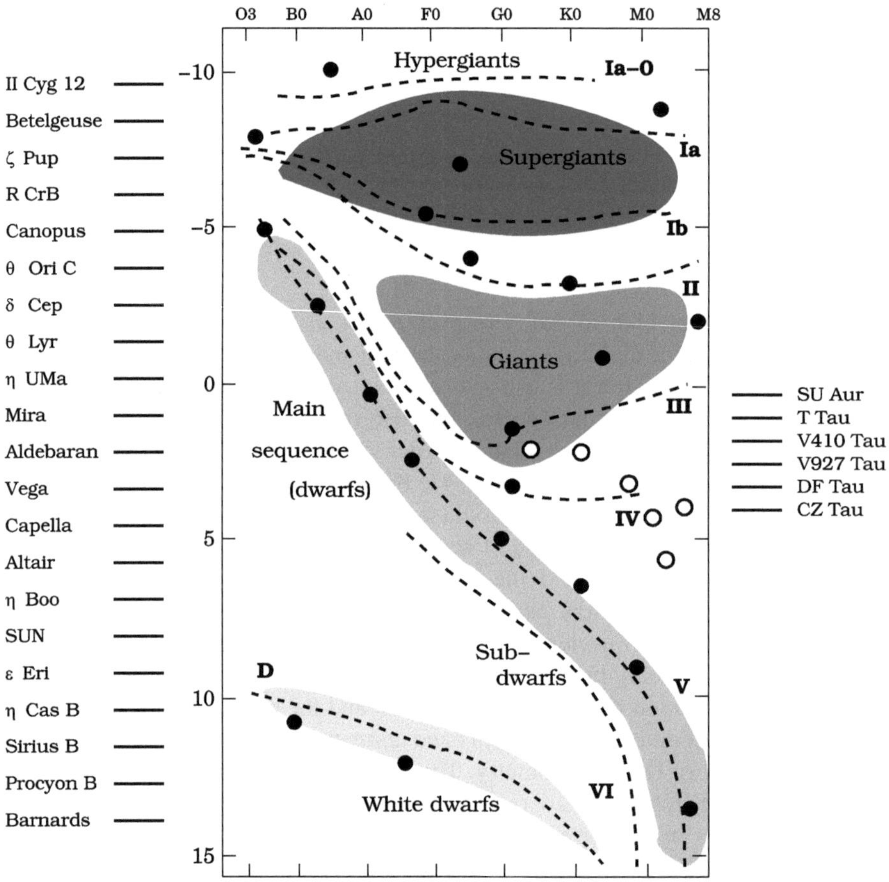

Fig. 2.6 Schematic HR-diagram plotting visual magnitude M_V and spectral classes from the Harvard classification. The plot identifies the various stellar sequences as well as luminosity classes from the MK classification. The filled circles are various examples of stars, which are identified on the left side. The open circles mark young T Tauri stars and are identified on the right side

as a G2 star on the *main sequence*. It was quite obvious that stellar properties require more criteria in order to account for their position in the HR-diagram. One of them is detailed line properties in optical spectra, such as the hydrogen Balmer lines. In 1943, W. W. Morgan and P. C. Keenan at the Yerkes Observatory [22] also introduced luminosity classes 0 to VI, which feature sequences from *class Ia-0 hypergiants* to *class VI subdwarfs* (see Fig. 2.6). The main sequence in this classification is populated by *class V dwarf stars* and represents the beginning of the evolution of fully mature stars once their energy source is entirely controlled by nuclear fusion. In evolutionary terms, young stars are then called *pre-main sequence stars* recognizing that they have not reached the main sequence. The term pre-main sequence has been introduced since the first calculations of evolutionary tracks in the HR-diagram in the mid-1950s.

2.2 The Quest to Understand the Formation of Stars

The historic reflections presented so far have had very little to do with stellar formation but more with general aspects relevant to the history of stellar astronomy. Still, they are important ingredients for putting the following sections into a proper perspective.

2.2.1 The Rise of Star Formation Theory

The first comprehensive account of the formation of the Solar System was formulated by the German philosopher Immanuel Kant (1724–1804). In his *nebular hypothesis* (in German *Allgemeine Naturgeschichte und Theorie des Himmels*, 1755) he postulated that the Solar System and the nebulae form periodically from a protonebula (in German 'Urnebel'). The swirling nebula contracted into a rotating disk out of which, in the case of the Galaxy, stars contracted independently, as did the planets in the case of the Solar System. In 1796 the French mathematician Pierre Simon de Laplace, based on many new detections of fuzzy nebulae in the sky, formulated a similar hypothesis, which was long after referred to as the *Kant–Laplace Hypothesis*. There is a distinct difference between Kant's and Laplace's postulates. While Laplace takes possible effects of angular momentum into consideration. Kant remained more philosophical and envisaged a more universal mechanism which includes not only the Solar System but also features rudimentary concepts of today's galaxies and concluded that formation is an ongoing and recurrent process [23].

This nebula hypothesis in its main elements was not well founded in terms of detailed physical descriptions and calculations. However, Kant could build on about 140 nebulous stellar objects known at the time through the works of C. Messier and F. Machain. The situation for Laplace was more comfortable as he could rely on F. W. Herschel in 1786, who created a list of over 1,000 such nebulae through observations with his giant telescope. For over 100 years not much happened with respect to stellar formation and most efforts were directed to deciphering the structure of these nebulae. Of course, the hundred-year period of spectroscopic and photometric advances contributed much to the physical understanding of stars (see above). The 19th century was dominated by the discovery of several thousands of these nebulae, and in 1864 J. Herschel published the precursor of the *New General Catalog* which contained over 6,000 entries. Though at the time it was unknown, most of these entries were actually galaxies. A small fraction of these nebulae appeared peculiar in that they revealed a spiral nature. It was not until after the turn of the century that in 1936 E. Hubble's book *The Realm of the Nebulae* finally put an end to the discussion about the nature of the nebulae.

In the 19th century only one reference to stellar formation can be found which involves gravitation and thermal physics and this is the attempt by H. Helmholtz and W. Thomson (Lord Kelvin) to explain the energy radiated by the Sun through slow gravitational contraction (see Sect. 2.2.2). The start of the 20th century marked the beginning of intense activity to apply the laws of thermodynamics developed in the 19th century. At the forefront were contributions by J. Jeans, A. S. Eddington, and R. Emden. In his paper *The Stability of a Spherical Nebula* in 1902 Jeans first formulated what is known today as the *Jeans Criterion* describing the onset of gravitational instability of a uniform sphere of gas. After A. Einstein formulated his theory of special relativity in 1905, it was Eddington in 1917 who realized that the conversion of mass to energy could be key to the Sun's luminosity [24]. The physics of uniform gas spheres, specifically the mass versus luminosity relation of stars, was much debated, specifically between Jeans and Eddington. While Eddington published a series of manuscripts like *The Internal Constitution of Stars* in 1926 [25] or *Stars and Atoms* in 1927 [26] giving a first account of modern astrophysics, Jeans' work entitled *Astronomy and Cosmogony* in 1928 [27] is considered by many to be the first book on theoretical astrophysics. On the observational side contributions by J. Hartmann, H. Shapley, E. E Barnard and many others shaped the perception of what is today referred to as the *interstellar medium* (see Chap. 3).

In the end it was Eddington who basically formulated the right idea, that the age of the Sun must be scaled by the ages of the oldest known sedimentary rocks and the main source of the star's energy must be subatomic. The formulation of thermonuclear fusion of hydrogen as the main energy source for stars by H. Bethe a decade later in 1938 confirmed Eddington's speculation [28]. Here S. Chandrasekhar's famous book entitled *An Introduction to the Study of Stellar Structure* in 1939 nicely summarizes these developments and is highly recommended for further reading [29]. Though the physical and mathematical ground work was set by Kant's Nebula Hypothesis, it was another 30 years before R. Larson (see Chap. 4) performed the first numerical calculations of a gravitational collapse. One decisive ingredient, though already identified in galactic clouds, was then introduced into the discussion: *dust*. The importance of dust in the context of stellar formation cannot be underestimated, a point emphasized by the title of this book.

2.2.2 Understanding the Sun

The Sun is the star everyone is most familiar with as it is directly involved in daily life (see Fig. 2.7). As the central star of our Solar System it provides all the energy for the Earth's biosystem. Without the steady inflow of light, all life would perish almost instantaneously and the Earth's surface would reduce to a cold icy desert. Radiation from the Sun regulates the oceans and the climate, makes plants grow, warms all living beings and stores its energy in the form of fossil resources. Sunshine, or the lack of it, even affects our moods.

The Sun is the closest star and its shape and surface can be observed with the naked eye once dimmed behind clouds or in atmospheric reflections during sunset. The earliest recording of dark spots on the Sun's surface are from Chinese records from 165 B.C., whereas the first sighting in Europe was recorded during 807 A.C. [15]. These records must have been forgotten and with great surprise, shortly after the invention of the telescope, Galilei and others observed these dark spots in stark contrast to what had been taught since Aristotle, that the Sun's surface is pure and without fault. The existence of these spots led to many theories and

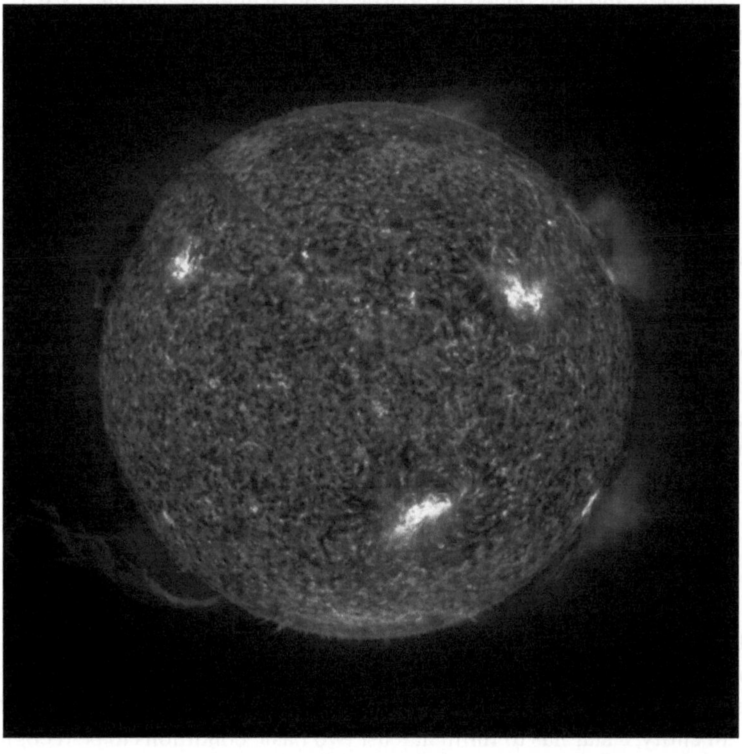

Fig. 2.7 The Sun as photographed by the solar observatory satellite SOHO at ultraviolet wavelengths. Credit: SOHO/ESA/NASA

speculations which ranged from interpretations such as solar mountain tops by G. Cassini to connections with climatic disasters like a devastating harvest in the 17th century by F.W. Herschel. However, the interest in Sun spots initiated a constant surveillance of the surface of the Sun, which lead to the discovery by S. Schwabe in 1826 that these spots appear regularly on the Sun's surface and their abundance and strengths underlies an eleven year cycle. The study of these spots not only showed that the Sun is rotating, it also led to the determination of its rotational axis and the discovery by R. Carrington between 1851 and 1863 that the Sun's rotation varies between the poles and the equator. This fact is enormously important for understanding the Sun and stars, as it is characteristic of rotating gas spheres with a dense core. Furthermore it is indicative of the dynamo in the Sun's interior. Today it is known that the sunspot cycle even correlates with plant growth on Earth and from carbon dated tree rings it can be concluded that this cycle has persisted for at least 700 million years.

This time span as well as the projected age for the Earth's existence of billions of years ascertained from geological formations, were in stark contrast with estimates of the Sun's age put forth by theories explaining the origins of solar energy. These theories evolved at a time when physicists like S. Carnot, R. Mayer, J.P. Joule, H. von Helmholtz, R. Clausius and W. Thomson formulated the laws of thermodynamics, and the concept of conservation of energy emerged [30]. Concepts like a meteoric bombardment or chemical reactions were dismissed on the basis of either problems with celestial mechanics or of the projection that the Sun could not have been radiating for more than 3,000 years. By the middle of the 19th century von Helmholtz suggested that the Sun's heat budget was derived from contraction of an originally larger cloud. This theory had two advantages. First, it would provide tens of millions of years of energy. Second, the contraction rate of the Sun would be too small to measure and thus the idea would stay around. One hundred years later, when the nuclear chain reactions were discovered that transform hydrogen into helium under the release of unprecedented amounts of energy [28], the secret behind the Sun's energy source was finally identified. However, although in the end the contraction theory proved to be incorrect in explaining of the Sun's heat, it marked the first time that the process of gravitational contraction of a gas cloud was formulated in terms of detailed physical laws. Thus the lifetime of a star through gravitational contraction is called the *Kelvin–Helmholtz timescale* (see Sect. A.9).

2.2.3 What is a Star?

Since the developments during the 19th century and specifically since the systematic spectroscopic studies at the beginning of the 20th century, it has been recognized that the Sun and the stars are bodies of a kind. Although there is no simple definition of such an entity, a star has to fulfill at least two basic conditions to be recognized as such. It has to be self-bound by gravity, and it has to radiate energy which it produces in its interior. The first condition implies that there are internal forces that prevent the body from collapsing under self-gravity. The body has to be approximately spherical as the gravitational force field is radially symmetric. Deviations towards obliqueness may arise through rotation. Internal forces can stem from radiation, heat motions, or lattice stability of solids just to name a few factors. If stability under gravity and symmetry were the only condition, then planets and maybe comets and asteroids could be stars as well. Comets and asteroids of course can be ruled out as modern imaging clearly show them to be non-spherical. Historically the view that these bodies are stars of some kind persisted for centuries, as expressions like 'wandering star' for planets or 'guest stars' in connection with comets still testify. It must be the second condition that makes a star: there has to be an energy source inside the star making it shine. Planets, although they appear as bright stars in the sky, cannot do that and their brightness is due to illumination from the Sun. Of course, no argument is perfect and the two largest planets in the Solar System, Jupiter and Saturn radiate more energy than the Sun provides, indicating that there is in fact an internal energy source. In this respect one has to add to the condition that the internal source also provides the energy to sustain the force against gravity.

2.2.4 *Stars Evolve!*

If the discovery that stars burn nuclear fuel had any consequence to star formation research it probably was the immediate realization that stars must evolve within a certain lifespan. To be more precise, while the star fuses hydrogen, it probably undergoes changes in its composition, structure, and appearance. As V. Trimble, an astronomer from the University of California, wrote in 2000 [31]:

Stars are, at least to our point of view, the most important building blocks of the universe, and in our era they are its major energy source. In 1900 no one had a clue how stars worked. By mid-century we had them figured out almost completely.

The developments in the 1930s from Eddington to Bethe indeed laid the ground work for an almost explosive development in the field of stellar structure and evolution. Specifically, stars had to evolve as a strong function of their mass. Key were hydrogen fusion lifetimes. From early calculations it was easily recognized that low-mass stars can burn hydrogen for billions of years. On the other hand, massive stars radiate energy at a rate which is many orders of magnitude larger than low-mass stars. Since their mass is only a few ten times larger this means that their fusion lifetime is much smaller. In fact, for massive O stars it is of the order of ten million years and less. Stars burn about 10% of their hydrogen and as a rule of thumb one can estimate the lifetime of stars by:

$$t_{life} = 7.3 \times 10^9 \frac{M/M_\odot}{L/L_\odot} \mathrm{yr} \tag{2.2}$$

where $M_{\odot}$ and and $L_{\odot}$ are the mass and luminosity of the Sun and M and L the same for the star [19]. Clearly, this development put an end to the perception that the Draper classification could resemble an evolutionary scale where early type O stars evolve into late type stars.

2.2.5 The Search for Young Stars

The short lifetimes of early type stars offers an opportunity to determine their birth places based on the argument that they could not have traveled far from their place of origin. As G. H. Herbig [32] remarks, it seems fairly surprising that the identification of stellar clusters containing OB stars as sites of on-going star formation did not occur until the early 1950s. By then A. Blaaw [33,34] determined proper motions of early type stars with median velocities of 5 km s^{-1} with some showing peak velocities of >40 km s^{-1}. Today the argument is even stronger as dispersion velocities of associated clouds have similar velocities (see Chap. 4) and net velocities of the stars relative to the parent cloud are usually less than 1 km s^{-1} (see [35]). An O star with 2 Myr of age would travel a distance of 2 pc; a fast one may reach 15 pc.

In 1947 B. Bok and his colleagues [36, 37] found small dark spots projected onto the bright H II region M8 and determined a size of less than 80,000 AU. The interpretation at the time was that these now called *Bok Globules* accrete matter from their environment and it is radiation pressure that forces them to contract. Bright H II regions, which of course contain O stars, would then have to be prime sites for stellar formation.

T Tauri stars were discovered in 1942 by A. H. Joy [38], but were not immediately identified as young stars. They got their name from *T Tau*, the first detected star of its kind. However, besides their strikingly irregular light curve, one obvious property was that they seem to be associated with dark or bright nebulosity [39]. Probably the first one to seriously suggest the notion that T Tauri stars are young, and not in these clouds by coincidence, was V. A. Ambartsumian [3] in 1947. It still took well into the 1950s until the fact was established [40, 41]. One of the earliest catalogs of young stars in star-forming regions was compiled by P. Parenago in 1955 [42]. But now help also came from the theoretical side. With the first calculations of positions of contracting stars in the HR-diagram in 1955 [43,44] it became clearer that properties of T Tauri stars matched these predictions. In Fig. 2.6 a few examples are shown. Generally T Tauri stars possess significantly higher luminosities for their identified spectral class. According to C. Lada, decades later, these studies revealed that mature low-mass stars emerged from OB associations and were formed at a much higher rate than massive stars [45]. It was also suggested that star formation is still ongoing, even in the solar neighborhood.

The strong emission lines of T Tauri stars, specifically the Balmer Hα line became their trademark in objective prism surveys, which scanned the sky up to the late 1980s (see Sect. 6.3.2). The most systematic optical catalog of T Tauri stars was compiled in 1988 [46].

Today searches for young stars are pursued throughout the entire electromagnetic spectrum (see Sect. 2.3.1) as a result of the accelerating development of technologies. To be more specific, searches today are performed in the Radio, Sub-mm, IR, and X-ray bands (see Sects. 2.3.2 and 2.3.3).

2.3 Observing Stellar Formation

The fact that most research today is actually performed outside the optical wavelength band is for a very good reason. Collapsing clouds fully enshroud the newborn star within the in-falling envelope of dust and gas thus blocking the observer's view. Thus, if it were only up to the human eye and the spectral range it is sensitive to, then the process of stellar formation would be unobservable. In fact, as the course of the book will show, even if we could take a peek inside the collapse there might not be much to see as most radiation in the very early phases of the collapse happens at much longer wavelengths. The following three sections briefly outline the challenges that had to be overcome to finally reach broadband capability as well as to demonstrate the current and future technical advances necessary for stellar-formation research.

2.3.1 The Conquest of the Electromagnetic Spectrum

Newton, in the middle of the 17th century, found out that light is composed of a whole range of colors and that this spectrum spans from violet to blue, green, yellow, orange and red. It implies a wavelength range from about 400 to about 750 nanometers. Not much was known about the nature of light until 1690 when C. Huygens showed that light has a wave character. Over 100 years later this scale rapidly expanded at both ends. F.W. Herschel found in 1800 that the largest amount of heat from the Sun lies beyond the red color in the spectrum. One year later J. W. Ritter detected radiative activity beyond the violet color of the spectrum. Both events mark the discovery of infrared and ultraviolet light. J. Maxwell postulated in 1865 that light is composed of electromagnetic waves, a theory that was proved to be correct by H. Hertz in 1887, who also detected electromagnetic waves with very long wavelengths and thus added the radio band to the electromagnetic spectrum. In another milestone in 1905 W. C. Röntgen detected his *Röntgenstrahlung*, the mysterious X-rays which obviously rendered photoplates useless and which were able to penetrate soft human tissue. Seven years later *M. von Laue* finally demonstrated that these X-rays are electromagnetic waves of extremely short wavelengths. By then the wavelength range of the electromagnetic spectrum spanned from about 10^{-10} to 10^{2} meters.

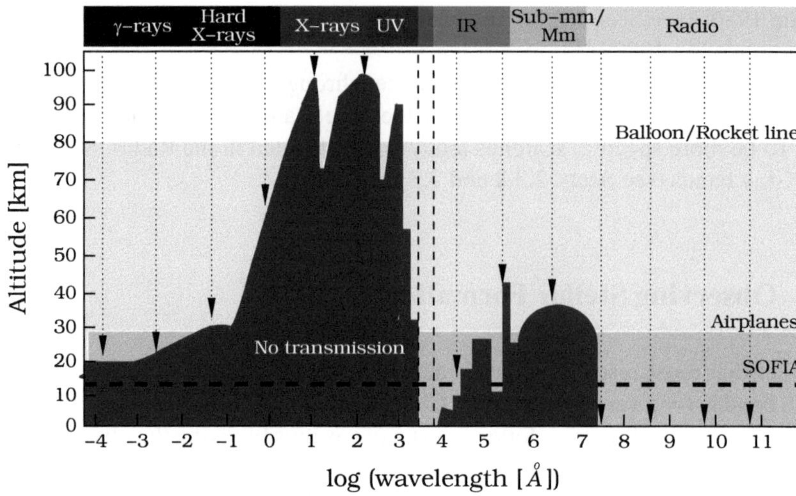

Fig. 2.8 The Earth's atmosphere does not allow for penetration of light from the entire electromagnetic spectrum to the ground-based observer. Throughout the spectral band, absorption in the atmosphere dominates over transmission. The dark region marks the wavelengths and heights through which light cannot penetrate. Some wavelengths, like sub-millimeter radiation or X-rays are absorbed only to an intermediate height and can be observed with high-flying balloons. Adapted from Unsöld [19]

But the study of the nature of light so far was entirely confined to laboratory experiments. A similar conquest in the sky is an ongoing challenge. It requires high-flying balloons, rockets, and satellites as well as extremely sensitive electronic equipment. Electromagnetic radiation from stars and other cosmic objects can penetrate the Earth's atmosphere only in limited fashion depending on its wavelength. Figure 2.8 shows schematically how only radio, optical, and some of the near-IR wavelengths have broad observing windows, while, for example, the UV range is almost entirely blocked out and X-rays can only be reached by high-flying balloons for short exposures. Sub-mm observations can be made from the ground at elevated heights, though most bands are entirely absorbed by the atmosphere's water vapor.

2.3.2 Instrumentation, Facilities, and Bandpasses

Today searches for young stars are pursued throughout the entire electromagnetic spectrum as a result of the accelerating development of advanced technologies specifically for focal-plane instrumentation. Instrumental development is an essential part of stellar research, and astronomy in general always motivates the creation of new technologies. The goal is to gain increased sensitivity, increased spatial and spectral resolution, and increased wavelength coverage. Of course, one would like to achieve all this throughout most essential parts of the electromagnetic spectrum.

For a short review readers are directed to J. Kastner's review on imaging science in astronomy [10]. In star-formation research, observing efforts today engulf almost the entire spectral bandwidth from radio to γ-radiation. In essence, it took the whole second half of the 20th century to conquer the electromagnetic spectrum technologically in its entire bandwidth for star-formation research. The following offers a very limited review of the use of instrumentation over this time period, and for various wavelength bands without providing technical specifications, which are outside the focus of this book. More detailed explanations and a description of instrumental acronyms in the following sections and chapters can be found in Appendix F.

Observations of young stars are predominately performed in medium- to short-wavelength ranges:

Near-IR :	3.5 μm	$\lesssim \lambda \lesssim$	0.8 μm	*thermal continua, vibrational lines*
Optical :	0.8 μm	$\lesssim \lambda \lesssim$	0.4 μm (4,000 Å)	*molecular/atomic lines*
Near-UV :	4,000 Å	$\lesssim \lambda \lesssim$	1,000 Å	*molecular/atomic lines*
Far-UV :	1,000 Å	$\lesssim \lambda \lesssim$	150 Å	*atomic lines, continua*
X-ray :	150 Å	$\lesssim \lambda \lesssim$	1 Å	*inner shell atomic lines*

There are strong signatures at longer wavelengths as well but they are normally not understood in the context of the protostar itself. Dust envelopes around protostars are only visible in the far-IR and the sub-mm. Long wavelengths exclusively identify very young protostars in the context of thermal emission from dust and molecules and some non-thermal emission in the radio band [47]. Ranges are:

Radio :	50 m	$\lesssim \lambda \lesssim$	5 mm	*non-thermal continua rotational lines*
mm :	5 mm	$\lesssim \lambda \lesssim$	1 mm	*rotational lines, dust*
Sub-mm :	1 mm	$\lesssim \lambda \lesssim$	0.35 mm	*rotational lines, dust*
Far-IR :	350 μm	$\lesssim \lambda \lesssim$	20 μm	*rotational lines*
Mid-IR :	20 μm	$\lesssim \lambda \lesssim$	3.5 μm	*vibrational lines*

Optical telescopes can be used from the near-UV to the mm-bandpass. Observing capability is dominated by the transmission properties of the atmosphere and thus only a few windows outside the optical band are really visible from the ground, with some becoming visible at higher altitudes (see Fig. 2.8). At even higher altitudes IR and sub-mm waves become accessible, though mirror surfaces are increasingly sensitive to daytime to nighttime temperature changes and require specific accommodations. Thermal background noise irradiated by instrument components is reduced detector cooling. Throughout the IR band, variable thermal emissionfrom the Earth's atmosphere is a problem growing with increasing wavelengths. Modern facilities thus use choppers or alternate beams to subtract atmospheric radiation and filter out the difference signal for further processing.

The 1940s and 1950s observed primarily older regions of stellar formation such as open clusters, OB, and T Tauri associations (see Sect. 2.2.5). The first systematic studies of young stars and star-forming regions were performed with optical telescopes. Though most of these studies occurred long after the 1950s when more advanced photomultipliers became available, hypersensitive photographic plates were used even into the early 1990s, which due to immense advances in photographic techniques had not much in common with the original plates [1]. To further enhance sensitivity the plates are sometimes submitted to a heating process (baking). The dynamical range of photographic plates is limited and so is their efficiency in comparison with photoelectronic devices. Thus some surveys used photographic plates specifically to cover wide fields, some used photoelectric detectors. Both methods produced uncertainties s ranging between 0.005 and 0.015 mag. Today large area charge-coupled devices (CCDs) with higher linear dynamic ranges and efficiencies have replaced most photographic plates. In order to obtain spectral information filter combinations [18] are applied. For higher resolution objective prism plates, objective grating spectrographs and slit spectrographs have been used [46, 48–53]. Objective prisms had already been used early in the century [21] and were specifically useful for scanning extended stellar fields. Today grating spectrographs are used with edged gratings for Cassegrain spectrographs and Coudé spectrographs that allow spectral resolutions of up to 100,000. Similar results can be achieved with lithographic reflection gratings in Echelle spectrographs [19, 54].

Many near-IR observations have been performed throughout the 1970s, notably [55–58], which together with optical observations provided a vital database for further studies of young stars. For wavelengths $\lesssim 3.5\ \mu$m nitrogen-cooled InSb-photodiodes are commonly used. Wavelengths $\lesssim 1.2\ \mu$m can be observed with optical photomultipliers, though nitrogen-cooled photo cathodes are needed.

Research in the 1980s also systematically began to survey star-forming regions in the mid-IR up to the mm-band as more advanced electronics became available. Specifically CO surveys provided a direct probe of molecular clouds and collapsing cloud cores (see Chaps. 3 and 4 for more details). For observations in the sub-mm and mm band nitrogen cooling becomes insufficient and superfluid helium cooling needs to be in place instead, forcing focal plane temperatures to below 2 K. S. Beckwith and collaborators, for example, used a He-cooled bolometer to measure 1.3 mm continuum emission with the *IRAM* 30 m telescope at an altitude of around 2,900 m on *Pico Valeta* in Spain [59]. The *IRAM* and *VLA* telescopes were used throughout the early 1990s to map molecular clouds, foremost the ρ Oph A cloud which hosts very recently formed stars with CO outflows [60]. Other surveys and observations of recently formed stars also involve dust emission from *Herbig Ae/Be stars* [61], objects in the ρ Oph cloud [62], H_2O emission in circumstellar envelopes of protostars [63], and from various collapsing cores with envelope masses $<5\ M_\odot$ [64], to name but a few out of hundreds of these observations performed to date.

Today many measurements are performed with *SCUBA* on the *JCMT*, which is a state-of-the-art facility located in Hawaii and which came into service in 1997.

Another even more recent sub-mm facility is the 10 m *HHT* on Mt. Graham in the USA. *ALMA* is a mm/sub-mm array of 50 antennas of 12 m diameter each located on the Chajnantor plain at 5,000 m altitude. Its design is driven by three key science goals:

- detect spectral line emission from CO or CII in galaxies at a redshift of $z = 3$
- image the gas kinematics in protostars and in protoplanetary disks in the nearest molecular clouds (150 pc)
- provide precise high dynamic range images at an angular resolution of 0.1 arcsec.

ALMA will become operational by 2012. Major recent, currently active, and future observing facilities like *CSO, BIMA*, and *OVRO* with short characterizations are listed in Appendix F.

2.3.3 Stellar Formation Research from Space

The benefits of space research with respect to stellar formation studies are undisputed. It should be noted, though, that space observatories are generally highly specialized, expensive, and usually suffer from much more limited lifetimes than their counterparts on the ground. In this respect modern star-formation research could not survive without the contributions of an armada of ground facilities equipped with instruments sensitive to ground-accessible wavelengths (see above). Specifically ground observations in the radio and sub-mm domain still contribute most to studies of molecular cloud collapse and very early phases in the star formation process.

On the other hand, there are domains where the exploration from space is not only highly beneficial, but is simply the only way to observe at all. The former is certainly true for bandpasses in the IR and sub-mm range, the latter is exclusively valid for the high energy domain. These two spectral domains have specifically been proven to be most valuable for investigating the properties of very young stars. Another advantage of in-orbit observations is that they provide long-term uninterrupted exposures as well as a much larger sky accessibility. And finally, observations from space offer the possibility to provide various deep and wide sky surveys on fast timescales. From the late 1970s, several observatories were launched into space which contributed to the understanding of early stellar evolution in major ways. Table 2.1 gives an account of all space missions with an agenda for star formation research that have been concluded or are still ongoing. It shows that there has been little X-ray coverage in the 1980s.

With the launch of the *X-ray observatory EINSTEIN* in late 1978 a new era in early stellar evolution began as X-rays were detected from young low-mass stars [65–67] soon after the observatory went into service and pointed at the Orion Nebula Cluster (ONC). X-rays from stars were not un-heard of, X-ray stars in binary systems were known to possess X-ray emission much more powerful than that of the Sun. Also at about the same time, X-rays were also detected from hot massive

Table 2.1 Space missions relevant for star formation research. The bandpasses are generally given in wavelength units, except for X-ray missions that did not have high spectral resolving power. Here it is more common to specify the range in keV.

Mission	Bandpass spectral range	Mission dates	Major science impact in star-formation research
IUE	UV 1150–3299 Å	Jan. 78–Sep. 96	UV spectra of young massive stars, Herbig Ae/Be stars.
EINSTEIN	X-ray 1–20 keV	Nov. 78–Apr. 81	Detection of X-rays from low-mass stars.
IRAS	IR 8–120 μm	Jan. 83–Nov. 83	All-sky survey, photometry of young stars.
HST (Hubble)	Optical/IR 0.3–3.5 μm	Apr. 90–	High-resolution images of star-forming regions.
ROSAT	X-ray 0.1–2.4 keV	June 90–Feb. 99	First X-ray all-sky survey; identifications, distributions and luminosities of young stars.
EUVE	Far-UV 70–760 Å	June 92–Jan. 01	All-sky survey catalog, coronal spectra of stars.
ASCA	X-ray 0.3–10 keV	Feb. 93–July 00	Hard X-rays from star-forming regions.
ISO	IR/Sub-mm 2.5–240 μm	Nov. 95–May 98	High-resolution images and spectra of stellar cores young stars.
FUSE	Far-UV 905–1195 Å	June 99–Oct.07	Molecular absorption lines in spectra of young stars.
Chandra	X-ray 1.5–160 Å	July 99–	High-resolution images and spectroscopy of star-forming regions, line diagnostics.
XMM-Newton	X-ray 6–120 Å	Dec. 99–	High-efficiency spectroscopy of young stars.
SST (Spitzer) 3–180 μm	IR/Sub-mm	Aug. 03–June 08	Imaging and spectroscopy of star-forming regions with high precision.
		July 08–	Warm Imaging mission
Herschel	Far-IR/Sub-mm	May 09–	Star formation in galaxies, interstellar medium.
SOFIA	2.2–26 μm	2011–	Permanent IR observatory flying at 13 km above ground.
GAIA	Optical	Launch 2013	3-D mapping of the Galaxy (ESA).
JWST (James Webb)	Optical/IR	Launch 2018	Images beyond HST.

stars [68,69]. Here a plausible model for the X-ray emission from hot stars was soon on the table [70], whereas X-rays from young low-mass stars remained a mystery for much longer and even today high-energy emission is not fully understood ([71], see Chap. 10).

In the 1990s *ROSAT* [72] was the main X-ray observatory available for stellar evolution research. A model of the satellite in-flight configuration is shown in Fig. 2.9. It not only produced a large number of long exposed observations of star-forming regions but also conducted the first X-ray all-sky survey (Figure 2.10). This first survey provides a valuable resource for current and future missions and allows us to study wide distributions of young stars [73]. The currently active X-ray observatories *Chandra* (see Fig. 2.11) and *XMM-Newton* add a new quality to previous X-ray missions. One of them is the capability to perform high-resolution X-ray spectroscopy at resolving powers of 300 to 1,200. The high spatial resolution of *Chandra* with 0.5 arcsec adds another dimension to X-ray studies as the cores of dense very young stellar clusters, such as the Orion Nebula cluster, are now fully spatially resolved.

Space missions in the IR and sub-mm band took a similar path with time. The launch of *IRAS* in 1983, though its mission was only 10 months long, provided researchers with the first ever all-sky survey conducted in space in the IR. *IRAS* generated a database which still keeps researchers busy today, specifically for its

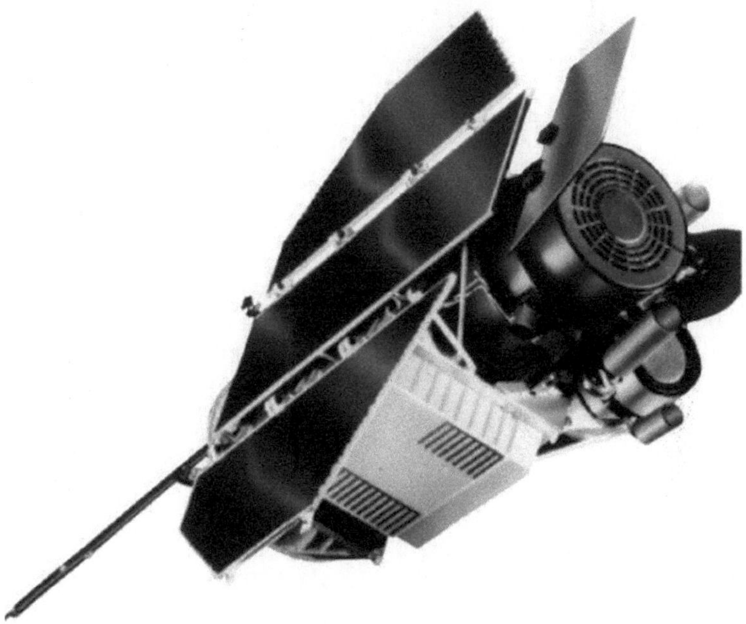

Fig. 2.9 Model of *ROSAT* in flight configuration. Throughout the 1990s *ROSAT* observed major star-forming regions in X-rays. One of its major accomplishments was the completion of the first X-ray all-sky survey (*RASS*). Credit: Max Planck Institut für extraterrestrische Physik

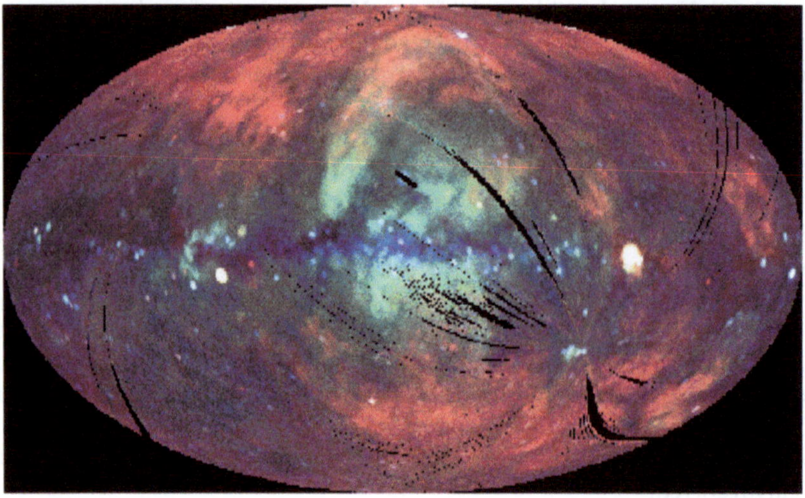

Fig. 2.10 The X-ray sky at energies between 0.1 and 2.4 keV as observed with the X-ray observatory *ROSAT* in the early 1990s. The colors represent X-ray energy, where red is near the low, blue near the high energy boundary. Credit: Max Planck Institute für extraterrestrische Physik

Fig. 2.11 Artist's impression of the *Chandra* spacecraft, which is orbiting Earth in a highly eccentric orbit (highest altitude is 139,000 km) since July 1999. Its currently projected tenure of operation leads into the year 2009. For more detailed information, see descriptions and links in Appendices E and F. Credit: NASA/CXC/SAO

wide field views. Missions with more powerful and higher precision instruments followed, like *ISO* in 1995 and *SST* in 2003. In it first 5 years in orbit the *SST* imaged the infrared sky with unprecedented precision at wavelengths between 3 and 180 micrometers with a large emphasis on star-forming regions providing the most detailed infrared data archive for decades to come.

This canonical development in the IR and X-ray bands is not accidental. Comparisons of *ROSAT* observations with IR images could demonstrate that observed activity in early evolutionary phases are quite complementary in these bands and their simultaneous study is highly beneficial. The vast improvements in observational quality in these two wavelength bands is demonstrated in Fig. 2.12. It shows four images of the Orion Nebula region centered on the *Orion Trapezium*

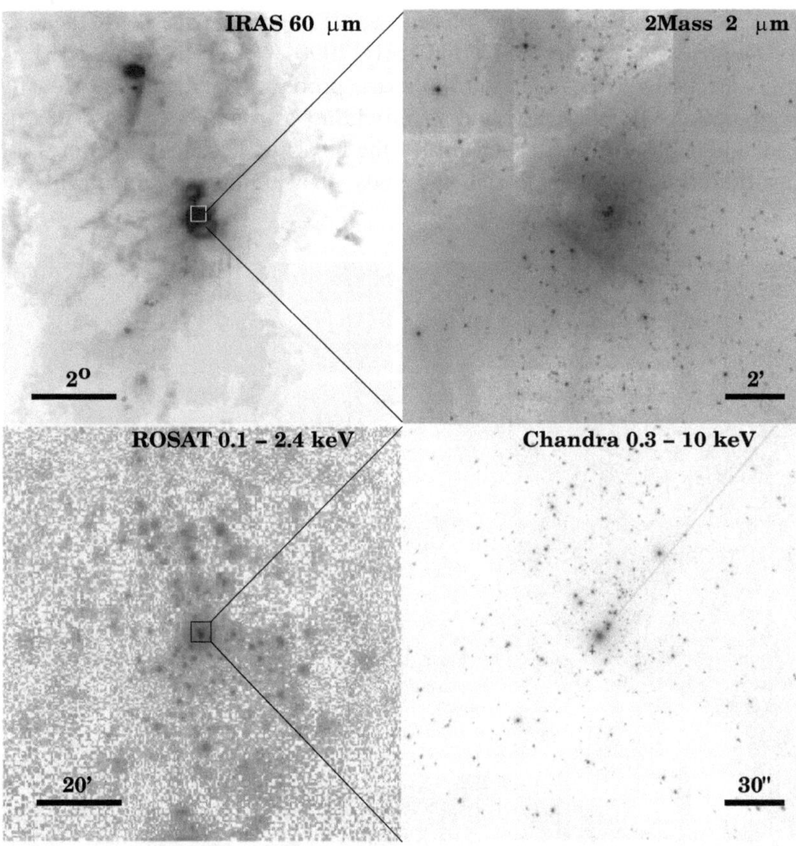

Fig. 2.12 (*top left:*) Wide view of the Orion region with *IRAS* at 60 μm. (*top right:*) Close-up view of the same region focusing on the center of the Orion Nebula with the *2MASS* observatory at 1–2 μm. (*bottom left:*) Wide view with the *ROSAT* PSPC in the X-ray band below 2.5 keV (above 5 Å). (*bottom right:*) Close-up view with *Chandra* ACIS below 10 keV (above 1.2 Å). The two comparisons illustrate the huge advances in the field of IR and X-ray astronomy, as well as the potential that exists to pursue wide-angle views as well as deep views

cluster. The very wide view of *IRAS* shows the large-scale structure of diffuse matter around the Nebula without being able to resolve many stars. The wide view of *ROSAT* shows that in X-rays only stars contribute to the emission. With a resolution of 25 arcsec of the *PSPC* (see Appendix F), however, *ROSAT* was still quite limited in resolving most point sources. In contrast the *2MASS* image shows highly resolved point sources as well as diffuse gas within the *IRAS* bandpass. The *2MASS* is a 2 μm on-ground survey observatory operated by the University of Massachusetts. Wide-view ground observations now always complement today's space observatories. Finally the *CHANDRA* view complements these IR observations with X-ray images of unprecedented resolution. All stars in the ONC with the exception of close binary systems are now fully resolved in the IR and X-ray domain providing researchers with a unique opportunity to study aspects of the early evolution of stars.

The future of star formation research from space is already well in progress. Table 2.1 mentions four exemplary missions, *SOFIA* as the successor of the airborne *Kuiper Observatory* of the 1980s and early 1990s, *JWST* (see Fig. 2.13) as the scientific successor to *Hubble*, and the space probes *Herschel* and *GAIA*. *SOFIA* began operations in mid-2010 and is providing astronomers access to optical, infrared, and sub-millimeter spectrum for the next two decades. It features a free-floating IR telescope mounted into the body of a jumbo jet, which will engage

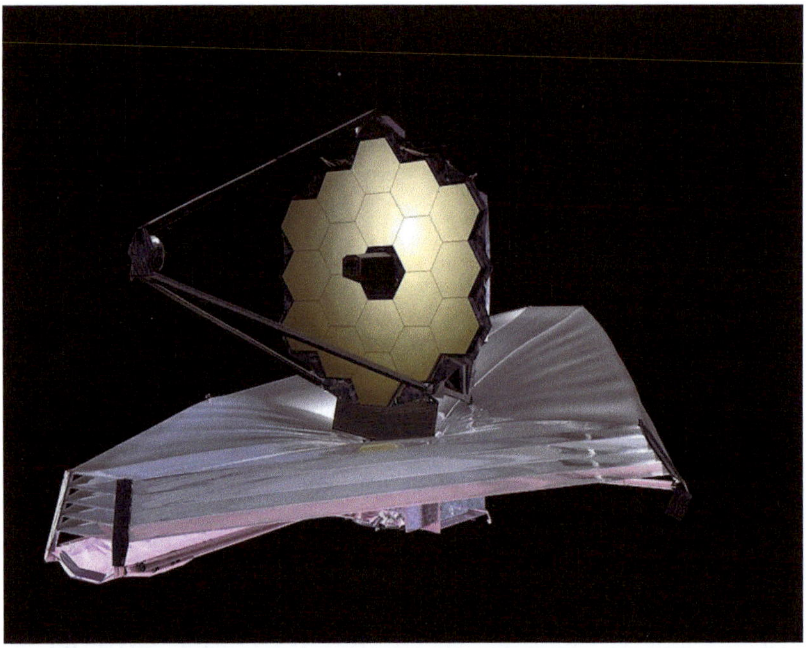

Fig. 2.13 Artist's impression of the *JWST* spacecraft, which will be launched around 2015 into one of the Earth-Sun Lagrangian points and will provide adaptive optics with a collecting area of 25 m^2. For more detailed information, see descriptions and links in Appendices E and F. Credit: NASA

in regular flights through the Earth's stratosphere, thus avoiding the absorbing effects of vapor clouds. *Herschel* was launched in 2009 and is to date the largest space telescope in operation featuring a 3.5 m Cassegrain telescope (see links in Appendices E and F). *Herschel* has the rather unique objective of studying the formation of galaxies in the early universe and thus provide information about the formation of stars in the early universe and clues toward cosmic star formation histories. *JWST* is to be launched during the second decade of the 20th century and with a collecting area of 25 m^2 in the near-IR band will be the largest telescope in space. Its primary scientific mission includes the search for light from the first stars, to study the formation and evolution of galaxies, stars, and planetary systems.

Chapter 3
Studies of Interstellar Matter

3.1 The Interstellar Medium

3.1.1 The Stuff between the Stars

Observational evidence that the space between the stars in our Galaxy is not empty came as early as the beginning of the 20th century with the discovery of stationary Ca II lines in spectra of the binary star δ Orionis by J. Hartmann in 1904. In 1922 E. Hubble concluded that dust clouds scatter the light of nearby cool stars, which are then detected as reflection nebulae, while hot stars excite the gas in emission nebulae. He concluded that diffuse clouds irradiated by a radiation field have kinetic temperatures of about 10,000 K and frequent collisions would establish Maxwellian velocity distributions for electrons and ions in space.

The realization that the spurious and sharp Ca II or Na I lines are the result of partially ionized layers of gas distributed throughout the entire Galaxy opened a new area in the field of celestial spectroscopy. Particularly important was the discovery that the interstellar medium is mostly hydrogen and thus quite similar to what is observed in stellar atmospheres – a fact that has wide implications in the context of stellar formation. In 1927 E. E. Barnard published a photographic atlas of the Milky Way which showed silhouettes of dark clouds shadowing the background starlight. R. J. Trumpler also showed in 1930 that interstellar extinction of light not only is a phenomenon created by dark clouds, but also has to be taken into account in objects that are observed at a distance of only a few hundred parsecs through empty space.

Progressive methods in optical data analysis revealed small velocity components in interstellar absorption lines which ultimately were linked to clouds in the interstellar medium as kinematic units. Such clouds were then also discovered in the Galactic Halo [74] raising the issue of the stratification of the *interstellar medium (ISM)* in the gravitational potential of our Galaxy. Perhaps one of the biggest breakthroughs in the understanding of the ISM came in 1957 when the idea was formulated that the ISM is directly connected to stellar evolution and that stars and ISM coexist and evolve together [75].

N.S. Schulz, *The Formation and Early Evolution of Stars*, Astronomy and Astrophysics Library, DOI 10.1007/978-3-642-23926-7_3, © Springer-Verlag Berlin Heidelberg 2012

Where does the ISM come from? To quite a large extent it is a remainder of the Big Bang, where once the universe had cooled down to sufficiently low temperatures to allow the formation of atoms. Here hydrogen and helium were formed with an abundance ratio of roughly 1,000:85, with only traces of other, light elements. The ISM as it appears today is thus in part debris from primordial star formation. In fact, its composition today is not very much different from what it was a few billion years ago when the ISM was formed. However it is not just primordial debris as it is under constant reformation through ongoing star formation, nucleosynthesis, and stellar mass loss.

The ISM is an extremely tenuous gas with an average density of 10^{-20} times the Earth's atmosphere near sea level and is mostly hydrogen with small amounts of carbon, oxygen, iron, and other higher Z elements. In this composition hydrogen contributes 90 percent of all interstellar matter in either its molecular, neutral, or ionized form. The ISM breaks down into the following components [76]:

- Hydrogen (H_2, H I, H II, e^-).
- Helium (He I, He II).
- Trace elements (C, O, Ne, Mg, Fe and others, including ions).
- Molecules (CO, CS, and others).
- Dust.
- Cosmic ray particles.
- Magnetic fields.
- Radiation fields.

Stars are born from ISM matter and during their lifecycle they return some of that material back into the ISM. Figure 3.1 illustrates this recycling process. Giant molecular cloud cores collapse to form stars or whole stellar associations, removing mass from the parent cloud and thus the ISM. During the stars' lifetime, matter is deposited back into the ISM. This can happen as a result of winds from massive stars or mass ejections from young or late evolutionary stages. Mass loss from massive stars ($M > 20\ M_\odot$) is especially important for replenishing the ISM with reprocessed material. Here, for example, we consider stars with mass-loss rates in excess of $10^{-7}\ M_\odot\ \mathrm{yr}^{-1}$, like W-R stars and most early supergiants. Also M supergiants are important and where they appear in larger concentrations in our Galaxy they actually return more material back to the ISM than W-R stars [77]. Mass-losing AGB stars return up to 3–6 $\times 10^{-4}\ M_\odot\ \mathrm{yr}^{-1}\ \mathrm{kpc}^{-2}$ [78]. When massive stars die they return a large portion of their mass to the ISM, enriched by nucleosynthesis. In general, it is estimated that about 20 percent of the mass used up in star formation is fed back into the ISM after the star is born. There are also indications that the Galaxy is still accreting some material from the *intergalactic medium (IGM)*, but at the same time is losing some of it into the IGM through diffusion processes or in catastrophic galaxy collisions.

All in all, it is hard to believe that we have equilibrium conditions under these circumstances. In fact the ISM is far from any dynamic, physical, and chemical equilibrium and is under constant change. Much of these conditions are a consequence of the localized structures of the ISM, where 'point-like' stars provide

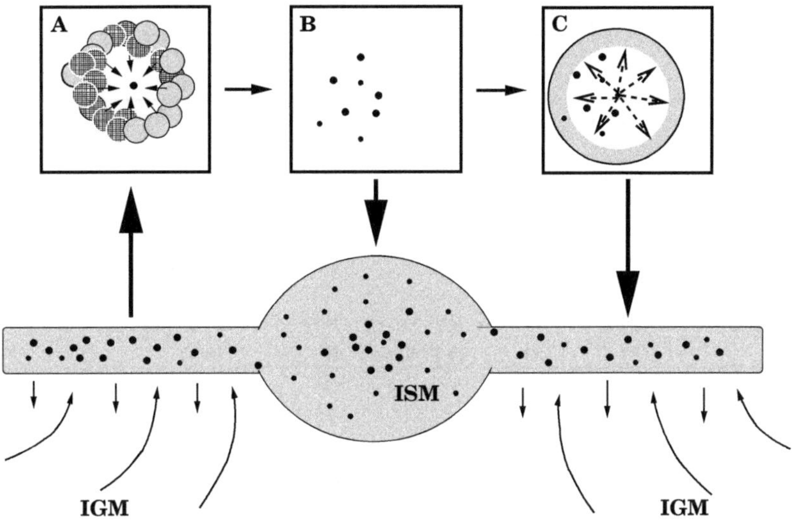

Fig. 3.1 The ISM is not just matter filling the space between stars, it is quite an active medium. Stars evolve from the ISM (A) and feed (in part) reprocessed matter back during their life cycle (B) and when they die (C). The ISM also accretes small amounts from, as some material diffuses into, the intergalactic medium (IGM). Adapted from an illustration by G. Knapp, *Sky&Telescope* 2001

the main energy sources. As a result the structure of the ISM is highly complex and temperature, density, and even velocity varies vastly across the Galaxy.

3.1.2 Physical Properties of the ISM

The stars are spread so far apart that the average interstellar space is actually very cold with a mean temperature of only a few degrees above absolute zero. Thus the mean density is just one hydrogen atom per cm^3. However depending on the location in space, the density of the ISM varies by many orders of magnitude. Table 3.1 summarizes these properties. Observed temperatures range from near absolute zero to over 10^6 K, and densities range from 0.001 to over 10,000 atoms per cm^3 in molecular cloud cores.

Molecular clouds are large condensations of cold gas with very high densities. They are quite ubiquitous thoughout the Galaxy. Their masses can reach up to one million Suns (see Chap. 4 for details). Most dominant throughout the ISM is atomic hydrogen which occupies about 50 percent of the volume. A large fraction of this hydrogen appears in cold clouds and warm diffuse gas. Cold atomic clouds are comparatively small, with sizes of around 10 pc and masses of about 1,000 Suns. The average distance between these clouds is about 150 pc. These clouds are embedded in a more tenuous and diffuse warm gas. The gas pressures of these

Table 3.1 Phases of the ISM ([82])

Medium	Phase	H state	n (cm^{-3})	T (K)	Heating signature	Comments
Molecular clouds, cores	Cold	H_2	>1,000	10–50	Cosmic rays	Icy dust
H I clouds	Cold	H	30	100	Dust	Diffuse ISM
Warm H I	Warm	H	0.1	8,000	Dust	Diffuse ISM
Warm H II	Warm	H^+	0.03	10^4	Photoionization	Faint
H II regions	Warm	H^+	>100	10^4	Photoionization	Transient, expanding
Hot ISM	Hot	H^+	10^{-3}	$10^{6.5}$	SNe shocks	Low mass
SNRs	Hot	H^+	Variable	10^7	Shocks	Dynamic

phases can be described by the ideal gas law; see Sect. A.2 (i.e., they are proportional to temperature and number density). Since these pressures turn out to be quite similar the two media can coexist in a fairly stable manner. The hot ISM appears either as a coronal gas with an extremely low density of 0.001 hydrogen atoms per cm^3 or as hot gas from shocks in supernova remnants (SNRs). Although its total mass is extremely small compared to the other phases, this hot ISM occupies almost 50 percent of the H I volume. In this respect the interplay of the three phases gives the ISM an almost 'sponge-like' structure.

3.1.3 The Local ISM

The Local ISM (LISM) is an example of how the ISM structure appears spatially distributed in the Galaxy. The LISM is the neighborhood of our Sun, within a radius of 460 pc (1,500 lyr). The Sun is about 8,500 pc away from the Galactic Center and is located in the Orion spiral arm of the Milky Way. Figure 3.2 shows the plane-projected locations of molecular clouds, stellar associations, and the diffuse LISM with respect to the Sun [79]. The Sun is surrounded by quite a number of gas clouds. These clouds, of course, are still thin enough so one can see right through them. The Sun is located in a void comprising coronal gas, which we refer to as the *Local Bubble* [80, 81]. This void borders into molecular clouds near the *Scorpius–Centaurus* stellar association, which extends into the Aquila-Rift, into the Gum Nebula and the Vela-Remnant, into the Taurus cloud, and into the Orion stellar association with its molecular clouds (see also Chap. 12). Part of the latter is the famous Horsehead Nebula located near Orion's belt. The Gum Nebula is a complex region of ionized hydrogen. Many surrounding clouds and associations are regions of fairly recent star formation (<10–50 Myr ago) and part of what is called the *Lindblad's Ring*, an expanding H I ring surrounding the Sun, and *Gould's Belt*. In fact the Local Bubble may not really be a bubble any more as there are indications that these bordering star-forming regions are pushing into an older void formed

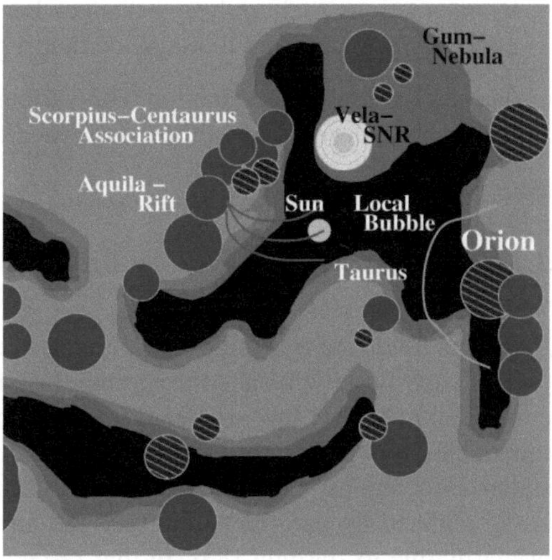

Fig. 3.2 A schematic view of the LISM. The baseline of the diagram is ∼1,500 lyr. The black areas represent hot, low-density regions, the gray areas warm, more dense diffuse gas. The dark circles are molecular clouds and the hatched circles major stellar associations. Adapted from an illustration by L. Huff (*American Scientist*) and P. Frisch (U. Chicago), 2001)

by the Lindblad Ring. On the other hand the LISM may also still be under the influence of recent star-forming activity. The filaments reaching towards the Sun in the diagram in Fig. 3.2 are warm, partly ionized gas shells resulting from star formation 4–10 Myr ago in the Scorpius–Centaurus association [79].

3.1.4 Phases of the ISM

Everything that fills the space between the stars within the Galaxy constitutes the ISM. This includes all radiation fields and magnetic fields, although this chapter will concentrate mostly on the matter part. There is gas, which consists of atoms, molecules, ions, and free electrons, and there is solid matter of a wide range of sizes, from microscopic dust particles to larger, more complex, dust grains. Of interest are ISM characteristics such as temperature, mass density, abundances of elements, and chemical composition of molecules. The ISM appears in three phases, a cold phase consisting of molecular and atomic hydrogen gas and dust, a warm phase with atomic hydrogen and ionized hydrogen gas, and a hot phase with shocked gas from supernova explosions as well as what is referred to as coronal gas (Table 3.1).

The phases of the ISM as decribed in Table 3.1 show temperature components ranging from a mere 10 K to 10^4 K for the bulk of mass, with over 10^7 K in the most extreme cases. This requires a variety of heating processes which purely for comprehensive reasons may be characterized in three categories:

- by heating from radiation fields; photoelectric absorption in the neutral ISM components, photodissociation in molecules, photo-ionization.
- by collisional heating from turbulent flows, supernova remnant shocks, and cosmic rays.

- by interactions such as heat exhanges between the dust and gas components, coupling of ionized gas to magnetic fields, gravitational collapse.

These processes in conjunction with various cooling rate components provide a semi-empirical description of the energy balance in the ISM.

3.1.5 Interstellar Radiation Fields

Radiation fields are a common property of the interstellar medium and significant photon emissivities emanate throughout the entire electromagnetic waveband from coronal sources of a few angstrom to the cosmic microwave background. Most of this radiation is highly anisotropic with significant variations at locations close to stellar associations, clusters, and near molecular clouds. Interactions with matter are dominated by the visible and UV radiation field in the diffuse ISM which is primarily due to starlight [83]. In that respect there is no average global field but one has to always normalize the field strengths to location, i.e. the normalizations in the solar neighborhood are different than at a location at several kpc distance [84].

The interstellar radiation field (ISRF) in the Far-UV is dominated by photospheric emission from hot OB stars, while in the optical and UV low-mass main sequence stars become important, mostly A- and F-type stars in the optical, and stars with M $< M_\odot$ in the red and near-IR bands. The strength of the FUV/UV radiation field varies and has anisotropic components. On average, OB stars produce up to a few 10^6 photons cm^{-2} s^{-1} Å^{-1} in the UV, normal stars about several 10^5 photons cm^{-2} s^{-1} Å^{-1} in the optical band. The integrated photon intensity from the hydrogen K absorption edge (911 Å) up to about 2,400 Å amounts to about 1.7×10^8 photons cm^{-2} s^{-1} on average. At regions near PDRs which are illuminated by close stars, fluxes can be over an order of magnitude larger.

Serious attempts to determine the ISRF stem back to the late 1960s, when H. Habing [85] used a large ensemble of nearby OB stars to recalculate the interstellar radiation density, later expressed in terms of the Habing field of 1.2×10^{-4} erg s^{-1} sr^{-1}, which corresponds to about 10^8 photons cm^{-2} s^{-1}. The average FUV ISRF G_o is then 1.7 Habing fields. The study also already realized that the albedo of interstellar grains is a considerable source of uncertainty in the UV and that intensity is considerably reduced below 1,100 Å due to anomalous interstellar extinction. More recently Henry [86] used over 118,000 stars in the *Hipparcos* catalogue to calculate the strength of the FUV/UV ISRF in detail emphasizing the importance of radiation scattering and absorption. Figure 3.3 shows the expected radiation field for the unrealistic assumption that all scattered photons remain part of the ISRF at an albedo, i.e. the amount of reflectivity of dust, of unity. The large break at 911 Å is due to photo-electric absorption of hydrogen. Inclusion of albedos below unity and interstellar extinction significantly reduces the FUV field.

In order to constrain properties of the local interstellar cloud at the heliosphere, Slavin and Frisch [87] modeled the background radiation field using local stellar

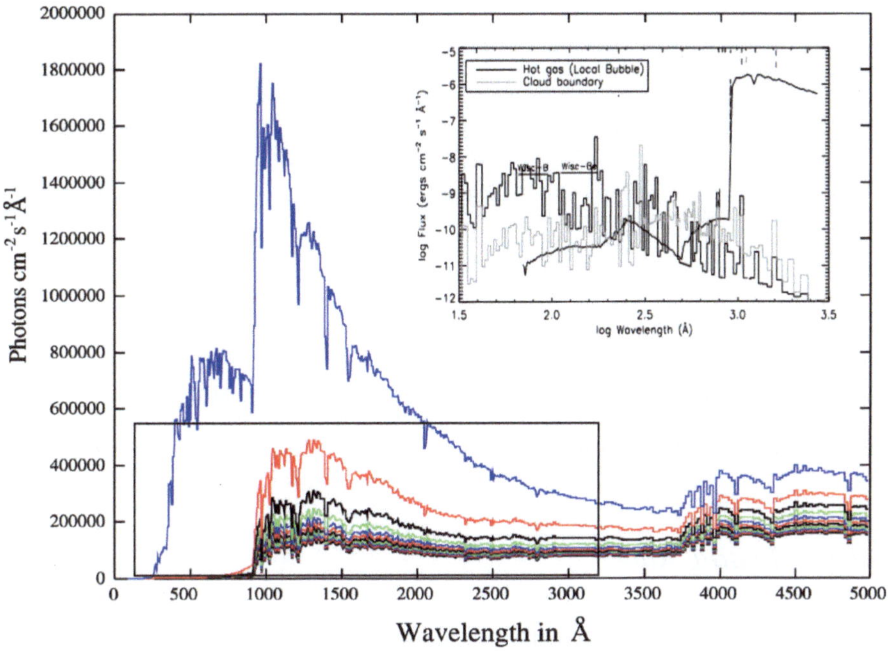

Fig. 3.3 The expected radiation field (*blue*) from the spectral distribution of the integrated radiation of all stars in the *Hipparcos* catalog and the assumption that all scattered photons remain part of the ISRF (from [86]). Albedos in this calculation range from 1 (*blue*), to 0.6 (*red*) to 0.1 (*black*). More realistic cases (*other colors*) appear when dust albedos of significantly less than one are introduced and some scattered radiation is absorbed by dust. The inset shows the regime indicated by the black rectangle in log scale computed for the solar neighborhood (from [87])

FUV/UV emission and the soft X-ray background. The local cloud within the LISM (see Fig. 3.2) is embedded in the low density and hot ($>10^6$ K) Local Bubble (see Sect. 3.1.3). Modeled IRSF emissions from this hot Local Bubble and the local cloud boundary are shown in the inset of Figure 3.3. These ultra-far UV and X-ray emissions are present thoughout the galactic ISM and part of the hot ISM. Absorptions by this hot medium are observed in X-ray spectra of bright X-ray sources (see Fig. 3.4).

In large part as a consequence of the blue (UV) ISRF there is diffuse thermal IR radiation from dust grains at wavelengths larger than 0.1 mm. First surveys showed that these emissions originate mostly from regions in the solar neighborhood [88, 89]. More detailed studies find that about 80% of the observed FIR/sub-mm flux comes from dust embedded in extended low-density H II regions [90]. The rest is likely associated with dust immersed in diffuse atomic hydrogen. The shape of the ISRF can be modeled using the MRN approach (see Appendix C.9) using a mixture of bare graphite and silicate grains with a size distribution $\propto a^{-3.5}$ and with column densities of atomic and ionized hydrogen primarily heated by OB stars. It has therefore been suggested that this FIR emission be used to estimate the total OB star formation rate in galaxies [90].

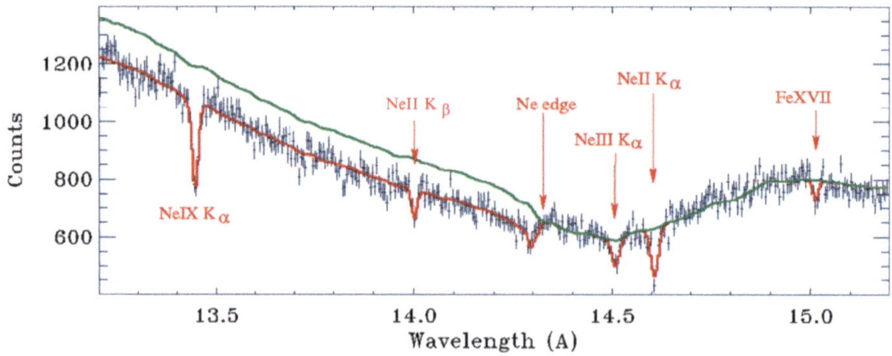

Fig. 3.4 Manifestation of cold, warm, and hot phases of the ISM in bright X-ray spectra. It shows the K photoelectric edge from cold neon, the II and III 1s-2p (1s-3p) absorption from warm neon ions, as well as the highly ionized absorption from hot neon IX and Fe XVII ions. Credit: Y. Yao, CASA

3.1.6 Heating Rates

The average kinetic heating within the phases involving radiation fields can be expressed through:

$$\int_{\nu_i}^{\infty} I_\nu \sigma_i(\nu) E_i \, d\nu = \frac{3}{2} k T_i \int_{\nu_i}^{\infty} I_\nu \sigma_i(\nu) \, d\nu \tag{3.1}$$

where I_ν is the mean photon intensity of the incident radiation field, ν_i corresponds to the photo-ionization threshold of a species i, i.e. a specific atom, molecule, or dust grain. σ_i, E_i, and T_i are the ionization cross-section, the kinetic energy of the photo-electron, and the average temperature, respectively. Using the constant field $G_{rad} = G_o$, heating rates $n\Gamma_x$ for specific processes x can then be reduced to the form:

$$n\Gamma_x = C_x \epsilon_x \langle n_x \rangle \langle G_{rad} \rangle \; [\text{erg cm}^{-3}\,\text{s}^{-1}] \tag{3.2}$$

where n is electron density and C_x a characteristic scaling factor for the specific process. The important physical dependencies are then summarized in the efficiency ϵ_x for each process. Table 3.2 summarizes these components for some of the more important cases.

The first category requires the presence of radiation fields, which in H I regions, i.e. outside H II regions, originate from lower mass stellar populations and are generally not energetic enough to overcome the H ionization potential of 13.6 eV. Exceptions are far-UV emissions from star-forming regions and X-ray emissions. With the bulk of the mass fraction in the neutral ISM, heating through photoelectric absorption in large molecules such as polycyclic aromatic hydrocarbons (PAHs) and dust grains provides an important contribution to the heating balance. The reason why PAHs are so effective in heating the ISM gas is because they are large and abundant, and contain a lot of carbon. The photoelectric heating efficiency in the

Table 3.2 ISM Heating Rates Components (Most inputs from [91])

Process x	region	C_x	$\langle n_x \rangle$	ϵ_x	$\langle G_{rad} \rangle$
photoelectric absorption	PAHs	10^{-24}	n_e	$\frac{1}{2} f_0 \frac{h\nu - E_{IP}}{h\nu}$	G_o
	dust	10^{-24}	n_e	$\sim Y \frac{h\nu - E_{grain}}{h\nu}$	G_o
photo-dissociation	H_2	1.4×10^{-23}	n_{H_2} $(n_e < 10^4)$	$f_\tau\ e^{-2.6A_V}$	G_o
photo-ionization	H I	2.2×10^{-22}	n_e	$f_C A_C\ e^{-2.6A_V}$	G_o
	H II	1	$4\pi\ n_i$	1	$\int_{\nu_i}^{\infty} G_{UV} h$ $\times(\nu - \nu_i)\sigma_i\ d\nu_i$
Dustgas heat exchange		$\sim 10^{-33}$	n_e^2	$T^{1/2}(T_d - T)$	1
Cosmic rays		$\sim 10^{-27}$	n_e	ζ_{CR}	1
X-rays		1	$4\pi\ n_e$	1	$\int G_{XR}(\nu)$ $\times e^{-\sigma_\nu N} \sigma_\nu\ d\nu$
Gravity	Cores	$\sim 4.3 \times 10^{-31}$	$n^{3/2}$	T	1
Turbulence	H I	2.8×10^{-30}	n_e		1
	Cores	3×10^{-28}	n_e		1
Mag. diffusion	Cores	$\sim 1.6 \times 10^{-30}$	$n^{3/2}$	T	1

f_0 is the neutral PAH fraction
E_{EP} is the ionization potential
Y is the yield defined by the dust opacity χ^d (see Appendix C.9)
E_{grain} is the energy of the grain
f_τ is a self-shielding factor by H_2 at optical depth τ from FUV excitation
A_V is the visual extinction
f_C and A_C are the carbon fraction and abundance, respectively
ζ_{CR} is the cosmic-ray ionization rate in fractions of 2×10^{-16}
G_{XR} is the mean X-ray intensity, σ_ν the X-ray absorption cross-section

case of PAHs can then be expressed as the fraction of ionizing energy with respect to the ionization potential (see Table 3.2), where $h\nu$ is the photon energy, E_{IP} the ionization potential, and f_0 the fraction of neutral PAHs. The latter follows from the trend that for very large or ionized molecules E_{IP} likely exceeds 13.6 eV. For typical PAHs in H I clouds f_0 is more close to unity, whereas in dense and more ionized environments such as photodissociation regions (PDRs), f_0 is significantly reduced.

The processes in dust grains are only slightly different. Far-UV photons absorbed by grains lose energy by cascading through various collisions. Those electrons overcoming the work function and electrostatic potential ($E_{grain} = W_{grain} + \phi_c$) of the grain during this diffusion process are injected back into the ISM with an estimated efficiency (see Table 3.2), where the yield Y is defined by the dust opacity χ^d (see Appendix C.9). In essence, for large grain sizes the escape probability for photoelectrons is low and the yield is small; for smaller sizes it can reach unity. The heating efficiency is reduced when the grain is charged, an effect which is negligible in diffuse clouds but important in dense PDRs.

The situation changes near dense molecular clouds where H_2 and other small molecules become increasingly abundant. In general, fragments of photodissociated molecules heat the ambient gas through transfer of residual kinetic energy. With the sheer abundance of H_2 in these regions as well as by its kinetic preference, photodissociation is most effective for this process. There are several aspects to this process depending into which vibrationally excited ground state the FUV excited molecule decays into. The efficiency with which the H_2 molecule entirely dissociates is $\epsilon_{dis} \sim 0.15$. The remaining ~85% decay back to some vibrationally excited gound state where they either cascade back to the ground state emitting IR photons or through primarily collisional de-excitations in very dense ($>10^4$ cm^{-3}) clouds.

Figure 3.5 shows some of the important heating rates over large density regimes in the ISM. At very low densities as in the warm neutral phase, photo-excitation,

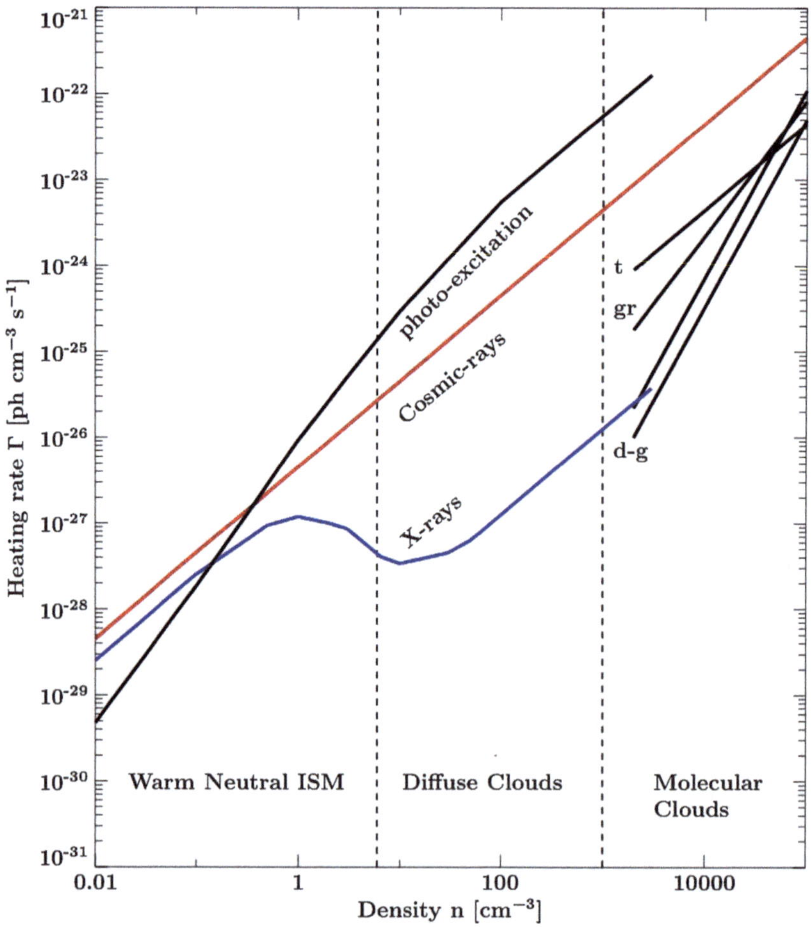

Fig. 3.5 Important heating rates over large density regimes in the ISM (numbers adapted from Table 3.2 and [91])

Cosmic ray, and X-ray heating are similarly important, in the immediate regime as in diffuse clouds, most X-rays become absorbed. At high densities as in molecular clouds photo-excitation is inefficient as radiation is absorbed, but many of the other contributions appear ranging from turbulence (t), gravitation (gr), and dust–gas exchange (d-g). Heating by cosmic rays is here the dominant process.

3.1.7 Cooling Processes in the ISM

Most if not all cooling in the ISM is radiative, either through spontaneous, or collisional de-excitations. How effective these processes are depends on the local physical environment, specifically for collisions. In atomic gas radiative cooling is provided through electronic de-excitation and recombination, in molecular gas cooling also involves rotational and vibrational states. A description of cooling in an interstellar gas is therefore complex as it depends on the conditions at a local level and the radiative transfer properties in that region. However, for many applications a description at low densities appears fairly adequate.

In general the cooling rate due to a single transition *fi* and species z is described by

$$n^2 \Lambda_z(h\nu_{fi}) = n_f A_{fi} h\nu_{fi} \beta(\tau_z) S'_z(h\nu_{fi}) \tag{3.3}$$

where n_f is the number density of species z in an excited state f, A_{fi} the spontaneous transition probability, τ_z the optical depth of species z, and $h\nu_{fi}$ the energy of the emitted photon. $\beta(\tau_z)$ is the escape probability of the photon from the cooling medium, $S'(h\nu_{fi})$ is the source function containing the statistical weights of transition fi, and a correction for the background radiation field due to the cosmic microwave background infrared dust emission.

In many situations a two-level system which considers both, collisional and spontaneous de-excitation provides a good description. Spontaneous emissions are simply described by the Einstein coefficient A_{fi} and level populations are defined by their statistical weights. The collisional de-excitation rate γ_{fi} is decribed through level populations given by the Boltzmann expression at the kinetic temperature of the gas [91]:

$$\gamma_{fi} = \frac{4}{\sqrt{\pi}} \left(\frac{\mu}{2kT} \right)^{3/2} \int_0^\infty \sigma_{fi}(v) v^3 \exp\left[-\frac{\mu v^2}{2kT} \right] \mathrm{d}v, \tag{3.4}$$

where μ is the reduced mass, σ_{fi} the collisional de-excitation cross-section, and v the particle velocity. In a two-level system with spontaneous and collisional de-excitation a critical density

$$n_{cr} = \frac{A_{fi}}{\gamma_{fi}} \tag{3.5}$$

allows us to relate collisional rates to all levels. If we further ignore the possibility of photon trapping (i.e. $\beta(\tau_z) = 1$) and heating background ($S'_x(h\nu_{fi}) = 1$), and acknowledge that $n_f + n_i = nA_z$, where A_z is the abundance of species z, then Eqn. (3.3) simplifies considerably and in the limit of low densities ($n < 3\times10^3$ cm^3) becomes

$$n^2\lambda \approx n^2 A_z \gamma_{fi} h\nu_{fi}. \tag{3.6}$$

Typical de-excitation rate coefficients for neutral collision particles are of the order of $3\times10^{-11}\,T^{1/2}$ cm^3 s^{-1}, for electrons with neutral particles 1×10^{-9} cm^3 s^{-1}, and for electron–ion collisions $1\times10^{-5}\,T^{-1/2}$ cm^3 s^{-1}. Ionized gases have a few other means of cooling through either radiative recombination through free-bound transitions of electrons or bremsstrahlung of electrons through free–free transitions of electrons; neither, however, is particularly effective. Here additional cooling is provided by low-lying electronic states of higher Z trace species such as C, N, and O below temperatures of 10,000 K. The lowest excited level of hydrogen corresponds to temperatures above about 120,000 K (10.5 eV).

The atomic interstellar cooling function for various ionization fractions x is shown in Figure 3.6. It shows this low-density case in dependence of temperature. Most of the interstellar medium outside denser regions such as molecular clouds and photodissociation region can be described by this function. It shows the effective cooling at low temperatures by neutral and ionized trace species. Above 10,000 K

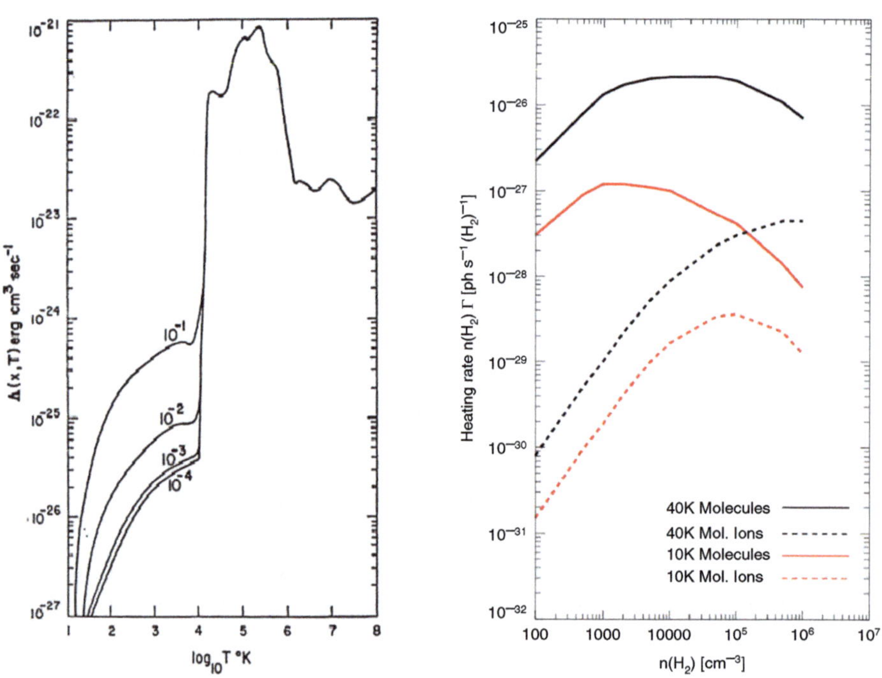

Fig. 3.6 *Left:* The cooling rate for (atomic) interstellar gas as function of temperature for the low-density case ($n < 3\times10^3$ cm^{-3}) and various ionization fractions (from Dalgarno & McCray [1343]). *Right:* The cooling rate per H_2 molecule for molecular gas as a function of density (numbers adapted from [91])

cooling is largely provided due to collisional Lyman α excitation of neutral H. Above this temperature collisional ionizations are highly effective and the cooling rate increases rapidly. Above 10,000 K hydrogen becomes fully ionized, at higher temperatures the other trace elements below Fe follow. Above 10^7 K most of the cooling then is due to bremsstrahlung.

Cooling rates in molecular gas are more effective because rotational and vibrational transitions are much more abundant and cooling occurs in a much larger wavelength range well beyond tens of micrometers. Molecular clouds are therefore generally not only high in density but also much colder. CO molecules are most abundant and provide most of the cooling; at higher densities O_2 and H_2O also contribute significantly. Even though large molecules such as PAHs are quite powerful cooling agents their abundance is simply too low to be of importance. Ionization fractions in molecular clouds are also generally low. However, ionized molecules possess large dipole moments for effective cooling also at very higher densities (Fig. 3.6) where cosmic rays persistently keep some ionization fractions.

Cooling times are very difficult to assess specifically for warm and hot media. The cool phase is relatively straightforward once equilibrium conditions apply where the cooling time equals the heating time. The latter then can be determined via ideal gas conditions where the integrated heat $\Gamma t_{th} = 3/2(1.1 + x)n_H kT$ in the isobaric case, which for very low ionization fractions can be approximated to

$$t_{th} \approx 10^5 \left(\frac{n}{[10^3\ \mathrm{cm}^{-3}]}\right)\left(\frac{T}{[100\ \mathrm{K}]}\right)\left(\frac{\Gamma}{[10^{-26}\ \mathrm{erg\ s}^{-1}\ \mathrm{cm}^{-3}]}\right)^{-1} \mathrm{yr}. \tag{3.7}$$

This means that for cold media thermal equilibrium can be reached in less than 10^5 yr. For warm and hot media the linear dependence on density and temperature is not correct specifically when conditions are well out of equilibrium. However, in the case where the media are on average close to equilibrium conditions, Eqn. 3.7 is not unreasonable. For typical warm medium parameters, i.e, $n \sim 0.1$, $T \sim 4 \times 10^3$, $\gamma \sim 10^{-27}$ ergs^{-1} cm^{-3} the cooling time is still of the order of 4×10^5 yr. For hot medium parameters cooling is quite short, i.e. $<10^4$ yr, but slows down again once temperatures fall below 10^4 K. The short cooling time again indicates that since we observe a sustained hot phase, it must be far out of equilibrium.

3.1.8 Heating and Cooling in H II Regions

In typical H II regions EUV radiation from an early type star, generally of type O3–B1, ionizes and consequently heats the surrounding gas within a finite radius. The residual of the balance between photo-ionization and radiative recombination determines the ionization fractions of this H II region. The actual temperature of the gas within the H II region is finally determined by heating and cooling through thermalized photo-electrons and collisional de-excitations. The EUV field ionizes its environment to a large degree and and the degree of ionization in a pure hydrogen nebula $x = n_p/(n_p + n_H)$ defines a neutral fraction $f_n = 1 - x$, which is usually of the order of 4×10^{-4}. In such a nebula the radially dependent photo-ionization heating can then be expressed by

$$n\Gamma_H(r) = f_n 4\pi n \int_{\nu_i}^{\infty} G_{UV}(\nu, r)\sigma_i h(\nu - \nu_i)\, d\nu \tag{3.8}$$

(see Table 3.2), where $h\nu_i$ is larger than 13.6 eV (911.6 Å). Since the local ionization balance in H II regions depends on the recombination rate coefficient of all atomic hydrogen levels with $n > 1$ $\beta_{n>1}(T_e) = 2.6 \times 10^{-13}(10^4/T_e)^{0.8}$ at an electron temperature T_e,

$$n_H 4\pi \int_{\nu_i}^{\infty} G_{UV}(\nu)\sigma_i\, d\nu = n_e^2 \beta_B(T_e) \tag{3.9}$$

and taking $3/2kT_H$ for the net average thermalized energy of the photo-electron the heating rate becomes

$$n\Gamma_H(r) = x^2 n^2 \beta_{n>1}(T_e)\frac{3}{2}kT_{bb}(r). \tag{3.10}$$

The average energy of electrons is still low with respect to the hydrogen ionization theshold of 13.6 eV and levels with $n < 2$ are not relevant. Also, the radiation field to first approximation is caused by the central star with a black-body temperature T_{bb}.

The cooling part is generally dominated by free–bound and free–free de-excitations, though in the case of hydrogen not very effective as its lowest excited level (10.2 eV) is still higher than the average energy of electrons (0.7 eV). Effective cooling is provided by trace elements providing an almost shell-like structure with distance to the ionizing star. In that respect the highest net heating rate is reached at the outer edge of the H II shell, which drops by a few factors inside due to cooling contributions of various trace ions. Figure 3.7 shows the temperature profile for a typical H II bubble, in the shown case using the radiation field of an O4 star and densities of 10^3 cm^{-3}.

The boundary of the H II region can be found from the ionization radius in which all ionizing photons are effective and engulfs the total recombination rate integrated over the entire nebula. The radius of the defining spherical volume is called *Strömgren radius* and can be written as:

$$\begin{aligned} R_{H\ II} &= \left(\frac{3}{4\pi\beta_{n>1}}\right)^{1/3} \mathcal{N}_{UV}^{1/3} n_e^{-2/3} \\ &\approx 0.23 \left(\frac{n_e}{[10^3\ \text{cm}^{-3}]}\right)^{-2/3} \left(\frac{\mathcal{N}_{UV}}{[10^{49}\ \text{photons s}^{-1}]}\right)^{-1/3} \text{pc} \end{aligned} \tag{3.11}$$

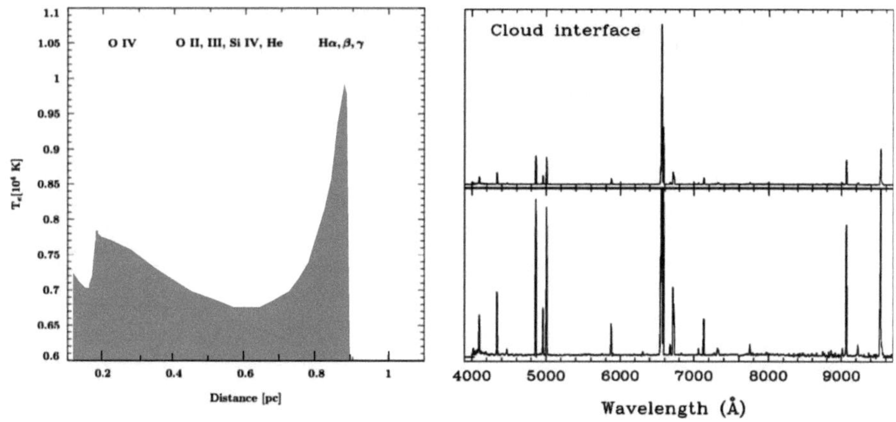

Fig. 3.7 *Left:* The calculated radial temperature distribution in an H II region (adapted from [91]). *Right:* The optical spectrum of the massive compact H II region IRAS 16362-4845 near the RCW108 molecular cloud in the Ara OB1 association [92]

where $\mathcal{N}_{UV}$ is the total rate of ionizing UV photons. For an O-star with a UV photon rate of 4×10^{49} photons s^{-1}, a typical temperature of 10,000 K for H II regions with a density of 10^3 cm^{-3}, this radius is about 0.9 pc as shown in Figure 3.7.

3.1.9 Interstellar Shocks

The ISM is under the constant influence of perturbations. Most events that cause these perturbations are quite violent impacts of *supernova explosions, stellar winds*, and *cloud collisions*. Once a certain pressure is reached these events drive *shocks* through the ISM and through even denser clouds. Other events that can increase the pressure in clouds are photo-ionization and photo-evaporation. The resulting shock wave is treated as a pressure-driven fluid-dynamical disturbance. As has been explained [93], these shocks are irreversible according to the second law of thermodynamics, as the kinetic energy of the gas is dissipated into heat thus increasing the entropy of the gas. In general, shock waves are always *compressive* and heat is actually dissipated. In dense neutral gas this can happen through molecular viscosity. In plasmas, depending on density, velocity gradients occur through either collisionless dissipation or friction. Specifically in partially ionized plasmas it can impose a so-called *ambipolar shock*, which is caused by a jump-like neutral-ion drift (see also Sect. 4.3.4).

$$\mathcal{M}_s = \frac{v_s}{c_s} \quad \text{and} \quad \mathcal{M}_A = \frac{v_s}{v_A} \tag{3.12}$$

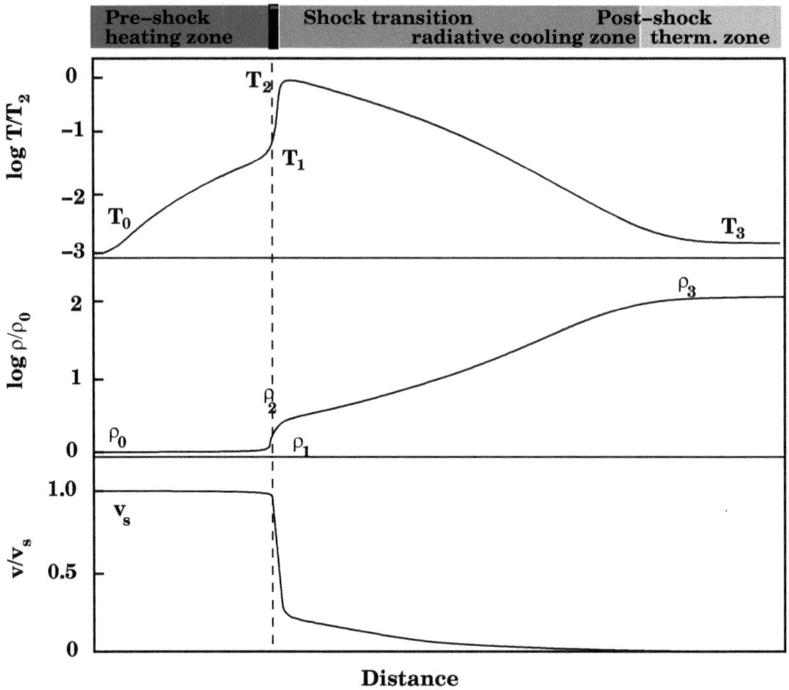

Fig. 3.8 Temperature, density, and velocity dependence of a strong shock in a single fluid flow distinguishing between three shock zones: the pre-shock with radiative heating, the shock transition zone with jumps in temperature, density and velocity, and the post-shock zone which is cooling radiatively. Adapted from Draine and McKee [93]

$\mathcal{M}_s$ is the ordinary Mach number and $\mathcal{M}_A$ is the Alfvén Mach number. For strong adiabatic shocks (i.e., $\mathcal{M}_s \gg 1; \mathcal{M}_A \gg 1$), which may be more relevant for stellar formation, the compression ratio reaches the limit of 4, while the post-shock equilibrium temperature at T_2 (see Fig. 3.8) for a pure and fully ionized hydrogen gas is

$$T_2 = 1.38 \times 10^5 \left(\frac{v_s}{[100 \text{ km s}^{-1}]} \right)^2 \text{K}, \tag{3.13}$$

while for a pure but neutral hydrogen gas is

$$T_2 = 2.87 \times 10^5 \left(\frac{v_s}{[100 \text{ km s}^{-1}]} \right)^2 \text{K}, \tag{3.14}$$

and for a mix somewhere in between (from [93]). For turbulent velocities that are typical of the ISM one could get shock temperatures of up to 10,000 K. In more violent environments like the acceleration zones of stellar winds where shock velocities reach over 1,000 km s^{-1} the temperatures will exceed 1 million K and the cooling radiation may be observed in the X-ray band.

Supernova explosions are a major source to heat the hot component of the ISM. Typical explosion energies of $E_{51} = 10^{51}$ ergs and initial average blast wave velocities of $\sim 10^4 (E_{51}/M_{ej})^{1/2}$ km s^{-1} provide initial shock temperatures $>10^8$ K. The shock wave propagating into the ISM during the free expansion phase may be approximated by the Sedov–Taylor blast wave solution which provides

$$r_s(t) = 1.5 \times 10^{19} \left(\frac{E_{51}}{n_{ej}}\right)^{1/5} \left(\frac{t}{[10^3\ \mathrm{yr}]}\right)^{2/5} \ \mathrm{cm}$$

$$v_s(t) = 1.9 \times 10^{3} \left(\frac{E_{51}}{n_{ej}}\right)^{1/5} \left(\frac{t}{[10^3\ \mathrm{yr}]}\right)^{-3/5} \ \mathrm{km\ s^{-1}} \tag{3.15}$$

$$T_s(t) = 5.3 \times 10^{7} \left(\frac{E_{51}}{n_{ej}}\right)^{2/5} \left(\frac{t}{[10^3\ \mathrm{yr}]}\right)^{-6/5} \ \mathrm{K} \tag{3.16}$$

where r_s is the distance of the shock, $v_s = dr_s/dt$ the blast wave velocity at r_s, T_s the shock temperature, and t the age of the blast wave. Once the shock velocity drops below a critical value of about 180 km s^{-1}, about one-third of the energy is radiated away and the adiabatic approximation of the Sedov–Taylor solution breaks down. This corresponds to a radius $r_s \approx 30$ pc. Now collisions become important and the blast wave fades away until its reaches its sound speed at $r_s \approx 60$ pc. At the end of the adiabatic phase the shock heats the ISM to about the temperature of the hot component. At a supernova rate of $\sim$50 yr^{-1} in the Galaxy and by comparing the remnant volume of a 30 pc radius to the total volume of the Galaxy this means that the same volume in the ISM occupied by the remnant will see another blast wave only after a few 10^8 yr. From this it is clear that the hot component of the ISM in the Galaxy has to be highly anisotropic.

3.1.10 Heating and Ionization by Cosmic Rays

All phases and components of the ISM are under constant bombardment of highly energetic particles which make up the bulk of what we call cosmic rays (see Sect. 3.1.11). It constitutes quite an important contribution to the overall heating rate. With a scaling factor C_x of the order of 10^{-27} (Table 3.2) cosmic ray ionizations are almost as effective as photoelectic absorptions, photodissociations and -ionizations which depend on local radiation field fluxes of much larger than 10^{-4} erg cm^{-2} s^{-1}. This scaling factor results from an average thermal energy transfer per ionization of about 9.6×10^{-12} erg (6.0 eV) in diffuse H I regions. In dense environments such as molecular clouds cosmic rays also provide the dominant contribution over turbulence, gravity, magnetic diffusion, and dust-gas exchange as the ionization rate for H_2 is about 1.6 times higher than for the atomic case (Fig. 3.5).

Most cosmic rays consist of a mixture of particles with a dominance of relativistic protons, next to α prarticles and some heavier elements. There are some electons as well, however with a low particle density. The bulk of the electrons primarily responsible for the heating and ionization come from proton collisions with interstellar matter. This produces a diffuse γ ray component in the ISM (see Sect. 3.2.5), while proton interactions are particularly effective in molecular clouds. Most common is ionization of H_2 and H I. Note, that even though H_2 dominates in molecular clouds, protons and electrons can scatter inelastically and excite H_2 molecules to the point of dissociation and consequently ionization of atomic H becomes relevant.

3.1.11 The Origins of Cosmic Rays

Cosmic rays are an important source of ionization in the ISM and specifically in molecular clouds (see Fig. 3.5 and Sect. 3.1.10). Their discovery dates back to 1912 when Victor Hess found that the ionization rate increased substantially with altitude which could neither be explained by the Earth's natural radioactivity nor being an effect coming from the Sun as the increase remained even during eclipse events [94]. The term "cosmic rays" itself was introduced by Robert Millikan who could prove that they were extraterrestrial in origin.

Gottlieb and Van Allen showed that the primary cosmic particles are mostly protons (H nuclei, 89%) with some α particles (He nuclei, 10%) and a small fraction of heavier particles. All of the rest of the elements make up only 1%, including very rare elements and isotopes in significant abundances such as Li, Be, B and ^{22}Ne, respectively. It should be noted that for measurements of the cosmic ray flux from Earth, the main source still is the Sun which produces significant fluxes uo the 10 GeV.

Cosmic Ray particles are accelerated to high energies with the bulk in the range between 0.1 and 10 GeV, which makes these particles highly relativistic with up to 0.996c. The highest energies can reach over 10^{20} eV (see Fig. 3.9, left). The cosmic ray flux up to 10^{15} eV has beem reasonably well studied and can be decribed as

$$\frac{\mathrm{d}N}{\mathrm{d}E} = kE^{-\alpha} \tag{3.17}$$

with $k \sim 5 \times 10^{-1}$ cm-2 s^{-1} sr^{-1} GeV^{-1} and $\alpha \sim 2.5$ above 1 GeV, and $\alpha \sim 0.2$ between 0.1 and 1 GeV. It is that last part of the spectrum which is relevant for the cosmic ray heating rate.

For a long time it has been speculated that cosmic rays are accelerated in the blast waves of supernova remnants. In these remnants expanding clouds of gas and magnetic fields can effectively accelerate cosmic rays. This has been shown in the case of supernova remnant RXJ1713.7-3946 [95]. Observations with the *Chandra* and *Suzaku* X-ray Observatories in conjunction with the High Energy Stereoscopic

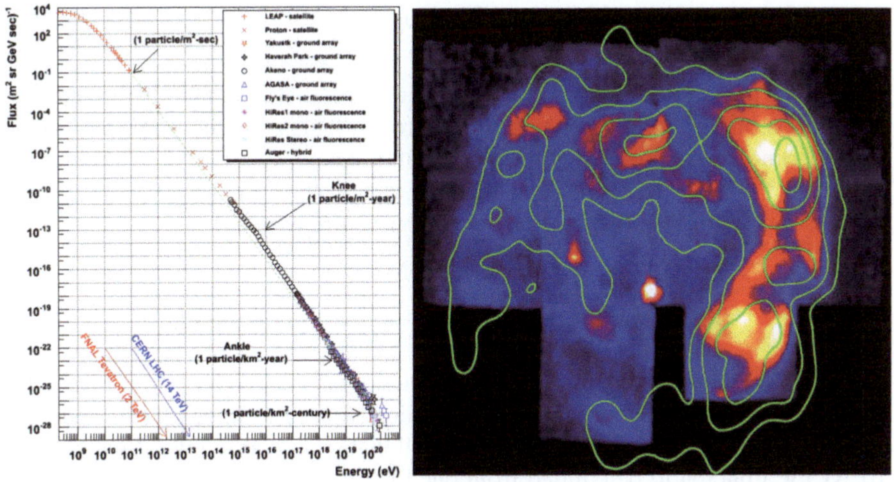

Fig. 3.9 *Left:* The cosmic ray flux spectrum in the range from 10^8 to 10^{20} eV sampled from various experiments. Indicated are also average particle detection frequencies (Credit: W. Hanlon, University of Utah). *Right:* An X-ray image of the 1,600 yr old supernova remnant RXJ1713.7-3946 from the *Suzaku* X-ray observatory. The contour lines show bright gamma-ray intensities as measured by the High Energy Stereoscopic System (HESS) located in Namibia (Credit: JAXA/Takaaki Tanaka/HESS)

System (HESS) telescope located in Namibia show the acceleration of Cosmic Ray particles was directly observed (Fig. 3.9). However, as much as this might explain the origin of particles in the GeV and TeV range, the generation of higher energies still remain mysterious.

3.2 Interstellar Gas

The ISM is largely gaseous and, as seen in the example of the LISM, appears in the form of clouds as well as gaseous streams. The physical properties of atoms and molecules are determined by the local temperature, density, volume and radiative environment. Dense H I clouds have to be cool, with kinetic temperatures usually between 50 and 200 K, in order to stay neutral. The warm, diffuse H I gas of the ISM has temperatures of over 1,000 K and thus appears as a very thin gas for the very same reason. The kinetic temperatures in this gas are for all atoms, not only hydrogen, as there always exists an equipartition of kinetic energy among other gas particles through elastic collisions. Modern research utilizes advanced diagnostic tools to analyze the distribution of neutral and molecular hydrogen. In the next sections large surveys are presented together with key conclusions.

3.2.1 *Diagnostics of Neutral Hydrogen*

Perhaps the most efficient means to survey the global distribution of hydrogen is by observation of the 21 cm line of neutral hydrogen. This emission originates within the hyperfine structure of the hydrogenic ground state:

$$\begin{aligned}\textbf{Total spin F} &= 1\ (p^+ \textit{ and } e^- \textit{ spin are parallel})\\ &\Downarrow \nu = 1{,}420.4\ \text{MHz}\\ \textbf{Total spin F} &= 0\ (p^+ \textit{ and } e^- \textit{ spin are antiparallel}).\end{aligned}$$

The energy difference between these two states is only $h\nu = 6 \times 10^{-6}$ eV and the corresponding line can be observed in the radio band at a wavelength of 21 cm. The oscillator strength of this transition is so small that it takes an average time of 1.1×10^7 yr for a photon to be emitted from one hydrogen atom. Such a rare transition can only be observed if there is an enormous abundance of neutral hydrogen.

Though H_2 is the most abundant molecule, it does not have any strong spectroscopic signatures in the radio or infrared band, which would allow us to survey its large-scale distribution, because of its lack of a dipole moment. Although there are molecular lines from H_2 in the ultraviolet, deeper observations are difficult because of extinction (see below). There is, however, an indirect way to measure the H_2 distribution. The carbon-monoxide molecule CO is the next most abundant molecule after H_2, with a relative number density between CO and H_2 of about 10^{-4}. The classical moment of inertia of the CO molecule is orders of magnitude larger than that for H_2 and thus its rotational transitions lie in the radio domain. Of interest is the transition from the first excited rotational state to the rotational ground state:

$$\begin{aligned}&\textbf{Rotational}\quad \textbf{quantum number J} = \textbf{1}\\ \textbf{115.27 GHz } (^{12}\textbf{CO})\ &\Downarrow \nu = \ 110.25\ \text{GHz } (^{13}\text{CO})\\ &\textbf{Rotational}\quad \textbf{quantum number J} = \textbf{0}\end{aligned}$$

The corresponding wavelengths are 2.60 mm for ^{12}CO (J = 1–0) and 2.72 mm for ^{13}CO (J = 1–0). Although the ^{12}CO isotope molecule is usually many times ($\sim 10^2$) more abundant, the transition from ^{13}CO becomes of interest once emission from ^{12}CO becomes optically thick in dense molecular clouds.

There are rotational and vibrational quadrupole (also called *forbidden*) transitions of H_2 in the infrared, which are related to the hot component of the H_2 gas in the ISM and are specifically of interest as they occur near regions of ongoing star formation.

The CO/H_2 ratio is relevant also to determine the dust to gas ratio and its dependence on metallicity and the local radiation field. For that one defines the quantity χ_{CO} which is that conversion factor from the integrated CO intensity to the column density of H_2. This quantity scales to one for the mass of molecular gas α_{CO} as [96]

$$\chi_{CO}[\mathrm{cm}^{-2}(\mathrm{K\ km\ s}^{-1})^{-1}] = 4.6 \times 10^{19} \alpha_{CO}\ \mathrm{M}_{\odot}\mathrm{pc}^{-2}(\mathrm{K\ km\ s}^{-1})^{-1}, \tag{3.18}$$

where values for the Milky Way yield $\alpha_{CO} \sim$ 4–9 $\mathrm{M}_{\odot}\mathrm{pc}^{-2}(\mathrm{K\ km\ s}^{-1})^{-1}$, whereas in the low-metallicity SMC, values of 90–270 $\mathrm{M}_{\odot}\mathrm{pc}^{-2}(\mathrm{K\ km\ s}^{-1})^{-1}$ have been suggested (see Leroy et al. [96] and references therein). A larger study Draine et al. [97] determined dust masses and PAH abundances for a large sample of morphologically similar galaxies and derived a value for $\chi_{CO} \approx 4 \times 10^{20}\ \mathrm{cm}^{-2}(\mathrm{K\ km\ s}^{-1})^{-1}$ for the CO-to-H_2 conversion factor. Figure 3.10 shows α_{CO} as a function of metallicity for galaxies of the local group with comparison to values usually obtained for the Milky Way and the one obtained in [97].

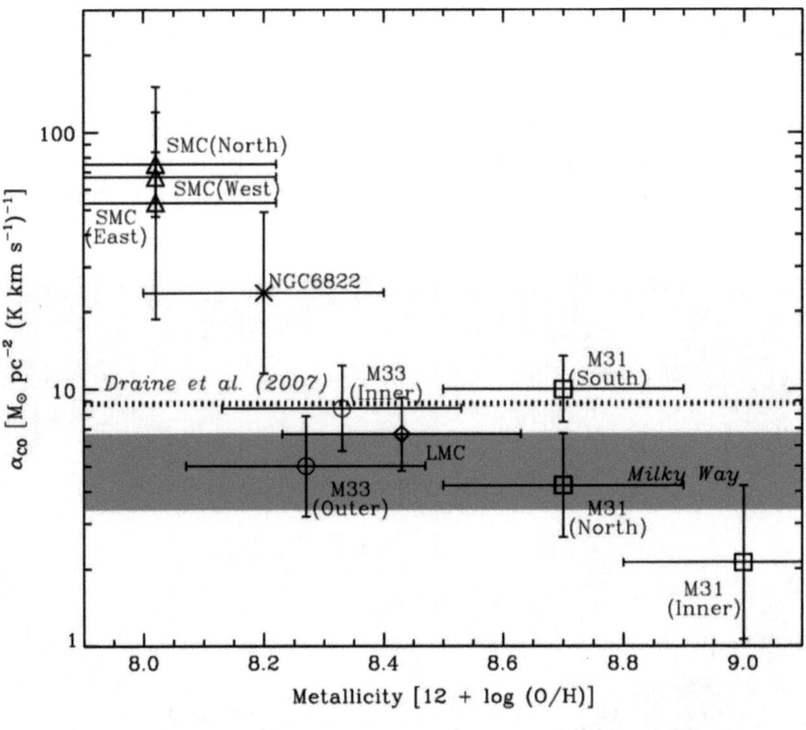

Fig. 3.10 The CO-to-H_2 mass ratio α_{CO} as a function of metallicity for the Milky Way and galaxies of the local group (from Leroy et al. [96])

3.2.2 Distribution of Neutral and Molecular Hydrogen

Early H I surveys date back to 1951 [98] when the 21 cm radio emission was discovered and just a year later the first large-scale survey was completed [99]. Since H I populates all portions of the ISM its 21 cm emission not only becomes detectable but can be used to reconstruct the basic spiral structure and kinematics of the Galaxy [100]. Advances in H I research are mainly based on large radio observatories such as the Arecibo Observatory, the *NRAO* 300-ft telescope, and, most recently, the *VLA*. Under the more recent activities are H I surveys by [101] and [102].

Figure 3.11 shows an all-sky survey of Galactic H I observed at a wavelength of 21 cm. The image center is also the Galactic Center and the Galactic plane runs horizontally through the image. Strikingly, the largest concentration of neutral hydrogen is within 250 pc of the Galactic plane, and the gas density is fairly constant between an inner radius of 4 kpc and an outer radius of 14 kpc around the Galactic Center. Below and above this plane the density is slowly decreasing. This is clearly in contrast to the total mass density. It could be shown [103] that the ratio of the dynamically determined Galactic mass density and the H I density is fairly constant outside a radius of 5 kpc from the Galactic Center, but the H I density is much lower inside a radius of 4 kpc. This fact may indicate that inside this radius the gas has been depleted more during stellar formation in the early phases of the Galaxy.

This depletion within the central regions of the Galaxy is also seen for molecular hydrogen between 1 and 4 kpc, although the overall distribution of H_2 is very different from the H I density distribution. The global H_2 distribution is deduced from CO measurements (see Fig. 3.12 and the section below). It shows a sharp

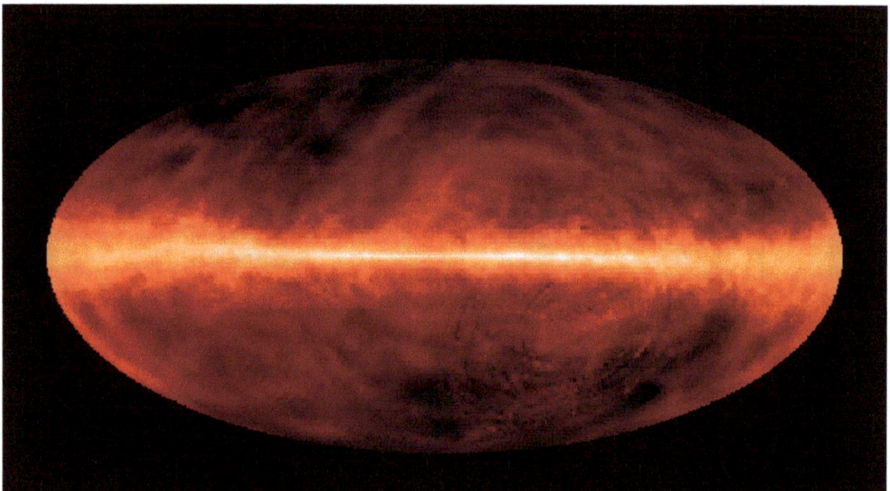

Fig. 3.11 H I distribution in the Galaxy observed at the 21 cm wavelength. Credit: J. Dickey (UMn), F. Lockman (NRAO), Skyview, 1998

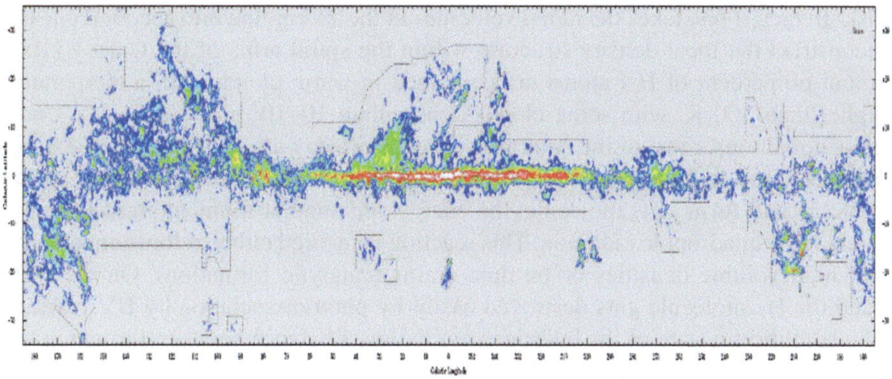

Fig. 3.12 A composite CO survey from data of many individual surveys obtained throughout the last decades containing 488,000 spectra. The horizontal scale is ±180 deg Galactic longitude, the vertical scale is ±30 deg Galactic latitude. Credit: from Dame et al. [104]

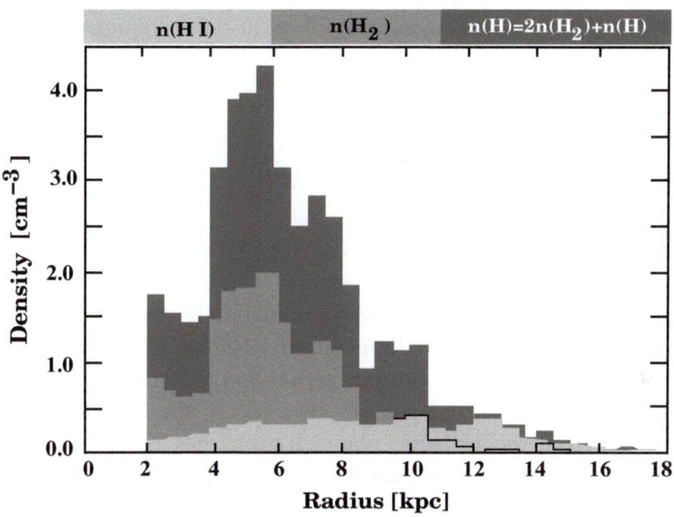

Fig. 3.13 Schematic representations of radial distributions in the plane of the Milky Way of the volume densities of atomic and molecular hydrogen. The light shade indicates n(H I), the medium shade $n(H_2)$, the dark shade all hydrogen. Adapted from Gordon et al. [103]

peak between a radius of 5 and 6 kpc from the Galactic Center and decays sharply towards larger radii. Beyond 12 kpc H_2 is only found at very low densities. Figure 3.13 illustrates the radial distribution of neutral hydrogen. Although their density distributions are quite different, the integrated mass of H I and H_2 in the Galaxy is about the same, $2.5\times10^9\ M_\odot$ for H I, and $2.0\times10^9\ M_\odot$ for H_2 [103].

Figure 3.11 also shows that diffuse H I clouds with sizes of hundreds of light-years across cluster near the Galactic plane. This gas is not isotropic but forms arches and loops indicating that it is stirred up by stellar activity in the Galactic disk. In fact, if one takes the radial velocities in the 21 cm line into account, one can reconstruct the local density structure within the spiral arms of the Galaxy [100]. About 60 percent of H I atoms are contained in warm clouds with a temperature higher than 500 K, with some clouds denser than 10–10^3 atoms per cm^3. Under these conditions, most of the neutral hydrogen will stay atomic for several reasons: a pure H I gas will always have a non-zero probability such that a collision of two H atoms will form a H_2 molecule; the energy and angular momentum so gained is released by quadrupole radiation. This reaction is favored either at low temperatures and high volume densities or on dust grains (catalytic formation). On the other hand the H_2 molecule gets destroyed easily by photodissociation by UV photons. The equilibrium state of the hydrogen gas is thus a balance between formation and dissociation of molecular hydrogen, and the relative amount between H_2 and H I densities can vary greatly within the ISM.

Large local variations from the overall equilibrium are expected to exist not only in very cool and super-dense regions, but also in regions of high density and high UV flux. So-called *photodissociation regions* have been identified in the Galactic ISM both in the general diffuse gas as well as in dense regions near hot young stars [105].

3.2.3 Distribution of Ionized Hydrogen

The Wisconsin Hα Mapper (WHAM) has surveyed the distribution and kinematics of ionized gas in the Galaxy above declination $-30°$. The WHAM Northern Sky Survey (WHAM-NSS) provides the first kinematically resolved map of the Hα emission from the warm ionized medium within ± 100 km s^{-1}. The map in Fig. 3.14 contains 37,565 spectra with Hα profiles revealing the ionized gas in the Galaxy. Distributions of massive ionized gas are revealed in the nearby spiral arms up to 1–2 kpc away from the Galactic plane. Ionized gas is also detected toward intermediate velocity clouds at high latitudes. Several new H II regions are detected around early B stars and evolved stellar cores [106].

3.2.4 Distribution of CO and CO_2

CO surveys of the entire Galaxy are rare. One of the first composite surveys was compiled in the late 1980s [107]. Many previous surveys concentrated more on specific areas in the sky. A new, more recent, large-scale survey integrates almost half a million individual surveys into one highly resolved composite all-sky CO map [104] (Fig. 3.12). The map shows (like Fig. 3.11 for H I) that most of the material is concentrated within 250 pc of the Galactic plane. However, in contrast to H I, the CO appears clumpy and concentrated within only half the radial extent around the Galactic Center. Although in the composite CO map there is a peak at the

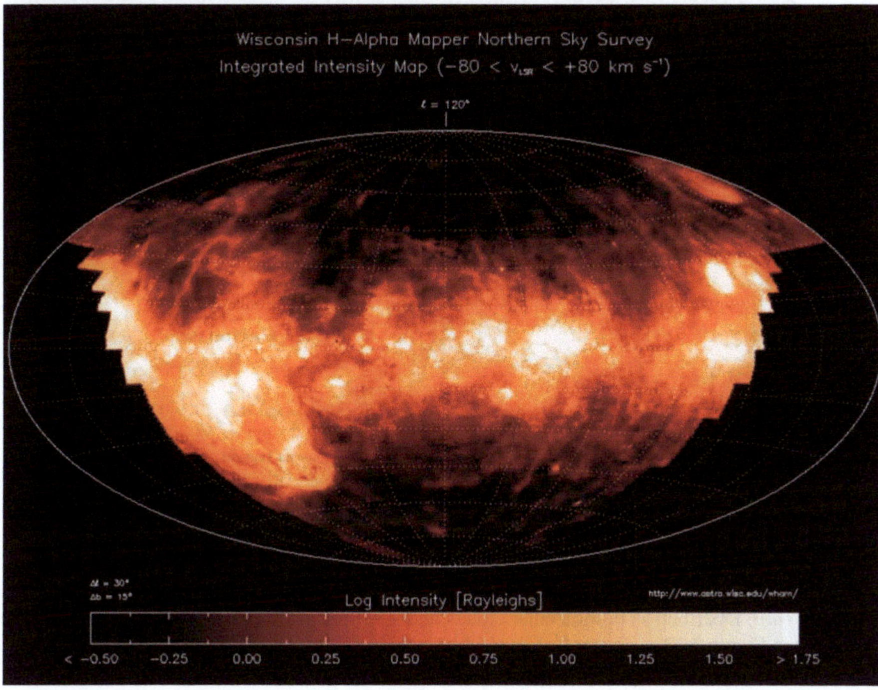

Fig. 3.14 Distribution of ionized hydrogen from the Wisconsin Hα Mapper (WHAM) in the northern sky (Credit:WHAM consortium)

Galactic Center (the dark spot at the center of the map is also the Galactic Center position), the radially integrated distribution also shows a peak around 5 kpc. Its shape looks similar to the one shown for H_2 [103] (Fig. 3.13).

The power of CO (2.6 mm) surveys to find dense regions in the sky cannot be underestimated. The appearance of CO molecules directly traces the H_2 molecule and therefore also indicates the densest regions in the ISM (i.e., molecular or giant molecular clouds). Although these clouds are discussed in more detail in the next chapter, Fig. 3.15 presents a finding chart for clouds and associations based on the composite survey by [104]. The top panel is the velocity-integrated composite map. Each region of clouds has its own peculiar motion on top of the Galactic rotation and depending on its direction, the observed line appears blue- or red-shifted indicating a radial velocity component. The map displays the locations of the most prominent star-forming regions in the Sun's neighborhood in Galactic coordinates. Figure 3.15 shows the radial velocity of these clouds with respect to Galactic latitude. It reveals the existence of a higher velocity molecular ring within 60 degrees of Galactic longitude that coincides with the peak in the hydrogen distribution.

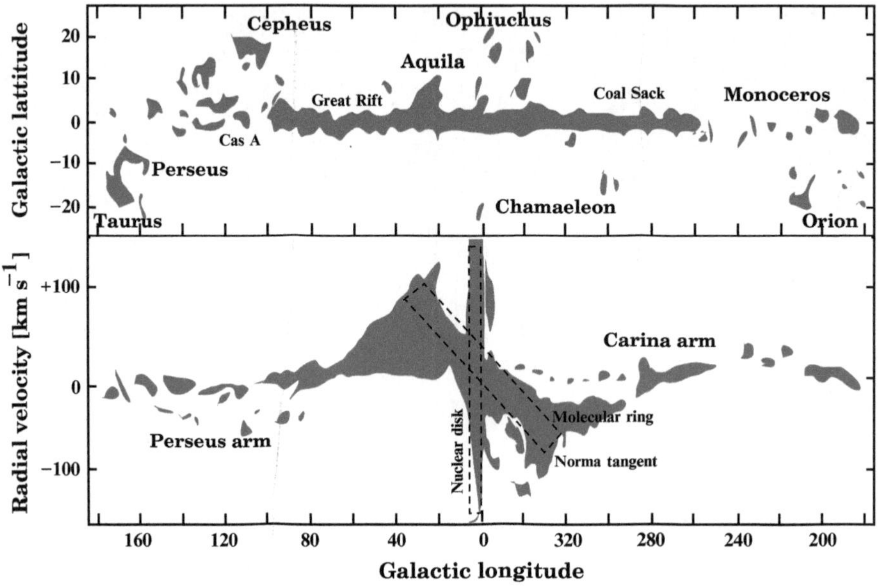

Fig. 3.15 (*top*) Distribution of molecular clouds and star-forming regions based on the composite CO survey in Figure 3.12. (*bottom*) Corresponding radial velocity distribution in the Galactic plane. Adapted from Dame et al. [104, 107]

3.2.5 *Diffuse High-Energy Emissions*

The peak in the radial density distribution as observed in CO surveys has its counterpart in the distribution of the diffuse γ-radiation in the Galaxy. Observations with satellites in the mid-1970s showed the existence of a broad emission region along the Galactic plane that is directed toward the Galactic Center with a peak in the radial distribution at 5–6 kpc [108–110]. The first full γ-ray survey of the sky was produced by the *EGRET* experiment on board the *Compton Gamma-Ray Observatory (CGRO)* for energies larger than 100 MeV. Figure 3.16 shows a more recent false-color image of the Milky Way obtained with *LAT* on board the *Fermi Observatory*.

Besides bright unresolved point sources there is a diffuse component in the Galactic plane. The most likely mechanism responsible for the high-energy emission is the decay of π^o mesons produced in interactions of cosmic ray nucleons with interstellar gas nuclei [111]. The peak at 5–6 kpc thus coincides with the high gas density of CO and H_2 in the Galaxy where the highest mass density is found. Variations in the cosmic ray flux are presumably too small to create this peak and such an explanation is not supported by the Galactic distribution of supernova remnants nor by the strength of the non-thermal radio continuum.

Fig. 3.16 Diffuse γ-ray emission observed with *LAT*, the high-energy telecope onboard *Fermi* for energies >100 MeV after 1 year of operation (Credit: NASA/DOE/Fermi LAT Collaboration)

3.3 Column Densities in the ISM

One of the centerpieces of ISM research is the analysis of absorption lines. The behavior of line equivalent widths on the *curve of growth (COG)* is one of the most powerful analysis tools (see Appendix D). The following sections present some applications to absorption spectra.

3.3.1 Abundance of Elements

Hydrogen, as the most abundant element, produces the strongest of all the absorption lines at the Lyman α line wavelength of 1,215.67 Å [113]. For most stars radiation damping is the dominating source for line broadening as long as the observed star is not too close to the observer. A comparison of the Lyman α measures of the hydrogen column density with those derived from observations of the 21 cm radio emission revealed a rather poor correlation [114]. One of the suspected reasons for this discrepancy is that most early-type stars used for Lyman α surveys only show the nearby distribution of hydrogen, missing rich concentrations of interstellar material throughout the rest of the Galaxy [113].

The relative abundance of elements with respect to hydrogen in the Galactic H I gas should be similar to solar (sometimes also referred to as 'cosmic') abundances (Table 3.3). A now classic study [113] determined the relative abundance of elements as measured in the direction of the early-type star ζ Ophiuchi. Early-type

Table 3.3 Abundance and mass distributions

Element	μ_Z (amu)	A_Z Sun	A_Z ISM	A_Z Cosmic rays	1-β_X (-)	$N_{H,Z}$ atoms/cm^2	M_{Gal} $M_\odot$
1 H	1	12.00	12.00	12.00	1.0	$1.0\ 10^{22}$	$4.5\ 10^9$
2 He	4	10.99	10.99	11.86	1.0	$9.8\ 10^{20}$	$1.8\ 10^9$
6 C	12	8.60	8.38	9.59	0.5	$2.4\ 10^{18}$	$1.3\ 10^7$
7 N	14	7.97	7.88	8.98	1.0	$7.6\ 10^{17}$	$4.9\ 10^6$
8 O	16	8.93	8.69	9.54	0.6	$4.5\ 10^{18}$	$3.2\ 10^7$
10 Ne	20	8.09	7.94	8.79	1.0	$8.7\ 10^{17}$	$7.9\ 10^6$
11 Na	23	6.31	6.16	8.01	0.25	$1.4\ 10^{16}$	$1.6\ 10^6$
12 Mg	24	7.59	7.40	8.83	0.2	$2.5\ 10^{17}$	$2.7\ 10^6$
13 Al	27	6.48	6.33	8.03	0.02	$2.1\ 10^{16}$	$2.5\ 10^5$
14 Si	28	7.55	7.27	8.73	0.1	$1.7\ 10^{17}$	$2.2\ 10^6$
15 P	31	5.57	5.42	7.36	0.6	$2.6\ 10^{15}$	$3.6\ 10^4$
16 S	32	7.27	7.09	8.06	0.6	$1.2\ 10^{17}$	$1.7\ 10^6$
17 Cl	35	5.27	5.12	7.28	0.5	$1.3\ 10^{15}$	$2.0\ 10^4$
18 Ar	40	6.56	6.41	7.76	1.0	$2.6\ 10^{16}$	$4.7\ 10^5$
20 Ca	44	6.34	6.20	7.93	0.003	$1.6\ 10^{16}$	$3.2\ 10^5$
22 Ti	48	4.93	4.81	7.82	0.002	$6.5\ 10^{14}$	$1.4\ 10^4$
24 Cr	52	5.68	5.51	7.76	0.03	$3.2\ 10^{15}$	$7.6\ 10^4$
25 Mn	55	5.53	5.34	7.54	0.07	$2.2\ 10^{15}$	$5.4\ 10^4$
26 Fe	56	7.50	7.43	8.62	0.3	$2.7\ 10^{17}$	$6.8\ 10^6$
27 Co	59	4.92	4.92	<6.88	0.05	$8.3\ 10^{14}$	$2.2\ 10^4$
28 Ni	59	6.25	6.05	7.19	0.04	$1.1\ 10^{16}$	$2.9\ 10^5$

stars are most favorable for such an analysis as they do not have too many intrinsic spectral lines interfering in the area of interest. Figure 3.18 shows a more recent study using *HST* data. Here depletions for various lines of sight and clouds with different radial velocities are compared. By comparing these measured abundances relative to an abundance distribution we observe in the atmosphere of the Sun (Table 3.3), it appears that most elements, most prominently Al, Ca, Ti, Fe, and Ni, suffer from an underabundance by a large factor. Unless a solar distribution of elements is more the exception than the rule, this indicates that these elements are depleted in the cold ISM gas because they have coagulated with interstellar dust.

The atomic hydrogen distribution also contains deuterium. D I was discovered by D. York with *Copernicus*, who measured the D/H ratio in the line of sight of the star β Cen to be $1.4 \pm 0.2 \times 10^{-5}$ [116]. The deuterium is of primordial origin as only ~50 percent of the amount created during the Big Bang has been destroyed in stellar interiors ([115] and references therein). There are still large uncertainties in the D/H ratio and measurements in different lines of sight yield different results. These differences do not correlate with the one expected from ISM reprocessing in stellar interiors and their origin remains unresolved to date [115]. Recent measurements with *FUSE* suggest a slightly higher value of D/H = $(1.52 \pm 0.08) \times 10^{-5}$ [117].

3.3.2 Absorption of IR, Optical, and UV Light

Neutral interstellar gas does not radiate in visible light. In the visible, it can be observed only through absorption of light from stars located behind the gaseous regions. Since interstellar atoms are usually in their ground state, most will not absorb photons at wavelengths larger than 3,000 Å [113]. However, ISM absorption from neutral atoms in the visible wavelength band are observed from Na, K, Fe, and Li, from (ionized) Ca II and Ti II, as well as from radicals of CN and CH.

There are various ways to detect such lines and distinguish them from stellar absorption lines. ISM lines are stationary as opposed to Doppler-shifted lines from, for example, stars in binary systems. Another characteristic is that interstellar absorption lines are extremely narrow compared to broad stellar lines. Figure 3.17 (left) shows an absorption spectral analysis performed in the oxygen K X-ray absorption band from spectra observed with *Chandra*. These observations are specifically useful for abundance studies. Similar studies in the wavelength range between 910 and 1,100 Å are under way with the *Far Ultraviolet Spectroscopic Explorer (FUSE)*. Bright continuum sources like compact binaries in the X-ray band and early-type stars in the FUV are well suited for such observations as they do not show interfering intrinsic lines in these particular wavelength ranges.

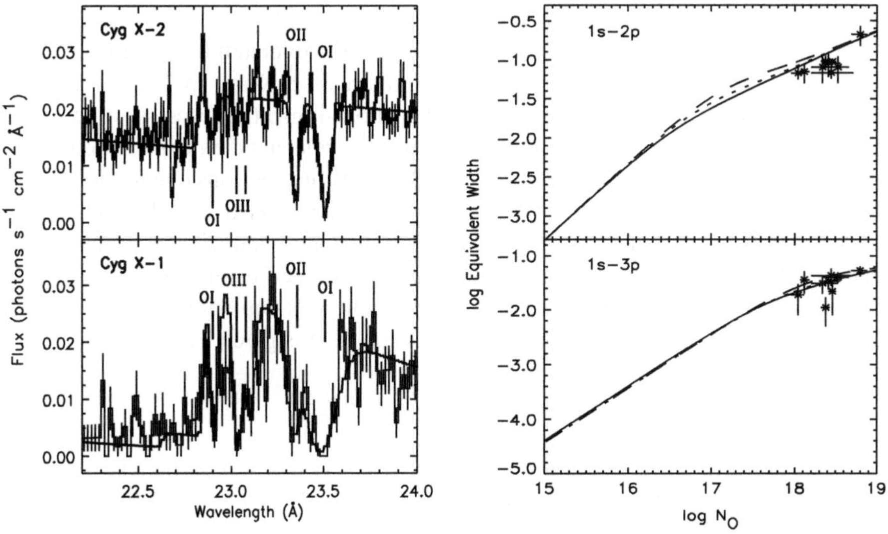

Fig. 3.17 The first curve of growth analysis performed in the X-ray band from high-resolution spectra obtained with *Chandra*. (*left:*) ISM K resonance absorption of O I, O II, and O III in two bright X-ray sources. (*right:*) Curve of growth analysis for O I 1s-2p and 1s-3p absorption. From Juett et al. [112]

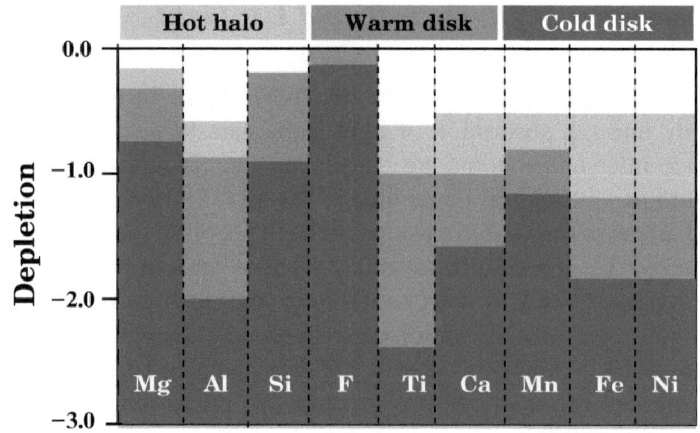

Fig. 3.18 Depletion of various elements in the direction of identified clouds based on HST data. The different shades show average measurements of element depletion (scale by factors of 10) for cold Galactic disk gas, warm disk gas, and hot Galactic halo gas. Adapted from Frisch [115]

3.3.3 Absorption of X-Rays

Knowledge of the element abundance in the ISM is crucial in many ways. X-rays, for example, get absorbed by material in the ISM, modifying the X-ray flux and spectrum from 100 eV up to 10 keV depending on the interstellar column density in the line of sight towards an X-ray source. In this respect X-ray spectra can be used as a diagnostic of ISM abundance. The total photo-ionization cross-section of the ISM is obtained by the sum over the contribution of relevant elements in the various phases of the ISM. In order to properly weigh these contributions one has to assume a certain abundance distribution of elements. Most widely used are solar abundances determined from the analysis of our Sun's photosphere and/or certain meteorites (see [118] and references therein). There also have been estimates for local ISM abundances [119] which include assessments of element depletion by dust.

The total photo-ionization cross-section of the ISM is given by:

$$\sigma_{ISM} = \sigma_{gas} + \sigma_{mol} + \sigma_{grains} \tag{3.19}$$

where σ_{gas} represent the cross-section of all gaseous phases, σ_{mol} the cross-section of all molecules (which is dominated by H_2), and σ_{grains} the cross-section of all solid material. Since hydrogen is by far the most abundant element in the ISM we are accustomed to normalize the number density of all elements with atomic numbers larger than unity to that of hydrogen. These cross-sections incorporate the basic photo-ionization cross-section modified by an abundance factor, ionization fractions, dust depletion factors, and assumptions about the chemical and physical composition of dust grains. X-ray absorption then can be described by an exponential law:

$$I_{observed}(E) = e^{-\sigma_{ISM}(E) N_H} I_{source}(E), \tag{3.20}$$

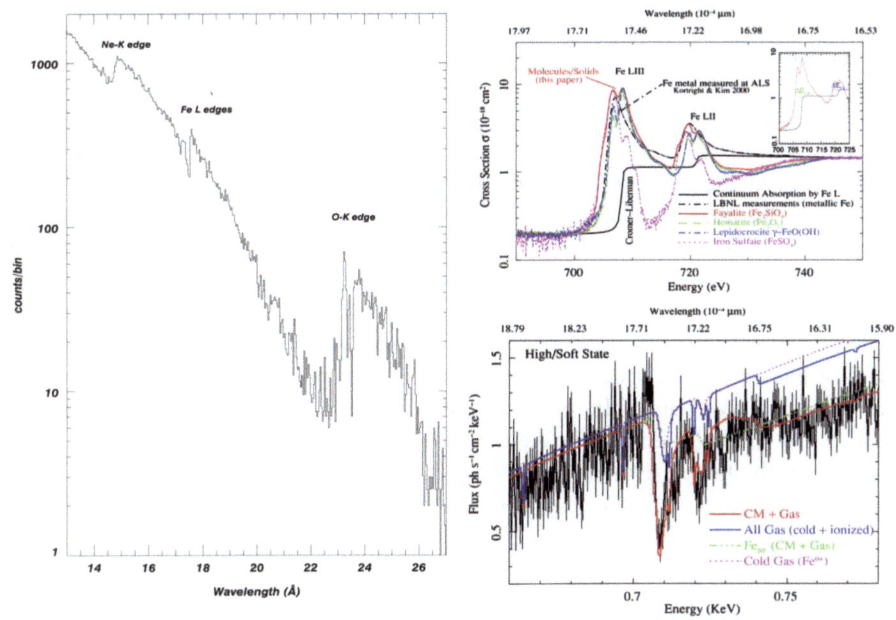

Fig. 3.19 **Left:** The absorbed X-ray continuum of the bright X-ray source Cygnus X-1. Absorbed by an equivalent hydrogen column density of 6.0×10^{21} cm^{-2}, it shows the absorption edges from O K, Fe L, and Ne K (from [123]). **Right:** (top) X-ray absorption near edge structure in the vicinity of the Fe L photoelectric edge, reveals that structures known as XAFS are distinct for different states of condensed matter. (bottom) The 700 eV Fe L photoelectric edge region as observed in the X-ray binary Cygnus X-1. It reveals the absorption components are decomposed to determine the quantity and composition of dust. The best fit (red) shows the combined absorption from all gas (blue) which includes line-of-sight cold gas (magenta) and XAFS signifying condensed matter (from [122])

where $I_{observed}(E)$ is the observed spectrum, I_{source} the original source spectrum, and E the energy of the X-ray photons.

N_H is the equivalent hydrogen column density in atoms per cm^2. The column density is therefore all the material integrated along the line of sight between the observer and X-ray source:

$$N_H = \int_0^D n_H \, \mathrm{d}l \tag{3.21}$$

where D is the distance to the source. In the X-ray band between 0.1 and 10 keV, absorption K and L edges are observed from the most abundant elements [120]. Figure 3.19 (left) shows how an X-ray spectrum is affected by the ISM (assuming there is no absorption intrinsic to the X-ray source) in the bandpass between 200 and 1,000 eV, which harbors the C K, N K, O K, Fe L, and Ne K edges. The optical depth of an edge is highly sensitive to the observed column density. Figure 3.19 (right) shows how the cross-sections of the Fe L3 and L2 edges vary from metallic Fe forms [121] to various molecular forms involving Fe [122].

Table 3.3 lists the most abundant elements with their molecular weights (μ_Z) and their abundance fractions A_Z, from which one can calculate column densities using:

$$N_{H,Z} = N_H \times 10^{A_Z - 12.} \tag{3.22}$$

Table 3.3 provides a solar and ISM abundance distribution, as given by [119], and a cosmic ray abundance distribution [124]. The latter is of some interest as the main contribution of the Galactic component of the cosmic radiation comes from supernovae which are replenishing the ISM with newly formed elements. Primordial element distributions determined for metal-poor stars [125] are of less relevance. The depletion factors β_X (see also Fig. 3.18) are taken from [119] again and serve as best guesses, as the exact amounts of element depletions are still not very well known. For a hydrogen column density of 10^{22} cm^{-2}, which is roughly the amount towards the Galactic Center, the element column densities listed in the second to last column of the table are obtained. With 4.5 $\times$ 10^9 $M_\odot$ of hydrogen in the Galaxy this leads to a total mass (in $M_\odot$) for each element.

3.4 Interstellar Dust

3.4.1 Distribution in the Galaxy

The density of dust between the stars is extremely low. In one billion cubic centimeters of space we find only a few dust particles. Dust particles usually have typical sizes of 0.01 to 1 μm, and they become apparent by observing the extinction of light through large volumes. Because of the small size of the dust particles, more blue light will be scattered than red light and, as a result, the images from such a dusty environment appear more red than they are without dust. This effect is known as interstellar reddening. Figure 3.20 is a collage of observations of the southern Milky Way obtained by A. Mellinger. It impressively shows the distribution of dust in the Galactic plane. Dust is generally cool and the stars provide the background illumination in Fig. 3.20. Dust absorbs and scatters part of the ISRF and radiates infrared light. In the visible band we thus only observe dark dust patches, arcs, and

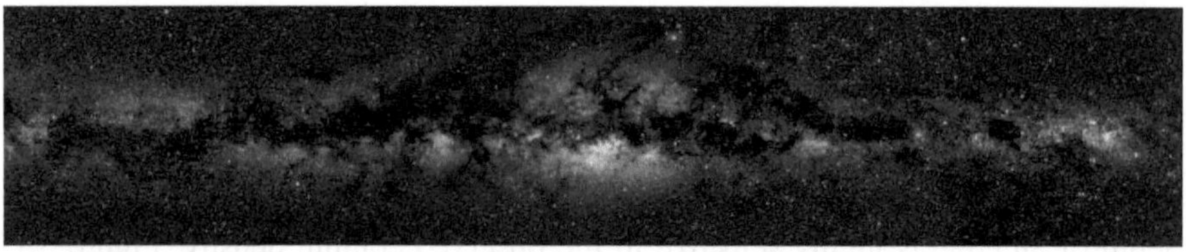

Fig. 3.20 Dust in the southern Milky Way. Credit: Photography by A. Mellinger; from http://home.arcor-online.de/axel.mellinger

loops of dust. Throughout the Galactic plane gas and dust are well intermixed at all scales. Regions of high gas density are usually also regions of high dust density. In this respect [126] showed that there is a close correspondence between H I and reddening per unit distance and that indeed gas and dust are likely to be physically associated. Taking into account all hydrogen components they also found that the average ISM column density per magnitude of optical extinction is of the order of 7.5×10^{21} atoms cm^{-2}. This number has been revised many times using different observational techniques and the mean gas-to-dust ratio for all lines of sight is now established to be closer to 5.8×10^{21} atoms cm^{-2}. A similar value has been suggested fitting the size of scattering halos of bright X-ray sources throughout the Galactic plane [127].

3.4.2 *The Size and Shape of Dust Grains*

The shape of dust grains is not well known, but has immense importance in stellar formation. Dust serves as a cooling agent during the stellar contraction phase (see Chap. 4). This cooling happens through radiation of light from a dust grain's surface and the cooling is more efficient the larger the surface is. By studying how dust absorbs, emits, and reflects light, one is able to deduce the geometrical properties of interstellar dust. This dust does not in the least resemble our normal house dust. Models of observed interstellar extinction [128] (see next section) over the wavelength range between 0.11 μm and 1 μm feature generalized particle size distributions of graphite, enstatite, olivine, silicon carbide, iron, and magnetite or combinations of these materials. The size distributions are power laws, which are monotonically decreasing towards larger sizes between 0.005 μm and 0.25 μm.

Figure 3.21 shows a grain of interplanetary dust caught by a high-flying U2-type aircraft. Its size is about 10 μm across and it is composed of glass, silicate minerals, and carbon. Models involving such shapes for dust grains are rare. Generally one approximates the shape of a grain to be spherical or cylindrical in order to more conveniently solve for electromagnetic field configurations. However, dust grains are likely not spherical. It is more likely that dust grains stick together, collide and shatter, leading to a rather random distribution of shapes. Figure 3.21 shows an attempt to illustrate the result of such a random conglomeration process. It assumes that dust grains are fractal, created from coagulation of smaller subunits called *monomers, dimers, tetramers,* and so on.

3.4.3 *Interstellar Extinction Laws*

Dust causes extinction of light. Extinction in the optical and near-IR is caused by absorption and scattering, which includes the reflection of light under various conditions. There is scattering by diffuse dust in the Galactic plane, reflection of

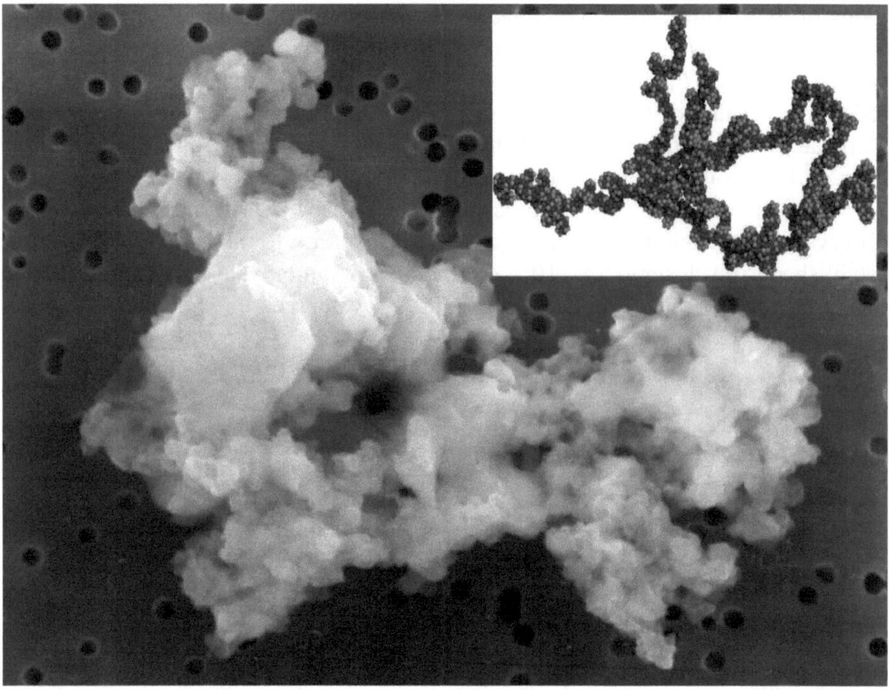

Fig. 3.21 A grain of interplanetary dust likely dating from the early days of our Solar System. It is 10 μm across and composed of glass, silicate minerals, and carbon. Credit: *NASA*, 2001. The inset shows, how interstellar dust, which is generally much smaller than 10 μm, could appear by means of fractals. Credit: E. L. Wright (UCLA), 1999

light in dust-rich nebulae, or scattering by more or less dense dark clouds. Thus, not only dark clouds are responsible for the effect, for extinction of light is present everywhere in the Galaxy as there is dust distributed throughout interstellar space at various density levels. In this respect each line of sight has its own level of extinction. To make matters worse, extinction depends on wavelength as well. In practice the effects of extinction are observed during photometric measurement as a change in color (or outside the optical band, as a shift of the spectrum). Commonly, since there is no unique law to express such a color shift, the ratio of two colors is used:

$$E(\lambda - V)/E(B - V) = (A_\lambda - A_V)/E(B - V) = (A_\lambda/E(B - V)) - R_V \quad (3.23)$$

where A_V is the visual extinction, i.e. the decrease of magnitude in the visual band, A_λ the same at a specific wavelength, and $E(B - V)$ the color excess as measured from the ratio of B (440 nm) and V (550 nm) fluxes. The color excess is the difference between the actual measured B-V color of the object and its true, intrinsic color. From photometric studies (mostly in the near-IR) of objects where the intrinsic color is known there is an average relation between color excess and visual extinction that is commonly used:

$$A_V = (3.1 \pm 0.1)E(B - V), \quad (3.24)$$

with $R_V = 3.1$. Unfortunately there is yet no unique perception of how the extinction of light by dust can be utilized to study not only the properties of dust itself, but also its concentration in dense clouds or in the diffuse medium. Many attempts are being made to study these properties continuously from the radio to ultraviolet wavelength bands.

The use of parameterized extinction data [129] led to the derivation of an average extinction law over the wavelength range of 3.5 μm to 0.125 μm [130]. The data included UV observations from the *International Ultraviolet Explorer (IUE)* and optical to near-IR observations for many lines of sight through the diffuse ISM, H II regions, and molecular clouds. Ultimately a relation was determined between R_V and the relative extinction A_λ / A_V. Figure 3.22 shows three cases of such a mean extinction law. The line of sight towards the star BD +56 524 is specifically of low dust density with $R_V = 2.75$, while the line of sight toward Herschel 36, the central star of the H II region Messier 8, is particularly dense with $R_V = 5.30$. A standard mean extinction law uses $R_V = 3.2$ for the diffuse ISM (see also [131]).

Common in all these extinction laws is an excess at wave number $\lambda^{-1} = 4.6$ μm^{-1}, which corresponds to a wavelength of 2,175 Å. The peak wavenumber seems quite stable with respect to R_V. Since the excess is very strong it has to be related to very abundant materials with large absorption strengths, like C, N,

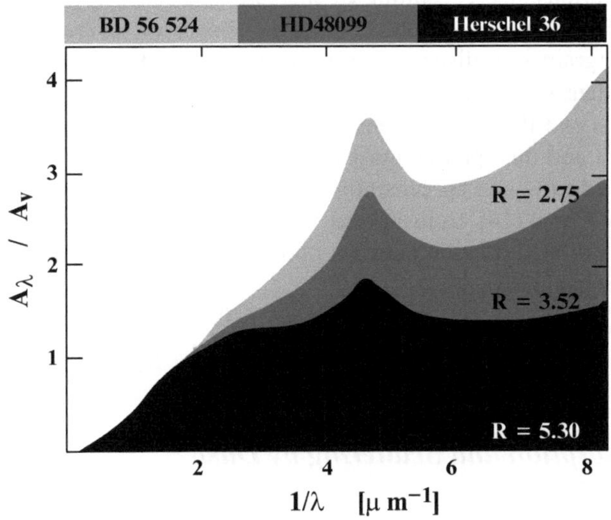

Fig. 3.22 Various extinction laws with respect to different lines of sight for a low-density case ($R = 2.75$), for close to the diffuse ISM ($R = 3.52$), and a high-density case ($R = 5.30$) from UV to IR. Adapted from Cardelli et al. [130]

O, Mg, Si, and Fe. Carbon seems very likely and it may not be a coincidence that graphite, a solid form of carbon, has a resonance close to 2,175 Å. The size of these graphite particles has to be smaller than 0.005 μm as larger sizes would shift the resonance peak. Observations suggested that amorphous carbon rather than graphite is injected into the ISM from stars [132]. This peak corresponds to a well-known and broad absorption band due to amorphous carbon in the UV spectrum. There are large variations in the UV absorption largely due to various sizes and morphologies of dust grains. Such variations may correspond to coagulation of small grains into larger grains, but also to partial or total destruction of some types of grains in shocked region. UV extinction curves are therefore quite sensitive to physcial properties of dust.

Another important feature is the fact that extinction in the near-IR seems to be independent of R_V. This implies that the size distributions of the largest particles are the same, independent of direction, which is a result also supported by interstellar polarization studies [133]. Near-IR photometry mostly uses the so-called Johnson filters, which are defined as the J, H, and K bands at 1.25, 1.65, and 2.20 μm, respectively. A single relation for extinction from diffuse dust and outer cloud dust is of great important for the analysis of near-IR observations of young embedded stars.

Extinction affects light in the far-UV and X-ray band as well. Under the assumption that grains are mainly composed of silicates and graphite, absorption cross-sections for particles of sizes between 0.003 and 1.0 μm for wavelengths larger than 300 Å have been evaluated [134]. The UV opacity has a maximum at 17 eV (730 Å) and vanishes near 100 eV (124 Å). At higher energies the absorption from dust is basically X-ray absorption on neutral atoms (see Sect. 3.1.4) as long as the absorbing atom does not get ionized. Some X-ray emission is scattered by dust particles of various sizes, resulting in characteristic halos to be observed around the source [135, 136]. Using *ROSAT* data the shapes of several halo distributions with specific grain size distributions were fitted leading to the determination of dust column densities [127]. They indicate that some correlation exists between the measured dust and the column densities obtained from X-ray spectra, confirming again that gas and dust phases must exist co-spatially. Furthermore, these results were found to be in good agreement with the relation of column density and color excess derived by [114] from optical measurements of hot stars. From the X-ray analysis a refined relation between optical extinction and equivalent hydrogen columns emerged [127]:

$$A_V = 0.56 N_H + 0.23 \tag{3.25}$$

where N_H is the equivalent column density in units of 10^{21} cm^{-2}.

3.4.4 Absorption and Scattering by Dust

Much of what we observe is affected by dust extinction, scattering, and absorption. The interaction of radiation with dust is highly complex and reveals itself not only through its traditional extinction properties. Various radiative absorption, emission, and scattering processes, as well as polarization, can indicate the presence of dust.

For example, some of the scattered light can be observed as diffuse light in the Galactic plane, as reflective light near nebulae, and even as a faint glow around dark clouds. The latter can be seen best at high Galactic latitudes contrasting against the dark, out-of-plane sky.

Throughout the near UV, visual, and IR spectra there are broad interstellar absorption bands. The most prominent band in the UV at 2,175 Å has already been mentioned in this context; it causes the distinct peak in the extinction law (Fig. 3.22) and is mostly due to graphite [137]. There are many other bands in the visual and IR, which are usually narrower and weaker, leaving them unrecognized in the context of the mean broad band extinction laws. The next strongest band appears in the visual spectrum at 4,430 Å at a width of 30 Å, more than a factor of 10 narrower than the one in the UV. The strongest bands, though, in the IR are at 3.1, 9.7, and 18 μm. There are over 200 of these diffuse bands known, and it is established that because of their width only solid particles and large molecules are responsible for them.

Generally the sum of dust absorption and scattering provides the total amount of extinction

$$C_{ext} = C_{abs} + C_{sca}, \tag{3.26}$$

where C_{ext}, C_{abs}, and C_{sca} are the total extinction, absorption, and scattering cross-sections, respectively. Scattering is a directional effect and usually does not change the frequency of the radiation except in regions of the UV band where the albedo is affected by absorption effect, i.e. the 2.175 Å UV band for example. Scattering can be separated from absorption through the albedo, i.e. the total reflectivity of the dust grains, defined as

$$A = \frac{C_{sca}}{C_{ext}}, \tag{3.27}$$

and polarization properties. The scattered flux is

$$F_{sca}(r) = \frac{F_0}{r^2} C_{sca} = \frac{F_0}{r^2}\frac{1}{k}\int_{4\pi} \mathcal{L}(\theta, \phi)\, \mathrm{d}\Omega, \tag{3.28}$$

where F_0 is the initial source flux, r the source distance, and $\mathcal{L}$ a dimensionless function decribing the change of direction of the scattered photons. However, most scattering by dust grains can be decribed by a single scattering albedo and a scattering phase function using measured intensities. The scattered intensity is then calcuated by Henyey and Greenstein method [138], where the scattered intensity is determined by

$$I_{sca} = A * \int_{4\pi} I_0 \Phi(\alpha)\, \mathrm{d}\Omega, \tag{3.29}$$

with I_0 being the initial, I_{sca} the total scattered intensity, Ω the solid angle, α the scattering angle. For most applications the Henyey and Greenstein approximation, which dates back to 1941, can be used to describe the scattering phase function

$$\Phi(\alpha) \approx \Phi_{HG}(\alpha) = \frac{1}{4\pi}\frac{1 - g^2}{(1 + g^2 + 2g\ \cos\alpha)^{3/2}}, \tag{3.30}$$

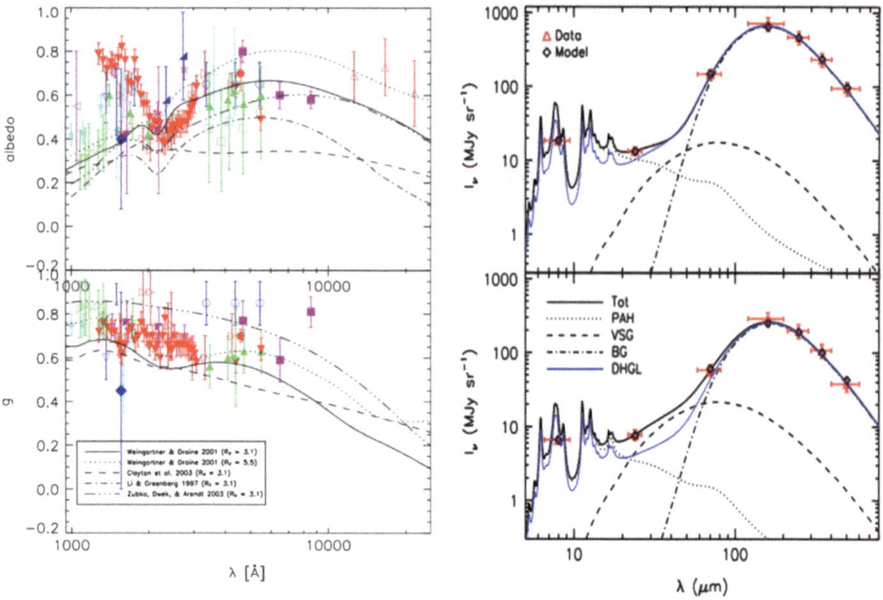

Fig. 3.23 *Left:* Albedos and g-parameters of interstellar dust from various measurements and models (from [139]. *Right:* Mean spectra from combined *Herschel* and *Spitzer* observations of the far-IR diffuse dust emission in a low excitation region of the high lattitude ($l \sim 59°$) interstellar medium (from [143])

where g is an anisotropy parameter expressed by

$$g = \int_{4\pi} \Phi(\alpha) \cos\alpha \, d\Omega. \qquad (3.31)$$

g varies from -1 to one, where -1 refers to complete back-scattering, 0 to isotropic scattering, and 1 to complete forward scattering. In a review of measurements and models K. Gordon [139] determined albedo and g-parameters as a function of wavelength for the reflection nebulae, dark cloud, and diffuse emission case. Figure 3.23 shows results from this review. It also shows results from various model attempts with reference to [83, 140–142]. The measurements show a rather high albedo with an average value of about 0.6 and strong variability in the UV specifically around the extinction peak at 2,175 Å and toward very low wavelengths which indicate enhanced absorption. At optical wavelengths the albedo appears rather constant. The rise in extinction at an A_V of 3.1 in this band is compensated by the strong rise in albedo at short wavelengths giving reflection nebulae their blue appearance. The g-parameter values show large scatter but are generally above 0 indicating some isotropy but a preference to a forward scattering direction.

3.4.5 Emissions from Dust

Emission from dust is as complex as its absorption properties. First there are the so-called unidentified IR bands, which occur at 3.3, 6.2, 7.7, 8.6, and 11.3 μm and could be caused by heating due to UV absorption by PAHs. There is also a red continuum from fluorescence emission that stems from the UV excitation of hydrocarbons. Many absorption processes heat the dust and its absorption efficiencies are most important to the radiative energy balance of dust:

$$\Gamma_{abs} = 4\pi\sigma_d \int_0^{\infty} Q(\lambda)J(\lambda)\,\mathrm{d}\lambda = 4\pi\sigma_d \int_0^{\infty} Q(\lambda)B(T_d,\lambda)\,\mathrm{d}\lambda = \Gamma_{em} \quad (3.32)$$

where $Q(\lambda)$ is the absorption efficiency, $J(\lambda)$ is the mean intensity of the radiation field. The absorption efficiency is generally complex and depends on grain size, shape, and temperature (see Sect. C.9 in Appendix C). For grain sizes $2\pi a$ smaller than λ, $Q(\lambda)$ for many cases can be approximated to $2\pi a/\lambda$ for $\lambda > 2\pi a$ and unity otherwise. This leads to dust temperatures

$$T_d \approx 9\left(\frac{a}{[1\ \mu\mathrm{m}]}\right)^{-1/5}\left(\frac{J}{[1.6\times10^{-3}\ \mathrm{erg\ cm^{-2}\ s^{-1}}]}\right)^{1/5}, \quad (3.33)$$

which results in values ranging from 9 K for 1 micrometer sizes to 22 K for 10 nm sizes. Larger grain sizes radiate like a blackbody and the temperature dependence of the absorption efficiency can be expressed as the Planck mean efficiency. Figure 3.24 shows some of the temperature dependencies for dust in general, but also larger dust particles such as silicates and graphite. The IR intensity then is given by

$$I(\lambda) = B(T_d,\lambda)\tau_d(\lambda), \quad (3.34)$$

where $\tau_d(\lambda)$ is the optical depth integrated over the line of sight, i.e.

$$\tau_d(\lambda) = \pi a^2 Q(\lambda)\int n_d(r)\,\mathrm{d}r. \quad (3.35)$$

Extinction specifically at lower wavelength may be considered by adding a factor $(1 - e^{-\tau(\lambda)})$ on the right side of Eqn. 3.34. Depending on the size and nature of the grains, temperatures can reach up to 100 K near hot stars in or near H II regions. The plot in Fig. 3.24 also shows the temperature dependence of dust in the case where it is near a more intense stellar radiation field. Inside H II regions temperatures can significantly exceed predictions because of additional heating through trapped Lyman α emissions. Note that in the same context grains start to evaporate at temperatures higher than 1,500 K. Dust in the diffuse UV radiation of the Galactic plane has temperatures of 50 K and less, and radiates at wavelengths longer than 100 μm [134].

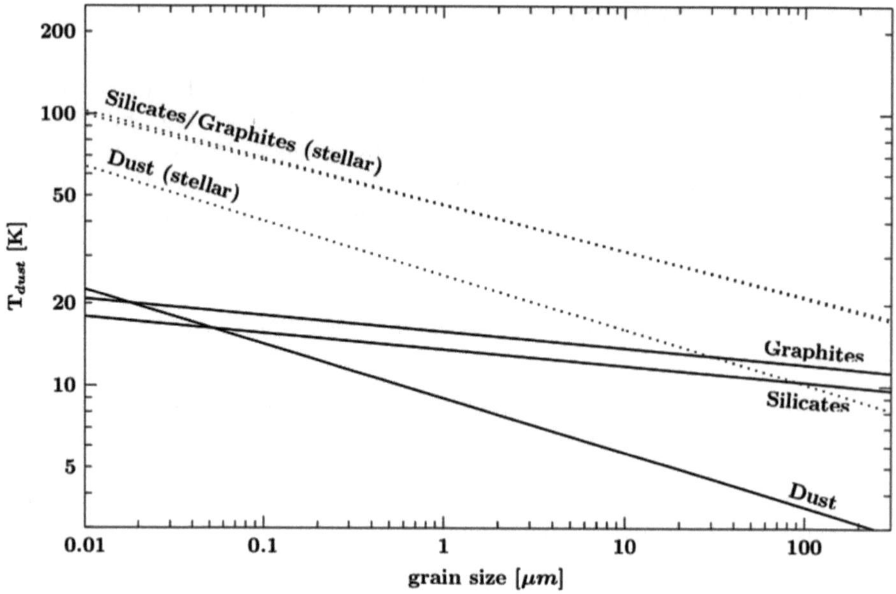

Fig. 3.24 Dust temperatures calculated from Eqn. 3.33 for small size dust (dust) and for silicates and graphite grains using Planck absorption efficiencies, The dotted curves are for the case of a strong stellar radiation field in the vicinity ($G_o = 4 \times 10^4$, distance = 0.2 pc) (inputs from [91])

Interpretations of the IR spectral energy distribution is more complex because in addition to dust emissions the spectra are overlayed by emissions from large molecules such as PAHs but also affected by extinction at smaller wavelengths. Figure 3.23 shows spectra obtained by the *Herschel* Observatory during their early mission phases covering high and low density areas of the high latitude diffuse ISM component rich in larger size dust populations ($\sim$0.1 μm) [143, 144]. The spectra are modeled by PAH emissions at lower wavelength, the background component at intermediate wavelengths, and the dust emissions at large wavelengths above 40 μm. Hydrogen equivalent column densities between the diffuse part (bottom) and dense part (top) vary between $\sim 2.5 \times 10^{22}$ cm^{-2} and $\sim 5.1 \times 10^{22}$ cm^{-2}.

3.5 The ISM in other Galaxies

3.5.1 Overview

The evolution of the ISM is best studied through observations of other galaxies, specifically spiral galaxies like our own Milky Way. Like our own Galaxy, the appearances of the chemical and dynamical states of the ISMs in other galaxies has roots in their star-forming histories. Important here is the relation between the existence of various stellar populations and interstellar gas and dust. The determination of abundances and metallicities give additional clues about the evolutionary stage of the galaxy. Observations of similar spiral galaxies, on the other hand, are helpful in understanding the evolution and star-forming history of our own Galaxy.

Galaxies can be classified into relatively few categories [145]. Classification schemes that exist today [146–148] are more or less based on Hubble's scheme. Some researchers, for example, describe physical parameters of the *Hubble* sequence of galaxies and investigate correlations [149]. One correlation is total mass with optical luminosity, as it reflects the variation of the amount of neutral hydrogen in galaxies. Specifically the hydrogen mass relates to the integrated 21 cm line emission as:

$$\frac{M_{HI}}{M_{\odot}} = 2.36 \times 10^5 D^2 \int S \, \mathrm{d}v \tag{3.36}$$

where D is the distance of the galaxy in Mpc and $\int S\,dv$ the 21 cm flux in Jy km s^{-1}. Note that sometimes the hydrogen mass to luminosity ratio M_{HI}/L_{bol} is determined, as well as the H I surface density σ_{HI}. Single (late-type) galaxies can have H I masses of over $10^{10} M_{\odot}$, while elliptical galaxies (E types) and S0 galaxies have considerably lower H I masses; E types usually have less than $10^7 M_{\odot}$. S0 types have quite a wide range of masses, between 10^7 and $10^9 M_{\odot}$, while others like Sc types have H I masses of around $10^9 M_{\odot}$.

Whether such a correlation implies a direct relation between H I mass and star formation is questionable, as it is molecular hydrogen that should show such a connection, and H I and H_2 have different spatial distributions. H_2, as observed through CO, also shows trends along the Hubble sequence, although these trends do not seem as obvious and direct as in the case of H I. Especially in the case of early-type galaxies [149], classification uncertainties add to the scatter in possible correlations. The mass ratio M_{H_2}/M_{HI} decreases with type from S0 to Sm, which seems mostly a consequence of the fact that M_{HI} increases and M_{H_2} is constant [150]. In this respect, if one considers the integrated mass of H I and H_2, then our own Galaxy would be quite abnormal as both appear at near-equal proportions relative to the total mass.

There have been several case studies of spiral galaxies in recent years putting the relationship between atomic and molecular hydrogen in a different perspective. As an example, the argument that an appreciable fraction of H I found in spiral arms of disk galaxies may be produced by dissociation of H_2 through UV light was introduced for *M83* [152] and *M51* [153]. In a more recent detailed comparison of the distributions of H I, H_2, and 150 nm far-UV (FUV) continuum emission in the spiral arms of *M81* (see Fig. 3.26) [151] could show that bright H α line emission maxima are always associated with a maximum in the FUV emission. The observed morphology can be understood if one postulates chimneys [154] which commonly expose H_2 to FUV from PDRs and areas of vigorous star formation [155, 156]. A cartoon sketch (Fig. 3.25) displays several cases of how various *Norman–Ikeuchi chimneys* could occur in a spiral arm of a galaxy [151]. The transverse size scale, which is likely powered by a cluster or association of O stars, may reach a few 100 pc. A study of the high-mass stellar population in *M51* registered over 1,300 Hα emission regions of sizes typically between 10 and 100 pc [157].

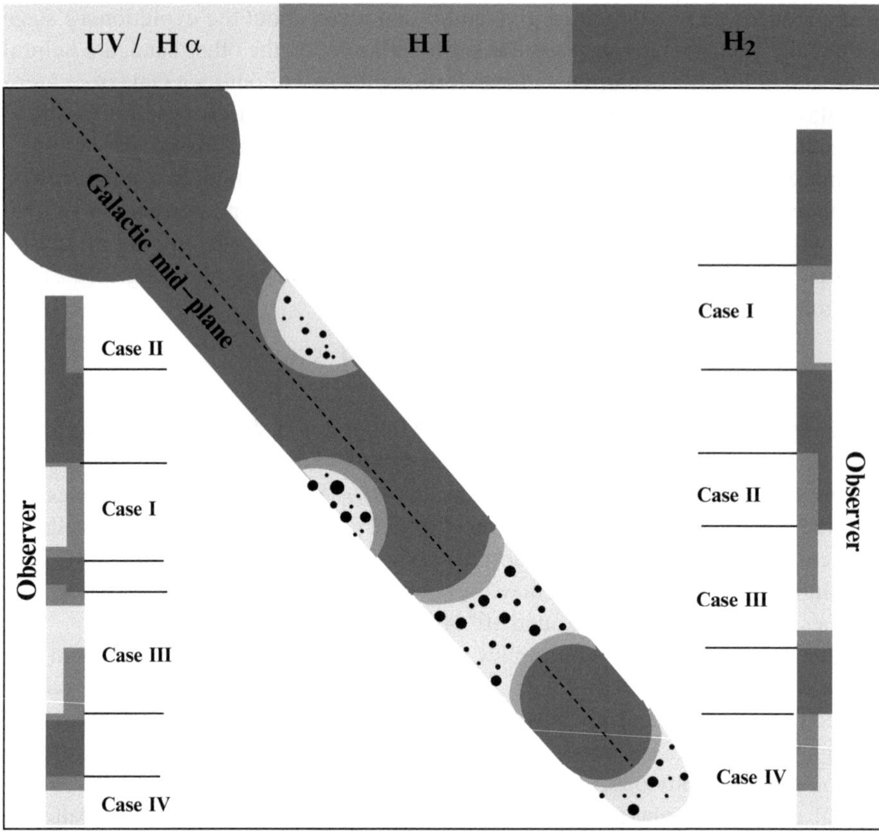

Fig. 3.25 Four possible cases of *Norman–Ikeuchi chimneys* in galactic disks show how a cluster or association of O-stars blow holes and cavities into the regions of dense H_2 [151]. The diagram also indicates the different PDR structure across the galactic surface viewed from opposite directions

Larger and more luminous regions seem likely to be blends of multiple regions. So even if the chimney picture offered by [151] is oversimplified, it is quite clear that photodissociation of H_2 plays an important role in the balance of H I and H_2.

The existence of significant amounts of dust in galaxies, as is impressively illustrated by the image of *M81* observed with *SST* in Fig. 3.26, is another important issue. Its physical properties and its interaction with light can be assumed to be quite similar to the ones observed in our own Galaxy and thus serves to illustrate how dust can act as an agent to catalyze the production of H_2, thereby countering photodissociation effects. This will result in a balance of H_2 production and the destruction in spiral arms of galaxies. Also, the energy removed by dust absorption in the optical and UV will be re-radiated in the IR and far-IR, making the assessment of mass through irradiated flux extremely difficult. Many studies dealing with dust opacities of, especially, star-forming galaxies are currently being pursued (see [158] for a review). These studies show that the dust contents of nearby galaxies depend not only on their morphologies but on their luminosities and activity levels as well.

Fig. 3.26 *SST/IPAC* image of the nearby (3.7 Mpc) Galaxy *M81* at IR wavelength. The Galaxy's dust at these wavelengths becomes the dominant source of emission, showing the entire extent of star-forming activity. Credit: NASA/JPL/Spitzer/IRAC/Caltech/S.Willner (CfA)

3.5.2 The Initial Mass Function

The distribution of stellar masses in a star formation event within a given volume of space is called the initial mass function (IMF). Estimates of the IMF for different stellar populations appears to hold for populations including present-day star formation in small molecular clouds, rich and dense massive star-clusters forming in giant clouds, as well as for ancient and metal-poor exotic stellar populations that may be dominated by dark matter [159]. First IMF estimates were published in 1955 when Edwin E. Salpeter produced a first IMF for stars in the solar neighborhood (see also Sect. 11.1.4). For stars with masses in the range 0.4 to 10 Myr the IMF could be described by a power-law form with an index of 2.35. This result implied that mass density diverged to very small masses, which in the early 1990s was speculated to be a result of faint stars or substellar objects being part of the mysterious dark matter. Studies of the stellar velocities in the solar neighborhood seemed to imply large amounts of missing mass in the disk of the Milky Way. During the last decades a large database has been established, including the *Gliese* catalog in 1969, which contain all known stars within the solar neighborhood with accurately known distance and fluxes down to very low masses [160–164]. Another, maybe more promising, method of measuring IMFs is to focus on young clusters. Here the statistics and observational uncertainties are generally worse, but systematic uncertainties can be much better handled through the observations of as many clusters as possible [165]. Specifically the IMF in the low mass range remains a significant source of uncertainty even though in the early 1990s the theoretical mass–luminosity relation of low-mass stars improved and observational biases due to unresolved binaries were further resolved [166, 167]. The lower boundary is due to the rather uncertain brown dwarf population which in any case hardly contributes to the total mass, the upper boundary is to about 150 $M_\odot$ [168], stars larger than that are speculative (see Sect. 9.2.3).

This database can be used to predict a current mass function of local young stars. Its elements would then contain the number of stars per pc^3 and the luminosity function within a magnitude interval and photometric color band, i.e. (see also Eq. (11.2))

$$g(M) = -f(L) \times \frac{\mathrm{d}m_V}{\mathrm{d}M} \tag{3.37}$$

where M and L are the stellar mass and luminosity, respectively, m the magnitude in the V band, $f(L)$ the luminosity function, $g(M)$ the mass function. Equation 3.37 shows that in order to obtain the mass function one needs the derivative of the stellar mass–luminosity (magnitude) relation (MLR) m_V(M). Thus any uncertainty in stellar structure and their evolution with time will alter this function considerably. The lack of knowledge about stellar age, chemical composition and metallicity, mass loss rates, and rotation will alter the solution in a significant way; it is still today major obstacle in the determination of proper mass functions.

Embedded in the mass function g(M) is the IMF $\xi(M,t)$ attached to a time modulation function $b(t)$ [159] accounting for the different lifetimes of stars (Eq. 2.2). With a timespan when the Galaxy formed ($t = 0$) until the current age of the Galaxy ($t = t_G$) the number of stars per pc^3 in the mass interval $[M, M + dM]$ that form in the time interval $[t, t + dt]$ is $dN = \xi(M,t)b(t)\,dM\,dt$. One boundary condition of this time history has to be that $g(M) = \xi(M, t_G)$ for $t_{life} \geq t_G$ and for $t_{life} < t_G$ the mass function then becomes

$$g(M) = \xi(m)\frac{1}{t_G}\int_{t_G - t_{life}}^{t_G} b(t)\,\mathrm{d}t \tag{3.38}$$

with

$$1 = \frac{1}{t_G}\int_0^{t_G} b(t)\,\mathrm{d}t$$

where t_{life} is the main-sequence lifetime (Eq. 2.2) and $\xi(m)$ the time-averaged IMF. For very small lifetimes as for OB stars, $\int_{t_G - t_{life}}^{t_G} b(t)\,\mathrm{d}t \approx \Delta t$ and $g(M) = \xi(M)\Delta t / t_G$. The mass-dependence of the time-averaged IMF is usually expressed as a power-law with

$$\xi(M) \propto M^{\alpha_i(M)}, \tag{3.39}$$

where the power-law index α_i varies in mass intervals i [169]. Figure 3.27 (left) shows measured index values for various clusters of different mass content and for various studies. The IMF (Fig. 3.27, right) in the Kroupa study is well determined at masses of 1 $M_\odot$ and higher with α values around 2.3, while lower values seem

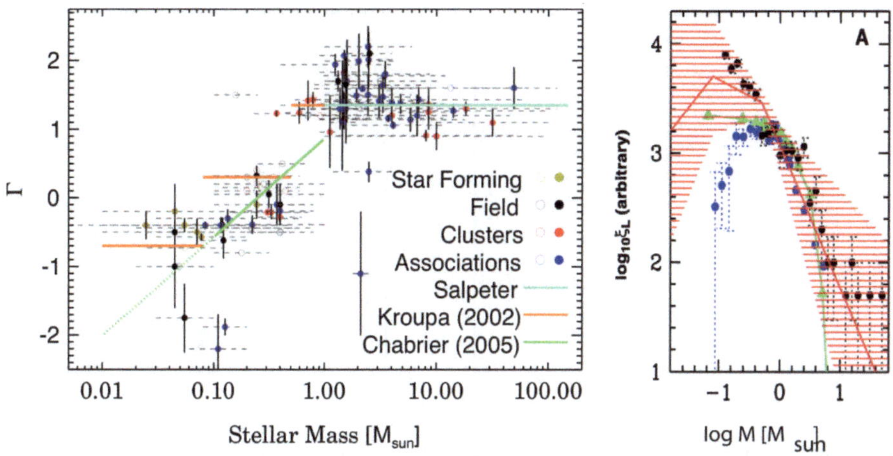

Fig. 3.27 *Left:* The "*alpha*-plot" representation. The derived index Γ of the initial mass function in clusters, nearby star-forming regions, associations and the field, as a function of sampled stellar mass (the dashed lines indicating the full range of masses sampled). Open circles denote studies where no errors on the derived index are given while filled circles are accompanied with the corresponding error estimate (shown as solid vertical lines) (from Bastian et al. [165]). *Right:* Measured stellar mass functions as a function of stellar mass using a variety of galactic star clusters [159]. The average galactic field-star (single) IMF is shown as the solid red line with the shaded areas as the associated uncertainty range (see also Sect. 11.1.4 and Fig. 11.4))

to apply for lower masses (see, however, Fig. 11.4). Note, that using a power law of a combination of power laws to describe the IMF is not the only way to decribe the IMF. [170] uses a power law and a lognormal part, while [171] apply a tapered power law (see Sect. 11.1.4).

This apparent universality of the IMF is a challenge for star-formation theory, because elementary considerations suggest that the IMF ought to systematically vary with star-forming conditions. It is also a question how starburst activity fit in this picture. The recent study by Weidner, Bonnell, & Zinnecker [172] (see Sect. 11.1.4) suggests a rather top-heavy IMF.

3.5.3 Star-Formation Rates

One of the key procedures to study properties and evolution of galaxies is the determination of star-formation rates. This rate is crucial in the assessment of content and evolutionary status of interstellar media in our and external galaxies. Galaxies show a wide range in young stellar content and star formation activity and observed variations are the actual basis for the *Hubble* sequence of galaxies (see [173] for a review). The star formation rate (SFR) is the rate at which galaxies, including the Milky Way, currently convert gas into stars. There are several types

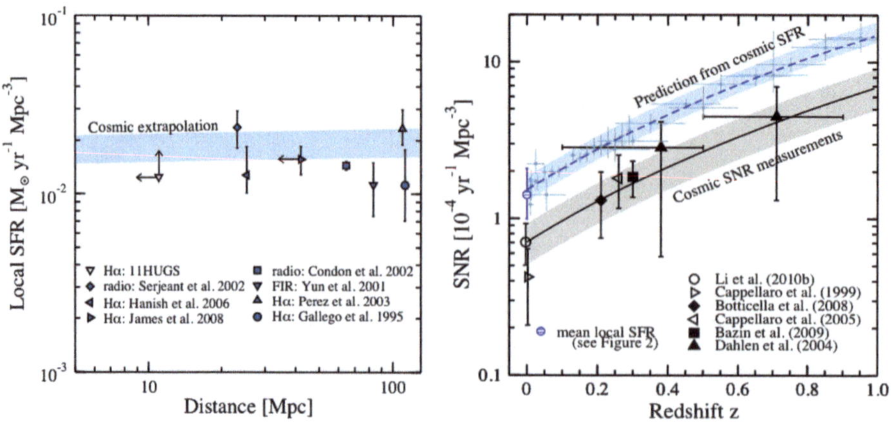

Fig. 3.28 *Left:* Local SFR density as a function of distance. *Right:* Comparison of the cosmic SFR to the cosmic SNR of all types of bright core collapses in the redshift range 0 to 1 (from [174])

of SFRs to consider. There is the local SFR which describes the rate for the solar neighborhood or at some location in the Milky Way and is usually given for the thin disk limit ($z \sim 100$ pc) in units of $M_{\odot}$ yr^{-1} pc^{-2}. The total SFR is the integrated value for the entire galactic disk and then has units of $M_{\odot}$ yr^{-1}. On cosmological scales a volume normalized density unit of $M_{\odot}$ yr^{-1}Mpc^{-3} is generally used.

A different but sometimes related number is the supernova rate (SNR), which decribes the number of supernovae occurring within one century in a galaxy and similar other units with respect to the SFR apply. The SFR and SNR describe the birth and death of stars, respectively, and given the short lifespan of massive stars undergoing core collapse, one would expect these numbers to be similar. So far, however, the cosmological SNR has been found to systematically lower than the cosmological SFR [174] (see Fig. 3.28) which is thought to be a measure of the number of intrinsically dark core collapses [175, 176].

There are several ways to determine SFRs and most of these methods involve the direct or indirect observation of massive stars. This has to do with the fact that very massive (OB) stars are shortlived in general (see Eqn. 2.2), they engulf the bulk of the converted mass into high Z elements, and their bright UV radiation is more easily observed, followed by their reprocessed dust glow in the far IR. These methods provide an account for the massive stellar content which then uses the IMF (see the previous Section) to extrapolate to a global SFR which includes the stellar mass content. SFR compilations include measurements from Hα, UV, FIR, and radio indicators. Several estimators are primarily used in studies, which involve

- Hα line luminosities from galaxies $z \leq 0.4$
- O II (3727Å) line luminosities from galaxies $0.4 < z \leq 1$
- UV continuum luminosity for $z > 1$ and 1,500 Å to 2,800 Å
- From far-IR luminosities at 60 μm for low IR background.

The usage of massive stars as beacons for star formation is as much a blessing as it is a curse. They can be easily observed over a large range of distances; however, one has to rely on massive stellar evolution models as well which introduce large systematic uncertainties toward the determination of the IMF.

The first item is a basic estimator and Kennicutt et al. [177] used the Hα line luminosities and UBV measurements of over 200 galaxies to determine

$$SFR_{H_\alpha} = 0.79 \times 10^{-41} \left(\frac{L_{H_\alpha}}{[\mathrm{erg\ s^{-1}}]} \right) [M_\odot \mathrm{yr}^{-1}], \tag{3.40}$$

assuming that $T_e = 10^4$ K and all ionizing photons are actually retained in the gas. For redshifts z exceeding 0.4 this line is shifted outside the optical range and it has been shown that the O II line doublet at $\lambda = 3{,}727$ Å can be used instead by scaling to an average Hα/O II line ratio [178, 179]. Note that in all these approaches the IMF is assumed to be universal with z (see Sect. 3.5.2):

$$SFR_{O\ II} = 1.4 \times 10^{-41} \left(\frac{L_{O\ II}}{[\mathrm{erg\ s^{-1}}]} \right) [M_\odot\ \mathrm{yr}^{-1}]. \tag{3.41}$$

At redshift z much larger than unity, most strong emission lines are shifted outside the optical range. The UV spectral continuum is dominated by the continuum emission of massive stars in young stellar clusters. Most studies [177] use evolutionary synthesis models of starbursts in which the UV luminosity per unit frequency becomes proportional to the SFR in the wavelength range 1,500 to 2,800 Å, i.e.

$$SFR_{UV} = 1.4 \times 10^{-28} \left(\frac{L_{UV}}{[\mathrm{erg\ s^{-1}\ Hz^{-1}}]} \right) [M_\odot \mathrm{yr}^{-1}]. \tag{3.42}$$

The wavelength range was chosen such that the integrated spectrum is nearly flat in flux for the Salpeter IMF. It is also free of contamination by the Lyman α forest and contribution from older stellar populations [173, 180]. However, SFR_{UV} is quite sensitive to uncertainties in the reddening correction.

A fourth SFR estimator makes use the fact that almost independent of the amount of reddening, a significant fraction of the ISRF (see Sect. 3.1.5) is absorbed by dust and re-emitted as thermal emission in the far infrared (>10 μm). The far-IR flux can then be used to scale the UV emissions of massive stars in galaxies. Studies conventionally use estimates of the 60 μm luminosity to scale the total far-IR luminosity in the range of up the 1 mm by $L_{fIR} \approx 1.7 L_{60\ \mu\mathrm{m}}$ [181,182]. The SFR can then be expressed by [173]

$$SFR_{fIR} = \frac{1}{\epsilon} 4.5 \times 10^{-44} \left(\frac{L_{fIR}}{[\mathrm{erg\ s^{-1}}]} \right) [M_\odot\ \mathrm{yr}^{-1}], \tag{3.43}$$

The numbers are from [183] using a Hubble constant of 70 km s^{-1} Mpc^1 for the flux to luminosity conversion. It has to be noted, that the use the SFR_{fIR} by itself is limited to rather obscured and dusty systems and for less dusty galaxies as the Milky Way or most of the galaxies in the local group should not be used as the only indicator.

Interstellar dust absorbs and reprocesses more than half of the radiation in the universe, and extinction produces the largest source of systematic error into measurements of SFRs in galaxies [173]. Even though the resulting systematic error in the SFR norm can be largely removed by applying a statistical correction for the dust attenuation, the attenuation within and between individual galaxies varies by several magnitudes causing random errors [173]. [184] cataloged the general properties and well-calibrated Hα, N II, and Hα continuum fuxes of 436 Galaxies. [174] used this catalog to determine the local SFR within 11 Mpc (see Fig. 3.28, left panel). Kennicutt et al. [185] then combined Hα and fIR tracers using 25, 60, and 100 μm fluxes to produce dust-corrected SRFs of a subset of galaxies within a distance of 3 and 30 Mpc. Here a scaled IR flux is used to calibrate the dust correction, i.e,

$$SFR_{H_\alpha} = 0.79 \times 10^{-41} \left(\frac{L'_{H\alpha}}{[\text{erg s}^{-1}]} + a_\lambda \frac{L_\lambda}{[\text{erg s}^{-1}]} \right) [M_\odot \text{yr}^{-1}], \tag{3.44}$$

where L'_{H_α} is now the uncorrected H_α line luminosity, a_λ a calibrated attenuation factor, and L_λ the luminosity at the fIR wavelength. Another source of uncertainty is the IMF. Most previous analyses use the Salpeter IMF whereas the Kroupa IMF with multi-powerlaw indices for different mass ranges is more appropriate (see Sect. 3.5.2). Using such an IMF, for example, in the SFR_{H_α} changes the scale factor from 0.79×10^{-41} to 0.55×10^{-41} in Eqns. 3.40 and 3.44 [185]. It also has been shown that the $SFR_{O\ II}$ and SFR_{UV} are systematically overestimated resulting from the effect of underlying stellar Balmer absorptions in the measured emission line fluxes [183].

Today it is emphasized that a smart choice of combinations of SFR indicators is the way of approach as star formation generally is obscured and the IR provides a complement to H_α and FUV determinations, which work great for unobscured objects. [186] and [96] provide additional calibrations for combined indicators.

It should also be realized that the indicators discussed above work well at large distance but not very well in the Milky Way or the local group. Most of the methods here rely on estimates of the ionizing photon production rate in combination with stellar population synthesis calculations. The advantage of using radio free–free continua is that one does not need to know anything about dust obscuration. The drawback is sensitivity as it only works feasibly in the Milky Way and its neighborhood. Murray & Rahman [187] use the total free–free emission in the WMAP foreground map as a probe of the massive star population, and derive a global SFR of 1.3 $M_\odot$ yr^{-1} for the Milky Way. Another local approach has emerged using *Spitzer* accounts of PMS stars [188]. Here population synthesis models determining the SFR were calibrated using the observed number of young stellar objects in the *Spitzer/IRAC GLIMPSE* survey of the Galactic plane resulting is a total SFR in the Milky Way of 0.68–1.45 $M_\odot$ yr^{-1}.

3.5.4 The Local Group and Beyond

The local group (LG) contains galaxies which share quite a number of common star formation properties but also appear surprisingly diverse. The group has currently 36 member galaxies of a large variety of sizes and shapes and is part of the Virgo Supercluster. With a radial extent of $\sim$1.2 Mpc it provides us with the closest volume of current star formation on cosmological scales (see Sect. 3.5.5). Its gravitational center is located between the Milky Way (MW) and the Andromeda Galaxy which are also the two most massive members of the group. The masses are still somewhat uncertain, depending mostly on the estimates of halo masses. Most recently [189] estimated the mass of the MW to be of 1.2–2.7$\times 10^{12}$ $M_\odot$ within a radius of 300 kpc, which makes the MW equally or slightly more massive than M31 with an extimate of 1.3–1.6$\times 10^{12}$ $M_\odot$ within a radius of 300 kpc. The mass equality gives the group a more dumbbell shape appearance as many of the LG members are satellites of the two big galaxies. The next massive members are the spiral galaxy M33 with $\sim 5\times 10^{10}$ $M_\odot$ and the irregular galaxy LMC with $\leq 10^{10}$ $M_\odot$. M32 is an interesting object as it has been classified a massive elliptical dwarf with a diameter of only 6,500 pc, even though claims exist that its distance has been greatly underestimated and it is a massive system outside the LG [190]. All galaxies together have an estimated total mass range of up to 2.6–4.5 $\times 10^{12}$ $M_\odot$. Of the 36 members 31 are dwarf galaxies with various morphologies which include ellipticals, irregulars, and spheroids. Almost all the mass is confined to the three spiral galaxies MW, M31, and M33 (>99%) and they thus dominate LG's overall star-formation rate and history. Figure 3.29 shows the M31 with some of its small galaxy satellites. Even though M31 is very similar in mass, it is classified as a SbI-II spiral. while the MW is to date is classified as barred SBc type. Detailed properties of local group galaxies can be found in [191–193].

Even though they contain only a tiny fraction of the LG mass, the existing dwarf galaxy population is of interest to study questions about whether their diversity is somewhat related to the star formation history and what the effects the absence of spiral density waves and interactions have on the star-formation process itself. Besides of smaller mass their dynamic development is determined by ram pressure and tidal stripping of their gas content [194]. Figure 3.29 shows the morphological segregation in the LG and other local groups which shows that gas-poor dwarfs concentrate around the nearest massive primary galaxy, likely an effect of tidal gas stripping, Furthermore, based on archival *HST* WFPC2 observations Dolphin et al. [193] find that in contrast to the irregular galaxies, the more luminous spheroidal and elliptical dwarf galaxies all show extended star formation episodes, while less luminous systems formed most of their stars over

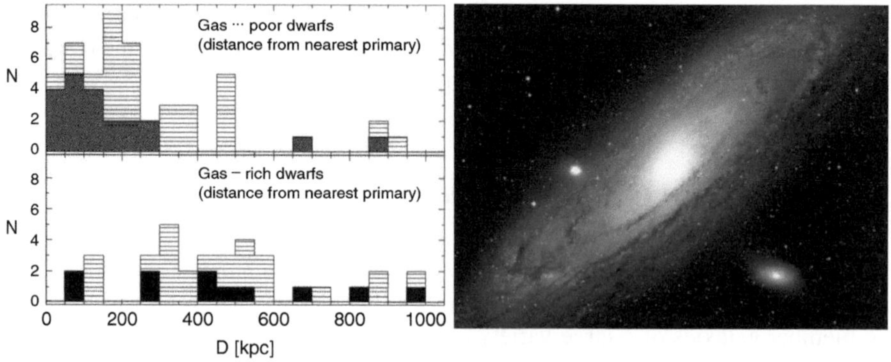

Fig. 3.29 *Left:* Morphological segregation in companion galaxies of local groups. The black histograms are for the local group, the hatched histograms for the M81 and Cen A groups (from [192]). *Right:* Gray scale image of the Andromeda Nebula with three of its companions

ten Gyr ago. A galaxy's classification as an irregular or elliptical/spheroidal is entirely explained by its star formation within the past Gyr or less. Most studies conclude that elliptical/spheroidal systems experienced such recent (<1 Gyr) star-formation episodes, and thus would have been indistinguishable from present-day irregulars prior to these episodes. Still the SFRs in all these small galaxies are very small, in elliptical/spheroidal systems it is practically zero, in irregulars $\leq$0.01. It thus appears likely that these low-mass dwarf irregular galaxies evolve into dwarf spheroidals [194].

The local ($z = 0$) star formation rate is usually the one within a radius of 100 Mpc. Some of the major recent studies on the local SFR are summarized in Fig. 3.28 which compares local SFRs in the range between 10 and 100 Mpc [174] yielding a mean rate (density) of 1.6×10^{-2} $M_\odot$ yr^{-1} Mpc^{-3} without significant variance over the radius of 100 Mpc. Gallego et al. [195] in the mid-1990s already determined a SFR density for the local universe ($z \leq 0.045$) to be 1.3×10^{-2} $M_\odot$ yr^{-1} Mpc^{-3}. For the local group this means that the current average SFR in the Milky Way is less than 10^{-2} $M_\odot$ yr^{-1}. Previous estimates of the Galactic SFR were much higher featuring values >2 $M_\odot$ yr^{-1} (see [188] for a review). The recent *Spitzer* GLIMPSE survey of young stellar objects finds values of 0.68–1.45 $M_\odot$ yr^{-1} [188] consistent with the one obtained by [187] using *WMAP*. These values for the MW now seem well established even though locally there might be larger differences as [196] recently determined the SFR in the Carina nebula (see Sect. 11.2.4) of 8×10^{-3} $M_\odot$ yr^{-1} even though as a lower limit only. This shows that using global rate densities to extrapolate to very local levels is dangerous as it ignores the density of galaxies over large distances. It also demonstrates that local rates from surveys of young stellar objects can be overestimates due to local fluctuations. Lada & Lada [197] calculate a lower limit to the nearby local rate from a subset of clusters within a distance 500 pc of ≥ 1–3×10^{-9} $M_\odot$ yr^{-1} pc^{-2} which integrated over the MW gives very high values.

3.5.5 Evolution with z

The star formation history of the universe is a fundamental problem in cosmology and here the *Hubble Deep Field (HDF)* observations have provided a key opportunity to study the rate of star formation over cosmic times. Section 3.5.3 showed that there are various methods at hand to determine cosmic rates at various different redshifts. In the early 1990s the group around Lilly et al. (see [199] and references therein) carried out deep redshift surveys, such as the *Canada–France Redshift Surveys I–V* which allow the direct study of galactic evolution and star-formation history for redshifts $0.0 < z < 1.3$ containing about 600 galaxies with median redshifts of $\sim$0.56. It was also noticed that the overall luminosity density significantly increased from the current ($z = 0$) universe to a ($z = 1$) universe indicating extensively higher star-formation activity in the past [195, 200].

Madau et al. [180] combined the data from [195,199] with the HDF and identified galaxies at redshifts $2.0 < z < 4.5$ which resulted in a significant variation of the SFR density over cosmic time. Figure 3.30 shows the SFR density over a redshift range between $0 < z < 4$ using data from [180, 195, 199] as well as models where the luminosity function of the galaxies follows the Schechter form [201]

$$\phi(L)\mathrm{d}L = \phi^* \left(\frac{L}{L^*}\right)^{\alpha} \exp\left(-\frac{L}{L^*}\right)\frac{\mathrm{d}L}{L^*}, \tag{3.45}$$

where L^* is the characteristic luminosity at some absolute magnitude, α the powerlaw index, and ϕ^* the SFR density at that point. These studies determined the evolution of the SFR until redshifts of $z \sim 4$ and the SFR evolution the peak rate was reached somewhere around $z = 1.2$ before it dropped rapidly until present day.

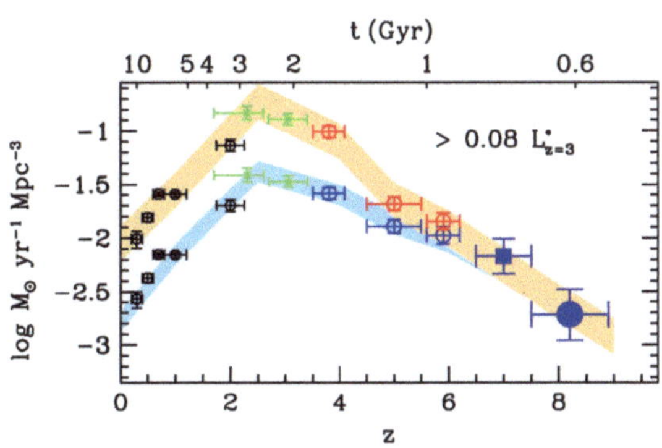

Fig. 3.30 Determinations of the UV luminosity density (light blue) and SFR density (light brown) for the $z \sim 8$ sample, as a function of redshift. The lower set of points (light blue region) show the SFR density determination inferred directly from the UV light, and the upper set of points (orange region) show the one inferred using dust corrections. The dust correction is zero at $z > 6$ (from Bouwens etal. [198])

The detection of galaxies in the *Hubble Ultra Deep Field (HUDF)* changed things dramatically. While in the previous *HDFs* further progress has been made characterizing galaxy populations even for $z \leq 6$, redshifts beyond that limit remained challenging. Until a few years ago only about two dozen of $z \sim 7$ candidates were available. The availabililty of the *HUDF* now allows detections up to $z \sim 8$ (see Bouwens et al. [198] and references therein). These authors utilize the *HUDF* to search for star-forming galaxies at $z \sim$ 8–8.5, which is only 600 million years past recombination. Figure 3.30 now shows the most recent compilation of SFR with redshifts up to ~8. The light blue curve shows the solution entirely derived from UV luminosities, the light orange curve marks the one inferred from dust corrections of the UV slopes. Two important observations can be made from this analysis and compilation. First, the peak of the SFR is now slightly beyond $z \sim 2$ and dust corrections are only important for redshifts of about $z < 6$.

Chapter 4
Molecular Clouds and Cores

Of specific interest for stellar formation are regions where matter in the ISM appears most dense and in the form of clouds. Although there is no particular definition of the term *cloud*, it describes regions with a density larger than about 10 to 30 atoms per cm^3. There are *diffuse* clouds, which do not reach these densities, and *self-gravitating* clouds in the range of the above densities. Some also define a third *unbound* category which are neither pressure nor gravity-supported and are probable clouds in the process of dispersal [202]. Clouds move as entities through the Galaxy with a peculiar velocity on top of the radial velocity due to the general rotation around the Galactic Center. Peculiar motions of such clouds usually have velocities of the order of 5 to 10 km s^{-1}. For a long time it was thought that two forces are at work to keep clouds stable: gravity to contract the cloud and internal (thermal) pressure and turbulence to disperse the cloud. However, it has been recognized in recent decades that this view may not be sufficient and magnetic as well as rotational properties of molecular clouds are also important. This chapter describes the global properties of clouds in the Galaxy, outlines relevant physical processes that affect these clouds and then finally discusses initial conditions of cloud cores before they undergo a gravitational collapse.

Almost all of the cold neutral atomic hydrogen exists in clouds. In the Galactic plane along a 1 kpc line-of-sight there are usually about 5 to 10 H I clouds as can be estimated from the number of line components in stellar spectra. With an average extent of 10 pc and densities of $\sim$30 atoms per cm^3 their average mass is less than 100 $M_\odot$. At a temperature of 100 K only clouds with masses of over $\sim$1,000 $M_\odot$ can become gravitationally unstable and thus most of these clouds are actually not very interesting for the process of stellar formation (see also Eqn. 4.21).

The Ophiuchus cloud complex (see Figures 4.1 and 11.10, bottom) serves in many ways as a prototype of molecular cloud structures. It is a nearby cloud at a distance of approximately 520 light-years and is comprised of filamentary clouds of low-density molecular gas with dense molecular cores interspersed. One of them, the westernmost cloud, contains a massive cloud fragment of size $\sim$1 pc $\times$ 2 pc and a total estimated mass of 500 to 600 Suns [203]. It is called the ρ *Oph* cloud, one of the few active nearby star-forming regions (see Chap. 11).

N.S. Schulz, *The Formation and Early Evolution of Stars*, Astronomy and Astrophysics Library, DOI 10.1007/978-3-642-23926-7_4, © Springer-Verlag Berlin Heidelberg 2012

Fig. 4.1 An exposure of the *Rho Ophiuchus* region in the Milky Way, which is full of filamentary molecular dark clouds. ρ *Oph* itself is near the upper right corner and is embedded in dark clouds and reflection nebulae. Credit: Photography by S. Pitt; from http://www.light-to-dark.com/sag.html

4.1 Global Cloud Properties

Clearly, the densest regions in the ISM are molecular clouds. The temperatures in these clouds are typically 10 to 50 K (see Table 3.1) and the result of a balance of heating and cooling mechanisms. The heating is provided by cosmic rays and nearby stars. The cooling proceeds through absorption and collision processes of dust and gas particles in the clouds. The energy is finally released as infrared radiation (see reviews by [113, 204]). This process is key to our ability to observe these clouds and detect a large variety of molecules that exists within them. They appear in quite variable sizes between 1 and 200 pc with masses as small as 10 to 100 $M_{\odot}$ and as large as $10^6 M_{\odot}$.

The formation of molecular hydrogen (H_2) is favored by the existence of dust grains. Ordinarily it is quite unlikely that molecular hydrogen would form through a simple merger of two atoms or radicals, as the newly formed molecule cannot radiate the gained binding energy away and will dissociate almost immediately. However, if this process happens near grains, this binding energy can be released without radiation onto a grain surface and the molecule remains stable. Also a shielding effect in cloud cores allows the fraction of H_2 to clearly dominate the amount of total hydrogen. Accordingly, H_2 can be found ubiquitously in clouds with large color excesses. At E(B-V) $>$ 0.08 mag the fraction of molecular hydrogen starts to become traceable, at E(B-V) $>$ 0.15 mag about 75 percent of measured stars show over 10 percent of H_2.

H_2 and other symmetrical molecules are hard to detect. They do not possess a permanent dipole field and therefore have no easily detectable rotational line spectrum at radio wavelengths. Some spurious lines observed in the infrared band originate from forbidden transitions. Infrared lines from H_2 hardly exist as the necessary high excitation energies are simply not available in molecular clouds. One notable exception is H_2 in shocks of jets and outflows. Ultraviolet lines cannot be observed at all as dust densities are too high in clouds. Densities in molecular clouds are thus determined indirectly though other molecules, foremost though collisional excitation of CO or CS with H_2 molecules (see [205]). Densities in molecular clouds vary between 10^2 and 10^6 H_2 molecules per cm^3.

A notable property of molecular clouds is that they are highly *irregular* structures. Many clouds have filamentary shapes with wispy structures that include clumps, holes, tunnels, and bows. These structures have been described in the past in many illustrative ways with expressions like *elephant trunks*, *coal sacks*, and *pillars of creation*, just to name a few. These names all represent various familiar images describing the shape of the clouds. Though no real communalities have been found so far other than the frequent occurrence of string-like structures, this fact testifies to the dynamic and turbulent nature of the matter involved in many of these clouds. The following sections give a phenomenological overview of various molecular cloud structures and star-forming activities.

4.1.1 The Observation of Clouds

Observations of molecular clouds are quite ubiquitous in literature and the reader will find observational accounts in most of the listed references. There are key diagnostical observation techniques that allow one to deduce specific physical properties. Thermal continua are related to temperatures below 100 K and observations are performed primarily at IR and sub-mm up to cm radio wavelengths (see Fig. A.2). Many molecular lines are observed at UV and optical wavelengths ($\sim$1,000–5,000 Å), and more lines appear at infrared and radio frequencies. They offer a wide range of spectral windows to study dense clouds. Several lines have even been amplified as *masers* and provide excellent probes to search for luminous embedded protostellar cores. Last but not least, some non-thermal emission in the form of synchrotron radiation may be seen throughout radio wavelengths. So far there is no significant amount of UV or high-energy emission from infalling cores. X-rays from star-forming regions become important at later evolutionary stages.

The global distribution of matter is also useful in studying clouds. Specifically Section 3.2.4 has discussed the composite CO survey shown in Figure 3.12 in some detail [104]. Other surveys have been performed during the same period [206–209]. These surveys locate most of the distributed mass and are resourceful data sets to map out the large-scale structure of molecular mass in the galaxy allowing us to study the global mass distribution of clouds [210, 211]. Here is a short overview of some of the major observational facilities used in the study of molecular clouds and collapsing cores.

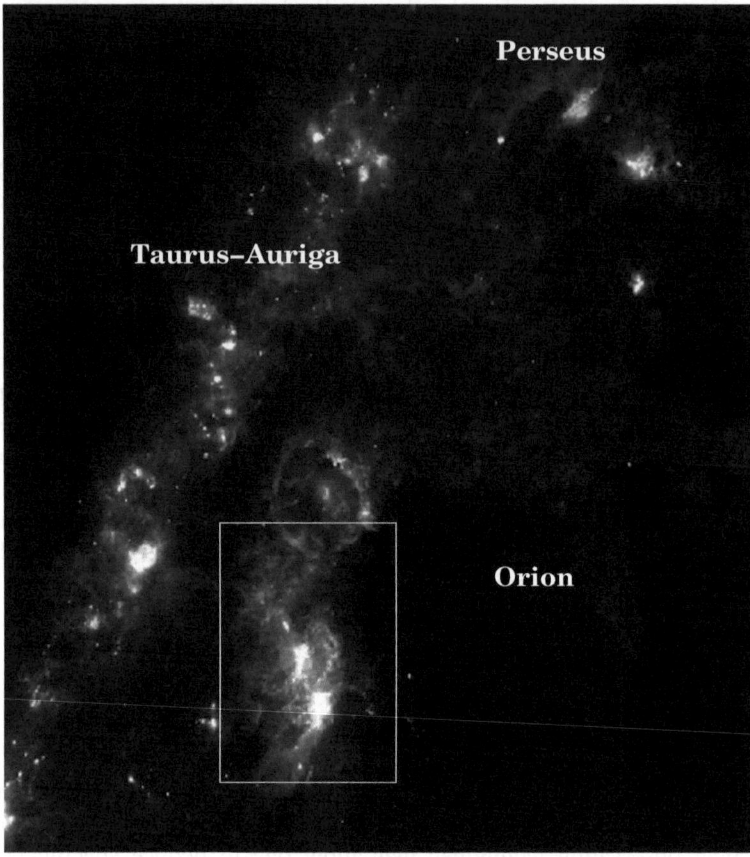

Fig. 4.2 A very large field of the sky that includes the Perseus molecular cloud complex (*upper right*), the Taurus–Auriga molecular cloud complex (*upper left*), and the Orion star-forming region (*lower center*), as observed with IRAS in the combined wavelength bands

The *Hubble Space Telescope (HST)* was launched in 1990 into a low Earth orbit to perform unperturbed sub-arcsecond resolution observations. Throughout the 1990s the two *Wide Field Cameras (WFCs)* were a major instruments used for observations relevant for stellar formation. Since 1997 they have been superseded by the *Near Infrared Camera and Multi-Object Spectrometer (NICMOS).*

The *Infrared Astronomical Satellite* (IRAS), launched in 1983, scanned more than 96 percent of the sky and provided the first high-sensitivity celestial maps at mid- to far-infrared wavelengths [212]. Though not a survey by definition, the IRAS observation archive is a rich heritage of IR data (see Fig. 4.2). The data from the observatory archive contain flux densities from wavelength bands centered at 12 μm, 25 μm, 60 μm, and 100 μm at a spatial resolution of about 1 arcmin. This, for example, allows us to determine the total measured infrared flux and thus luminosity and optical depth via:

Fig. 4.3 The Very Large Array (VLA) is a collection of 27 radio antennas located at the NRAO site in Socorro, New Mexico. Each antenna in the array measures 25 meters (82 feet) in diameter and weighs about 230 tons. The VLA is an interferometer, which means that the data from each antenna can be combined electronically so that the array effectively functions as one giant antenna. Credit: NRAO/AUI/NSF

$$S = \int_0^\infty B_\nu(T)\tau \, d\Omega \, d\nu = \int_0^\infty F_\nu \, d\nu \tag{4.1}$$

where τ is the optical depth and $d\Omega$ the solid angle subtended by the source. F_ν may be obtained by fitting some regression function to the data. IRAS found over 250,000 sources, which to a large extent are identified as emissions from distant galaxies and also as a large fraction of interstellar dust clouds and globules. This catalog allows for many types of surveys relevant for stellar-formation research.

Two other prominent infrared observatories include the *Infrared Space Observatory (ISO)* which was launched in late 1995 and the *Kuiper Airborne Observatory (KAO)*, created in 1974. ISO conducted nearly 30,000 observations at wavelengths between 2.5 and 240 μm. It was equipped with many instruments allowing for very high spectral resolution at wavelengths $\leq$45 μm. The KAO performed astronomical observations at a maximum height of about 14 km ensuring it to be above the atmosphere's water vapor. The KAO terminated its service in October 1995 and will be superseded by *SOFIA* in 2005.

The *Two Micron All-Sky Survey (2MASS)* is a more recent and ground-based near-infrared digital imaging survey of the entire sky conducted at 1.25 μm, 1.65 μm and 2.17 μm wavelengths. A key science goal of the survey is the determination of the large-scale structure of the Milky Way.

The *National Radio Astronomy Observatory (NRAO)* is a long lived working facility for radio astronomy. It operates many facilities around the world including the new upcoming *Atacama Large Millimeter Array (ALMA)*, the *Green Bank Telescope (GBT)*, the *Very Large Array (VLA)* (see Fig. 4.3), and the *Very Long Baseline Array (VLBA)*.

More details can be found in Table 2.1 and Appendix F. The latter also includes a comprehensive list of websites which offer many more details on these observatories as well as many more observational sites and facilities that allow for the performance of detailed imaging and spectroscopy which did not find any mention in this section.

4.1.2 Relation to H II Regions

Many dark molecular clouds are seen in connection with bright regions of ionized hydrogen, which are also active regions of star formation. Like H I clouds, H II clouds are much too hot and in the end not massive enough to become gravitationally unstable. Thus, they are not likely original sites of star formation. On the other hand, the reason why they glow bright in the sky is because they harbor young, and at their centers likely massive, stars. These stars not only ionize their environment giving rise to bright H II regions, they also erode the interface between this region and the dark molecular cloud through photoevaporation and photoionization.

This behavior can be demonstrated with a rough time scheme that allows for mainly three configurations as has been suggested by *C. Lada* in the 1980s and as is illustrated in Fig. 4.4:

- *Configuration A*: no H II region is associated with the molecular cloud. Depending on the mass and dynamics of the cloud this will remain the case as along as the star-forming activity does not produce massive stars (see Sect. 4.1.3).
- *Configuration B*: star-forming activity in the molecular cloud has begun and has produced one or more massive stars which ionize their environment forming blister-like structures. Massive stars have already reached the main sequence after a few times 10^4 yr and usually burn out their nuclear fuel on timescales of 10^7 yr. The *ionization age* of the H II region thus is short and most of the star-forming activity is still ongoing in the cloud. In this configuration the massive stars will carve out a large ionized region and expose the interface between the blister and the neutral cloud (see Sect. 4.1.8).
- *Configuration C*: Most of the molecular cloud has condensed into cluster stars and O stars, maybe even in different generations, ionizing the entire region leaving a few remnant cloud patches and globules. The system may now also be part of an association of early-type stars which also provide ionizing power from outside. Most of the initial star-formation activity has ceased and cluster stars evolve towards the main sequence (see Sect. 4.1.9).

Although this simple view covers many observational features it is controversial. The only dynamics involved in this picture are rather static gas masses affected

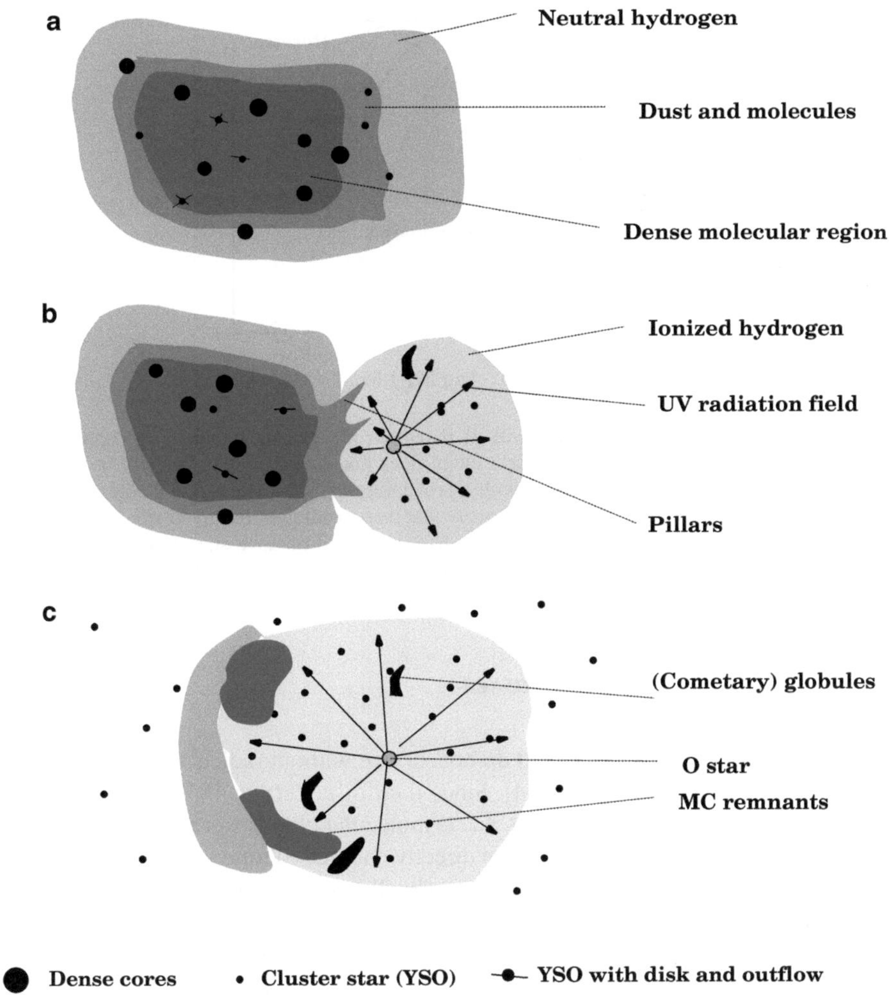

Fig. 4.4 Three cloud configurations showing GMCs (configuration **a**), a GMC with an H II blister (configuration **b**), and a giant H II region with molecular fragments (configuration **c**) with stars forming in all of them. Such a sequence of configurations was proposed in order to understand the formation of OB-associations [45]

only by progressing condensation into cluster stars as well as photoionization and photoevaporation effects from massive stars. Any details from ionization of the interstellar radiation field, internal cloud turbulence and velocity fields, and angular momentum and magnetic fields have been neglected. This clearly only affects molecular clouds at the top of the mass spectrum (see below). Molecular clouds with masses of less than $10^5 M_\odot$ are not likely to form massive stars. The limiting probability that a molecular cloud of mass M does not contain an O-star is shown in Fig. 4.5 [210]. Ultimately these configurations may serve as a formation scenario for OB-associations [45].

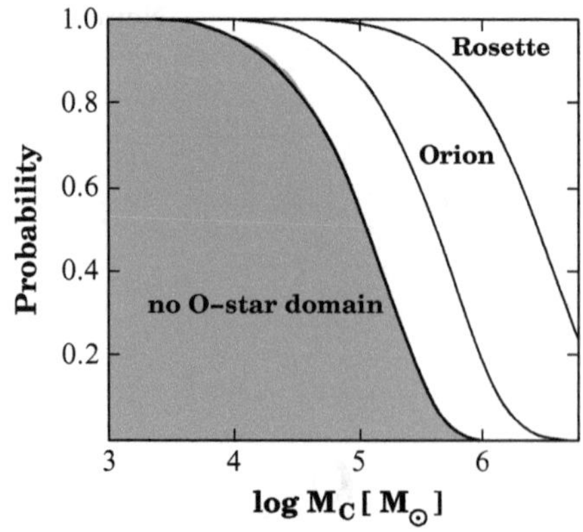

Fig. 4.5 Molecular clouds associated with H II regions are usually found on the upper end of the mass spectrum. The probability that a cloud does not contain a star as early as 09.5-type limits the cloud mass to about $10^5 M_\odot$ with a probability of 50%. The Orion Nebula Cloud has a mass of about 1×10^5 and thus the probability that there would not be any O-star is 87%, in the Rosette Nebula region it is 99%. These clouds have produced massive stars in the past and thus these numbers project the likelyhood that they will produce new ones. Adapted from Williams and McKee [210]

4.1.3 Molecular Cloud Masses

Clouds are transient structures and do not survive without radical changes for more than a few times 10^7 yrs [213, 214]. Since their discovery in 1970 through the 2.6 mm radio emission of CO (see Sect. 3.2.1) molecular clouds were seen as birth sites for stars and their lifetimes seem to be directly coupled to time spans of early stellar evolution. Some of the clouds have masses well in excess of 10^4 $M_\odot$. These *giant molecular clouds* (GMCs) are among the most massive objects in our Galaxy even rivaling Globular Clusters. Our Galaxy hosts about 4,000 GMCs (see Table 3.3) accounting for a substantial fraction of the mass in the Galaxy.

Configuration A in Fig. 4.4 comes closest to the structure of GMCs. They are embedded in neutral hydrogen gas within the ISM and should be considered as dense condensations in a widely distributed atomic gas. The interface of dust and molecular gas often represents a rapid transition from the molecular gas to the surrounding atomic gas providing effective shielding from the interstellar radiation field. This transition from H to H_2 depends in a sensitive way on the pressure and ambient radiation field to such an extent that slight changes in either values can alter a whole cloud population [202]. It also (see Sect. 4.2.6) depends on the column density of material that provides shielding from UV radiation (see Sect. 4.2.6). Importantly, there is no requirement that a cloud is gravitationally bound and, therefore, the molecular content is not particularly an indicator of whether the cloud is gravitationally bound or not.

The average density inside such a cloud is a function of the cloud mass and external temperature and pressure conditions. Fig. 4.6 shows their relation for a

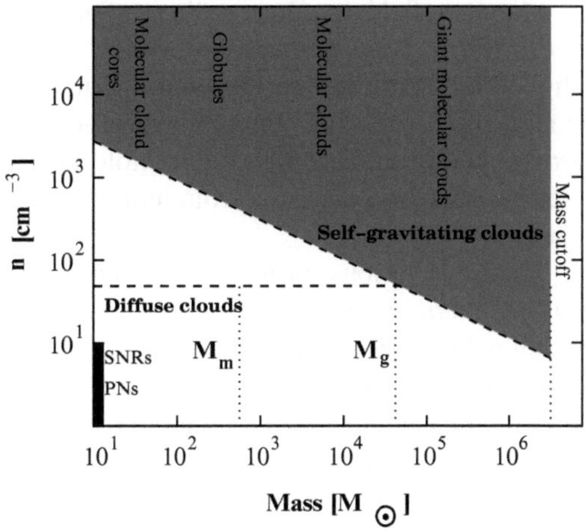

Fig. 4.6 The distribution of cloud types in the mass–density diagram. Adapted from Elmegreen [202]

given temperature and a pressure. Self-gravitating GMCs and MCs have a higher mass at lower densities. Cloud cores and globules are self-gravitating as well but because of their lower total mass require higher densities by an order of magnitude. Above a mass limit M_g even diffuse clouds become self-gravitating. From several CO surveys [208, 209, 215] accounting for about 40 percent of the molecular mass in the Galaxy, the mass function of GMCs was determined to be a power law with a -1.6 in slope. From these fits the cumulative number of molecular clouds within our solar neighborhood was modeled as [210]:

$$\mathcal{N}_c(>M) = 105\left[\frac{M_u}{M}^{0.6} - 1\right]. \tag{4.2}$$

The normalization of this function is still uncertain and depends on the galactic neighborhood. The upper end of the mass function is cut off at $M_u \sim 6 \times 10^6\ M_\odot$. Probable origins for this upper mass limit could be the tidal field of the Galaxy or massive stars within the clouds. Other more recent models based on a fractal interpretation of the ISM structure find similar values for the slope [211].

4.1.4 Magnetic Fields in Clouds

Radiation from magnetized environments is likely to include some fraction of polarized light. Partial polarization from molecular clouds can be observed in absorption and emission. In this section only continuum emission and absorption is considered. As will be outlined below, magnetic fields play a major role in the dynamics of molecular clouds and ultimately in the stellar formation process and the observation of magnetic configurations is vital to the understanding of these processes.

Magnetic fields produce grain configuration changes, which can be observed in two different ways through absorption of a background continuum by preferably aligned dust particles and through thermal emission of such dust particles. There is also grain scattering which, however, is more applicable to circumstellar nebulae around young stellar objects than molecular clouds (see [216] for a review). Also, thermal polarized emission from individual protostellar objects will be discussed in Chap. 6.

There are various forms of alignment of dust particles with magnetic fields. The *paramagnetic alignment* of thermally rotating grains is based on the dissipation of rotational energy perpendicular to the magnetic field, causing the angular velocity component in this direction to vanish. This type of alignment is also called *Davis–Greenstein alignment* and depends strongly on the magnetic properties of grains as well as the gas–grain temperature ratio [217, 218]. Alignment of supra-thermally rotating grains is called *Purcell alignment* and is based on the fact that many grains rotate significantly faster than expected from the kinetic gas temperature. This alignment is thus independent of the gas–dust temperature ratio. Other mechanisms under study include *radiative* and *mechanical* alignment [219]. The latter authors suggested that dust grains preferably align with the local magnetic field with the shortest axis of rotation.

A. A. Goodman [220] compiled a few factors determining the polarization of background light. For example, when it comes to polarization, grain size matters. Grains with sizes of the order of $\lambda/2\pi$ are the most effective polarizers. Other sizes add towards extinction, but do not enhance polarization levels [76, 221]. Thus drawing information from polarization maps of molecular clouds is still very difficult and the process suffers from many selection effects. To observe polarization due to absorption of light from background stars restricts the measurement to optical and IR wavelengths (i.e., at wavelengths $\ll$100 μm). Observation of polarized thermal emission in clouds with temperatures of less than 50 K requires extreme column densities in line with the simple argument that the more extinction, the more dust and the greater polarization percentage. With currently available instrumentation one needs extinction values of $A_V \sim 25$ mag and higher (i.e., column densities of the order of 10^{23} cm^{-2}) for detection of polarized emission at 350 μm. However, most interesting for star formation are dense dark clouds, which usually have a range of extinction around 0.5 mag $< A_V <$ 10 mag and where higher values are rare due to their limited size. Thus, measurements are almost entirely restricted to background light from stars rather than intrinsic thermal emission. Even in this case the sensitivity for polarization depends on the level of extinction and it could be shown that in dark clouds with extinction higher than $A_V = 1.3 \pm 0.2$ polarization from background light seems insensitive to magnetic fields [220, 222, 223].

This does not mean that optical and infrared polarimetry is entirely limited to low extinction. On the contrary, it was demonstrated in the late 1970s [224] that one can observe extinct areas by up to 40 mag over large scales. However, at shorter wavelengths one is subject to confusion along the line of sight as well as limited in flux at such high extinctions. At longer wavelengths (i.e., $>$100 μm)

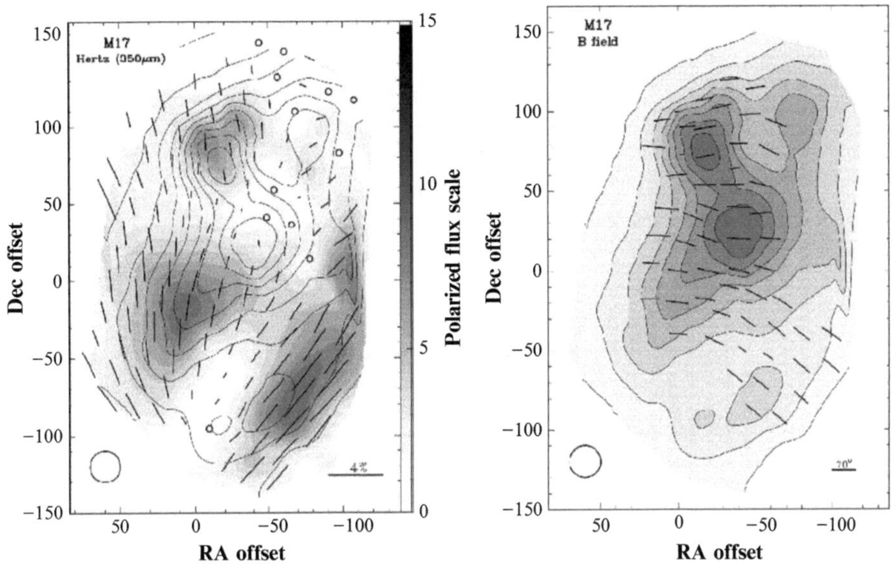

Fig. 4.7 (*left*) Polarization map of M17 at 350 μm. The center position is at R.A. = 18h17m314, Dec. = −16° 14'250" (B1950.0). The contours show the continuum flux levels. The gray scale gives the polarized flux according to the scale an the right. The thick lines illustrate the direction of the polarization vectors. The large circle is the beam width. (*right*) Orientation of the reconstructed magnetic field in M17. The vectors show the projected orientation. The length of the vectors represent the viewing angle as scaled in the right corner giving some 3-D information. From Houde et al. [234]

the absorption and scattering cross sections are small and one observes exclusively thermal emission from aligned dust grains.

The molecular cloud in M17 at a distance of about 2.2 kpc [225] is one of the best studied objects with respect to optical, infrared, and far-infrared polarimetry. Others are L1755 in Ophiuchus [220] and the Orion Molecular Cloud (*OMC-1*) [226,227]. M17 is associated with a PDR region and various condensations and young stellar objects resembling configuration B in Fig. 4.4. Optical polarization studied by [228] shows a quite incoherent pattern of polarization vectors uncorrelated with the nebula. At far-infrared wavelengths between 100 μm and 800 μm [229–232] polarization vectors seem to smooth out. A combination of Zeeman line splitting [233], polarimetry maps of 350 μm, and ion-to-neutral molecular line widths [234], have recently been used to determine the magnetic field configuration in M17. The result is shown in Fig. 4.7. The polarization map (left) shows a smooth distribution of polarization vectors with strong polarization in fair but not full complement of continuum flux. The right part of the figure shows the resulting magnetic field configuration with respect to the continuum flux. Clearly visible is the fact that there seems to be a poor correlation between the dense condensations and the overall cloud (see above).

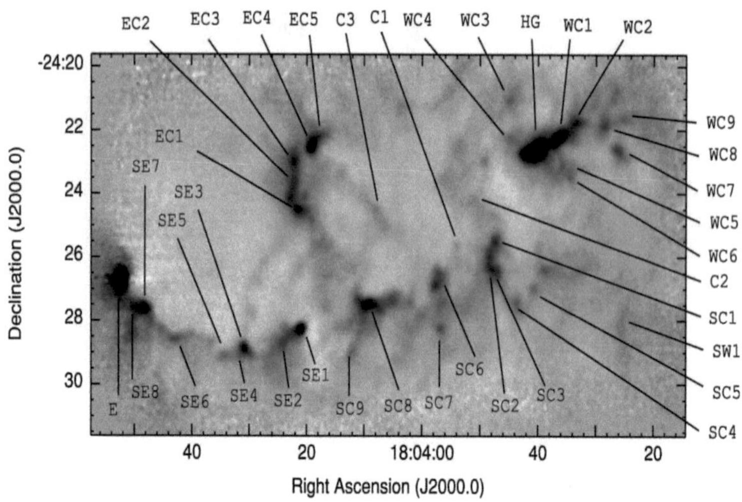

Fig. 4.8 Sub-mm mapping of clumps in the Lagoon Molecular Cloud near M8. The pointers with labels are identifiers of detected cores and clumps. From Tothill et al. [244]

4.1.5 Clump Mass Functions

Dense cores in dark clouds have been extensively observed by P. C. Myers and colleagues and the results have been well documented in a series of key publications throughout the 1980s and 1990s [235–237]. Molecular clouds are highly fragmented [213] and can be described in terms of *clumps* and *cores* (see Fig. 4.8). On a large scale this clumpiness satisfies a self-similar description of all ISM clouds [238] (i.e., with a mass spectrum that has a somewhat similar slope but a lower mass range). *Cores* are regions out of which single stars or binaries form and thus are assumed gravitationally bound. However, a large fraction of the GMC mass is in *clumps*, which are generally larger structures than cores. Specifically the more massive gravitationally bound clumps are thought to form stellar clusters. However, this also means that a large fraction of the mass occupies a relatively small volume compared with the total volume of the cloud, raising the the question: how is this mass actually bound? For regions that are in virial equilibrium, and which can be assumed to be gravitationally bound, [214] showed that observed internal dispersion velocities correlated tightly with the radii and masses in the form of power laws. For a large range of cloud radii a power law index of 0.38 was determined; others found indices of more like 0.5 (see [239] and references therein). This dependence is not so different from the turbulent ISM and it was asserted that the clumpy structure is at least in part the result of supersonic motions (see also Sect. 4.2.4). It was shown [239], on the other hand, that clumps are not in virial equilibrium in many other GMCs and instead are confined by the pressure of the surrounding medium.

It also seems that the mass distribution is not really an indicator for star formation. In a comparative study of two GMCs, the Rosette Molecular Cloud

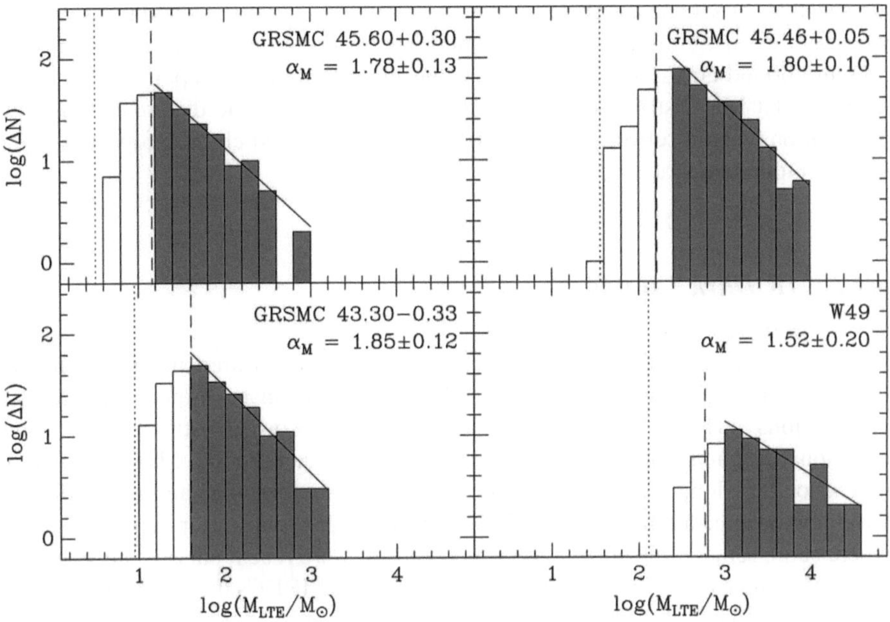

Fig. 4.9 Clump mass spectra of the four clumps. The *dashed line* marks the limit clump mass sampling is incomplete (*white bins*). The *solid line* is the powerlaw fit to complete bins with slopes α_M (from Simon et al. [247])

with a large ionizing luminosity [240] and high levels of star-formation activity, and G216-2.5 with a very low luminosity and no star-forming activity [241], almost 100 clumps were found in each of the clouds with a very similar mass distribution [242]. Thus, the shape of the mass distribution of clumps and cores in a GMC does not solely determine stellar formation; not every clump leads to a stellar cluster. In fact all clumps in G216-2.5 and many clumps in the Rosette Molecular Cloud turn out to be not gravitationally bound and thus would not form stars [243].

Clump mass functions in that respect are not expected to resemble cluster mass functions (see Sect. 11.1.4) or the stellar IMF (see Sect. 3.5.2). Clump mass functions have been derived from CO line surveys [245–247] resulting in clump functions

$$\Delta N/\Delta M \propto M^{-\alpha} \tag{4.3}$$

with slopes α between 1.6 and 1.8. which is considerably flatter than the Salpeter slope of 2.35 and the stellar IMF in general. Figure 4.9 shows clump functions for four clouds. Note that for the most luminous case W49 the slope appears much flatter (see Sect. 9.3.3).

This result seemingly contradicts measurements in other star-forming regions using dust observations in the far-IR and sub-mm range or extinction mapping which do produce slopes close to the stellar IMF [248–250]. The difference in power-law mass distributions is attributed to the different density regimes the traces methods are sensitive to. CO observations are more sensitive to the diffuse and transient gas whereas the dust emission/extinction methods trace the denser, gravitationally or pressure-bound cores. This implies that the mass functions steepen continuously from the flat MC distribution towards the stellar IMF and the different powerlaw distributions would reflect a structural change of the cloud/clump properties from transient to bound structures [251].

4.1.6 IR Dark Clouds

IR dark clouds (IRDCs) were first recognized in mid-infrared images from the Infrared Space Observatory (ISO) and the Midcourse Space Experiment (MSX) in emissions $\geq$15 μm [252]. IRDCs appear highly absorbed with $A_V > 25$ and in silhouette against the galactic mid-infrared (MIR) background. Morphologically they also show long filamentary knots and shapes (see Fig. 4.8) and many IRDCs have now been cataloged for ISO [253] and MSX [254] sub-mm/radio observation where some remarkable correlation was noted. A more recent survey using *Spitzer* data at 24 μm and above find in more than three dozen IRDCs that even though they remain dark, many of the cores are associated with bright 24 μm emission sources. Figure 4.10 shows mid-IR, far-IR, and sub-mm observations of the IRDC G028.53-00.25. The 24 μm image is entirely dark with respect to the IRDC, while the 850 μm

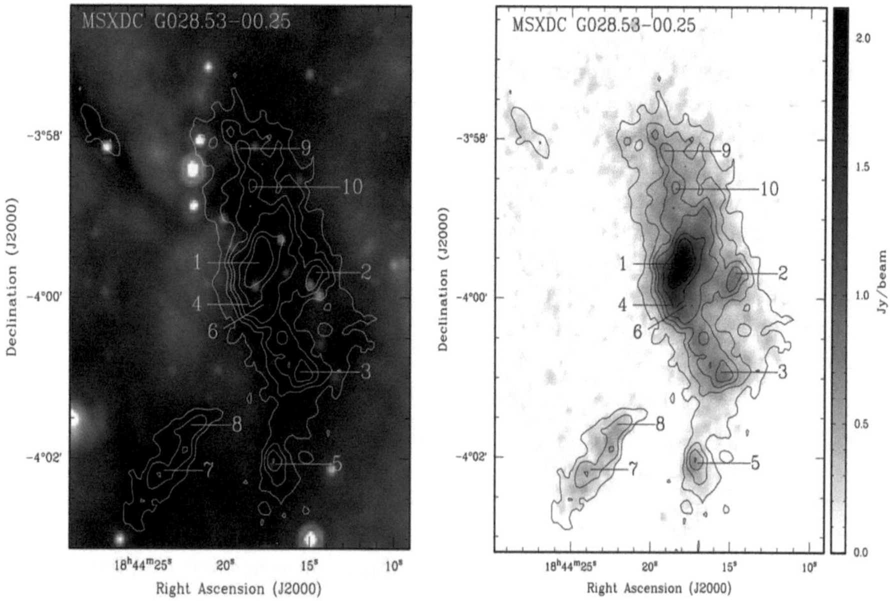

Fig. 4.10 Example of mid-IR (24 μm, *left*), far-IR (850 μm, *right*), and sub-mm (1.2 mm, contours) continuum observations of the IRDC G028.53-00.25 (from Rathborne et al. [255])

and the 1.2 mm correlate. This clearly suggests that they contain one or more embedded protostars and and are now considered to be precursors of star formation. Even though in many cases these dark cores are sites of low and intermediate star formation, the most massive ones are considered candidates for the very earliest stages in the high-mass stars-formation (see Sect. 9.3.1).

4.1.7 High-Latitude Clouds

Although most of the molecular clouds and dust concentrate close to the galactic equator due to the finite scale height of the galactic plane there are clouds at high ($> \pm 20$ deg) galactic latitude [256–258]. These clouds behave like atmospheric clouds with wispy structures. They are highly transient and disperse as rapidly as they form. In this respect these galactic *cirrus* clouds have long been thought to be unfit for stellar formation as they provide strong turbulent support against self-gravity. Searches for young stars associated with these clouds remained unsuccessful for a long time [259].

Recently, this view has somewhat come under scrutiny, specifically since the detection of young stars with *ROSAT* in soft X-rays far away from any active star-forming region [260, 261]. More evidence for gravitationally bound cores in molecular cirrus clouds have come from CS observations [262–264] and emission lines from HC_3N [265]. These findings raise the question of how much turbulent support is important for stellar formation.

4.1.8 Photodissociation Regions

The observation of cloud cores in the form of *elephant trunks* has gained huge popularity through spectacular observations by the *HST*. One of the most famous observations in this respect is the heart of the Eagle Nebula (see Figure 4.11), which hosts column-like structures of dust illuminated by the light of background stars as well as by newborn stars [266]. More popularly labeled as *pillars of creation* these structures are generally not visible to the naked eye as they are normally deeply embedded. They are formed at the interface between the bright H II region and the molecular cloud through the process of *photoevaporation* (see Sect. 4.2.7). These *pillars* are thus huge dust columns hosting *evaporated gaseous globules (EGGs)*, which are molecular core remnants in the intense UV radiation field. Clearly, the evaporation effect is interrupting the star-formation process in *EGGs*; however it is entirely unclear to what degree. In fact it is not obvious how stable pillars should exist at all. So far, none of the existing models can reproduce the observed dust and gas velocities and densities [267].

PDRs (see also Sect. 3.5) are an integral part of the ISM of galaxies. These are predominantly neutral regions in which heating and chemistry are regulated by UV

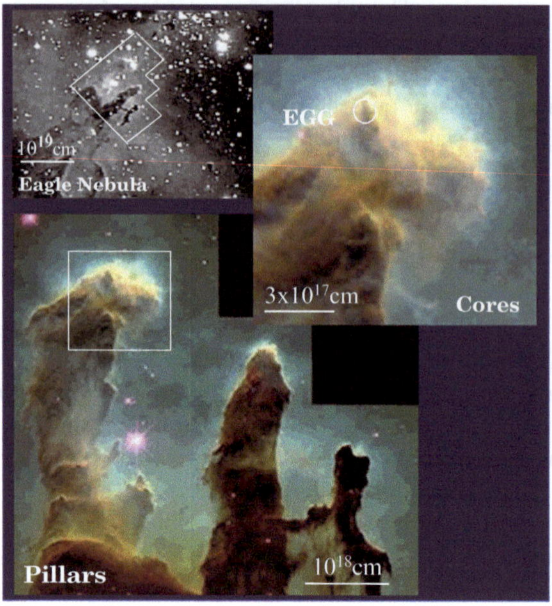

Fig. 4.11 The heart of the Eagle Nebula was observed with the WFPC camera onboard the HST in 1995. (*left*) A ground exposure of the nebula and its stellar environment. The HST exposure targeted the core of the Nebula as indicated by the white frame. It shows *pillars* consisting of dense dust columns which appeared as the strong UV radiation from nearby O-stars evaporated the condensed molecular environment. The diameters of these pillars reach up to 70,000 AUs. They harbor molecular cores which are the birthplaces of stars. About 70 *EGGs* have been identified in these pillars. Credit: J. Hester and P. Scowen, NASA

photons. An in-depth review of PDRs and their properties can be found in reviews by D. Hollenbach and A. Tielens [155, 156]. Whether it is the neutral hydrogen gas around MCs, EGGs, and pillars, the effects of radiation fields from massive stars as well as even stars in their youngest evolutionary stages (see Chaps. 6 and 10) are affecting the processes of stellar formation in many ways. Though on a global scale the entire neutral ISM, specifically the cold and the warm neutral phase (see Table 3.1), can be described as PDRs, models traditionally are applied to interstellar and molecular clouds, circumstellar disks and outflows, or even planetary nebulae, to name but a few. PDRs are primarily observed at IR and sub-mm wavelengths. These observations include lines of C and C^+, O, rotational–vibrational lines of H_2, rotational lines of CO, and features from PAHs [156].

4.1.9 Globules

The first *globules* were detected in 1947 and named after their discoverer: *Bok Globules*. Today two groups of globules can be distinguished. The first group are the Bok Globules, originally described as small dark spots projected onto bright

Fig. 4.12 Composite SST image of the bright globule *IC 1396A*. The image combines the *red color* for 24 μm, the *green color* for 5.8 and 8 μm and the *blue color* for 3.6 and 4.5 μm [269]. Credit: NASA/JPL-Caltech/W. Reach (SSC/Caltech)

H II regions with sizes of the order to 80,000 AU and less. The original assumption by Bok and Reilly that they accrete matter from their environment and collapse under the pressure of the ambient radiation field is incorrect. It seems more likely that they are fragmented remnants of the evaporation of an original GMC. As old cloud fragments they are probably not directly involved in the star-forming process but may host cores that form stars. The globule *TC2* in the Trifid Nebula is one of a very limited number of globules where star-formation is ongoing while being evaporated by the radiative continuum of the early type stars central to the nebula. From the molecular emission it is suggested that the star-formation process was probably initiated much less than 1 million years ago and since the ionization time scale of this globule is of the order of 2 million years, the star inside will have enough time to evolve before the globule is entirely evaporated [268].

A subset of Bok Globules are the so-called *cometary globules*, which form when the recombining downstream side of the ionization front behind the irradiated cores is shielded against direct irradiation leaving the globule with a fragmented tail [270, 271]. The giant H II region IC 1396 hosts a large number of globules and cometary globules. One of the most spectacular examples of a cometary globule is *van den Bergh 140 (IC 1396A)*. Fig. 4.13 shows a correlation between an optical image of globules *IC1396 A* and *B* overlaid by X-ray contours from a *ROSAT* observation [272]. Though the center of *IC1396A* hosts the young star double LKHα 349, *IC1396B* near the *IRAS* source *IR21327+5717C* does not identify with an X-ray source. Figure 4.12 shows a recent SST image of the globule. Infrared colors

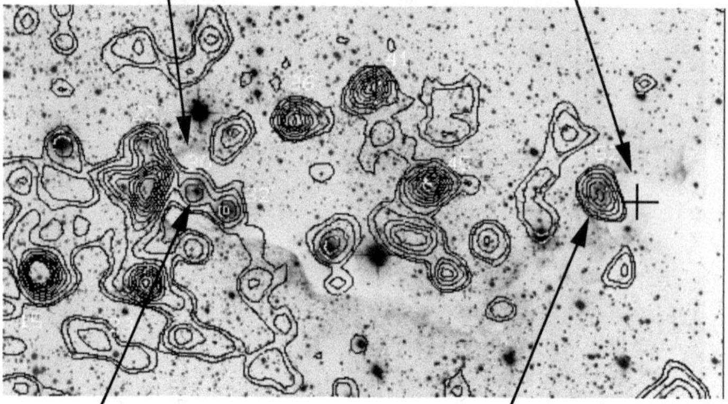

Fig. 4.13 A view of the cometary globules *A* and *B* in *IC1396*. The contours are X-rays from very young stars identified in the region. Globule *A* hosts the young stellar double LkHα 349, while globule *B* to the right (near but not associated with the source *IR21327+5717C*) does not sem to host a young star (from Schulz et al. [272]). An IR image of this globule observed by *SST* is also shown in Figure 4.12

of up to 24 μm detect several very young stellar sources, specifically of class I and 0 type (see Chap. 6). It is estimated that about 5 percent of the globule's mass is presently in protostars [269]. In general, Bok Globules do not necessarily have to contain an evolving star. However, recent observations of previously assumed starless globules with the *SST* revealed impressive signatures of hidden protostars. Examples are shown in Figure 2.1 and *L1014* [273].

A second group of globules are not associated with bright H II regions and appear as dark regions in the sky only noticeable to the eye by extinction of the stellar background. These dark clouds have extents from a few 0.1 pc to maybe a few pc and contain 10 to a few 100 $M_{\odot}$. In contrast to Bok Globules these globules have a young star formation history and contain collapsing cores.

4.2 Cloud Dynamics

Molecular clouds may be formed by gravitational instabilities in a more or less clumpy ISM [274] and experience rapid changes with time due to their sensitivity to the local radiation field [202]. An immediate indicator of a high level of dynamics is their short lifetime of around 10^7 yr. Evidence comes not only from their obvious link to stellar formation but also from the fact that they appear chemically relatively unevolved. Models of gas-phase chemical dynamics [275–277] show that many commonly detected molecules reach their abundance peaks at timescales of $1–3\times10^5$ yr before reaching a steady state within around $10^7–10^8$ yr. Most observed abundances are not consistent with steady-state conditions. In this respect it seems warranted to assume that these clouds are young [278] and that they are subject to a dynamical evolution with frequent chemical recycling. It has been suggested that clouds undergo turbulent cycling between their interiors and exteriors [279] with dense clumps being formed and dispersed [280].

Most of all, the large mass and relatively compact appearance of molecular clouds pose a problem for the simple picture of static clouds. Supported only by thermal pressure these clouds notoriously exceed their thresholds for gravitational collapse (see Sect. 4.3.1) and thus they should all contract on a rapid scale producing stars at a far higher rate than observed. In reality the star-formation rate is only about 5 percent of this rate [210] and thus there has to be additional support against self-gravity. Thus, molecular clouds and cloud cores are not static entities, they are dynamic and subject to instabilities and velocity dispersion.

4.2.1 Fragmentation

The process that governs the global structure from large-scale properties of the ISM to stars is called *fragmentation*. It is a hierarchical process in which parent clouds break up into subclouds, which may themselves break into smaller structures. It is a multiscale process [281] that ranges from scales of:

- 10 kpc – spiral arms of the Galaxy;
- 1 kpc – H I super clouds;
- 100 pc – giant molecular clouds;
- 10 pc – molecular clouds;
- 0.1 pc – molecular cores;
- 100 AU – protostars.

Stars are the final step of fragmentation (see Fig. 4.14). The first four items are large-scale structures in terms of fragmentation and their stability is determined by thermal, magnetic and turbulent support. The last two items are small-scale structures and are predominantly supported by thermal and magnetic pressure. Open issues relate to questions of whether magnetic fields can really prevent or at least delay gravitational collapse. Others consider the role of cloud rotation towards the stability of clouds and the impact of turbulence on steller formation. The following sections lay out the underlying ideas that help to find some answers to these questions. Although the discussion will be focussed on fragments scaling 100 pc and less it has been long speculated that conclusions are also more or less valid at larger scales [211].

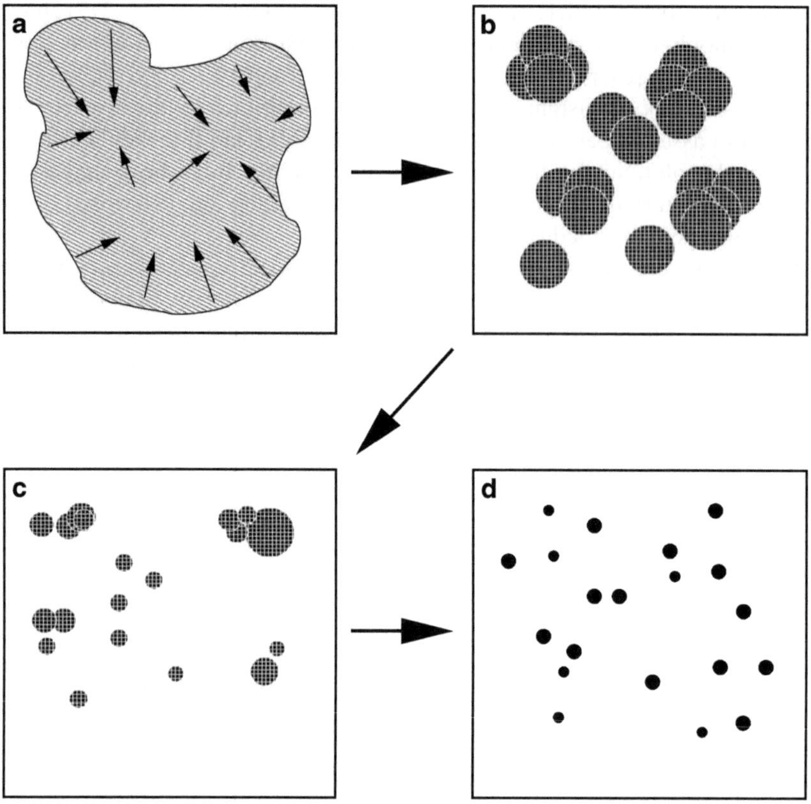

Fig. 4.14 A schematic illustration of the main processes that lead from molecular clouds to stellar entities. In (**a**) parts of the cloud fractionalize into cores, which then begin to collapse (**b**) into protostars (**c**). The latter evolve (**d**) towards the main sequence. Note that the illustration is idealistic as not all the cloud material winds up in newly formed stars

4.2.2 Pressure Balance in Molecular Clouds

The condition for a cloud (which is same for clumps and cores) to be gravitationally bound is inferred by the virial theorem (see Sect. A.7). An exhaustive treatment of basic cloud and subcloud theory can be found in [239,282,283]. The energy equation can be written as:

$$\frac{1}{2}\ddot{I} = 2\mathcal{T} + \mathcal{M} + \mathcal{W}. \tag{4.4}$$

Here I is a function describing the cloud's inertia which reflects changes in the cloud's shape. This term becomes important in the case of turbulence. $\mathcal{T}$ is the net kinetic energy of the cloud reduced by the energy exerted by the pressure from the ambient gas, $\mathcal{M}$ is the net magnetic energy with respect to the ambient field (see Sect. 4.2.3), and $\mathcal{W}$ the gravitational binding energy of the cloud under the condition that the field is dominated by self-gravity only. All these terms are net quantities with respect to the ambient medium and in this respect they are somewhat dependent on the location in the Galaxy. Under the assumption that the cloud is in virial equilibrium, the left-hand side becomes negligible. Similarly, in the case of zero magnetic fields, $\mathcal{M}$ vanishes as well. Using the gas laws ($\mathcal{T} = \frac{3}{2}\overline{P}V_{cl}$, where $\overline{P}$ is the mean total (thermal and turbulent) pressure, P_0 the ambient pressure, and V_{cl} the cloud volume) the virial theorem can be written in pressure terms:

$$\overline{P} = P_0 + \frac{3\pi a}{20} G\Sigma^2. \tag{4.5}$$

Here the gravitational energy density:

$$\mathcal{W} = \frac{3}{5} a \frac{GM^2}{R} \tag{4.6}$$

has been expressed by the mean projected surface density $\Sigma \equiv M/\pi R^2$. The numerical factor a measures the effects of non-uniform density distributions and the deviation from sphericity. Equation 4.5 shows that the mean pressure $\overline{P}$ inside the cloud (clump, core) is the sum of the surface pressure P_0 and the weight of the material inside the cloud (clump, core).

The balance works the same with clumps and cores. The difference now is that the surface pressure of the these entities is the mean pressure inside the GMC and thus the virialized pressure of the clump is proportional to the sum of the squares of the surface densities of the GMC and clump. This implies that in the case where surface densities of clumps are much smaller than the surface density of the GMC, these clumps are confined by pressure rather than gravity. Only the most massive clumps eventually have surface densities exceeding that of the GMC and are able to form stars. Throughout all these considerations it should be kept in mind that equilibrium conditions is an idealized assumption. Clumps and cores do not trap energy as efficiently as a protostar and the equilibrium assumption may be a weak one (see Chap. 5).

4.2.3 *Non-Zero Magnetic Fields*

It has been long speculated that magnetic pressure offers a feasible solution to explain slow star-formation rates. Magnetic pressure efficiently aids thermal pressure against gravity. The standard theory formulates a scenario of star formation in which interstellar magnetic fields prevent the self-gravitation and eventual collapse of clumps and cores with an insufficient mass-to-flux ratio [284, 285]. The results are cores in magnetohydrostatic equilibrium.

The negligence of magnetic fields in the pressure balance is only justified for weakly magnetized clouds. In the case of stronger fields the magnetic energy term $\mathcal{M} = \frac{1}{c}\mathbf{j} \times \mathbf{B}$ has to be evaluated. This is a very complex process and the following offers only a schematic treatment. For the case that the ambient medium is of low density and the field is integrated over a large enough volume V_a, so that the field has dropped to the asymptotic value of the ambient medium B_{ism}, the net magnetic energy due to the cloud reduces to:

$$\mathcal{M} = \frac{1}{8\pi} \int_{V_a} [(\overline{B}^2 - B^2_{ism}]) \, \mathrm{d}V \qquad (4.7)$$

where $\overline{B}^2$ denotes the mean field strength. This expression emphasizes the physical significance of the term as the total change in magnetic energy due to the pressure of the cloud, which is stored both in the cloud and the ambient medium [282].

The actual field configurations in clouds and cores are unknown and 3-D reconstructions from observations have been rare [234, 286, 287] (see Fig. 4.7). However, in the virialized frame (i.e., time-averaged and over large spatial regions) a *poloidal* field configuration is the simplest configuration, and in case of no ambient field $\mathcal{M}$ scales roughly similar to R as the gravitational binding energy (4.6):

$$\mathcal{M}_{cr} \equiv \frac{b}{3} \overline{B}^2 R^3 = \frac{b}{3\pi^2} \frac{\Phi^2}{R}. \qquad (4.8)$$

The numerical factor b in the magnetic case, as for a in the gravitational case (see Sect. 4.2.2) is of the order of unity and contains information about the shape of the cloud and field topology. For a sphere $b = 3/4$ and for a thin disk $b = 1/\pi$, under the condition that $\mathcal{M}_{cr} = |\mathcal{W}_{cr}|$, one can deduce a magnetic critical mass independent of R that is directly related to the magnetic flux threading the cloud as (see Fig. 4.15)

$$M^2_{\Phi} = \frac{5b}{9\pi^2 a} \frac{\Phi^2}{G} \quad \text{and} \quad M_{\Phi} \sim 0.21 \frac{\Phi}{G^{1/2}}. \qquad (4.9)$$

The right-hand numerical term was evaluated for the spherical case. Of course, reality is more complex and T. C. Mouschovias and L. Spitzer [288] have derived a lower and more realistic mass-to-flux ratio, which can be expressed as:

$$\frac{M_{\Phi}}{\Phi} < \left(\frac{M_{\Phi}}{\Phi}\right)_{cr} = c_{\Phi} G^{-1/2}. \qquad (4.10)$$

Here c_{Φ} is 0.13 for spherical clouds and 0.16 for a cylindrical region. In general these criteria hold for axisymmetric configurations. It has been shown that further developed they also apply for non-axisymmetric disks [289]. The ratio of the critical magnetic mass to the actual mass (i.e., $[(M_{\Phi}/M])^2$ [289]) is an invariant as long as the magnetic flux is frozen to the matter (i.e., the field lines move with bulk motions of the cloud mass). There are two cases to consider:

- For $M < M_{\Phi}$ the cloud is considered *magnetically subcritical* and the magnetic field dominates the stability of the cloud. As long as the field is frozen into the cloud there cannot be a gravitational collapse.

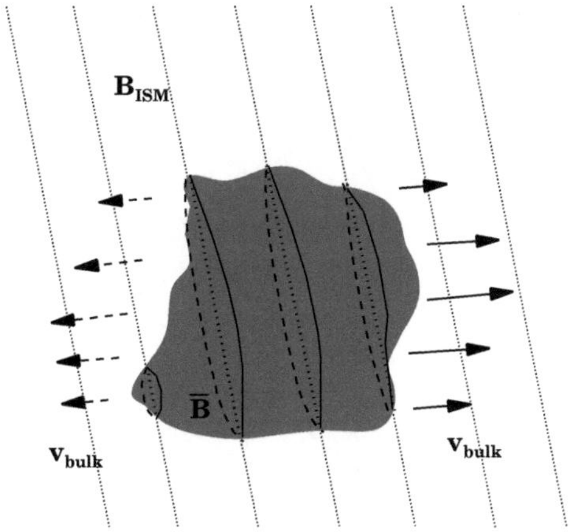

Fig. 4.15 A schematic illustration of a cloud fragment with a poloidal magnetic field configuration. The dotted lines resemble the asymptotic field of the ISM or the intercloud gas. For simplicity it has been plotted as poloidal as well. If the cloud is magnetically subcritical, the magnetic field is frozen in (*solid and dashed field lines*) and moves with the bulk motion of the cloud, which is stable against gravitational collapse. In the supercritical case (*fat dotted lines*) this may not be the case and unless thermal pressure cannot counter self-gravity, it will collapse

- For $M > M_\Phi$ the cloud is considered *magnetically supercritical* and either thermal pressure or gravitation dominates the stability of the cloud.

Equation 4.10 can be expressed more comfortably in terms of the mean density of the cloud ρ_{cl} and field strength B_{cl} of the cloud as [290]:

$$M_\Phi \sim 3.5 \times 10^{-3} \frac{B_{cl}^3}{G^{3/2} \rho_{cl}^2}$$
$$\sim 100 \left(\frac{B_{cl}}{[100\,\mu\mathrm{G}]} \right)^3 \left(\frac{N_{H_2}}{[10^4\,\mathrm{cm}^{-3}]} \right)^{-2} M_\odot \tag{4.11}$$

where the cloud is entirely composed of H_2. For typical cloud densities a 1 $M_\odot$ cloud needs magnetic fields of at least 20 μG to remain subcritical.

More importantly these criteria also hold once thermal pressure is included. External pressure (P_g) leads to a maximum stable mass M_J to balance gravity (see Sect. 4.3.1). Thus the critical mass exceeds the critical magnetic mass due to the additional external pressure. As a result the ratio of magnetic to gravitational energy is now the square of the ratio of the magnetic critical mass to the total critical mass and the virial theorem can be written as:

$$\overline{P} = P_0 + P_g \left[1 - \left(\frac{M_\Phi}{M_{cr}} \right) \right]. \tag{4.12}$$

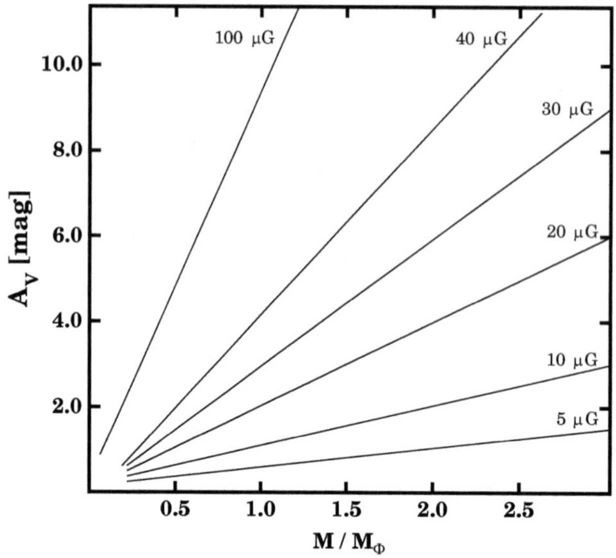

Fig. 4.16 A relationship between critical magnetical mass and extinction magnitude as it appears in (4.13)

There are several stable equilibria involving M_Φ and M_{cr}. These are equilibria for $M < M_{cr}$ and for $M_\Phi < M < M_{cr}$. There are none for $M > M_{cr}$ [291].

Furthermore, the value of the field can be evaluated by relating the ratio M/M_Φ to $\Sigma/\overline{B}$ and finally $N_H/\overline{B}$ [283]. This yields:

$$\overline{B} = 5.05\left(\frac{N_{21}}{M/M_\Phi}\right)\,\mu\text{G} = 10.1\left(\frac{A_V}{M/M_\Phi}\right)\,\mu\text{G}. \tag{4.13}$$

Figure 4.16 shows tracks of magnetic field values for various combinations of cloud mass and extinction. A typical value for cloud fields is 30 μG in contrast to the diffuse fields in the ISM, which are an order of magnitude smaller.

4.2.4 *Turbulence*

The view that self-gravitating clouds are supported magnetostatically is not uncontested and magnetic fields are certainly not the only means to aid pressure support of clouds. Turbulence and compressibility in interstellar clouds was first pointed out by C. F. von Weizäcker in 1951 [292]. By simply comparing the average thermal velocity of molecules (1–10 km s^{-1}) to the much larger observed velocities of interstellar clouds (10–100 km s^{-1}) he concluded that the motion of the interstellar gas in general must show the characteristics both of turbulence and compression.

Interstellar gas when viewed on a global scale, even with molecular clouds dispersed within them, appears scale-free and the gas dominates the dynamics [293]. However, when viewed in molecular surveys with intensity contours it appears like a collection of clouds. This creates a distinct bifurcation in modeling of intercloud dynamics [211]. Scale-free models involve turbulence and self-gravity (see [294] for a review), while others feature collisions and cloud compression [209, 295].

In the simplest sense turbulence is caused by irregular fluctuations in velocity and compression means that these velocity fluctuations will cause density fluctuations in the gas. Compressibility of turbulence is a key issue. Turbulence in most terrestrial applications is considered incompressible. However, supersonic flows in a highly compressible ISM can create density perturbations that even lead to collapse events. The treatment is quite difficult as it has to include dissipation through shocks (see Sect. 3.1.9), non-uniformity, inhomogeneity, and the fact that the ISM is magnetized. Critical is the supersonic nature of turbulence. Former calculations of density spectra in turbulent magnetized plasmas applying the theory of incompressible MHD turbulence [296, 297] include the effects of compressibility and particle transport for Mach numbers of much less than unity. The Mach number is defined as:

$$\mathcal{M}_{rms} = \frac{v_{rms}}{c_s} \tag{4.14}$$

where v_{rms} is the mean velocity of the flow and c_s the speed of sound as defined in Sect. A.8. For supersonic flows (i.e., $\mathcal{M}_{rms} \gg 1$) reliable analytical models of strongly compressible MHD turbulence are still underdeveloped and most conclusions are drawn from numerical calculations. It was proposed [294] that stellar birth is regulated by interstellar supersonic turbulence through the establishment of a complex network of interacting shocks. This allows flows to converge and generate regions of high density, which eventually assume the form of clumps and cores. Observational evidence for supersonic flows in molecular clouds comes from large line-width measurements (see [246] for a review). Mach numbers here turn out to be 10 or higher.

An important parameter is the decay time of supersonic turbulence. If it is shorter than the free-fall time during stellar collapse, follow-up shocks will disperse the clump before it can form protostellar cores. It can be shown [298, 299] that this decay time τ_d can indeed be smaller than the free-fall time τ_{ff} in dense molecular clouds independent of the existence of magnetic fields:

$$\frac{\tau_d}{\tau_{ff}} \simeq 3.9 \frac{\lambda_d/\lambda_J}{\mathcal{M}_{rms}} \tag{4.15}$$

The ratio of the driving wavelength λ_d of the flow [299] and the Jeans wavelength λ_J (see Sect. 4.3.1) of the clump is smaller than unity to maintain gravitational support [300].

Recent numerical models show in this respect that compressible supersonic turbulence may be able to regulate the star-formation efficiency in molecular clouds

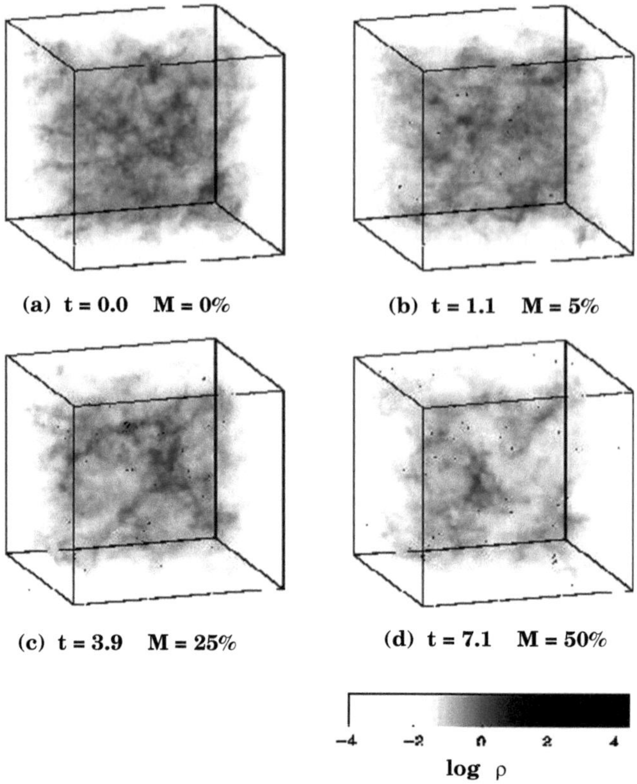

Fig. 4.17 Simulation of local collapse events in a globally supported cloud. A sequence of density cubes applying a compressible supersonic turbulence model. In (**a**) the full hydrodynamical solution without self-gravity is shown. Once gravity is added to the cube (**b**) density fluctuations are generated through converging and interacting shock fronts. The central regions of some high-density clumps have undergone gravitational collapse. With increasing time (in units of τ_{ff}) up to 50 percent of the gas mass has collapsed and been accreted by stellar cores in (**c**) and (**d**). See also Figure 4.18. Credit: R. Klessen, AIP [301]

(see [294, 302]). Figure 4.17 (see also Figure 4.18) shows a simulation [303] in which, on a global scale, the presence of high-speed turbulence inhibits stellar collapse by supplying support to average density structures. However, collapse cannot be prevented entirely on a local scale. Only in the case that the r.m.s. velocity of the flow is high enough to offer support for high-density shocked regions can local collapse be prevented. Another criterion is that the driving wavelength of the flow has to be smaller than the Jeans length. The full range of these effects and their implications still have to be investigated. Specifically intriguing are effects of enhanced star-forming activity inflicted by cataclysmic events like supernova explosions as is now observed with *SST* (see Figure 4.19).

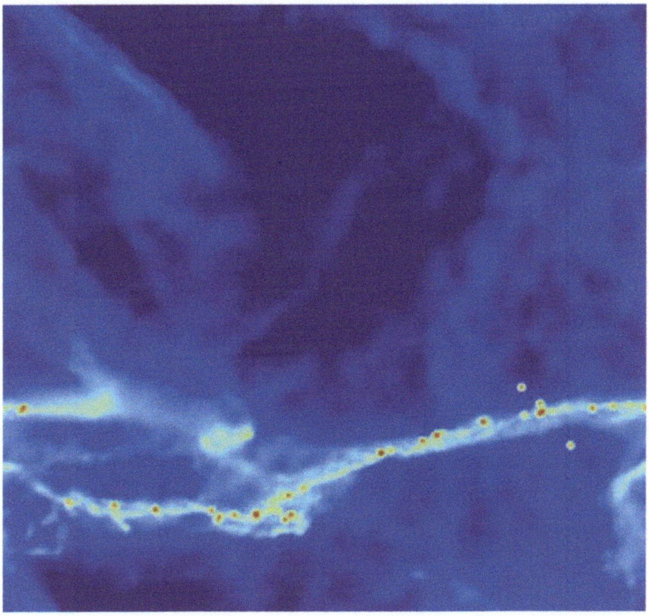

Fig. 4.18 A simulation of local collapse events in a globally supported cloud. The image shows a 2-D projection of a density cube (see Fig. 4.17). String-like gravitational fragmentation can occur as a result of supersonic turbulence. Credit: R. Klessen, AIP

Fig. 4.19 Events of triggered star formation likely through supersonic turbulence inflicted by a supernova explosion a few million years ago. The image was obtained with the *SST* and shows star formation in *Heinze 206* in the *LMC*. Credit: NASA/JPL-Caltech/V. Gorjian (JPL)

4.2.5 Effects from Rotation

Molecular clouds carry angular momentum which originates from the ISM and ultimately from the rotation of the Galaxy. The coupling to the ISM and the intercloud medium is mostly achieved by magnetic fields, though some angular momentum transport may also be achieved through shocks [304]. First studies of rotational equilibria considered spherical isothermal clouds embedded in a tenuous intercloud medium [304–306]. Just as there is a maximum stable mass for magnetized (M_Φ) and self-gravitating (M_J) clouds, there is also a maximum stable mass due to rotation:

$$M_{rot} = 5.1\left(\frac{\sigma J}{GM}\right) \tag{4.16}$$

with angular momentum J and the velocity dispersion $\sigma^2 = \overline{P}/\overline{\rho}$. The calculations [304] showed that thermal support is essential for stable unmagnetized rotating clouds and therefore the rotational mass M_{rot} is limited by a few times M_J. The total critical mass involving self-gravity, magnetic fields and rotation can be expressed as:

$$M_{cr} \simeq [(M_J + M_\Phi)^2 + M_{rot}^2]^{1/2} \tag{4.17}$$

in fair agreement with numerical results (see [283, 306]).

A common way to characterize the rotation of a cloud of radius R by the ratio of centrifugal to gravitational force is:

$$\alpha_\Omega \equiv \frac{\Omega^2 R^3}{GM}. \tag{4.18}$$

A sufficient criterion for clouds to be primarily supported by gas pressure rather than rotation is $\alpha_\Omega < 0.5$ (i.e., $M_{rot} < M_J$). This parameter, although valid for spherical clouds only, seems to be fairly robust to some oblateness and differences are expected to be less than 20 percent [307]. Yet the importance of rotational support in molecular clouds is still under debate as some observations indicate larger values of α_Ω, while some do not ([308], see [309] for a review), and further observations are needed to determine the fraction of clouds that appear rotationally supported.

Typical angular velocities of molecular cloud cores are of the order of $\Omega \sim 10^{-14}\ \mathrm{s}^{-1}$, which is consistent with the α_Ω and thus a reasonable choice. Since angular momentum is a conserved quantity this causes a problem once the core engages in free-fall collapse. A simple estimate illustrates the dilemma. During the course of the collapse (see Chap. 5) the radius of the cloud decreases by a factor 10^7, which, since angular momentum scales with R^2 and only with Ω, implies that the final star would rotate within a few seconds of a period. Thus, even if rotation may not contribute much to the stability of the cloud, it has to be dealt with during collapse and contraction.

4.2.6 Ionization Fractions

The term 'molecular' usually implies that constituents are considered to be neutral. In reality this is hardly the case as a substantial interstellar radiation field is constantly pounding the molecular gas. This implies that these clouds have substantial *ionization fractions*. A quick glance at (A.53) shows that molecular clouds with temperatures of 50 K and below are hardly collisionally ionized. Therefore the fractions have to come from external and internal radiation fields. In general, ionized material is hotter and less compressible and thus works against gravitational collapse. However, interstellar clouds are not ionized isotropically and under the absence of effective convection the ionizing radiation can only penetrate the cloud on a limited scale giving rise to a layer of predominantly ionized hydrogen H I around the cloud (see Fig. 4.4 and [246]). Such a layer, once established, provides effective shielding from the interstellar radiation field. Its depth primarily depends on the strength of the radiation field as well as the gas density, but also on the amount of dust in the layer. The Strömgren radius defined in Equ. 3.11 may be used

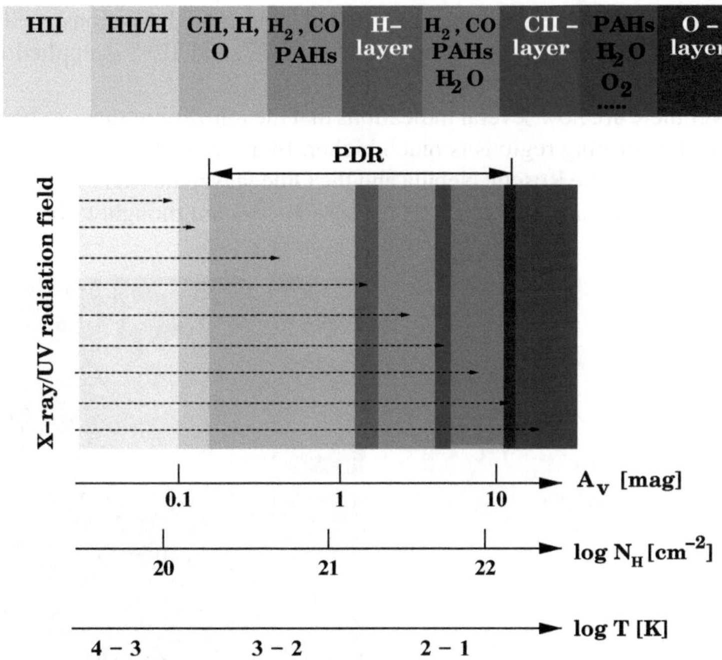

Fig. 4.20 The possible structure of a PDR [156]. The PDR is illuminated through a radiation field penetrating the region. The dissociative impact at various increasing cloud densities creates a layered structure of predominantly existing atoms, ions and molecules. Specifically in the vicinity of a massive star the PDR is preceded by an H II layer which creates a thin atomic (ΔA_V < 0.1) H II/H interface. Depending on depth there appear thin characteristic interfaces, where atomic hydrogen coexists with molecular hydrogen, atomic carbon with CO, and atomic oxygen with molecular oxygen. The density scales convert through (3.25). All scales, specifically the temperature scale are approximate. Adapted from Hollenbach and Tielens [156]

to estimate the penetration of an ISM radiation field. With radiation rate of a factor 100 lower than in the O-star case, a gas temperature of 10 K, and a density of 1,000 cm^{-3}, this results in a penetration of about 0.1 pc.

On the other hand, the penetration of UV radiation into PDRs likely creates a stratification of dissociated layers. Figure 4.20 demonstrates this schematically on a slab of molecular gas which is illuminated by a typical radiation field from a massive star. Depending on column density, different penetrated layers constrain specific elements. Also depending on the flux-gas density ratio, PDRs are characterized by a foremost neutral layer of mostly neutral atomic hydrogen (see also Fig. 4.4) which extends to a depth of $N_H \sim 2\text{–}4 \times 10^{21}$ cm^{-2}, and which in many cases can be preceded by H II gas and a thin H II/H I interface [156]. The layer is followed by an H I/H_2 interface and layers of C, C^+, CO, H_2, O and O_2 with increasing depth, respectively. In this respect the PDR includes gas where hydrogen exists predominantly in a molecular form and carbon is mostly present in the CO form.

Inside the molecular cloud the total ionization fraction is very low. Observations of HCO^+ and CO molecules indicate a fraction of 10^{-7} to 10^{-8} in some molecular clouds [311], and a limit of 10^{-6} may be inferred from deuterium fractionization reactions [312]. Using mechanisms discussed in Sect. C.8, charged grain reactions and ionization of incident cosmic rays [313] determined fractions well below 10^{-7}. For most considerations a range between 10^{-8} and 10^{-6} is applied in most cases [285].

However there are now several indications that the ionization fraction specifically in massive star-forming regions is much higher. In a recent study of two prototype H II blister regions, the Rosette Nebula and the Omega Nebula (M17), diffuse X-ray emission was detected [310] (Fig. 4.21). These X-rays are thought to be caused by

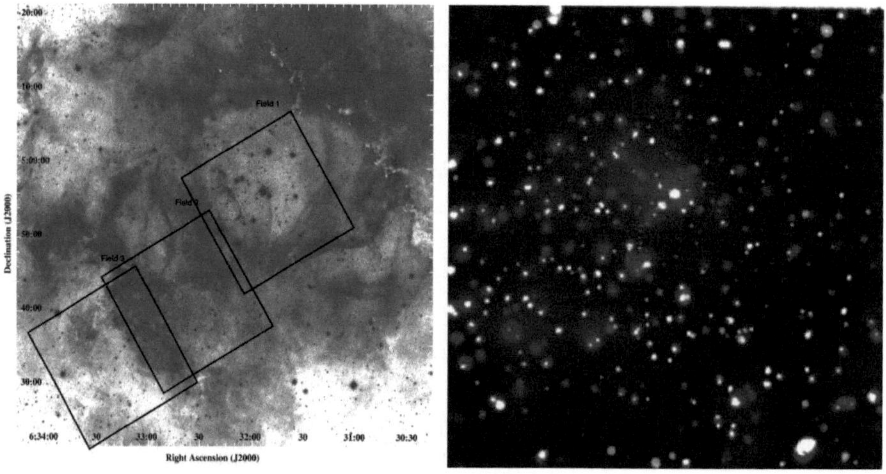

Fig. 4.21 (*left:*) Optical image of the center of the Rosette Nebula. The squares mark the fields covered by X-ray observations. (*right:*) X-ray image of field 1, the central region of the Rosette, obtained with the Chandra X-ray Observatory. Besides the many bright X-ray sources in the field of which many are O-stars there is strong diffuse emission present indicating the existence of hot plasma and thus high ionization. From Townsley et al. [310]

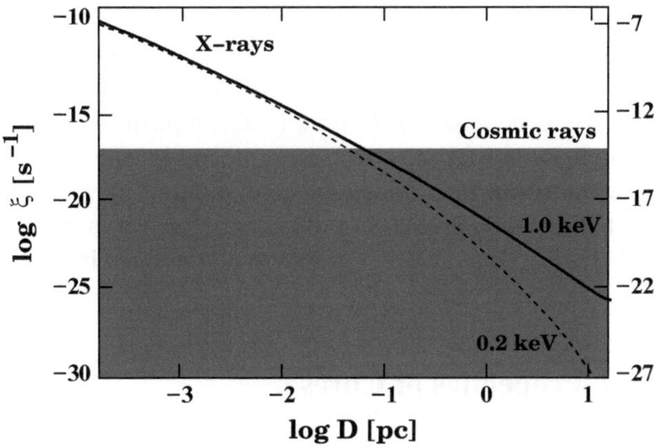

Fig. 4.22 Ionization rates in the vicinity of a T Tauri star as calculated by F. Palla [314]. Shown is the ionization rate (*left scale*) as well as a lower end projected ionization fraction for low cloud densities of 10^3 cm^{-3} against the distance from the T Tauri star in pc. The X-ray luminosity was assumed to be 10^{30} erg s^{-1} for both cases, 0.2 and 1.0 keV spectra (see Chap. 10)

thermalization of O winds through either wind–wind collisions or a termination shock against the surrounding matter. Since only small portions of these winds were recovered it is suspected that most of the energy flows without cooling into the low-density interstellar medium. This has two consequences. The notion of low ionization fractions may therefore not be true everywhere and the radial dependence of the ionization fraction may show a more stratified shell-like structure. Clearly, ongoing stellar formation in or near molecular clouds can increase the ionization fraction. Figure 4.22 indicates that the ionization rate within 0.1 pc of T Tauri stars is as high or even higher than the one from Cosmic rays [314]. For stellar densities of over 10^3 as they occur in embedded stellar clusters (see Chap. 10) this implies that ionization rates from X-rays can lead to a substantial contribution to the ionization fraction in the region.

4.2.7 Evaporation

The process of *photoevaporation* of molecular material and dust by the strong ultraviolet radiation field of O-stars has already been discussed in the mid-1950s [315]. It is the inevitable fate of the PDR. As shown in Fig. 4.20 the UV/X-ray radiation from a newly born O-star heats the dense cloud it is facing from less than 100 K to up to 10,000 K. This leads to a transfer of kinetic energy onto the dense material increasing the pressure and thus leading to high acceleration of cloud material [156]. The result is an evaporation effect as can be seen in the example of M16 (Figure 4.11).

The effective ionization lifetime of an early-type star is of the order of 10^6 yr. It has been shown that photoevaporation is actually a quite inefficient process and a single generation of O-stars can only affect a small fraction of the molecular cloud [316]. Calculations showed that the ionizing radiation can remove about 3–5 × 10^{-3} $M_\odot$ yr^{-1} which in the case of GMCs would take to a few times 10^7 yr and thus a several generations of O-stars to accomplish [317]. For small clouds (i.e., where the ionization lifetime becomes comparable with one stellar lifetime), modeling of the transition to a cometary configuration in which the embracing ionization front balances the pressure of the neutral gas was applied to the Rosette and Gum Nebula [318]. More about evaporation is presented in Chap. 8.

4.3 Dynamic Properties of Cores

In the sections above much has been said about the structure and dynamics of molecular clouds, and the sub-structural components and physical concepts that govern them. The following takes a closer look at the dynamical properties of self-gravitating cores and physical processes that will eventually lead to the initial conditions for gravitational collapse.

4.3.1 Critical Mass

The structures of non-magnetic, isothermal clouds were studied by W. B. Bonnor (1956) and R. Ebert (1955). The so called *Bonnor–Ebert spheres* are in equilibrium for a given external pressure up to a maximum mass M_{cr}. Similarly, at a given mass there is also a critical density that renders the sphere unstable, and as long as the consistency of the gas is not subject to change, the sphere can remain isothermal up to a large range of densities. Here traditional Bonnor–Ebert spheres have the trademark that $\mathrm{d}T/\mathrm{d}r = 0$. Some recent concepts allow for modified Bonnor–Ebert spheres and acknowledge the existence of a kinetic temperature gradient [319]. The concept of Bonnor–Ebert spheres is also very often applied to describe cores under collapse. Formally this is actually not quite accurate as the calculations for these conditions apply for a hot and much thinner gas in H I clouds rather than for cold and very dense gas in molecular cores.

Effectively the critical mass of a Bonnor–Ebert sphere, thus sometimes called the Bonnor–Ebert mass, is equivalent to the *Jeans mass*. By applying the virial theorem using (A.23), (A.18), and (A.24) J. Jeans (1902) defined a critical wavelength (called *Jeans length*) where an infinite sized cloud is unstable to density perturbations:

$$\begin{aligned}\lambda_J &= \left(\frac{\pi c_s^2}{G\overline{\rho}}\right)^{1/2} \\ &= 0.2\left(\frac{T}{[10\ \mathrm{K}]}\right)^{1/2}\left(\frac{n}{[10^4\ \mathrm{cm}^{-3}]}\right)^{-1/2}\ \mathrm{pc}.\end{aligned} \tag{4.19}$$

Thus for a sphere, this length corresponds to a critical radius $R_J \equiv \frac{1}{2}\lambda_J$ and a critical Jeans mass of:

$$\begin{aligned} M_J &= \frac{4}{3}\pi R_J^3 \overline{\rho} \\ &= \frac{\pi}{6}\overline{\rho}\lambda_J^3 \\ &= 1.6\left(\frac{T}{[10\text{ K}]}\right)^{3/2}\left(\frac{n}{[10^4\text{ cm}^{-3}]}\right)^{-1/2} M_\odot. \end{aligned} \tag{4.20}$$

Here $\overline{\rho}$ is the average uniform density. At a temperature of 10 K (the speed of sound then is about 0.2 km s^{-1}) and a number density for a cloud of hydrogen of $n \sim 10^4$ cm^{-3} the Jeans radius is 0.2 pc and the Jeans mass is 1.6 $M_\odot$. Thus typical small dense cores are always close to the onset of gravitational instability unless there are other influences. This is specifically true for large clouds of sizes of 100 pc, where the average density is much smaller and the temperature is higher, and the Jeans mass is around 250 $M_\odot$. GMCs therefore could not be stable if it were for thermal pressure and gravity alone. On a separate note, the Jean criterion holds in the case of rotation, that is, it provides the same length scale for the growth of perturbations. However, what changes is the rate at which density fluctuations grow in dependence of the angle towards the axis of rotation, forcing the infalling sphere into a disk.

Real clouds also feel some finite external pressure P_{ext}, which includes contributions from thermal pressure, turbulence and magnetic fields. Under these conditions the critical value of the gravitational stable mass can be expressed as [320, 321]:

$$\begin{aligned} M_{BE} &= 1.18\frac{c_s^4}{P_{ext}G^{3/2}} \\ &= 0.35\left(\frac{T}{[10\text{K}]}\right)^2\left(\frac{P_{ext}}{10^6 k_b}\right)^{-1/2} M_\odot. \end{aligned} \tag{4.21}$$

4.3.2 Core Densities

Critical are assumptions about the core densities. The singular isothermal sphere has a density profile that has a singularity at the center of the sphere (see Fig. A.4). A spherical gas cloud close to equilibrium between thermal pressure and gravity has a typical density profile of [322]:

$$\rho = \frac{c_s^2}{2\pi G r^2} \tag{4.22}$$

and thus the density grows to infinity at the center. The self-similar description in the standard model of stellar formation (for a review on self-similarity see [284] and [290]) identifies density regions of the infalling envelope once a hydrostatic core has developed at the center. There is a free-infalling envelope with a radius of $r_{inf} \equiv c_s t$ and a radial density profile of $\rho \propto R^{-3/2}$ due to a constant mass infall rate and a quasi-static outer envelope with $\rho \propto R^{-2}$ (see Fig. A.4).

The shape of initial density profiles plays a vital role in models for protostellar collapse. Such profiles have been suggested [323,324] and there are many reasons to question the density profile deduced from the collapse of an isothermal sphere. One reason is that artifacts such as that the central density seems to approach infinity, or that the core radius itself extends to infinity as well. Observed density profiles suggest a flatter profile at the center with a power law index between 0.5 and 1.0, while the outer envelopes have indices between 2 and 4 (see [319] for a review). Others apply either more complex dependencies [325] or versions of Bonnor–Ebert spheres [326–328].

4.3.3 Magnetic Braking

One of the concepts studied to regulate the rotation in molecular clouds is the effect of *magnetic braking*. The basic process is based on the likely situation that the angular velocity of the cloud is larger than the one of the ambient medium. At the same time, the magnetic field strength in clouds is significantly larger with respect to the ambient medium as well. Combined this causes the launch of toroidal Alfv́en waves into the cloud and the ambient medium (see [285] for a basic summary). These waves are able to transport angular momentum which eventually brings the cloud into co-rotation with the ambient medium. The observational account of the importance of this effect for molecular clouds is ambiguous, and as mentioned in Sect. A.11, rotation in large clouds may not seem very important. However, in collapsing cores it can be an efficient mechanism to transport angular momentum and is thus an important ingredient of initial collapse conditions.

The characteristic timescale for magnetic braking depends on the mass-to-critical-magnetic-mass ratio M/M_Φ [307, 329]. For a spherical cloud embedded in a uniform magnetic field the ratio of the braking time t_b to the free-fall time t_{ff} is:

$$t_b/t_{ff} \simeq 0.3 \frac{M}{M_\Phi} \left(\frac{\overline{\rho}}{\rho_0} \right)^{1/2} \tag{4.23}$$

where ρ_0 is the density of the ambient medium, and $\overline{\rho}$ the average density of the core. Magnetically subcritical cores which are also not very dense have thus very short braking timescales. Magnetically supercritical clouds which are considerably more dense have long braking times. In the case that the magnetic field substantially exceeds the one in the ambient medium the density dependence basically vanishes and the braking time is entirely dominated by M/M_Φ [307,330]. Magnetic braking in molecular cloud cores is quite effective and can reduce the specific angular momentum by two orders of magnitude [324].

4.3.4 Ambipolar Diffusion

The fact that molecular clouds have non-zero ionization fractions is an essential condition for magnetic fields to connect with the cloud gas. The magnetic field is directly felt by charged particles only. To invoke sufficient magnetic support within the cloud (see Fig. 4.15) a substantial number of charged particles pinned to magnetic field lines are required. Pinned matter drifts with field lines relative to the neutral particle population. The drag force that moves the charged particles with respect to neutral particles is proportional to the densities of the two populations. This process is called *ambipolar diffusion*. Thus for low ionization fractions there is a small drag force, high fractions would imply a large drift. The drag is also directly proportional to the difference of bulk velocities of the neutral and ionized particle population, which is called *ambipolar* drift velocity. According to [76] this velocity w_D relates to the magnetic field B as:

$$w_D = -\frac{1}{8\pi n_i n_H m_H \langle u\sigma \rangle} \nabla B^2 \tag{4.24}$$

where the bulk of the neutral matter is assumed to be hydrogen. Electrons and ions are co-moving with a similar velocity distribution (see Sect. A.12). The factor $\langle u\sigma \rangle$ is the slow-down coefficient through collisions incorporating the the random velocity u of the ions and the momentum transfer cross section σ. Various values for $\langle u\sigma \rangle$ are tabulated in [76]. Most appropriate are C^+–H collisions which give a value of 2.2×10^{-9} cm^{-3} s^{-1}. Thus, while the neutral gas contracts under gravitation, the pinned ion gas will expand relative to this contraction. For a simplified cylindrical configuration of contracting gas and magnetic field one can then define a characteristic *diffusion time* t_D as the ratio radius of the gas cylinder r to the drift velocity w_D, which then gives:

$$t_D = 5.0 \times 10^{13} \frac{n_i}{n_H} \text{ yr} \tag{4.25}$$

Ambipolar diffusion acts like a fine-tuning device for collapsing cloud cores. For low ionization fractions, as they are believed to exist in cloud cores, magnetic support comes hand in hand with ambipolar diffusion (i.e., as soon as ions begin to couple with magnetic field lines, there will be a drift and vice versa [285]). The fact that the magnetic field is frozen-in actually makes it difficult for low-mass stars to form. This can be seen through (4.8) and (4.9). At a typical magnetic field in the range of 10 μG (see Fig. 4.16) for a core of solar mass and a core radius of 0.1 pc, the critical magnetic mass is substantially higher than the core mass and the core is strongly magnetically subcritical. Ambipolar diffusion slowly weakens the magnetic field and eventually the core mass becomes supercritical. Key is that the hydrostatic equilibrium is established within the diffusion time. Once the density of the core and thus the drag force becomes large enough matter decouples from the field and collapse can pursue. There is only a limited range of ionization fractions where this picture can actually work. For fractions much larger than 10^{-7} the diffusion time becomes very large and star-formation timescales would be uncomfortably long. In fact [331] observed much more rapid time scales in the Orion Nebula cluster of the order of 1 Myr. More details about ambipolar diffusion is presented in Appendix D.

Chapter 5
Concepts of Stellar Collapse

One of the most fascinating aspects of stellar formation is the enormously vast change of physical parameters during the process of gravitational cloud collapse. A comparison of the initial density in a molecular cloud core with the average density in mature stars, for example, reveals a difference of twenty orders of magnitude. The typical density of such a cloud is $\sim 10^{-20}$ g cm^{-3}, whereas the average density of a star like our Sun is of the order of 1 g cm^{-3}. A change of this magnitude cannot be achieved by external forces but only through the self-gravitational collapse of matter. During this collapse the size of the object reduces from light years across to a few hundred thousand kilometers in radius with internal temperatures rising from a few K to 30 million K. The following sections outline the physics involved in the stellar-formation process from the onset of the collapse to what is generally considered to be a protostar and further describe its first approach towards the main sequence.

The process of gravitational collapse is part of the sequence illustrated in Fig. 5.1 that starts from giant molecular clouds which fragment into dense cores and then collapse. The result are protostars evolving over time toward the main sequence. The first step concerning the stability and fragmentation of a molecular cloud of critical density was the subject of the previous chapter. The last step, the evolution towards the main sequence, is the subject of Chap. 6. This chapter looks at the collapse of a dense cloud into a stellar core and highlights the physical processes involved.

5.1 Classical Collapse Concepts

Several researchers in the 1960s stated the complexity of stellar-formation processes and recognized the many phases involved. It was essential that at some point the collapse comes to a halt and internal pressure can counter the infall of matter. The resulting stellar core may then be in quasi-hydrostatic equilibrium (see Sect. A.7). It was soon realized that this simplistic view had fundamental problems. Early computations led to claims that for protostars greater than one-tenth of a solar

N.S. Schulz, *The Formation and Early Evolution of Stars*, Astronomy and Astrophysics Library, DOI 10.1007/978-3-642-23926-7_5, © Springer-Verlag Berlin Heidelberg 2012

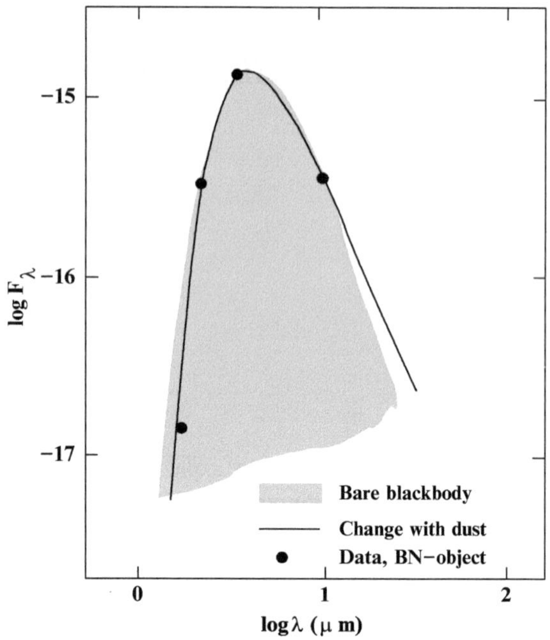

Fig. 5.1 Early comparison of the calculated luminosity of a collapsing protostar including dust opacities with infrared measurements of the Becklin–Neugebauer object in Orion obtained by R. B. Larson in the late 1960s [333]. The shaded region shows a spectrum that would only represent a simple blackbody

mass hydrostatic equilibrium should not be possible [332], because all the energy is radiated away rather than contributing to the temperature required to build up the necessary internal pressure.

Fig. 5.1 shows the spectral energy distribution emitted by a collapsing core. The energy initially radiated from the infalling gas is absorbed by dust grains suspended in an optically thick core and then re-emitted as thermal radiation at IR wavelengths. The result seems to match well with IR observations of the Becklin–Neugebauer object in Orion. Clearly, absorption and scattering events specifically involving dust must play a pivotal role during the collapse.

5.1.1 Initial Conditions and Collapse

One of the first full calculations of the gravitational collapse of such a core was performed by R. B. Larson in 1969 [334] considering a non-rotating, non-magnetic, spherically symmetric gas cloud. It has to be kept in mind that a collapse never occurs in isolation and external pressure and external radiation fields have to be considered. As long as the temperature in the collapsing sphere is of the order of the external temperature, the energy gained through the compression is irradiated away by dust particles in the gas flow and does not add to the thermal pressure. As a consequence the collapse happens on a free-fall timescale (see Sect. A.9) and, as has been pointed out in Chap. 4, depends on the initial core density distribution.

In addition there is external pressure exerted by the matter surrounding the core, which initially does not take part in the collapse but provides a pool of material for subsequent accretion.

Typical mean properties of molecular cloud cores at the onset of collapse are (see Chap. 4) :

- Mass: 1 $M_{\odot}$.
- Radius: $\sim$ 0.1 pc.
- Temperature: 10 K.
- Mean core density: 10^{-19} g cm^{-3}.
- Ionization fraction: $\sim 10^{-7}$.

These initial conditions indicate the collapse towards a low-mass object. Generally one considers *low-mass* objects as anything less than about 2 solar masses. This restriction stems from the fact that most protostellar calculations performed to date have been done most consistently for low-mass systems. Cloud cores are assumed to be isothermal and their density profile from the interior to the outer boundary has a functional dependence of r^{-2}. Such a core is unstable against gravitational collapse and its phenomenology is fairly independent of initial boundary conditions. Thus, many calculations with somewhat varying initial conditions (e.g., [284, 323, 334–337]) lead to similar scenarios.

One qualitative description that has been introduced in this respect uses the perception that the gravitational collapse is highly *non-homologous*. The central parts of the sphere collapse much faster than the outer parts introducing pressure gradients almost instantaneously. While the inner parts collapse first, the outer parts follow only after being reached by an outward propagating expansion wave at the speed of sound [284]. The concept was developed by F. Shu in the late 1970s and is commonly referred to as an *inside-out-collapse* scenario. The model allowed an analytical treatment of stellar collapse that led to some estimates of the accretion rate as well as the size of the collapsing sphere. After a short time a small core of 10–20 K forms and accretes with a constant rate:

$$\dot{M}_{acc} = 0.975 \, \frac{c_s^3}{G} = 4.36 \times 10^{-6} \left(\frac{T}{[20 \, K]} \right)^{3/2} M_{\odot} \, \text{yr}^{-1} \tag{5.1}$$

with an infall radius of:

$$r_{inf} = c_s t \tag{5.2}$$

In this picture r_{inf} is the boundary between the collapsing inner envelope and the remaining outer envelope. One of the trademarks of this model, but likely also one of the flaws, is the constant accretion rate. In this respect there are some applications of this model applying a time-dependent modification of the form of [338]:

$$\dot{M}_{acc} = \frac{c_s^3}{G} e^{t/\tau} \tag{5.3}$$

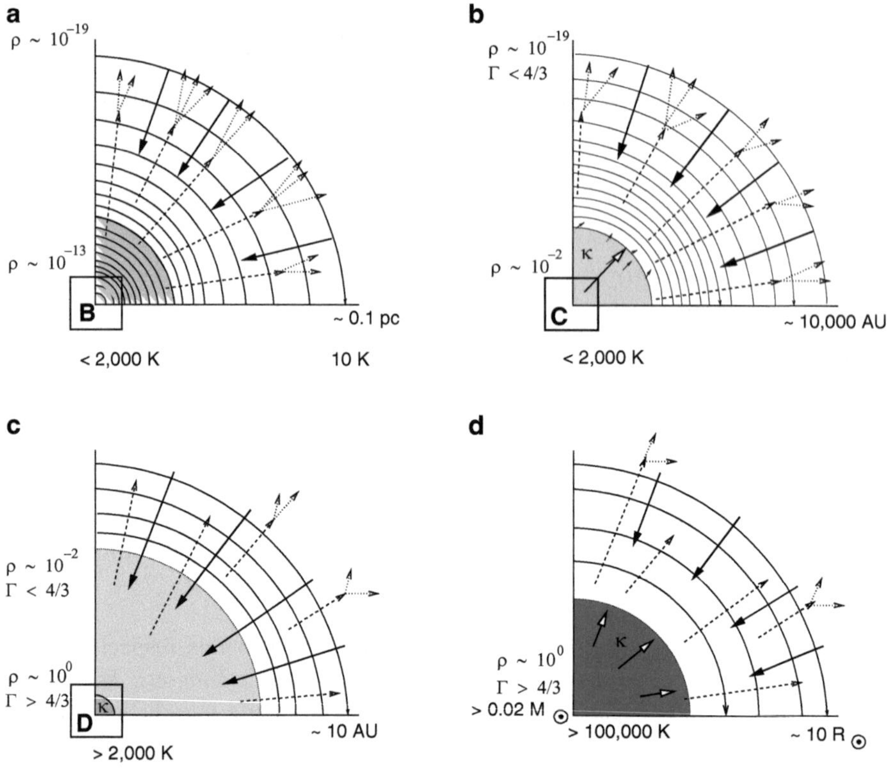

Fig. 5.2 Sketch of a collapse of an isothermal sphere which may be divided into roughly four phases. A free-fall phase in which density increases isothermally (**a**), a first core phase (**b**) where parts of the sphere become optically thick for the first time, a dissociation and ionization phase (**c**) where the collapsing core experiences a variety of critical changes in optical thickness, and (**d**) the accretion phase where the protostellar core finally remains optically thick and evolves towards the main sequence. The inward-pointing arrows illustrate inflow, the dotted outward-pointing arrows emission and re-emission, the white arrows at the centers represent trapped radiations, the circles represent mass density growing inwards, and the shades represent various opacities

with τ being a time constant for the case when the protostellar mass has reached its main-sequence value.

The concept of an outward propagating expansion wave is unique to Shu's standard model of star formation. Numerical treatments as persued by Larson are different and here one can identify several characteristic phases in the dynamical evolution of the collapse (see Fig. 5.2).

Free-fall phase (A). At central densities below 10^{-13} g cm^{-3} the collapse is fairly isothermal and matter falls in on a free-fall timescale. The central density rises most rapidly in the center of the collapsing sphere and less rapidly in the outer layers. The cooling is done effectively by mostly hydrogen molecules and grains radiating at IR-wavelength. As long as the temperature stays below ~200 K most hydrogen molecules are still in the rotational ground state ($J = 0$).

First core phase (B). Once the central density exceeds 10^{-13} g cm^{-3}, the collapse is no longer isothermal as some inner layers of the sphere become optically thick and compressional energy cannot be radiated away. Internal temperature and thus pressure increases. Eventually a first stable core of a few astronomical units forms containing some tiny fraction of the initial cloud core mass. At temperatures of about 1,000 K most grains evaporate and thus cannot absorb and re-radiate energy and hydrogen molecules are fully rotationally and vibrationally excited.

Opacity phase (C). Once the internal temperature reaches 2,000 K, hydrogen molecules start to dissociate causing the first adiabatic exponent to drop below its critical value of 4/3. This causes a second collapse as the internal pressure gradient is not steep enough to counter the increasing gravitational load. Once a density of 10^{-2} g cm^{-3} is reached, the ionization fraction of hydrogen has increased sufficiently and the first adiabatic exponent drifts back over its critical value. Several contraction cycles may follow due to ionization of He and higher Z atoms. What is left is a core of about $\sim 0.01\ M_{\odot}$.

Accretion phase (D). The remaining core now becomes optically thick and enters its main accretion phase. This is usually referred to as the *zero age* of a star. At some point it will develop an accretion disk. It is now that observations indicate the existence of a protostar. Low-mass stars will cease accretion before they reach the main sequence and the 'naked' star continues contracting until it reaches the main sequence and hydrogen burning ignites. Intermediate and massive stars probably reach the main sequence while still accreting.

Often phases B and C are combined and known as the *adiabatic phase* while phase D is sometimes also called the *hydrostatic phase*. There are many views and solutions to the collapse problem and there has been much discussion about the validity of collapse calculations of singular isothermal stellar cores.

5.1.2 Basic Equations

The collapse toward a protostar is a hydrodynamical process and the collapsing gas is treated like a continuous fluid. Basic functions involved are:

- the mass density ρ;
- the gas pressure P;
- the gas temperature T;
- the internal energy U;
- the gravitational potential ϕ; and
- the radiated energy flux L.

These functions depend on parameters like the specific heat μ and opacity κ, which contain the information about chemical composition and ionization states.

The generalized forms of these equations, specifically the ones for the energy balance and the energy transport, are quite complex and are found in various forms depending on what the initial assumptions are [334, 339, 340]. For the simple case of spherical symmetry and constant opacity the equations can be written using only one spatial coordinate. This may either be the mass M_r, which is the mass included within a radius r, or the radius r itself. It is only then that the gravitational potential, defined by the *Poisson equation*:

$$\nabla^2\phi = 4\pi G\rho \tag{5.4}$$

can also be expressed analytically.

The *equation of continuity* :

$$\frac{\mathrm{d}r}{\mathrm{d}M_r} = \frac{1}{4\pi\rho r^2} \tag{5.5}$$

describes the conservation of mass. The collapsing sphere does not gain or lose mass.

The *equation of motion*:

$$\frac{\mathrm{d}^2r}{\mathrm{d}t^2} = -4\pi r^2\frac{\mathrm{d}P}{\mathrm{d}M_r} - \frac{GM_r}{r^2} \tag{5.6}$$

is the expression for momentum conservation. It contains two external force terms, the force due to gas pressure and the gravitational force. Other terms can include viscous, radiative, and magnetic forces.

The *energy equation* resembles the 1st law of thermodynamics, that is:

$$\frac{\mathrm{d}L_r}{\mathrm{d}M_r} = -\frac{\mathrm{d}U}{\mathrm{d}t} - P\frac{\mathrm{d}(\rho^{-1})}{\mathrm{d}t} \tag{5.7}$$

It includes the total energy flux L_r through the surface of the sphere containing M_r, the change in the specific internal energy U and the work performed due to contraction or expansion. If energy were being conserved for the protostar, which is only possible if L_r is equal to zero, then all the work from compression or expansion would account for the change in internal energy.

Thus energy cannot be conserved for the protostar and one has to add an expression for the radiation field in order to obtain a full set of equations. The equation for the *energy flow* across the spherical surface, which is the irradiated luminosity [341], is given by:

$$L_r = -\frac{16\pi acr^2T^3}{3\kappa_R}\frac{\mathrm{d}T}{\mathrm{d}r} \tag{5.8}$$

(see also Appendix C). Here κ_R is the Rosseland mean of the absorption coefficient, c the speed of light, and a the radiation density constant ($4\sigma/c$).

5.2 Stability Considerations

The process of stellar formation has many phases which are defined by specific modes of stability – all for the purpose of countering the force of infalling matter. In this respect, internal pressure has to depend strongly on density and temperature and changes in this dependence have profound impact on the dynamical behavior of the gas under collapse conditions.

5.2.1 *Dynamical Stability*

Once macroscopic motions are introduced the stability behavior of the system has to be investigated. Introducing a small perturbation into the balance between gravitational attraction and internal pressure will either force the system to counter the perturbation and to return to hydrostatic equilibrium or it will destroy the equilibrium. In a classical sense this is similar to a positive and negative feedback cycle around an equilibrium point. The response to such a perturbation is called *dynamical stability*. The repetitive process of contraction (expansion) and response back to equilibrium is called *quasi-hydrostatic equilibrium.*

A radial perturbation means that there is a small compression or expansion. In order to obtain core stability the first adiabatic exponent Γ_1 has to fulfill the relation:

$$\Gamma_1 > \frac{4}{3} \tag{5.9}$$

independent of internal degrees of freedom. This condition is of immense importance for any stable core during the protostellar evolution. The internal pressure needs to grow sufficiently in relation to the mean density in order to provide enough force to counter the compression. Equation 5.7, on the other hand, limits Γ_1 to:

$$\Gamma_1 \leq \frac{5}{3} \tag{5.10}$$

For the polytropic index n the above limits then require a range between 1.5 and 3.

5.2.2 *Dynamical Instabilities*

Changes of the ionization structure of the gas have profound impact on the stability of a gas. Figure 5.3 illustrates the development of the first adiabatic exponent Γ_1 as a function of gas temperature and density under exclusion of radiation effects. It illustrates various valleys of instabilities as the internal temperature increases from 10 to 2×10^6 K. The figure shows the temperature dependence for the case

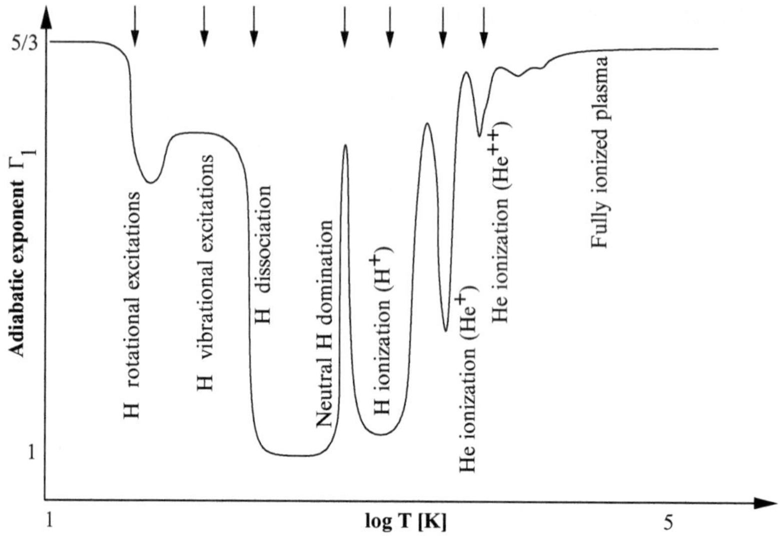

Fig. 5.3 The first adiabatic index Γ_1 as a function of temperature from calculations during the first collapse phases. The arrows mark the onset of specific physical processes that change the index most dramatically causing instability in the newly formed stellar core. Adapted from data from Wuchterl [342]

of $\sim 10^{-14}$ g cm^{-3}. For densities also varying from left to right (i.e., from 10^{-14} to 1 g cm^{-3}), the peaks and valleys in the diagram remain but with slight changes in their depths and height. For example, at higher densities the large Γ_1-valleys due to H_2 dissociation and H ionization become somewhat shallower.

At 10 K the gas is assumed ideal and Γ_1 is equal to 5/3. The first shallow valley of instability is caused by rotational and vibrational excitation of H_2 molecules (and roughly corresponds to phase B in Fig. 5.2). At around 2,000 K the dissociation of H_2 molecules sets in causing a large second valley of instability followed by a large third valley due to the ionization of H and He (this happens through phase C in Fig. 5.2). One therefore has to realize that each time a particle species is excited, dissociated, or ionized, the density structure of the gas is changed. This alters the polytropic description of the gas and thus its stability.

5.2.3 *Opacity Regions*

It is crucial that opacity sources are available to create internal layers that are optically thick for radiation. If there are no opacity sources the internal energy and thus temperature to build up internal pressure cannot increase sufficiently to even reach hydrostatic equilibrium. Many calculations of the Rosseland mean opacity for a variety of gases have been carried out since the 1970s and include

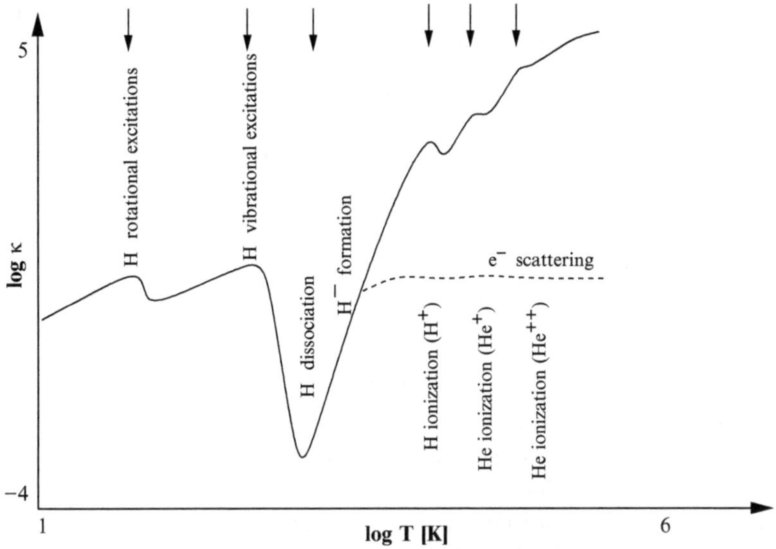

Fig. 5.4 The Rosseland mean opacity κ_R as a function of temperature from the same calculations shown in Fig. 5.3. It shows the behavior of opacity parallel to the changes in the first adiabatic index and thus how it can counter drifts into instability regions. Adapted from data from Wuchterl [342]

about 60 million atomic, molecular transitions from neutral elements and radicals including hydrogen, helium and various heavy elements, specifically iron [9, 343–345]. Absorption and scattering of various dust grain sizes are included as well. The international *Opacity Project (OP)* and the *Opacity Project at Livermore (OPAL)* (see [346] and references therein) now provide extended tabulations for temperatures above 8,000 K. Figure 5.4 illustrates the development of the Rosseland mean opacity κ_R as a function of temperature and density [342]. At low temperatures and densities, changes in κ are due to the melting of the ice coatings of silicate dust particles and their eventual evaporation leaving only molecules as opacity sources. This causes a big drop in opacity until the ionization of alkali metals sets in, and H_2 dissociates. The large rise in opacity is then due to the ionization of H and He. This example shows once again the critical impact of the dissociation of H_2 molecules as well as ionization processes (see also Appendix D).

Rapid changes of opacity in connection with increasing density and temperature causes what is called *secular* instability. Another form of instability occurs once κ_R increases upon adiabatic compression, which is called *vibrational* instability. The evaluation of these instabilities is not as straightforward as in the case of dynamical instability which involved only the first adiabatic exponent Γ_1. The conditions for secular and vibrational involve a complex interplay between the opacity gradients at constant temperature κ_P and pressure κ_T, the pressure gradients at constant temperature χ_T and density χ_ρ, and the temperature gradient at at constant entropy.

In this study [342] other regions of secular instability for a core of density of 10^{-2} g cm^{-3} have been identified as well.

5.3 Collapse of Rotating and Magnetized Clouds

The previous sections considered the classical case, where thermal and gravitational pressures alone define the dynamics of the collapse. While the treatment of an unrotating and non-magnetized sphere provides the theoretical groundwork it falls short in accounting for many observed phenomena (see Chap. 6), which feature accretion disks, matter outflows, and a variety of magnetic field effects. The following sections discuss several cases which include rotation or magnetic fields or both in the collapse process. The inclusion of these effects requires abandoning perfect spherical symmetry. Nearly all calculations now adopt axisymmetric configurations and provide numerical solutions, because analytical solutions for most magneto-hydrodynamic equations are not known. The benefits are that these calculations introduce structures into the theoretical treatment that do not appear in the classical approach: the formation of accretion disks and even the possibility of mass loss.

The classical collapse phases as described in the previous sections are valid with or without rotation and magnetic field support of the the collapsing core. In simple terms this means rotation and magnetic fields eventually cannot prevent the collapse once the core reaches critical initial conditions. The collapse will reach the *first-core* phase and its dynamical contraction will proceed as well. What will differ substantially is the matter flow onto the core as well as the dynamics of the infalling gas. The following sections put specific emphasis on characteristic length scales under various scenarios. It also has to be kept in mind that because of the vast parameter space that has to be covered most treatments of the problem do not cover all collapse phases. Calculations covering a density range between 10^{-22} and 10^{-16} g cm^{-3} [347–349] describe the problem during quasi-static cloud contraction and possibly the onset of the *free-fall* phase thus covering strictly isothermal phases of the cloud. Some calculations go a little further toward the formation of a first core [324, 350, 351] but still assume isothermality in the collapsing cloud. Many authors thus make the distinction of pre- and post-*point mass formation (PMF)*. Some quite recent calculations include rotation and magnetic fields and are specifically valid for phases after PMF [352–354].

5.3.1 Collapse of a Slowly Rotating Sphere

The detailed impact of rotation on the gravitational collapse is quite complex. One of the difficulties lies in the fact that for most cases the simplification of spherical symmetry is no longer valid and axial (cylindric) symmetry has to be assumed instead. This makes analytic treatments almost impossible, because the gravitational potential cannot be easily expressed in analytic form. Some basic considerations, however, can be made by studying the case of slow rotation. It has been argued that, if the rotation is slow enough so that the cloud envelope rotates only a small fraction of a full turn during the collapse, then one can identify two regimes: an inner region that is distorted due to rotation and the outer envelope that stays roughly spherical [355]. Here the collapse is always in the pre-PMF phase.

The inner region is limited by the centrifugal radius r_c where centrifugal support of the cloud becomes important. This radius r_c depicts the location of a centrifugal shock at which the rotational velocity is equal to the velocity of the infalling material. At radii smaller than r_c the core is significantly distorted; at larger radii rotation may be considered a small effect. This radius has been calculated to be [355]:

$$\begin{aligned} r_c &= 0.058\Omega^2 c_s t^3 \\ &\sim 0.44\left(\frac{\Omega}{10^{-14}\ \mathrm{s}^{-1}}\right)^2\left(\frac{c_s}{0.35\ \mathrm{km\ s}^{-1}}\right)\left(\frac{t}{10^5\ \mathrm{yr}}\right)^3\ \mathrm{AU} \end{aligned} \tag{5.11}$$

where the time should be small, more like a fraction of the free-fall timescale. Inside this radius the infalling matter encounters the centrifugal barrier in the equatorial plane and therefore this radius may define the dimensions of the seed accretion disk. The mass of the accreting protostar after an accretion time of 10^5 yr may then be expressed as:

$$\begin{aligned} M_{star} &= \left(\frac{16Rc_s^8}{G^3\Omega^2}\right)^{1/3} \\ &= 0.23\left(\frac{R}{3R_\odot}\right)^{1/3}\left(\frac{\Omega}{10^{-14}\ \mathrm{s}^{-1}}\right)^{-2/3}\left(\frac{c_s}{0.35\ \mathrm{km\ s}^{-1}}\right)^{8/3} M_\odot \end{aligned} \tag{5.12}$$

Since the radius r_c grows with angular velocity Ω^2 and $r_{inf} = c_s t$, it is clear that angular velocities cannot be too much larger than $\sim 10^{-14}\ \mathrm{s}^{-1}$, because r_c becomes of the order of r_{inf} and the free-fall envelope (see Fig. 5.5) cannot be assumed to be spherical any more. The region outside the free-fall envelope is stationary and isothermal, limited by the boundary r_{out} where the sound speed is equal to the rotational velocity (i.e., $c_s = \Omega r_{out}$). Cores that do not yet contain protostars should rotate at the same angular velocity as the outer envelope with typical values of $\Omega \sim 10^{-14}\ \mathrm{s}^{-1}$. Thus r_{out} may also be considered as the boundary between the collapsing sphere and the external medium.

For typical values of various parameters (M = 1 $M_\odot$, $\dot{M}_{acc} = 10^{-5}\ M_\odot\ \mathrm{yr}^{-1}$, $c_s = 0.35\ \mathrm{km\ s}^{-1}$, $\Omega \sim 10^{-13}\ \mathrm{s}^{-1}$ [356]) one finds for length scales $r_c = 6.6 \times 10^{14}$ cm (44 AU), $r_{inf} = 1.1 \times 10^{17}$ cm (7,400 AU), and $r_{out} = 3.5 \times 10^{17}$ cm (23,400 AU).

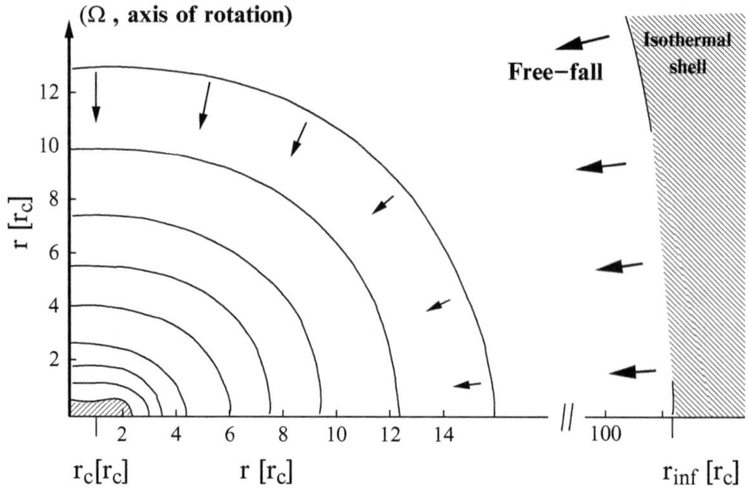

Fig. 5.5 Illustration of a slowly rotating, collapsing core. The scales of the horizontal and vertical axis are in units of the centrifugal radius r_c. The horizontal axis is also artificially extended to the approximate location of the outer isothermal envelope r_{inf}. The contours are approximate isodensity curves, which close to r_{inf} are roughly spherical, close to r_c strongly non-spherical. The density between r_{inf} and r_c varies from 10^{-19} to 10^{-13} g cm^{-3}. Adapted from Terebey et al. [355]

5.3.2 Collapse of Magnetized Clouds

Collapse calculations which include magnetic fields in the initial conditions have been performed by many authors [350, 357–360]. Of some interest is one example as it helps to illustrate several characteristic conditions in the magntized case [358, 359]. Here an unstable cloud core with a density distribution of an isothermal sphere is threaded by a uniform poloidal magnetic field (initial conditions as described in Chap. 4). The problem now is a MHD one and the basic set of equations has to be extended in order to account for magnetic field effects. This affects the system of equations in several ways. Magnetic forces act on the ionized material only and the matter density has to be specified for neutral and ionized material. The initial ionization fraction in the collapsing cloud is quite small (i.e., 10^{-7}) and forces due to gravitation and internal or external pressure can be neglected for ionized material. However, it is large enough to cause an ambipolar drift that weakens the coupling of matter to the magnetic field (see Sect. 4.3.4). Thus what is left is to add a generalized equation of motion for ions and neutrals as well as the induction equation and source equation for the magnetic field [358].

The dynamics of the collapse remains fairly unaltered. It still is non-homologous, that is it qualitatively happens in an inside-out fashion. However, it is not symmetric any more as the outward moving expansion wave travels faster in the direction parallel to the magnetic field. The result that the accretion rate $\dot{M}_{acc}$ hardly changes is a consequence of a rather unrealistic balance. The Lorentz force acts against

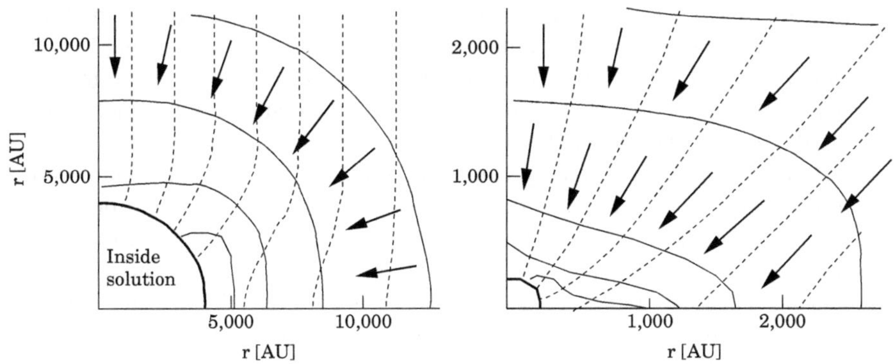

Fig. 5.6 Illustration of a collapse of a magnetic spherical, isothermal sphere fixed at about 2×10^5 yr after the start of the collapse and at the end of the *free-fall phase*. The *solid lines* are contours of equal density, the *dashed lines* are magnetic field lines, the thick arrows show the velocity field. The horizontal and vertical axes are in AU. The diagrams show the self-similar solution for the outside (*right*) and inside (*left*) case. Adapted from Galli and Shu [358, 359]

gravity and effectively slows the gas flow. However, the expansion wave, though now not spherically symmetric, propagates faster outwards. As a result the outer envelope collapses faster. In the end the value for $\dot{M}_{acc}$ remains roughly unchanged in this picture.

Quite dramatic are the changes in symmetry and thus in the gas flow. Magnetic pinching forces deflect the infalling gas toward the equatorial plane to form what is called a *pseudo disk* around the central protostar (see Fig. 5.6). Since there is no rotation it is not really an accretion disk which would be centrifugally supported. In this respect effects from rotation and pinching magnetic fields are not quite the same. However, in analogy to the centrifugal radius in the case of slow rotation, a magnetic radius can be defined as [359]:

$$\begin{aligned} r_b &= k_b G^{2/3} B_o^{4/3} c_s^{-1/3} t^{7/3} \\ &\sim 540 \left(\frac{B_o}{30\ \mu\text{G}} \right)^{4/3} \left(\frac{c_s}{0.35\ \text{km s}^{-1}} \right)^{-1/3} \left(\frac{t}{10^5\ \text{yr}} \right)^{7/3} \text{AU} \end{aligned} \tag{5.13}$$

where B_o is the initial uniform magnetic field and k_b is equal to 0.12, determined numerically. Another characteristic radius is defined where magnetic equals thermal pressure in the initial state:

$$\begin{aligned} r_m &= \frac{2c_s^2}{B_o \sqrt{G}} \\ &= 2.1 \times 10^4 \left(\frac{c_s}{0.35\ \text{km s}^{-1}} \right) \left(\frac{B_o}{30\ \mu\text{G}} \right)^{-1} \text{AU} \end{aligned} \tag{5.14}$$

This radius indicates when the magnetic pressure dominates over thermal pressure in the outer envelope. Fig. 5.6 illustrates the collapse environment in [358, 359] for roughly 2×10^5 yr into the collapse. The left-hand part of the diagram shows magnetic field lines, contours of equal density, and the matter inflow pattern for the self-similar treatment [358] and for the inside numerical treatment in [359], showing an increase of r_b to approximately 1,000 AU. As an interesting effect the infalling material actually pulls the magnetic field toward the central core and practically aligns with the infalling gas. Ambipolar diffusion is present but insignificant in these illustrations. For typical values of c_s, B_o, Ω, M, and $\dot{M}_{acc}$ the radius r_b always exceeds the centrifugal radius r_c by one or two orders of magnitude. This nourishes the idea that in the case of additional rotation a smaller, centrifugally supported, accretion disk is fed equatorially through an outer pseudodisk.

5.4 Cores, Disks and Outflows: the Full Solution?

The cases discussed so far have more illustrative than astrophysical value as they show unique characteristics with respect to rotation or magnetic fields, but do not properly account for other characteristics such as *ambipolar diffusion*, *magnetic braking*, *turbulence*, and, as has been mentioned before, a realistic change of accretion rate with time. The importance of a proper treatment of ambipolar diffusion and magnetic braking has been emphasized by Mouschovias and colleagues as this process likely dominates the coupling of matter to the magnetic field during inflow [324, 351, 361, 362] . On the other hand, it seems that if, for example, magnetic braking is very efficient it may favor the formation of single stars [11], whereas observations of stars and clusters feature a substantial amount of YSOs in binaries and clusters and thus fragmentation. This issue is currently under discussion specifically with respect to the influence of turbulence and core interactions [363] during cluster formation (see [302] for a review). The following examples, though, concentrate more on issues that concern the formation of single stars.

In realistic collapse scenarios both rotation *and* magnetic fields are important. The two separate cases show that in both instances disks will form but on different length scales. However, neither of the two cases create any form of outflow. Thus, accretion disks and outflows are likely produced by the cooperation of magnetic *and* centrifugal forces. Many solutions of a gravitational collapse of rotating, isothermal, magnetic molecular cloud cores have been produced recently [324, 347, 352–354], where the ones that also include post-PMF phases are specifically of interest. In the following three, examples of the most recent calculations are presented, attempting to reproduce many observed phenomena that will later become important in early stellar evolution.

5.4.1 Ambipolar Diffusion Shock

The concept of ambipolar diffusion in molecular cloud cores has been introduced in Sect. 4.3.4. Sizeable ionization fractions in the presence of magnetic fields will lead to an ion-neutral drift force that can grow strong enough to decouple the matter from the magnetic field creating a shock. A semi-analytical scheme incorporates such an ambipolar diffusion shock and magnetic braking before and after PMF [353]. The limiting case of a non-magnetic rotational collapse simply results in a centrifugal shock (see also [364]) at a centrifugal radius of about 80 AU. The application of parameter values as they were used in Sect. 5.3.1 indicates consistency with such a solution. However the model remains limited to the pre-PMF phase unless one postulates a mechanism to remove angular momentum from the core like a stellar wind [365]. In a magnetic rotational collapse, angular momentum is not only removed by a wind but also reduced due to magnetic braking which ultimately allows for the formation of a single central mass.

The infall is magneto-hydrodynamic in the outer regions only until it passes through an ambipolar diffusion shock and subsequently a centrifugal shock. After 10^5 yr the centrifugal radius grows even beyond 100 AU with a central mass growing to 1 $M_\odot$ and a disk mass of somewhat less than 10 percent of the central mass. Figure 5.7 shows the development of self-similar infall velocities (left) and surface densities (right) for non-magnetic rotation and MHD infall with ambipolar diffusion. In the first case infall basically stalls at the centrifugal shock. In the second case

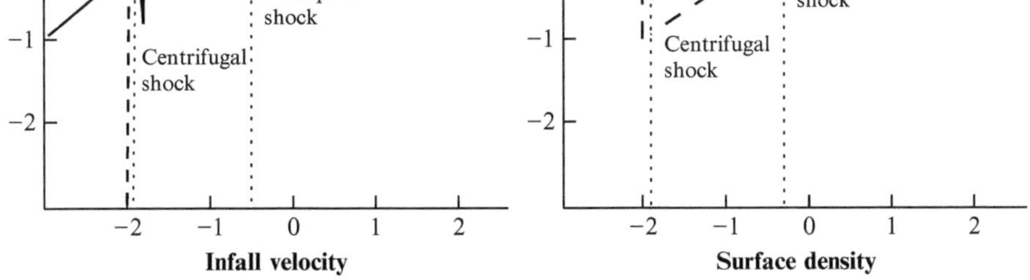

Fig. 5.7 Self-similar solution of a collapse of a magnetic and rotating isothermal cloud at some time into the collapse. Drawn are unitless self-similar variables. The horizontal axis is the self-similar radius ($r/(c_s t)$) of the collapsing sphere. (*left*) The development of the infall velocities. (*right*) The corresponding surface densities. The dashed line shows the solution for a non-magnetic rotating sphere, the solid line one solution including magnetic braking and ambipolar diffusion. Adapted from Krasnopolsky and Königl [353]

the centrifugal shock region allows the creation of a strong, rotationally supported accretion disk and an increase in surface density towards the central core. The established disk will drive centrifugal outflows transporting angular momentum and mass.

5.4.2 Turbulent Outflows

While in the above model gravitational torques and instability-induced turbulence are not relevant, these turbulent outflows become more important in some numerical calculations [354] (see Fig. 5.8). The generation of possible outflows here can be qualitatively understood by the following argument. By introducing rotational motion into what is assumed to be a poloidal magnetic field configuration, toroidal fields are generated simply by the generation of poloidal currents through Ampere's law. The coexistence of toroidal and poloidal magnetic fields creates torques that allow angular momentum transfer along the magnetic field lines, which eventually lead to the ejection of matter. Axisymmetric MHD simulations result in a quasi-static first core that is separated from an isothermal contracting pseudodisk through developing accretion shocks. Magnetic tension forces transport angular momentum to the disk surface where matter finally gets ejected through bipolar outflows. Two

Fig. 5.8 The matter flow at the core of the infall after PMF in K. Tomisaka's calculations. The shades in the upper part of the diagram represent the ratio of the toroidal to the poloidal magnetic pressure, in the lower part the density distribution. Light shades are high values, dark shades low values. The *solid lines* are magnetic field lines and the arrows velocity vectors of the matter flow. From Tomisaka [366]

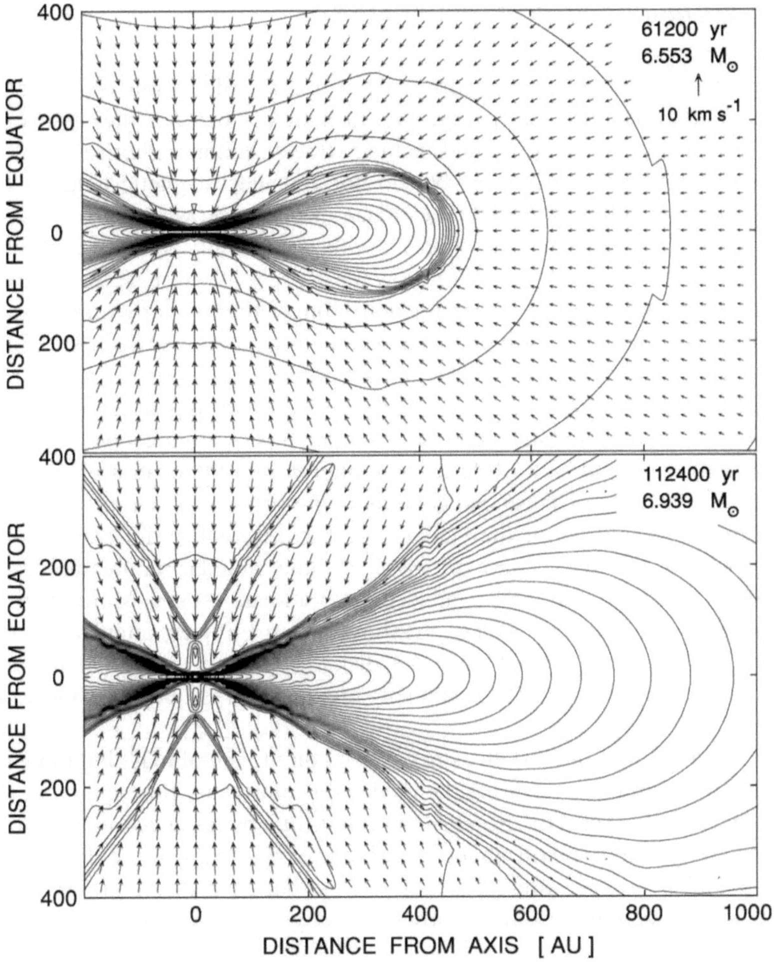

Fig. 5.9 Density and velocity structure during the growth of a protostellar disk calculated for an initial mass of 10 $M_\odot$. The stellar core has grown to about 6.5 $M_\odot$ after 60,000 yr. After about twice the time, the stellar core has not significantly grown, but the size of the accretion disk has more than tripled. From Yorke and Bodenheimer [352]

types of outflow can be distinguished. A U-shaped outflow forms when the poloidal field energy is comparable to the thermal energy in the contracting disk. The gas is accelerated by the cooperation of centrifugal forces and pressure from the toroidal field. A more turbulent I-shaped outflow is generated in regions where magnetic field lines and velocities are more randomly distributed. In both cases the outflow is launched by centrifugal forces.

5.4.3 Formation of Protostellar Disks

Another example of a recent hydrodynamical 2-D calculation is presented by H. Yorke and colleagues [352] (Fig. 5.9). In their calculations they focus on the evolution of a rotating protostar and include effects such as *radiative acceleration* and *angular momentum transport* but do not consider magnetic fields. The fact that angular momentum transport takes place has been shown by many numerical models [367–370] (see Chap. 8). Starting with a uniformly rotating and centrally condensed sphere their calculations follow mass accretion over typical protostellar lifetimes of the order of 10^7 yr, significantly longer than in earlier attempts [371, 372]. The model includes traditional disk parameterizations (see Sect. 8.1.1) as well as similar mechanisms for the introduction of *gravitational torques*. Another novelty in calculations of this extent is the inclusion of stellar masses up to 10 $M_{\odot}$. After PMF a warm, quasi-hydrostatic disk surrounding a central unresolved core forms, and grows in mass and size while it feeds matter onto the central object. One of the important aspects of these calculations is the growth in size of the accretion disk relative to the mass of the stellar core for different initial conditions. The radial growth of the disk is an expression of radial angular momentum transport which is achieved through tidally induced gravitational torques. More specifically, the radial growth of the disk is an expression of the insufficient angular momentum transport, and mass is loaded onto the disk faster than traditional momentum transfer mechanisms can handle. Eventually gravitational torques relax the transport problem.

Thus the growth of the disk does not depend on the initial mass but more on the initial angular momentum. During the modeled evolutionary time span quasi-static disks in excess of several thousand AU in radius can be produced. For a low mass star after 10^7 yr there is still about 35 percent of the mass in the disk. For high mass stars this is the case after a few times 10^4 yr but the disk still grows in size for another few 10^4 yr without significantly increasing the mass of the central star.

Chapter 6
Evolution of Young Stellar Objects

This chapter decribes the evolution of stellar objects after the PMF phase when a young stellar object forms via accretion from its circumstellar cloud. It is difficult, if not impossible, to draw a line between the process of protostellar formation and the evolution of a protostar toward a mature star on the main sequence. The dense central object that forms during the collapse of a cloud core is not yet a star, but it is also not a molecular core anymore. One should define what is meant by the terms *protostar*, *PMS star*, and *young stellar object* as these expressions are common in the literature:

- *Protostar*: the optically thick stellar core that forms during the adiabatic contraction phase and grows during the accretion phase.
- *Protostellar system*: the entire young stellar system with its infalling envelope and accretion disk. The nomenclature includes binary and multiple systems. The extent of the system is fairly well descibed by its Jeans length.
- *PMS star*: the premature star (PMS = pre-main sequence) that becomes visible once the natal envelope has been fully accreted and which contracts towards the main sequence.
- *PMS stellar system*: the PMS stars with their accretion disks and their star/disk interaction regions.
- *YSO*: the entire stellar system (YSO = young stellar object) throughout *all* evolutionary phases. However, it is specifically used during the collapse phases and in the case of massive systems, where evolutionary phases cannot easily be distinguished.

In the light of the collapse phases introduced in Sect. 5.1.1 the birth of the protostar may be placed somewhere near the end of the adiabatic phase, when the core becomes optically thick toward radiation and enters the accretion phase. Section 6.1 thus describes the evolution of the protostar during its accretion phase. Early stellar evolution can be generically pursued and observed through the evolution of effective surface temperature and bolometric luminosity; their evolutionary aspects are reviewed in Sects. 6.2 and 6.11. Theoretical studies on protostellar evolution do not usually explicitly make a distinction between before

N.S. Schulz, *The Formation and Early Evolution of Stars*, Astronomy and Astrophysics Library, DOI 10.1007/978-3-642-23926-7_6, © Springer-Verlag Berlin Heidelberg 2012

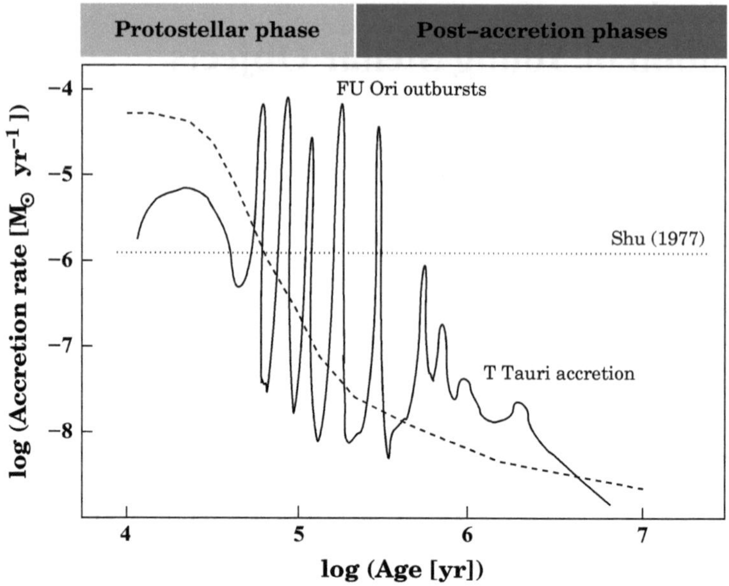

Fig. 6.1 The diagram schematically shows the time development of the accretion rates as they appear in various calculations and in relation to observed phenomena. The dotted straight line resembles the result from standard star formation [284], which does not depend on time and is based on infinitely large radius and mass and thus appears as a flat line throughout the entire time interval. The hatched curve represents an approximation to the results of the many more realistic calculations. The thick line was adapted from Hartmann [290] and shows additional features such as outbursts of FU Ori type protostars and T Tauri disk accretion. All rates apply for typical low-mass ($\sim M_{\odot}$) stars only

and after accretion since the only difference between the two phases is the fact that the protostar ceases to accrete more matter from its primordial envelope, while it continues to contract quasi-hydrostatically. To address a common misunderstanding up front, 'after accretion' does not mean that the matter flow towards the protostar has stopped entirely. It simply means that the accretion disk discontinues to get predominantly fed by the envelope. In fact, accretion at very low rates continues well after, which may have some importance with respect to high-energy emission of PMS stars (see Fig. 6.1).

6.1 Protostellar Evolution

The previous chapter dealt with the gravitational collapse beyond the first PMF phase into the adiabatic phase. The calculations discussed there have established the order of magnitude of the mass accretion rate that now determines the growth properties of the protostar. During the adiabatic phase the collapsing system evolves into two regimes based on their opacity properties. In the core is the optically thick protostar and nearby infalling matter prevents any radiation from escaping. This protostar is nearly in hydrostatic equilibrium and continues to grow. Its mass, originally as low as $\sim$0.001 $M_{\odot}$, increases quite rapidly by almost a factor of 10 to form an 'embryo' star (see phase D in Sect. 5.1.1). Since infalling matter is optically thick any accretion shock will heat the outer layers causing them to expand thereby increasing the size of the protostar to a few solar radii. During this expansion the opacity outside the accretion shock drops and photons can escape. The radiative energy loss prevents the protostar from further expanding and the radius of the star remains fairly constant throughout the accretion phase [373].

6.1.1 Accretion Rates

Details of the amount and variability of accretion rates specifically during adiabatic and accretion phases are still highly uncertain and probably vary greatly with time. From Chap. 5 it seems unclear how much rotation or magnetic fields affect the accretion rate. It is established, though, that they determine the geometry of the accretion flow which leads to the formation of an accretion disk. The time dependence of the accretion rate is likely a matter of angular momentum transport (see Chap. 8). More relevance may also be given to the density of the infalling envelope and the infall speed relative to the sound speed to determine the rate. It is sometimes useful to scale accretion rates by c_s^3/G (see (5.1)), which is the time independent solution obtained by Shu's standard model [284]. For an isothermal infalling envelope of $\sim$10 K this would lead to a constant accretion rate of about 1.6×10^{-6} $M_{\odot}$ yr^{-1}. Accordingly, it then takes about $\sim 7 \times 10^5$ yr to form a 1 $M_{\odot}$ star. Though this is not unreasonable more detailed calculations for spherical collapse show [374–376], (see also [377] for a review) that typically the rate varies with an initial burst of rapid accretion lasting about 10^4 yr. Half of the envelope is accreted within 10^5 yr and nearly all of it within 10^6 yr. Axisymmetric collapse scenarios that include rotation and magnetic fields do not really change much of this picture. Calculations predict initial accretion rates of up to 50 c_s^3/G, which then rapidly and steadily decline with time [364, 378–381]. Observations show in very late protostellar stages accretion rates of only 0.06 c_s^3/G, corresponding to rates of less than 10^{-7} $M_{\odot}$ yr^{-1}. Figure 6.1 schematically illustrates the dependence of accretion rates with time. In general, observations so far are inconclusive about the order of magnitude of the accretion rate. A recent study of accretion in a large number of young and very low-mass objects from a variety of star-forming regions derives estimates of the accretion rate in very late spectral types between 10^{-9} and 10^{-12} $M_{\odot}$ yr^{-1} with a clear trend of decreasing accretion rate with stellar mass [382].

6.1.2 Matter Flows

Concepts that include magnetic fields, ambipolar diffusion, and rotation (see Chap. 5) predict various types of matter flows onto and from the protostar (Fig. 1.3, see also [354]). Flow motions can be seen as nearly spherical infall of the outer envelope and accretion from an inner accretion disk. The physics of the boundary where the inflow from the accretion disk connects with the growing protostar is highly uncertain. Recent detections of X-ray emission from these stars hint at the existence of strong magnetic activity in at least some of the protostars [383]. Observationally, the existence of accretion disks in very young protostars is difficult to verify as most of the time the star and its disk are hidden by dust and infalling gas. *IRAS* observations established that about 10 percent of the observed PMS population are protostars likely in their accretion phase [384–386].

Figure 6.2 also shows that there are outflows in the form of winds and jets. The calculations by [354] strongly suggest what has been suspected all along: these outflows are based on magnetic acceleration from the surface of a Keplerian disk.

6.1.3 Deuterium Burning and Convection

Deuterium is one of various elements that engages in nuclear fusion at temperatures between 10^6 and 10^7 K, and among all others (^{6}Li, ^{7}Li, ^{9}Be, and ^{3}He) it is the most abundant one. A protostar steadily accretes mass and though the pressure balance can be treated quasi-hydrostatically, its internal structure evolves with the amount of mass accreted [388–390]. Deuterium burning occurs at temperatures larger than 10^6 K. For low-mass protostars ($M \leq 1\ M_\odot$) this temperature is reached

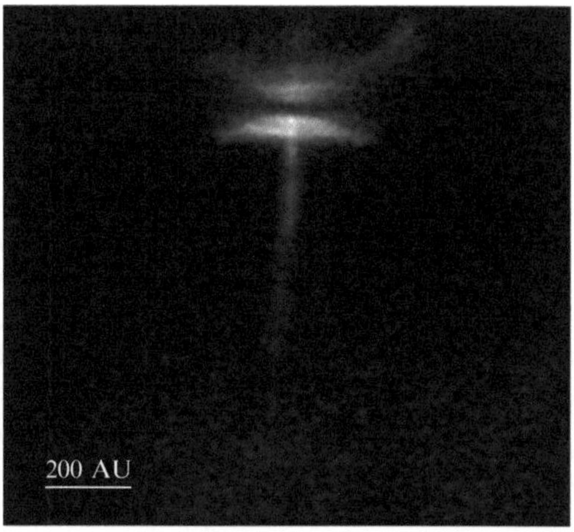

Fig. 6.2 A spectacular image of the disk and jets of the YSO HH30 with high outflow activity in the jets. The outflow velocities are several hundreds of km s^{-1}. Adapted from Burrows et al. [387]

at its center setting off the process of sustained fusion of deuterium. The quantity of the mass accretion rate is therefore not only important for the star's final mass, it also determines the supply of deuterium to the hydrostatic core which leads to modification of the energy balance inside the star. A fast decline in the accretion rate with time will also lead to an early depletion of deuterium in the core assuring a short burning phase. In stars more massive than 2 $M_\odot$ accretion rates stay high long enough that deuterium burning will continue beyond a phase of central depletion, resulting in shell type burning [390]. Thus, in these stars the interior is never thermally relaxed and quasi-static contraction may not proceed homologously [346].

In low-mass stars the amount of deuterium burned is of the order of the amount that is accreted from the outer envelope. This can only happen if the accreted deuterium is transported efficiently to the center of the star. This mechanism is *convection* and is effectively driven by deuterium burning. A treatment of convection in terms of a single element of turbulence (*eddy*) in adiabatic regimes can be found in [391]. More complex models involve entire distributions of eddies [392, 393]. These models parameterize convection in terms of a *mixing length* or *scale length* scaled to the solar radius and temperature. Once convection is established, accreted deuterium is rapidly transported into the stellar interior. The produced luminosity is proportional to the mass accretion rate and is given by:

$$L_{D_o} \simeq 12 L_\odot \left(\frac{\dot{M}_{acc}}{10^{-5} M_\odot \mathrm{yr}^{-1}} \right) \tag{6.1}$$

where the magnitude is mainly determined by the deuterium abundance and the energy produced per reaction. It is important to realize that the amount of energy available from deuterium is of the same order of magnitude as the total gravitational binding energy. Deuterium thus regulates the mass–radius relation of the star. The proportionality assumption, however, only holds for a narrow regime of accretion rates and for very low-mass stars. Generally, more deuterium is consumed than is provided through accretion and thus the produced luminosity is:

$$L_D = L_{D_o} \left(1 - \frac{\mathrm{d}(f_D M)}{\mathrm{d}M} \right) \tag{6.2}$$

where M is the mass of the star, f_D the fractional deuterium concentration and the product $f_D M$ the total deuterium content [346]. Under the assumption the star is fully convective, Fig. 6.3 shows the change of the fractional deuterium concentration f_D with protostellar mass for various accretion rates. It shows that for high accretion rates the drop in deuterium concentrations is shifted towards higher stellar masses.

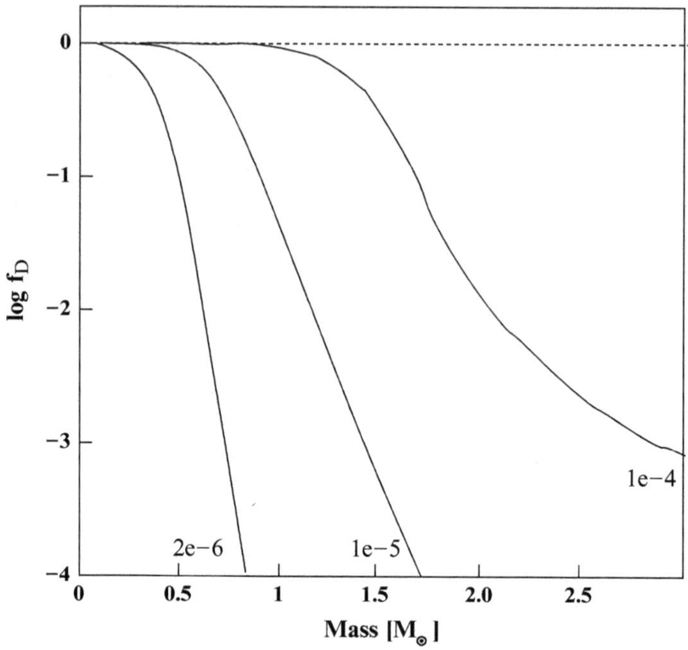

Fig. 6.3 The decrease in fractional deuterium concentration in low-mass protostars for various accretion rates. Adapted from Palla and Stahler [346]

6.1.4 Lithium Depletion

The interstellar abundance of [Li/H] $\sim 10^{-9}$ is about $\sim 7 \times 10^{-7}$ times smaller than the relative abundance of deuterium. Thus, it is clear that these elements do not contribute to the intrinsic energy balance of the star and are structurally unimportant. However their abundance can serve as tracers to test the internal structure of PMS stars as well as their evolutionary status. A good example of this is the depletion history of Li in low-mass stars [394]. Surface abundances provide a critical test for convection, rotation, and mass loss in young, but also evolved, stars. Often important is the ^{7}Li isotope which is an order of magnitude more abundant than ^{6}Li – the main reaction under study is ^{7}Li(p,α)^{4}He at temperatures of about 2.5×10^6 K [395].

A special case in this respect is the diagnostics of the Li abundance in *brown dwarfs*. Though these are objects at the final stage of stellar evolution, their study is critical for the study of PMS populations as well as for stars at the lowest end of the mass spectrum. They are stars with masses below 0.08 $M_{\odot}$ down to about a few tens of Jupiter masses, which due to their low mass never engage in nuclear hydrogen fusion and thus never reach the main sequence. Some of the more massive ones manage to burn deuterium, but they also hardly produce temperatures hot enough to ignite Li burning in their interiors. The mass cut can be placed at around 0.06 $M_{\odot}$. Atmospheres of brown dwarfs are thus Li rich as there is no convective motion to transport the element into the interior for eventual destruction.

6.1.5 Mass–Radius Relation

The evolution of radius and mass of a protostar is central to our understanding of the early evolution of stars. From (5.12) one can infer a simple relation for the mass-to-radius ratio which predicts protostellar masses of the order of less than 0.35 $M_{\odot}$ with radii between 2 and 5 $R_{\odot}$. The future evolution is dominated by deuterium burning and previous histories will converge at the onset of deuterium burning [396]. Figure 6.4 shows the mass–radius relation for a protostar accreting at $\dot{M} = 10^{-5} M_{\odot}$ yr^{-1} as adapted from [396] and [397]. The curves represent snapshots at a late stage in the protostellar accretion phase. The left part of the figure highlights the mass range below 1 $M_{\odot}$. It shows that all mass–radius relations prior to deuterium burning converge to a point at roughly 0.3 $M_{\odot}$ and 2.5 $R_{\odot}$. From this point the radius starts to expand to about 5 $R_{\odot}$ as the mass grows to 1 $M_{\odot}$.

Development toward more massive stars is complex and still to a high degree uncertain. The grey area in Fig. 6.4 highlights the mass range in which the stellar core can be assumed to be fully convective due to deuterium burning. Toward more

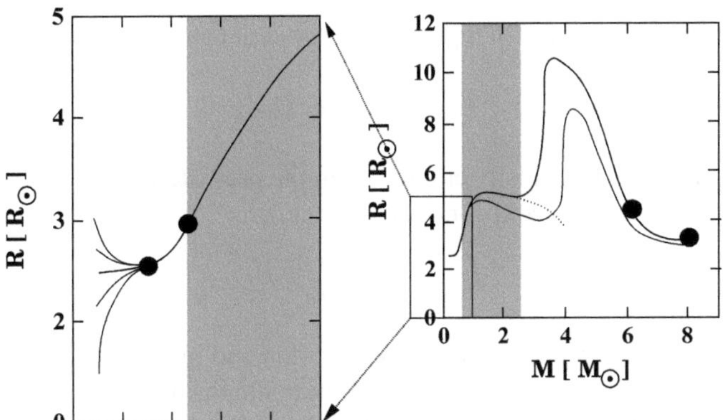

Fig. 6.4 A snapshot in the evolution of the protostellar radius with mass at a steady accretion rate of $\dot{M} = 10^{-5} M_{\odot}$ yr^{-1}. The evolutionary state of stars following this mass-radius relation can be placed near the time when the stars become visible due to their surface luminosity. For example a 2 $M_{\odot}$ star at that stage has a radius of about 5 $R_{\odot}$ and, an 8 $M_{\odot}$ star has less than 4 $R_{\odot}$. The left-hand diagram highlights the protostars of less than 1 $M_{\odot}$, the right-hand diagram for the entire mass range up to 8 $M_{\odot}$. The gray area indicates the range where stars are fully convective. The black circles (from left to right) mark the onset of deuterium burning, the point where stars are fully convective, the ignition of hydrogen (CN cycle) in the core, and the arrival at the ZAMS (see Sect. 6.2.2). The dotted line in the right-hand figure represents the case where no deuterium shell burning occurs and the thin line the mass–radius relation using different boundary conditions. Adapted from Stahler [396] and Palla and Stahler [397]

massive stars a radiative barrier develops which eventually leads to deuterium shell burning giving rise to a steep increase in the stellar radius of up to 10 $R_\odot$. Without deuterium shell burning, this rise would not occur and a more massive star would simply contract at that point (dotted line). The protostar is then in a range where it is radiatively dominated with convective zones near its stellar surface. After reaching the peak, more massive stars contract towards the main sequence, which is reached at 8 $M_\odot$. This means that massive stars reach the main sequence well into their protostellar phase.

The extent to which these predictions depend on initial boundary conditions, specifically for more massive stars, is also illustrated in Fig. 6.4. In the case of spherical infall and the development of an accretion shock the evolution is represented by the thick line. Whereas assuming photospheric conditions dominated by disk accretion one sees a slight delay in the onset of deuterium shell burning and thus less radius expansion.

6.1.6 Protostellar Luminosities

The total emitted luminosity of a protostellar system can be written as:

$$L_{tot} = L_{star} + L_{acc} \tag{6.3}$$

where L_{star} is the net luminosity emitted from the protostellar surface and produced by deuterium burning L_D, and contraction or expansion L_{gc}:

$$L_{star} = 4\pi R_{star}^2 \sigma T_{eff}^4 \tag{6.4}$$

L_{acc} is the luminosity produced by mass accretion and infall and includes contributions from energy dissipation of infalling matter onto the accretion disk and onto the central star and radiative energy losses throughout the envelope and the disk.

In observations of YSOs it is difficult to determine L_{star} for a large fraction of its evolution, because the protostar remains embedded in its natal envelope. Thus for L_{star} in protostars one depends heavily on theoretical models. This fact is nicely illustrated in Fig. 6.5 [398]. For up to about 4 $M_\odot$ L_{acc} is the dominating source of the total luminosity. Note, L_{star} is all the luminosity the star's surface can radiate under steady-state conditions as long as the excess luminosity from deuterium burning goes into radius expansion. Luminosity from radiative energy transport L_{rad} is insignificant unless the mass of the star exceeds about 2 $M_\odot$. Eventually the radiatively carried luminosity is the dominating source. At the condition where $L_{rad} = L_{acc}$ the surface temperature is also high enough to radiate all radiatively carried energy. From this point the star contracts homologously towards the main sequence [398].

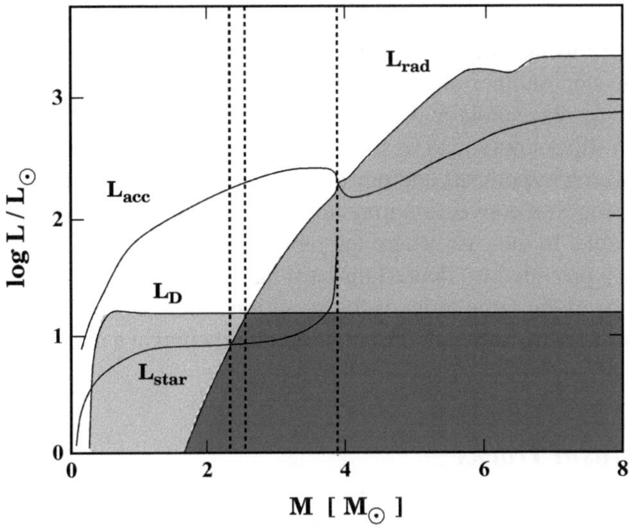

Fig. 6.5 The mass dependence of luminosities within a protostellar system, for an accretion rate of $\dot{M} = 10^{-5} M_{\odot}$ yr^{-1}. For low-mass stars L_{acc} continuously dominates L_{star}. The luminosity from deuterium burning is also much higher than L_{star}. This situation causes the stellar radius to expand. The vertical dotted lines mark the intersections of the radiatively carried luminosity L_{rad} with the functions for L_{star}, L_D, and L_{acc}. Adapted from Stahler [396] and Palla and Stahler [398]

6.2 Evolution in the HR-Diagram

It had been recognized early on that young stars can be identified by their location in the HR-diagram. The HR-diagram is a valuable tool in understanding stellar evolutions. Comprehensive introductions can be found in standard textbooks. The HR-diagram takes advantage of the fact that in the course of stellar evolution only specific paths of luminosity and effective temperature occur. The first calculations of pre-main sequence radiative tracks for contracting stars of less than ~2 $M_{\odot}$ were published in the 1950s [43, 44], although it was not yet obvious that T Tauri stars would meet these conditions [32].

The basic equations for stellar structure have already been introduced in Chap. 5. Ingredients discussed so far include opacities, deuterium burning, lithium depletion, convection, and mass accretion rates. As important as these ingredients are, evolutionary projections need boundary conditions as well as initial assumptions. An obvious condition at the center of the protostar is that both mass and luminosity have to vanish. Another set of conditions determines the dynamic properties of the photospheric surface of the star in terms of pressure and luminosity, i.e.:

$$M = M_{star} \rightarrow P_e = \frac{2}{3} \frac{GM_{star}}{R^2_{star}\kappa} \qquad L = 4\pi R^2_{star} \sigma T^4_{eff} \tag{6.5}$$

These conditions reflect some implicit assumptions that are central to how these calculations are to be interpreted. One is that there is thermal equilibrium at the surface of the star, another one is that the star should be fully convective. The first condition seems reasonable and fairly consistent with observations though events like accretion shocks may lead to some departure from LTE. The second condition is more a working hypothesis and presumably not always valid. In stars much more massive than the Sun convection may not be as efficient as the star becomes more radiatively stable. In the case where the radiative luminosity L_{rad} is of the order of the luminosity provided by deuterium burning, the contraction timescale t_{KH} (see (A.38)) may be of the same order as the accretion timescale (see (A.39)) for a small range of masses [346], although observations clearly hint at a more complex picture.

6.2.1 Hayashi Tracks

C. Hayashi [399] already emphasized that in order to describe the evolution of protostars, its is useful to follow the rate of energy change at every point in the density–temperature diagram. He demonstrated that any star of a certain mass and radius has a minimum effective temperature. Simple arguments, for example as introduced by [399–401], show that there are zones of instability in the HR-diagram, which constrain the evolution of protostars. The boundary of these zones of instability is then determined by the evolution of a fully convective star.

PMS stars of masses $M < 2\ M_\odot$ are considered fully convective. The HR-diagram can be used to display the development of the equation of state of the protostar. The zones of instability in this respect are areas where the first adiabatic exponent Γ_1 (see (A.9) and Sect. 5.2) drifts out of the range of stability. In convective zones this temperature gradient is close to adiabatic. Significant deviations from that condition will not maintain quasi-hydrostatic equilibria. From (A.7) and (6.5), as well as the above boundary conditions, one finds a relation between luminosity and effective temperature that traces a line in the HR-diagram for a specified stellar mass, called a *Hayashi track*. Though not true evolutionary paths, they represent *asymptotes* to these instability zones, also sometimes referred to as forbidden zones, in the HR-diagram. In Fig. 6.6 this region is located to the right of the tracks between lifetimes 1 and 2.

Of course, these evolutionary tracks should connect early with the outcomes of the collapse phases discussed in Chap. 5 with timescales discussed in Sect. A.9. The beginning of the Hayashi track thus should be at the adiabatic phase C and the track point mostly downwards. However, in the case of the Hayashi tracks it is mandatory that from the collapse of a gravitationally unstable and isothermal cloud a fully convective structure emerges. Under the assumption that all available energy released in the gravitational collapse was absorbed in the dissociation of H_2 and ionization of H and He, Fig. 5.4 indicates that the average internal temperature of the protostar in quasi-hydrostatic equilibrium is between 10^4 and 10^5 K almost independent of stellar mass and that the opacity is very high. Here the protostar

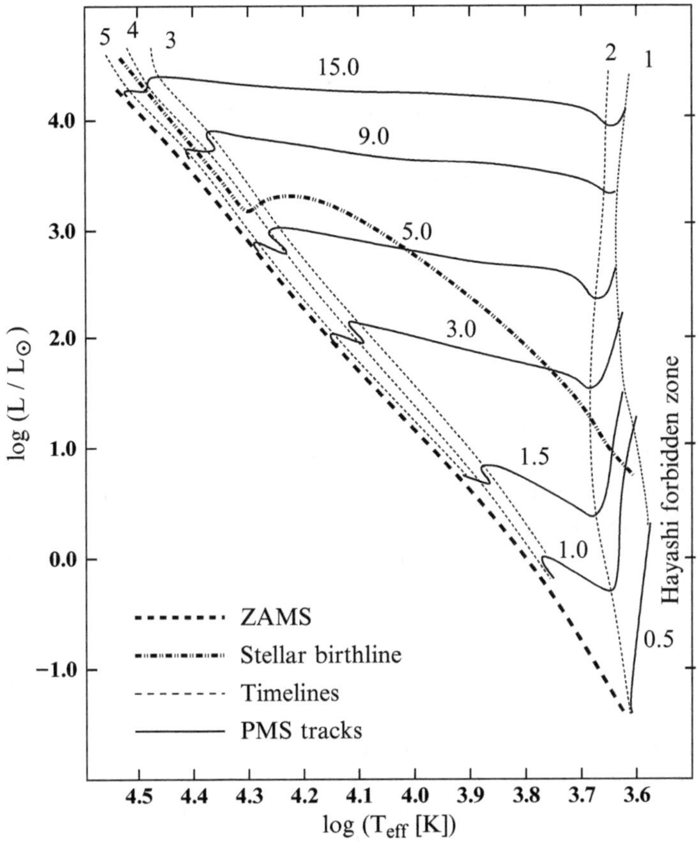

Fig. 6.6 Evolutionary paths in the HR-diagram for stellar masses ranging from 0.5 to 15 $M_\odot$ (solid tracks, adapted from Iben [402]). These paths are marked by thin hatched lines marking time periods labeled 1 to 5. The thick hatched line to the left approximately indicates the location of the ZAMS. The line across the tracks is the stellar birthline approximated from [403] for an accretion rate of $\dot{M}_{acc} = 10^{-5} M_\odot$ yr^{-1}

may indeed be considered fully convective and this is usually where Hayashi tracks have their starting point (timeline labeled 1 in Fig. 6.6). When the opacity drops the internal temperature rises and the convective zone recedes from the center. This causes the evolutionary path of the star in the HR-diagram to move away from the Hayashi track toward higher effective temperatures. The protostar is now on the *radiative track* of the HR-diagram (timelines 2–5, see also Table 6.1).

The details of Hayashi's argument on convectivity and stability have come under some more scrutiny in recent years [404]. The problem is not with Hayashi's argument itself, but with the assumption that stellar collapse phases automatically lead to *fully* convective conditions. These authors stress the importance of including the entire star-formation history in order to account for the true starting point of early stellar evolution (see Chapter 12). In other words, the situation that the initial conditions for early stellar evolution have to be redefined once evolution steps beyond the stellar collapse phases may be fundamentally flawed. Differences in the collapse and accretion history can lead to a significantly different temperature and luminosity evolution. M. Forestini [405] also suggested that a different mixing length in the treatment of convection leads to a shift of the Hayashi track (see below).

6.2.2 ZAMS

The *zero-age main sequence (ZAMS)* is the endpoint of PMS evolution. At the end of the evolutionary track it represents the first time when energy generation by nuclear reactions in the stellar core *fully* compensates the energy losses due to radiation from the stellar photosphere. The onset of sustained fusion defines the end of PMS evolution. The interior of such an evolved PMS must get hot enough not only to ignite hydrogen burning, but to establish a dominant CNO cycle, which requires temperatures above 15 million K. It then marks the end of contraction and the point where C^{12}, C^{13}, and N^{14} reach their equilibrium [390]. The time it takes a PMS star to develop this equilibrium is substantial. Although (see timeline 4–5 in Fig. 6.6) progress in the HR-diagram in terms of luminosity and T_{eff} is marginal, it takes stars of 1 $M_{\odot}$ 16 million years to achieve this.

For massive stars the concept of a ZAMS is specifically intriguing since theories predict that the star should be still accreting. What eventually halts the influx of matter is most likely an emerging wind in these stars, whose existence is a known property of mature massive stars ($M \geq 8\ M_{\odot}$), but whose time of emergence is fairly unknown [390, 406, 407]. Recent calculations (see [346]) are shown in Fig. 6.7 and clearly show very short contraction timescales of massive PMS towards the ZAMS. Results for stars of higher mass are not yet available. Given the steep decline of time towards the ZAMS, stars of about 8 $M_{\odot}$ will have extremely brief contraction times to the ZAMS [390].

Table 6.1 Classical evolutionary PMS lifetimes as calculated by [402] together with more recently derived lifespans for low-mass stars [346]

M $M_{\odot}$	1–2 (Myr)	2–4 (Myr)	4–5 (Myr)	Δt_D (Myr)	t_{rad} (Myr)	t_{cZAMS} (Myr)	t_{ZAMS} (Myr)
0.5	160	–	–	0.25	8.3	160	98
0.8	–	–	–	0.015	2.5	92	52
1.0	8.9	25	16	<0.010	1.4	50	32
1.5	2.4	8.1	3.0	–	–	18	20
3.0	0.21	1.2	0.28	–	–	2.5	2.0
5.0	0.029	0.35	0.068	–	–	0.58	0.23
6.0	–	–	–	–	–	0.42	0.040

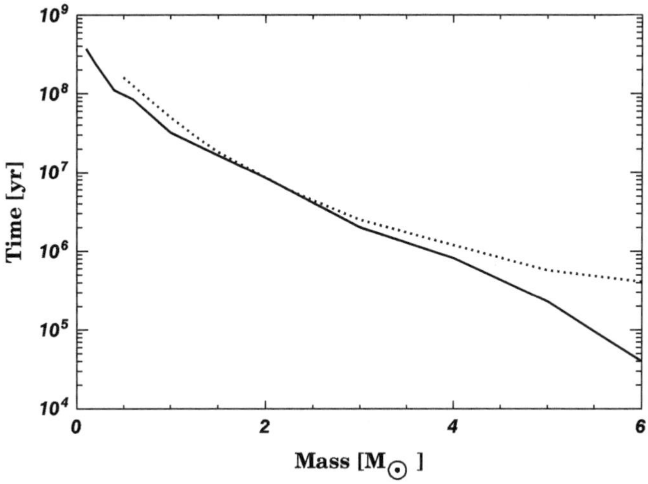

Fig. 6.7 The time t_{ZAMS} it takes a PMS star to reach the ZAMS with respect to the final mass of the star. Classical calculations are shown by the *hatched line*, more recent ones are shown by the *solid line*. Specifically towards higher masses recent calculations show much shorter contraction times (see also Table 6.1)

6.2.3 *The Birthline*

When young low-mass protostars leave the accretion phase and start to quasi-hydrostatically contract towards the main sequence they become optically visible. The location in the HR-diagram where this happens is called the *stellar birthline* [408]. This is not to be confused with the *zero age* of a star introduced in Chap. 5, which is the *beginning* of the accretion phase (phase D). The birthline thus is the first light (i.e., $\tau = 1$ at its surface for $\lambda \sim 5{,}000$ Å) of what is generally referred to as a PMS star. The problem is that there are many exceptions to this concept as will become clear in the further course of the book.

The determination of the stellar birthline has two aspects. The first one is that it marks the end of the stellar collapse phase at which the protostar more or less has reached its final mass. It is thus sensitive to the mass accretion rate. The second aspect has to do with the fact that observationally there should not be any PMS star visible with effective temperatures and luminosities above this line [346].

Figure 6.6 shows the approximate position of the birthline according to [398] and [403] for a thermal (constant) accretion rate of 10^{-5} $M_\odot$ yr^{-1}. For PMS stars with masses up to around 2 $M_\odot$, the birthline cuts through the Hayashi track. These PMS stars are fully convective when they are first observed. This location marks the separation between the T Tauri and Herbig Ae/Be birthline [390]. Higher mass PMS stars already have radiatively stable cores. Calculations show [403] that at masses of 2.6–2.75 $M_\odot$ sufficiently high temperatures are reached to allow for the appearance of a radiative barrier. For stars with masses above 4.3 $M_\odot$ the star at the birthline is thermally relaxed and becomes fully radiative. Finally, at above about 7.1 $M_{\odot}$the birthline attaches to the ZAMS indicating that these stars do not possess a PMS phase and may already be on the main sequence once they become visible.

The birthline also depends on the mass accretion rate. As can be seen below, an accretion rate of 10^{-5} $M_{\odot}$ yr^{-1} seems quite appropriate for low-mass stars. This may not be true for intermediate and higher mass stars. In general, a lower average accretion rate would move the birthline below the one indicated in Fig. 6.6; a higher rate would move the curve above this line [398]. Interesting in this respect is the point where the birthline touches the ZAMS. Young massive stars that would appear to the right of the ZAMS then may indicate higher accretion rates.

6.2.4 PMS Evolutionary Timescales

Knowledge about intrinsic properties of YSOs relies almost entirely on theoretical stellar models. From these models ages, masses, and stellar radii are derived given that the brightness, distance, and effective temperature of the YSO is known. The so-determined ages constitute the only available evolutionary clocks for tracing the development of the star towards the main sequence [409]. Thus, it has to be stressed that there is no simple estimate of various PMS lifetimes and one always has to rely almost entirely on calculations. Classical PMS timescales [402] predict lifetimes on the Hayashi track between well above 100 million yr for very-low-mass stars and only a few 100 yr for stars more massive than 10 $M_{\odot}$. While the latter is simply a reflection of the fact that massive stars are more effective in stabilizing radiatively, timescales of several 100 million yr, however, indicate high convective stability (see Fig. 6.7).

6.2.5 HR-Diagrams and Observations

Evolutionary paths in the HR-diagram play a vital role in the modeling of the contraction of a PMS star toward the main sequence and it is therefore warranted to engage in a few more details. Most pioneering work in recent times has been performed by F. Palla and S. Stahler. Various differences in theoretical calculations will have an impact on the comparison with observations. Besides the already presented classical [402] and recent [390, 397, 398, 410] calculations, other computations. apply various differences in the treatment of composition, opacities, convection, mass range and other physical parameters [403–405, 409, 411–413]. These differences in general either shift PMS tracks or slightly change their length and as a result move the ZAMS, the birthline and isochrones (lines of similar age from the birthline). The spacing, for example, of evolutionary tracks in Fig. 6.6 varies across the $[T_{eff}, L_{bol}]$ plane causing significant differences for predicted masses in different domains. Corrections of PMS ages can be substantial for intermediate mass objects, and for all stars younger than a few million years [410]. Many issues concerning the comparisons of evolutionary predictions with observations will also appear in Chap. 11 which mostly deals with observational details of active star-forming regions.

Observations first produce color–magnitude ($[E - V, M_V]$) diagrams (see Fig. 2.4) which are then transformed into the $[T_{eff}, L_{bol}]$ plane of the HR-diagram. In general this can be done from any applicable photometric band. Typical errors of well observed quantities in nearby star-forming regions translate into typical uncertainties in [346]: for T_{eff} they typically amount to ± 300 K and for $log(L_{bol}/L_{L_\odot})$ to ± 0.25. In many other cases uncertainties are worse. Misidentifications as single, binary, or even multiple star systems occur quite often and are sources of systematic errors. Unidentified stellar binaries may imply too high bolometric luminosities shifting the data upwards in the HR-diagram. Similarly, the bolometric luminosity can lead to incorrect identifications if various emitting components contribute to the observed luminosity. These issues have implications on the determination of the birthline and age of the stars. The sketch in Fig. 6.8 demonstrates an example of possible ambiguities in the comparison with results under very good conditions. In general the comparison of computations to observations is difficult and it is wise to compare observations with many different calculations before drawing conclusions. It has to be cautioned that the birthline concept is not unbiased with respect to observations. The problem is that other

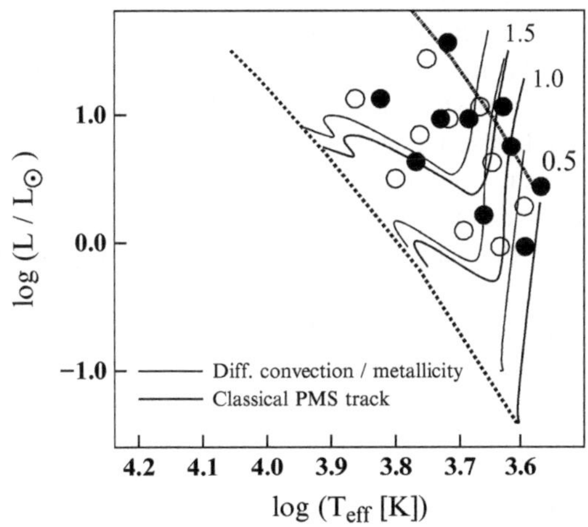

Fig. 6.8 Classical evolutionary tracks for masses 0.5, 1.0, and 1.5 that have been shifted according to [405] for different treatments in convection and metallicity. The ZAMS and the birthline are drawn as in Fig. 6.6. Overplotted is a set of arbitrary data points (open circles). The solid data points show the same sources with their color indices systematically underestimated and assuming that 60 percent of the sources were unknowingly binaries. Not only would the measured birthline of stars be in a different place, but discrepancies in mass determination would amount to 0.5 $M_\odot$ for a solar mass star

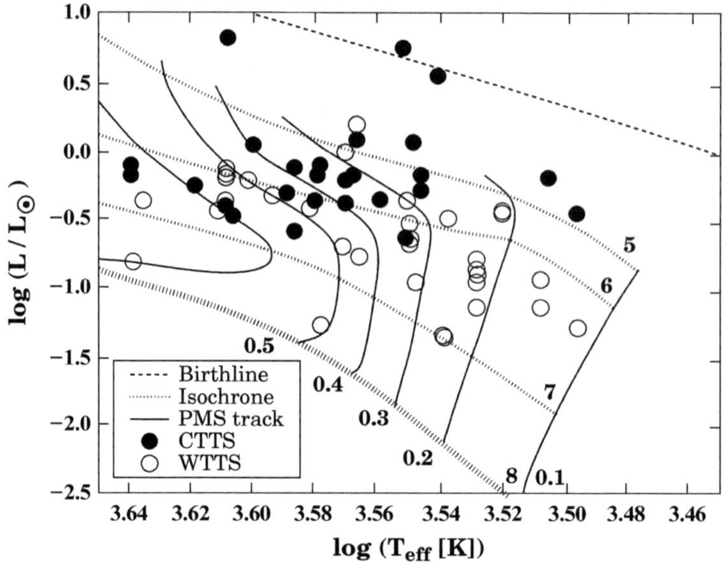

Fig. 6.9 T Tauri stars from the *Taurus-Auriga* region in the HRD diagram (data from [414]). The diagram shows stellar surface temperature and bolometric luminosity of T Tauri stars with respect to the stellar birthline [398] and theoretical PMS tracks [411]

effects can affect the visibility of PMS stars through obscuration by its parent molecular cloud or its host embedded cluster.

PMS stars should not be visible before they reach the birthline, which is pretty much the case for T Tauri stars as well as Ae/Be intermediate mass stars [398]. It is particularly emphasized that a birthline computed for a accretion rate of $10^{-5}\ M_{\odot}\ \mathrm{yr}^{-1}$ represents the best placement for the visibility cut-off in the HR-diagram. Fig. 6.9 illustrates this for the case of T Tauri stars in the *Taurus–Auriga* region [414]. Though the data have not accounted for possible misplacements due to binaries in the sample, all data points lie below the birthline within their uncertainties. The diagram also shows that most T Tauri stars line up around the isochrone representing an age of 1 Myr.

6.3 PMS Classifications

Classification schemes are never really accurate, though they are useful to test theories. Observers try to classify large amounts of data using statistical communalities of their properties. In 1984 [415] investigated *spectral energy distributions (SEDs)* of IR sources observed in the 1–100 μm wavelength band in the core of the Ophiuchi dark cloud. They found that these sources could be divided into morphological classes based on the shape of their spectra. In 1987 C. Lada identified

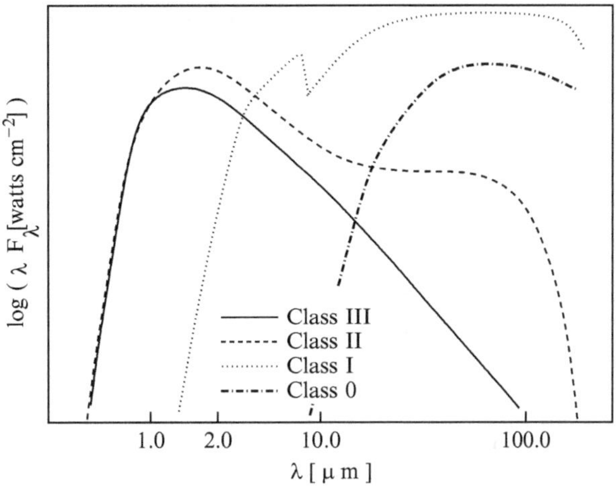

Fig. 6.10 Proposed classification of PMS stars based on SEDs in the 1–100 μm wavelength range

three classes for which he proposed a general IR classification scheme. The basis for this classification is the level of long wavelength excess with respect to stellar blackbody emission (see Fig. 6.10). Quantitatively the spectral index is defined as:

$$\alpha_{IR} = \frac{\mathrm{d}\,\log(\lambda F_\lambda)}{\mathrm{d}\,\log \lambda} \tag{6.6}$$

between 2.2 and 25 μm. *Class I* sources have very broad energy distributions with spectral indices of $\alpha_{IR} > 0$, *Class II* sources have $-2 < \alpha_{IR} < 0$ and *Class III* sources have $\alpha_{IR} < -2$. The large IR excesses are attributed to thermal emission from dust in large circumstellar envelopes and Class I sources are likely to be evolved protostars. In 1993 [60] discovered embedded sources that remained undetected below 25 μm indicating significantly larger amounts of circumstellar material than in Class I sources and they proposed a younger *Class 0* of YSOs. UPDATE on Class O stars: Greaves&Rice2011.

Much more moderate IR excesses than in Class I sources are seen in Class II sources and even less in Class III. Figure 6.11 presents a rough overview of the possible physical properties of these classes, which will be discussed in more detail in the next few sections. This illustration tries, on a best effort basis, to synchronize these IR classes with physical evolutionary phases of YSOs. Reality is more complex than that, and the separations with respect to the phases are much fuzzier. In other words, the evolution of a protostar itself is not traceable because its environment is not transparent for radiative signatures. In fact, many of the physical mechanisms producing emissions are decoupled from the properties of the central young star. Specifically, the morphology of IR spectra can only give a qualitative measure of the amount of circumstellar material, because at shorter wavelengths (i.e., <100 μm) disk emission is optically thick leading to poor mass estimates [62].

The SEDs cover a mass spectrum of YSOs between 0.2 and 2 $M_\odot$, which certainly includes the majority of YSOs in the Galaxy. YSOs of masses outside that range may show similarities but many other properties need to be addressed additionally.

	Infalling protostar	Accreting protostar	Contracting PMS star		MS star
YSO properties	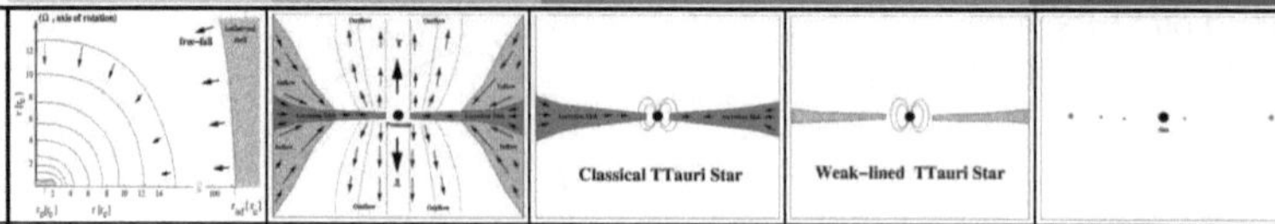				
Phase	adiabatic (A,B,C)	accretion (D) deuterium burning onset of convection	convective radiative onset of nuclear burning		convective radiative full nuclear burning
Matter flows	mostly infall disk & outflows form	some infall mostly accretion outflows, jets	low accretion	?	—
Envelope/ disk size	< 10000 AU	< 1000 AU	< 400 AU	~ 100 AU	—
Infall/ accretion rate	10^{-4}	10^{-5}	10^{-6} -- 10^{-7}	?	—
Age	10^4 - 10^5yr	10^5 yr	10^6 -- 10^7 yr	10^6 -- 10^7 yr	—
Emission bands (except IR)	thermal radio X-ray?	radio X-ray	radio optical strong X-ray	non-therm. radio optical strong X-ray	non-therm. radio optical X-ray
Classes	Class 0	Class I	Class II	Class III	ZAMS

Fig. 6.11 The IR classification in the context with evolutionary phases and matter flow parameters

6.3.1 Class 0 and I Protostars

The notion that the SED classes are part of an evolutionary sequence was suggested early on by [416] and [417]. It was argued that Class I sources are low-mass protostars building up mass by the *accretion* of infalling matter [45]. Thus what is observed is not the total luminosity of the YSO, L_{tot}, but the accretion luminosity of the circumstellar envelope. It is assumed that Class I objects are heavily obscured and thus invisible. The accretion luminosity can be estimated by:

$$L_{acc} = G\frac{M_{star}\dot{M}_{acc}}{R_{star}} \tag{6.7}$$

For accretion rates of the order of 10^{-6} to 10^{-5} $M_\odot$ yr^{-1} the accretion luminosity from this estimate is of the order 10 – 50 $L_\odot$. Observed integrated SEDs of protostars yield luminosities around 1 $L_\odot$[414], which is more than an order of magnitude lower than predicted. According to [290], one possibility for this discrepancy could be that most of the accretion goes into disk growth rather than onto the star, which at some point gets rapidly released onto the star causing outbursts as seen in *FU Orionis stars* (see Fig. 6.1 and Sect. 6.3.5). Another possibility is that the luminosity is radiated through other mechanisms (see Chap. 8).

One of the most luminous protostars known is YLW 15 in the ρ Ophiuchus Cloud [62,418] and from luminous X-ray emission Montmerle and colleagues [383] argue for a braking mechanism involving a magnetic field and the rotation of the protostar. However it is not clear whether this is a common property of protostars or how much it really affects the discussion about accretion rates.

Class I protostars should be rather *evolved* protostars and Class 0 sources are their younger cousins. This perception comes from two observational results. One is that Class 0 sources seem to have significantly higher envelope masses than Class I sources. The other is that different outflow rates and morphologies exist in these sources.

Thermal emission from dust in circumstellar disks is optically thin and proportional to the total particle mass [419]. P. André and T. Montmerle [62] estimate the mass of circumstellar mass using the flux density over the entire observed source extent and relation for the emissivity under LTE conditions:

$$M_{envelope} = \frac{S_\lambda d^2}{j_\lambda} = \frac{S_\lambda d^2}{\kappa_\lambda B_\lambda(T_{dust})} \tag{6.8}$$

Here κ_λ is the dust opacity, and $B_\lambda(T_{dust})$ the blackbody spectrum at a dust temperature of T_{dust}. Sub-millimeter observations of a large number of YSOs in the ρ *Oph* cloud yield an average dust opacity of $\sim$ 0.01 cm^2 g^{-1} [420] and a dust temperature around 30 K. Circumstellar masses for:

- Class I objects are between $\sim$0.015 and $\sim$0.15 $M_\odot$; and
- Class 0 objects are between $\sim$0.14 and >2.8 $M_\odot$.

In Class I objects, a large fraction of the envelope has already been accreted onto the protostar. Fig. 6.12 shows that measured envelope masses [421] place Class I sources just beyond the expected birthline [396], whereas Class 0 sources have not reached the birthline. It should be noted that there is some uncertainty in the determination of envelope masses from millimeter flux densities mainly because contributions to the dust opacity are not well understood (see Appendix F). The birthline [396] seems to favor an accretion rate of 10^{-5} $M_\odot$ yr^{-1}, which is higher than predicted in the standard collapse model [284], but not unreasonable when compared to more recent detailed calculations (see also [323]).

Another observational result indicates that outflow activity, though still quite potent, appears more dilute with larger opening angles and lower mass-loss rates. Class 0 stars are clearly younger and contain much larger circumstellar envelopes. Though the detailed mechanisms that drive jets and outflows in YSOs are quite poorly understood, it is predicted that while matter is still infalling at a high rate and while accretion flows are substantial, significant outflow activity is created. In this respect *all* Class 0 and I sources should show outflow activity. It has been pointed out [422] that in some cases Class I sources may be misclassified due to an overestimate of the bolometric luminosity. This may be for several reasons. Either these objects are embedded in large and dense clouds, or lie behind those clouds, or likely fail the prerogative that all observed luminosity is indeed due to

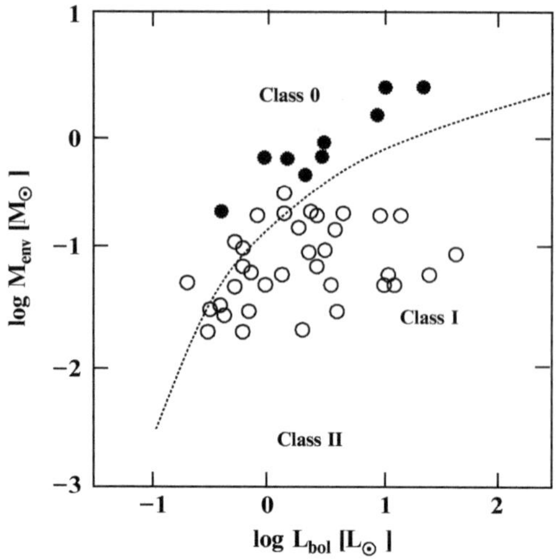

Fig. 6.12 Circumstellar envelope masses for different SED classes versus the stellar luminosity. The hatched line represents the birthline. Adapted from [421]

accretion from the envelope. From the J = 2–1 line of ^{12}CO observations shown in Fig. 6.12 [421] also suggest that Class I sources possess significantly lower outflow momentum as a result of the decay in the mass accretion rate.

These properties can be made visible specifically in IR color–color diagrams. The sensitivity towards disk and envelopes is particularly emphasized in the bandpasses available to the *IPAC* on board the *SST* due to their close placement towards a 10 μm silicate emission feature, which is strong in the presence of large amounts of circumstellar material. The *IRAC* color–color diagram shown in Figure 6.13 distinguishes between stars with only disk emission (see below) and with circumstellar envelopes [424]. The models involve mass accretion rates ranging from 10^{-6}–10^{-9} $M_\odot$ yr^{-1} and disk radii between 100 and 300 AU.

The detection of high-energy radiation from Class I and maybe even Class 0 sources is a rather new phenomenon (for more details see Chap. 10), but of enormous importance as it extends the detectability of these objects into domains that are optically thick even for sub-mm and mm observations [71].

6.3.2 *Classical T Tauri Stars*

The original detection of young low-mass stars was based on objects that exhibited strong hydrogen Balmer (Hα) line emission and were associated with reflection nebula and molecular clouds [39]. These stars showed similar physical characteristics to stars like T Tau in the Taurus cloud and thus were named *T Tauri* stars. Even now, 60 years after their detection, they are still prototypes of what is referred to as *classical*

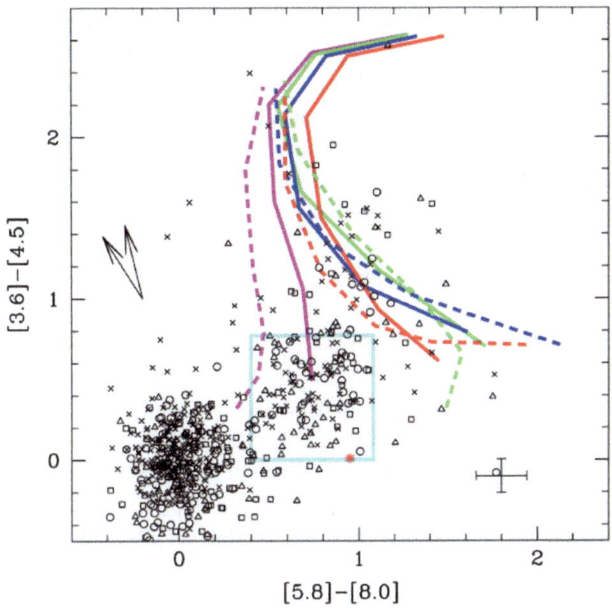

Fig. 6.13 IR colors of YSOs from various star-forming regions observed with *SST (IRAC)*. The data are from various surveys (i.e., *S140* (*circles*), *NGC 7129* (triangles), *Cep C* (*crosses*)) [423]. Class III sources accumulate at the lower left corner of the diagram, class II sources are marked by the light blue square. Class I and younger sources scatter in the middle of the diagram. The color lines depict results from various disk and envelope models for class 0 and I stars [424]. Shown are model tracks for two envelope radii at 50 AU (solid lines) and 300 AU (dashed lines) as well as a range of central source luminosities at 0.1 (*magenta*), 1 (*green*), 10 (*blue*), and 100 (*red*) $L_\odot$. From Allen et al. [424]

T Tauri Stars (CTTS). Other early objective-prism surveys were performed through the 1950s and 1960s [40, 425–427]. T Tauri stars are generally low-mass PMS stars of spectral types F to M and surface effective temperatures of 3,000 K to 7,000 K. Other more sensitive Hα surveys using objective prisms have been performed since then providing a huge database for the study of optical and near-IR properties of PMS stars (see also Chap. 11). This is in clear contrast to the detection of Class 0 and I sources, which required more advanced IR and sub-mm instrumentation to be detected and which therefore have only been identified comparatively recently. Throughout the 1980s the term *T-association* for young low-mass stars in star-forming regions was created in analogy to the *OB-associations* of massive stars. Notable surveys during this era were [51, 52, 428–434] and in the 1990s [435–439].

CTTS are optically visible as they only possess a very small fraction of their natal envelope (Fig. 6.12 and beyond the birthline on the Hayashi track in the HR-diagram). Their Kelvin–Helmholtz timescales are significantly larger than their accretion timescale and therefore all contain low-mass PMS stars. In the SED classification CTTS are Class II sources and their energy spectra peak between

1 and 10 μm with moderate IR excesses reaching beyond 100 μm. Several large IR and sub-mm surveys of T Tauri stars have been carried out in the last fifteen years [59, 440, 441] and established that classical T Tauri stars have strong circumstellar disks [442] with accretion rates that in some cases may still be as high as 10^{-5} $M_{\odot}$ yr^{-1} for a short amount of time, but otherwise are generally much lower than 10^{-6} $M_{\odot}$ yr^{-1} (see Fig. 6.1). In the IR color–color diagram (see Figure 6.13) these sources place themselves more towards the lower left corner marked by the colored square.

The transition from a protostellar to a PMS stellar system involves two radical changes which do not happen on the same timescale. The first one is the onset of deuterium burning during the protostellar phase and the star at some point becomes convective, which together with stellar rotation provide conditions that under various circumstances generate an internal dynamo leading to external magnetic fields (see Sect. 8.3.1). The second one is the drop in accretion rate at the end of the accretion phase causing a variety of physical processes which contribute to the total luminosity of the CTTS. These events do not happen synchronously and some processes, specifically ones that relate the stellar magnetic field to matter flows should occur in Class 0 or I sources as well (see above). Some of the principal topological features of CTTS are illustrated in Fig. 6.14. Somewhat more detailed descriptions will follow in the upcoming chapters.

Accretion disks in CTTS have masses between 0.0003 and 1 $M_{\odot}$ with a typical median of $\sim$0.01 $M_{\odot}$. Disk radii are much less than 400 AU, typically around 100 AU [59, 443]. There are several physical processes generating disk emission. Though relatively small in total mass, dust in the outer, optically thin circumstellar disk efficiently absorbs optical and UV light from the central region and irradiates this energy in the form of reprocessed long-wavelength photons throughout the radio, sub-mm and far-IR bands. The wavelength, in in this respect, correlated with disk radius. Far-IR bands probe disk radii of $\sim$1–5 AU, while sub-mm and radio bands reach into the outer edge of the disk [444]. Dust on the surface of the optically thick central and inner parts of the disk get heated up to 1,000 K at the inner edge of generating IR-radiation near 3 to 10 μm. Viscous dissipation would contribute to the IR emission, though radiation from the central star seems to be the dominant process [414]. Therefore mass accretion rates are determined from accretion flows near the stellar surface rather than from disk emissivity [439].

Matter that moves closer to the stellar surface (i.e., beyond an inner radius of 3–4 stellar radii) gets heated to higher temperatures where it dissociates and ionizes. The size of this inner region does depend on the stellar mass (see Fig. 6.4), but is usually of the order of 0.1 AU. It is entirely dominated by the *stellar magnetosphere* which disrupts the inner disk and forces the material flow into *accretion streams* and *accretion columns*. Accretion streams are the production site of the optical broad emission line spectrum containing the strong Balmer Hα line [445, 446]. Accretion columns are the parts of the accretion stream where the flow is directed down onto the stellar surface basically in free-fall. Here *accretion shocks* in the free-falling flow with velocities of 150 to 300 km s^{-1} may heat plasma up to 10^6 K giving rise to soft X-ray emission [447]. In addition, excess emission is thought to arise from

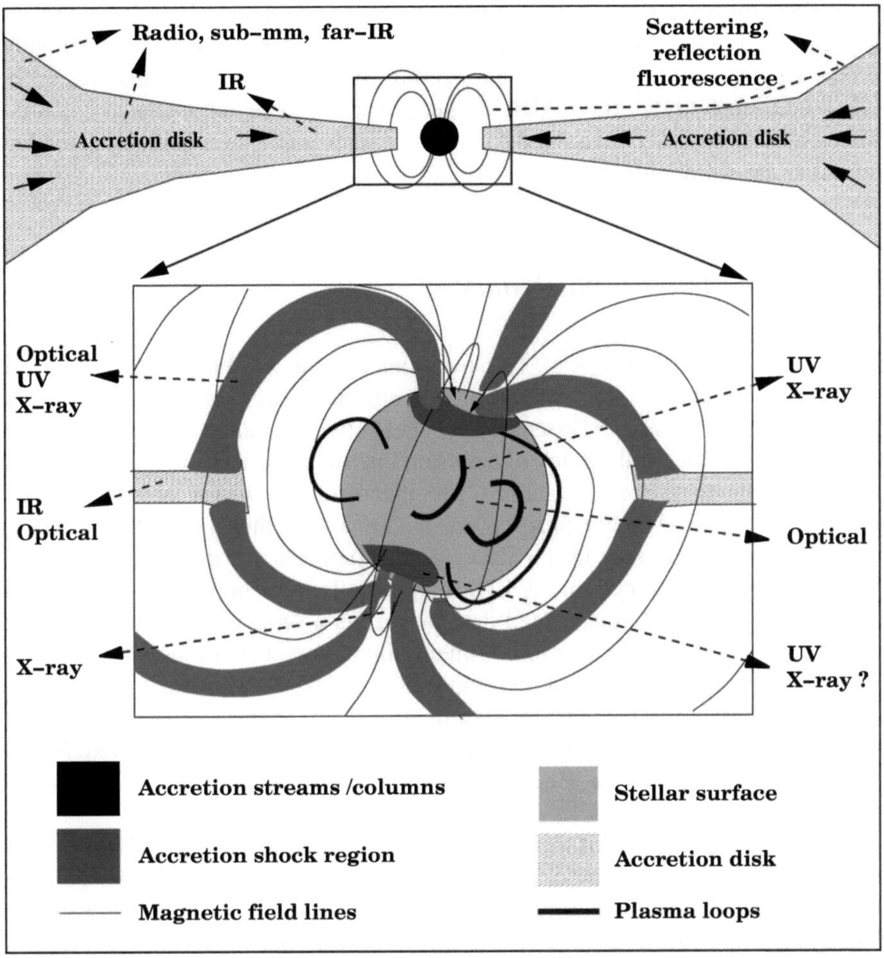

Fig. 6.14 Schematic impression of a classical T Tauri stellar environment. Dominant in size (~100 AU) is an accretion disk. The circumstellar disk emits at radio, mm, sub-mm, and far-IR wavelengths. The accretion disk itself radiates at IR, specifically near 1–5 μm. The radiation from the central source at shorter wavelengths may reflect and scatter at the circumstellar disk. The local environment of the PMS star is characterized by the interaction of the turbulent magnetic field with the inner accretion disk. The stellar magnetic field not only disrupts the inner disk (at ~0.1 AU), but creates accretion streams and columns onto the stellar surface, which may produce accretion shocks. Accretion streams will emit broad-line emission from optical (i.e., H Balmer lines) to X-rays (i.e., O, Ne, Mg, Lyman α). Hot accretion shocked regions may produce continuum emission. The latter has no effect on the effective temperature of the surface, which thus radiates an optical blackbody spectrum. The near stellar magnetic field creates a hot stellar corona, similar to the one observed in the Sun, but with X-ray fluxes that are 1,000 times stronger. Explosive high-energy emission has its origin in magnetic reconnections, which could be caused by rotational twists of the turbulent magnetic field lines

the regions of *accretion shocks* at the stellar surface, which is thought to be the origin of UV continuum emission at temperatures of 10^4 K [290, 439]. Most likely, X-rays are generated through coronal heating, similar to X-ray production in the Sun, but at much higher fluxes [71]. In general it has to be acknowledged that the processes responsible for heating the magnetospheric gas and for generating the line emissions are still highly uncertain.

6.3.3 Weak-lined T Tauri Stars

Many of the optical surveys of T-associations recognized the fact that there are a significant number of stars which, though still associated with the star-forming region, do not show strong Hα emission. These stars are called *weaklined T Tauri stars (WTTS)*. In the third catalog of emission line stars of the Orion population [46] applied a distinction criterion against CTTS in terms of an Hα line equivalent width of less than 10 Å. Another property of WTTS is that they generally no longer show IR excesses and thus in the SED classification scheme are Class III sources. Consequently it is assumed that WTTS is a generally older population of stars than CTTS. Today it is widely accepted that these stars no longer accrete matter from a circumstellar disk. For quite some time a young post-T Tauri stellar population was discussed in the same context and called *naked T Tauri stars (NTTS)* and strong X-ray emission helped find and identify many of these objects [431, 433]. Today it is clear that all NTTS are WTTS, though not all WTTS behave like NTTS, which resonates the fact that Class II and III sources span a wide range of physical phenomena [448].

The current perception of the nature of WTTSs is summarized in Fig. 6.15 and goes as follows. The remnant or inactive disk becomes at some point optically thin at all wavelengths and thus any type of IR excess is orders of magnitude smaller than seen in CTTS. Note that before the disk disperses or evolves further into protoplanetary disks, there may still be a remnant disk in WTTS. However, this disk is clearly decoupled from the central region and no accretion takes place. Herein lies the main difference between WTTS and CTTS. Though from the evolutionary standpoint this difference is not a unique one [449] as there are possible PMS stars that never had an accretion disk beyond the birthline. Another possibility would be that accretion only temporarily ceased, although this has not been observed yet.

The fact that the optical lines, specifically the Balmer Hα line, are weak and show line broadening of much less than 100 km s^{-1}, which is in contrast with the fast accretion streams in CTTS, indicate the different dynamics. Optical lines of low-velocity widths are produced near the stellar surface in the chromosphere of the star [290]. Another consequence is that there is no hot continuum from accretion shocks. This of course does not mean that WTTS are weak emitters. On the contrary, most of the emission comes from the stellar surface and from coronal activity, and specifically high-energy emissions are on the rise.

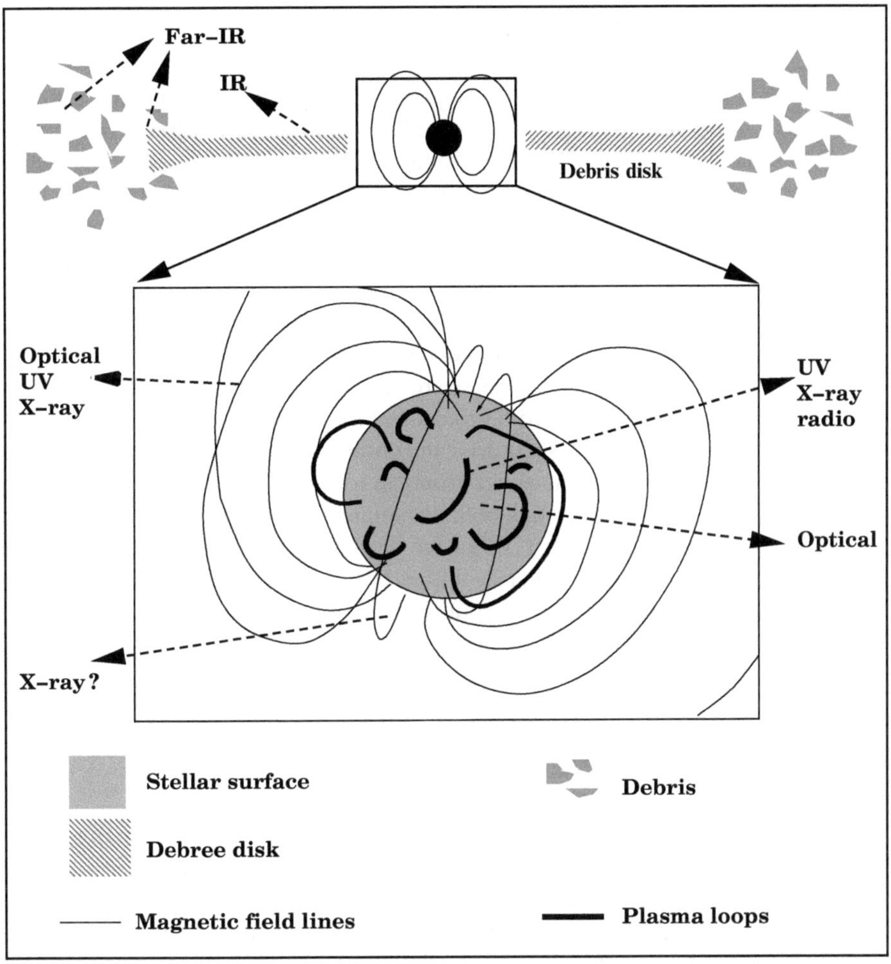

Fig. 6.15 Schematic impression of a weak-line T Tauri stellar environment. The near stellar magnetic environment is now dominant, while the accretion and circumstellar disk may still exist in debris form. Thus the IR excess is orders of magnitude lower than in CTTS, likewise for mm and sub-mm emission, though the circumstellar debris still contains large amounts of radiating dust. This causes emission from molecules (e.g., from ice mantles around dust and debris). Since there are no or only very weak accretion streams, broad H Balmer α is also weak – a fact from which the PMS stellar system historically got its name. The stellar surface radiates its optical blackbody spectrum, and the near-stellar magnetic field still creates a hot stellar corona with strong X-ray fluxes. Through an effect of further contraction the magnetosphere may be a bit more compact and magnetic reconnection events causing X-ray flares may become less frequent. Generally the origins of high-energy emission in CTTS and WTTS are highly uncertain and currently heavily under study

A large fraction of WTTS are actually detected in X-rays as they are, similarly to CTTS, strong X-ray emitters (see also Chap. 10, [67]). With the launch of *ROSAT* [72], wide fields specifically in the Taurus–Auriga region were scanned in the wavelength band between 0.1 and 2.5 Å [73, 450–452].

6.3.4 Herbig–Haro Objects

The detection of *Herbig–Haro (HH)* objects [453–456] happened soon after the discovery T Tauri stars. Quickly it was realized that the nebula found around the star T Tau and HH objects have remarkable similarities. HH objects, though, are associated with much higher mass outflow rates. HH objects are not stars but rather are produced by high-velocity material ramming into the ambient ISM producing powerful shocks [457]. The source producing these shocks are likely jets and winds emanating from a central star. Though in many cases the nature of the central star is uncertain, it is presumably either a protostar or T Tauri star. Fig. 6.2 shows an example of such an object, other examples can be found in [458,459] and references therein. Today hundreds of such objects are cataloged and many of them are also associated with *collimated jets* from protostars [460]. The jet phenomenon itself is directly related to HH objects as jets drive the observed outflows. In some cases both the collimated jet as well as the shock fronts in to the ISM are visible, sometimes only the ISM shock is seen. Though both phenomena are associated with embedded IR sources, not all protostars or CTTS have been associated with HH objects and show jets. From the point of stellar evolution HH objects are important as they on the one hand, allow one to identify young star-forming regions, and, on the other hand, aid the study of outflow dynamics in young stars.

6.3.5 FU Orionis Stars

FU Orionis stars [461] came under special scrutiny when it was realized that these systems had some remarkable properties [462]. These stars undergo optical outbursts of several magnitudes, which classified them as specifically variable young stars. That FU Ori stars are young comes from their spatial and kinematic association with star-forming regions. Their optical spectra indicate spectral types of late F to G with effective temperatures of $\sim$6,000–7,000 K, though their IR characteristics indicate K to M supergiant atmospheres of around 3,000 K. The morphology of these outbursts are quite remarkable as they show rise times of $\sim$1 yr and decay time constants of 50–100 yr. Thus not too many objects of this type are known; so far the best studied objects are FU Ori, V1057 Cyg, V1515 Cyg and BBW 76. The latter object has been found to be as bright as it is today for over 100 years [463]. A most comprehensive account of these stars is given in [290] and only a short summary is presented here. Similar outbursts were known to occur in close binaries as a result of accretion disk instabilities.

Though the FU Ori phenomenon is not of binary nature it was realized that these are very young stars with accretion disks which undergo episodes of increased mass accretion rates [464]. The bright optical emission during outburst episodes is thus not from the stellar surface but from the rapidly rotating inner accretion disk witnessed through the observation of double-peaked absorption lines [464]. To explain the disk temperature one needs to invoke very high mass accretion rates of 10^{-4} $M_\odot$ yr^{-1}. In an evolutionary sequence they are likely to be classified as very young PMS stars as indicated in Fig. 6.1. Recent observations of *FU Ori* and *V1057 Cyg* challenge this perception [465]. It was concluded that a rapidly rotating star near the edge of stability loosing mass via a powerful wind [466], can at least equally account for the observed line emission. Although a disk or stellar-wind phenomenon as the cause is thus still under discussion, very recent observations of line profiles from the FU Ori star *V1057 Cyg* indicate good agreement with disk models. Residual discrepancies were qualitatively explained due to turbulent motions in the disk caused by the magnetorotational instability [467].

6.3.6 Herbig Ae/Be Stars

Intermediate mass PMS stars have masses between 2 and 8 $M_\odot$ and are called *Herbig Ae/Be (HAEBE) stars*. They are of spectral types between B0 and early F and show emission lines and were discovered in the early 1960s lying in obscured regions and illuminating reflection nebulae [468]. *HAEBE*s appear above the MS in the HR-diagram and are the more massive analogs to T Tauri stars [469, 470]. It has to be emphasized that though all *HAEBE*s are young, not all young A or B type stars are *HAEBE*s [471]. Specifically with the advent of *IRAS* the original characterizations of these stars were adjusted to the point that they are stars with IR excess due to hot and/or cool circumstellar dust [472, 473]. This adjustment clearly excludes classical Ae and Be stars [474]. A limitation of *HAEBE* stars to luminosity classes II to V also eliminates confusion in comparison with evolved Be supergiants, which are massive post-sequence stars with strong equatorial outflows.

HAEBE stars generally share similar long-wavelength properties with the low-mass T Tauri stars, though there are some compelling differences caused by various intrinsic physical properties. The t_{KH} timescale is shorter and thus *HAEBE* stars spend much less time contracting towards the main sequence. On the other hand they spend a relatively long time in the prototellar phase before they become visible. This has severe consequences for their intrinsic evolution [398]. Stars in the range between 2.5 $M_\odot$ and 4 $M_\odot$ are only partially convective. Their lower mass limit with respect to T Tauri stars approximately marks the mass above which the stars have already developed a radiatively stable core as they begin to engage in quasi-static contraction. Unlike their low-mass cousins, which are fully convective and have had time enough to relax thermally once they appear optically, *HAEBE* stars undergo a rather critical phase of global relaxation. As a consequence these stars appear *underluminous* at first and increase in luminosity, and, as is illustrated in Fig. 6.4, undergo significant radius expansion before contracting towards the MS. This is shown in Fig. 6.16. While the Hayashi phase is skipped almost entirely, the stars move upwards in the HRD [346]. More massive stars are intrinsically fully radiative. Above stellar masses of about 4 $M_\odot$ stars are intrinsically already fully

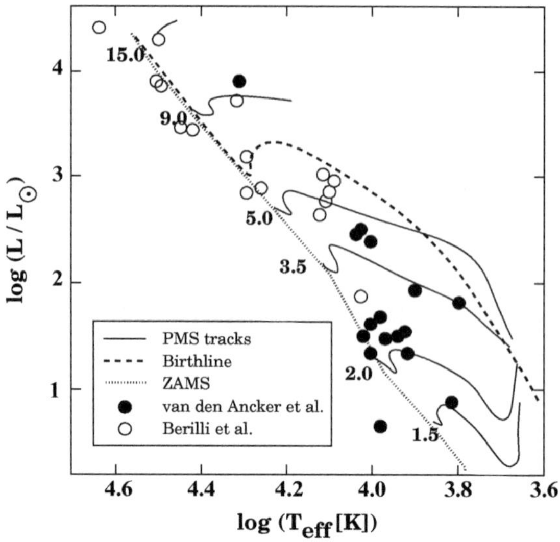

Fig. 6.16 HRD of Herbig Ae/Be stars for two different data sets. The solid symbols are stars with parallaxes measured recently by *Hipparcos* (from [475]). The circles are a few hot and massive stars from a sample from [476]. The PMS tracks are from [398] for stars from 1.5 to 15 $M_{\odot}$, the birthline (approximate) represents an accretion rate of 10^{-5} $M_{\odot}$ yr^{-1} and was drawn from [403]

radiative and start to contract towards the main sequence under their own gravity and all the above happened on a somewhat faster scale before the star becomes visible.

Above about 8 to 10 $M_{\odot}$ stars should burn hydrogen before they could become visible, although here the entire birthline concept becomes obsolete. Stars more massive are almost impossible to predict as there is hardly any way to identify the stars' arrival at the birthline not to speak of arriving at the ZAMS. The study of *HAEBE* stars is in this respect an important link to understand more massive stars and with intermediate masses they serve as an interface between low- and high-mass star formation. There are many issues between theory and observation that are clearly not understood. For example, from the birthlines derived [398, 403], the data in the HRD would not require more than moderate pre-PMS mass accretion rates in line with the rates needed for low-mass stars, which is also corroborated by theoretical considerations [346]. In contrast, from the modeling of SEDs it is indicated that *HAEBE* and more massive stars should have accordingly higher accretion rates [469]. In many respects, further studies of *HAEBE* stars should prove fruitful in understanding the early evolution of massive stars.

Chapter 7
Multiplicity in Star Formation

The previous chapter formulated the characteristics of the formation of single young stars not taking in account effects of fragmentation. That fragmention in critical stages when seed stars form is likely to occur throughout the Galaxy. This implies that the vast majority of protostars and PMS stars are found in multiple systems. Mergers might occur in locations of very extreme stellar densities only as capture cross-sections are hopelessly inefficient (see Sect. 9.4.6, [477]). In a statistical study Zinnecker [478] compared various modes of star formation and concluded that some form of fragmentation process is the dominant mode of binary formation. It is general knowledge that in the solar neighborhood about 60 percent of the stars are part of either a binary or multiple system and there is no reason to believe that this is exceptional throughout the Galaxy. This means that many protostars and PMS stars are multiple systems. In fact, recent surveys of star-forming regions have concluded that this is indeed the case [479,480]. Summaries of a variety of observational results about the binary nature of PMS stars have been compiled [480, 481]. In summary, the ubiquity of binarity suggests that PMS multiplicity must be taken in account in any theory attempting to describe star formation (see, e.g. [482, 483]).

7.1 Observational Account

7.1.1 *Binary Frequency*

Comparisons between PMS and MS binaries are useful indicators of binary evolution. The main categories for such a comparison are frequency, orbital parameters, and mass ratio. To date such comparisons are still relatively rare. An early study in the late 1960s [484] suggested that multiple star systems appear hierarchical which now supported by formation concepts (see below). Measured mass ratios for PMS multiplets are yet rare compared to MS systems and so are binary periods, though the latter seem to have very similar distributions MS stars in some studies [485].

N.S. Schulz, *The Formation and Early Evolution of Stars*, Astronomy and Astrophysics Library, DOI 10.1007/978-3-642-23926-7_7, © Springer-Verlag Berlin Heidelberg 2012

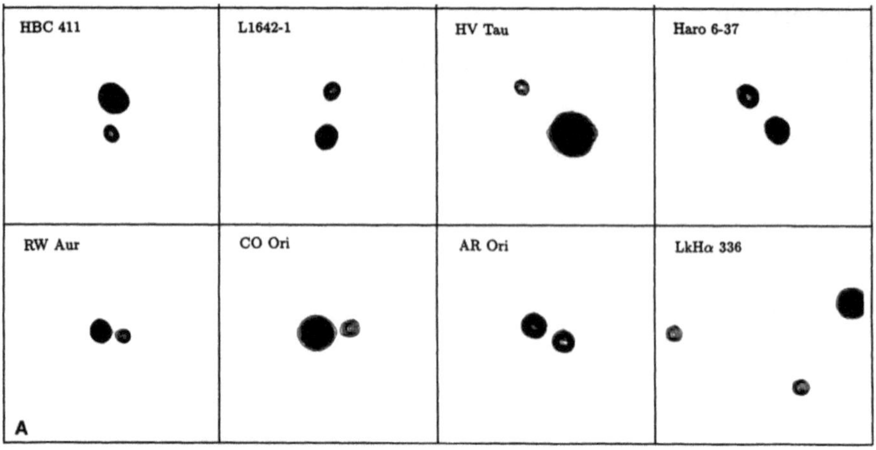

Fig. 7.1 A subgoup of CCD images of visual PMS binaries at 0.9 μm. The size of each field is 13 × 13 arcsec (from [490])

Accounts of the detection of T Tauri star binaries are plenty [486]. In the early 1940s, Joy & van Biesbroeck detected binary companions in young stars such as RW Aur, UX Tau, and S CrA. Up to the mid-1980s several statistics about binarity and multiplicity were available [487–489]. Until the mid-1990, surveys of several hundred T Tauri stars had been published [490–492]. Estimates of binarity in the Taurus–Auriga T association even predict a frequency of nearly 100% [485], twice the amount observed in the solar neighborhood. A study by [490] surveyed visual PMS binaries in the solar neighborhood of the over 200 stars observed almost 25% could be visually resolved. Figure 7.1 shows a some of the images of these visual binaries at 0.9 μm. The binary components exhibit a wide range of separation and luminosity ratios.

There are several different measures to quantify multiplicity. The study by [493] uses what is called the multiplicity frequency f_{mult} which accounts for all subgroups containing singles (s_1), binaries (s_2), triples (s_3) and so on. The sum of all multiplets then is $s_{mult} = \sum_{i>1} s_i$ and the multiplicity frequency is

$$f_{mult} = \frac{s_{mult}}{s_1 + s_{mult}}. \tag{7.1}$$

Studies find (see [493, 494] and reference therein) that values for field G dwarfs can exceed 0.6, while the brown dwarf multiplicity frequency is substantially lower unless they reside in a very tight brown dwarf pair. It is also suggested that roughly 20% of multiple systems are triple systems and higher. [490] investigated the multiplicity frequency in many nearby molecular clouds and found that in many of them, F_{mult} averages about 0.13 ± 0.3, but some at almost 0.35 ± 0.10. The result might indicate a weak relation between binarity and the size of a population of young stars in a cloud. Other studies [495–498] found variations in multiplicity frequency in different star-forming environments. In that study, young binary stars

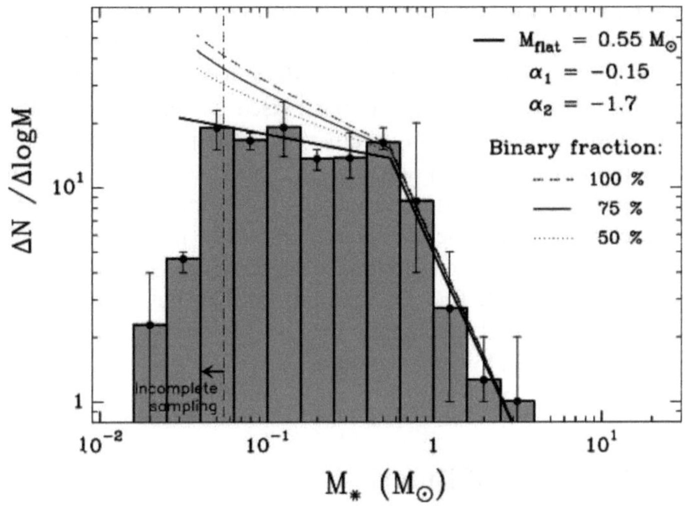

Fig. 7.2 The mass function in the ρ Oph cloud and the effect of binarity to the low-mass slope (from [500])

in large-scale regions appear to be twice as frequent as in confined and clustered regions. The one in clustered regions appears in good agreement with MS binaries. [499] also find in the ONC that the companion star frequency of low-mass stars is comparable to that of main-sequence M-dwarfs, but of half that for solar-type MS stars, and even by several factors lower than in the Taurus–Auriga and Scorpius–Centaurus star-forming regions. The study also found that the binary frequency of low-mass stars hardly varies through the cluster. This contrasts predictions that the low binary frequency in clusters is caused by the disruption of binaries due to dynamical interactions. This could mean that either the initial binary frequency in the ONC was lower than in Taurus–Auriga in the first place. It could also mean that its proto-cluster was much more dense and dynamically active.

The multiplicity frequency also has consequences in the determination of the cluster mass function. [500] quantified such an effect during an analysis of ISOCAM observations of the ρ Oph cloud. Figure 7.2 shows the mass function of 123 Class II YSOs. While neither the break point nor the primary mass function >0.55 $M_\odot$ is effectively altered by the includion of binarity, the low-mass slope significantly steepens with binary fraction and contributes considerably to the uncertainties in the determination of mass functions.

7.1.2 *Binary Separation*

The question how these properties, multiplicity frequency and binary separation of PMS stars, compare with MS stars in different star-forming regions is of interest as any significant variation would impact binary formation scenarios and early binary evolution. PMS binary separations in the solar neighborhood have been measured to be between 15 and and 1,800 AU [490, 495]. It has been acknowledged that the real account of close binaries is rather biased as separations <15 AU, even in regions as close as *Taurus–Auriga*, are very difficult to resolve [480]. In that respect the account of true binary separations is still incomplete, not only because in most cases orbital eccentricities are unknown, but mainly because the majority of PMS binaries have sub-arcseond separations, even in the nearest clouds, The existing surveys [490] find a distribution of separations heavily skewed towards separations <150 AU, [494] find a wide and flat distribution with a mean of 30 AU for G-dwarfs, others find specifically for very low mass stars and brown dwarfs distributions towards much lower separations.

7.2 PMS Properties of Binaries

Most PMS binaries components behave like CTTS and WTTS. Binary components possess circumstellar disks likely even allowing for planet formation. Most binaries are *coeval* [501]. Figure 7.3 demonstrates this by plotting six examples into the HR-diagram. Besides the visible bias introduced by an unresolved binary nature it shows that pairs of primary and secondary stars within measurement uncertainties align well with isochrones. In a previous study only 60 percent of stars in the

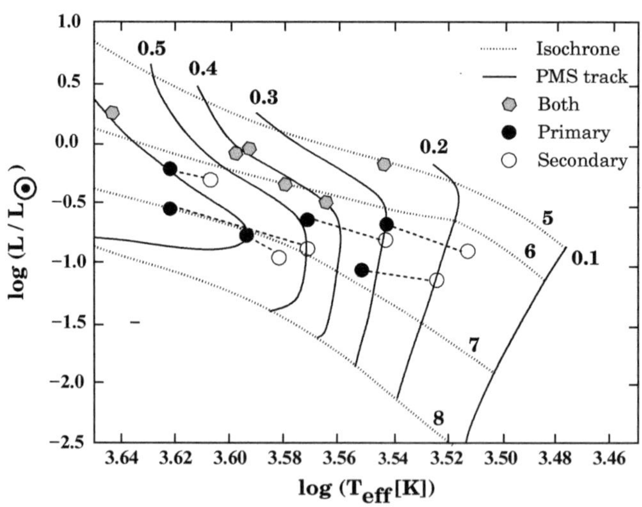

Fig. 7.3 Measured binary stars in the HRD. PMS tracks range from 0.1 $M_{\odot}$ to 0.5 $M_{\odot}$, isochrones from log t [yr] or 5 to 8. The unfilled and solid circles show the HRD positions of the primary and secondary binary members once resolved. The hexagons show the measurement results of the previously unresolved systems

sample were recognized as coeval [502]. Here, however, the study included wide and very wide binaries which are not likely physical but just optical pairings. Surveys of protostellar binaries do not exist yet and only a few such entities have been found, though some of them are of higher multiplicity [503–506]. The separations in surveys so far found a range from 15 AU to 2,800 AU (see below). Radial velocities from *RASS* sources near *Taurus–Auriga* indicate a strong excess of short-period binaries, a result which, however, in part is biased due to a selection effect with respect to X-ray brightness [507]. The following examples illustrate the wide range of properties exhibited by PMS binaries and triples.

Recent studies of members of the *TW Hydra Association (TWA)* recognize the existence of large-separation PMS binaries and one might speculate that there are more such systems to be discovered in other local, sparse and maybe older PMS groups. Two loosely bound binaries have been identified with very large separations in *TWA*, *TWA 11AB* and TWA 11C with a projected separation of ~13,500 AU, and *TWA 1* and *TWA 28* with ~41,000 AU [509]. The very low-mass *TWA 30A* and *TWA 30B* star/disk systems are also weakly bound with a separation of the sky of ~3400 AU [508]. *V4046 Sgr* is a very close (a ~2.4 $R_\odot$) binary of twin 0.9 $M_\odot$ stars in an orbit of 2.4 days and which at an ago of 12 Myr are still accreting [510,511]. Here Kaster et al. (2011, in prep.) now also find a third body at a separation ~13,000 AU interacting with *V4046 Sgri AB* best described via Kozai cycles (see also Sect. 13.3.6). Fabrycky & Tremaine [512] argue that the orbits of a large fraction of binary stars shrank by an order of magnitude after formation through secular perturbations from a distant companion star and tidal friction.

7.2.1 Hen3-600 (TWA 3)

Some of the most IR-bright sources with low luminosity stars are found in the TW Hydra association (TWA) where most stars exhibit optical depths of approximately 0.1 [514], which is an order of magnitude larger than usual in these stars. The triple system Hen3-600 is classified as a CTTS still actively accreting similar to the main single CTTS TW Hya in the association [515]. The conclusion is based on the ballistic infall signature of their broad Hα emission profiles. This is a bit surprising as TWA is about 10 Myr old. Estimates of the disk mass accretion are orders of magnitude lower than in 1 Myr old CTTS. Hen3-600 shows a complex system of dust disks around its components. [516] found that component Hen3-600A, classified of M3 type, harbors a disk of circumstellar dust which may be truncated by the secondary and provides tentative evidence for a central disk hole. [513] (and references therein), however, find from the overall IR SED that Hen3-600A is a spectroscopic binary, surrounded by a dust disk; it orbits Hen3-600B, classified as M3.5, which then again is surrounded by a circumstellar disk. The SED and deduced configuration is illustrated in Fig. 7.4. Based on this assessment Hen3-600 is a hierarchical triple in a classical sense with two disk structures, one associated with the spectroscopy binary component A and a size of about 70 AU, and a

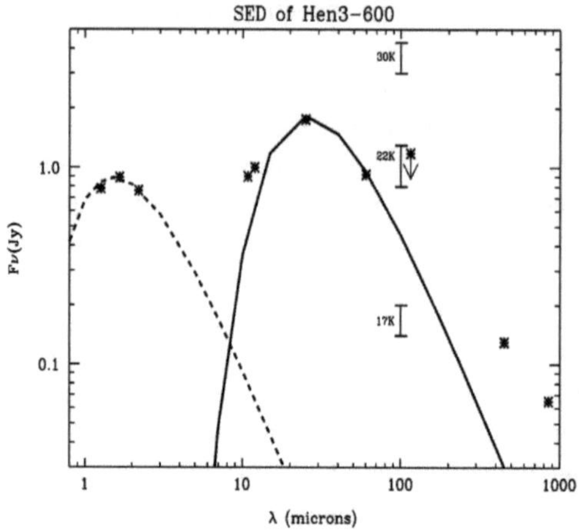

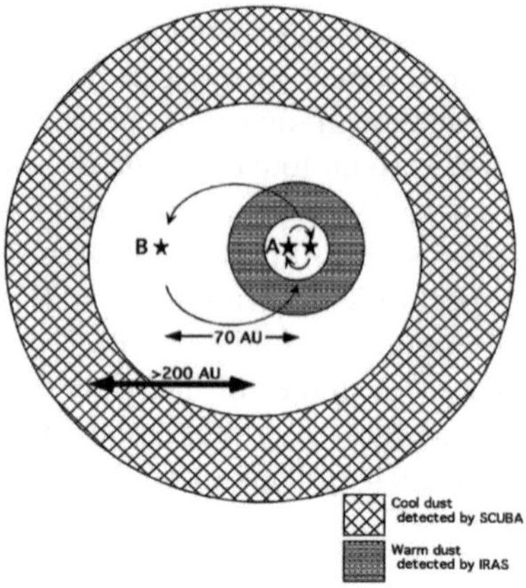

Fig. 7.4 (*Left*) The SED of Hen 3-600. The stellar photosphere is a 3,350 K blackbody (dashed line); a 200 K blackbody is fitted to the IRAS 25- and 60 μm observations. The 450 and 850 μm fluxes measured with SCUBA lie well above the emission expected from the 200 K grains and much cooler grains must also be present. (*Right*) A sketch of the dust distribution in Hen3-600 inferred from the SED analysis. Sizes of the various regions are not to scale. IRAS and mid-IR observations primarily detect the warm dust around the primary, which is a spectroscopic binary. SCUBA reveals the cool dust which, likely, orbits the visual binary (AB) (from Zuckerman [513])

circumstellar disk around the triplet with a diameter around 10 times larger. The inner disk harbors the warm dust which existence had already known since its *IRAS* detection. The outer disk ring is cooler dust observed with *SCUBA* at 450 and 850 μm. The origin of the dust is still debated, but its still relatively young age, its lack of debris signatures (see Sect. 12.1.2), as well as the high-energy emissions argue for a primordial origin. However, [517] imaged the system with the high-resolution *Submillimeter Array (SMA)* at 880 μm to specifically detect thermal emissions from circumstellar dust disks and find no evidence for this outer disk ring. Its broadband emission characteristics from the X-ray, UV, optical (H_α), to the IR bands indicate that these young stars, in particular the spectroscopic binary component A, are in transition between rapidly accreting CTTS and non-accreting WTTS [518].

7.2.2 *DQ Tau*

The DQ Tau binary is a typical Class II system consisting of two almost equal mass ($\sim$0.65 $M_\odot$) and equal radius ($\sim$1.6 $R_\odot$) stars of spectral types in the range from K7 to M1. The young stars have a fast rotational period of about 3 days and are locked in an orbital period of about 15.8 days and its SEDs can be also fit by a large circumbinary disk of about 0.002–0.02 $M_\odot$. In general there is nothing special about this system, except that its orbit is highly eccentric with $e = 0.556$ and the periastron separation of the two stars is only 8 stellar radii. [519]. This has a few interesting consequences. One consequence is that as long as the two stars are not near periastron passage, they indeed behave like moderately active CTTS. Once near periastron passage things change. For more than 65% of periastron or nearby encounters, the system experiences optical brightenings and spectral hardening as a result of variable and irregular accretion [519,520] with recurrences almost identical to the orbital period. Unlike observed in HEN3-600, the IR SED also does not indicate an inner disk clearing and warm material is present near the stars, possibly through inflow and outflow activities. The optical brightening is interpreted as a variable accretion rate which is regulated by the binary orbit. Models also show that circumbinary disk material can stream across an entire binary orbit at certain orbital phases resulting in a pulsed accretion flow onto both stars [521].

These circumstances seem to explain these periodic outbursts solely by accretion [520]. However, it has to be noted that the periastron separation is smaller than the sum of the magnetospheric radii of the two stars and induced magnetospheric interactions have been proposed. [522] argue on the basis of mm flares expected to induce magnetospheric interactions and propose just that based on the rare detection of a flare at 3 mm from DQ Tau at scales and magnetic field strengths comparable to those inferred for a sample giant X-ray flares observed in Orion [523]. The paradigm was put to the test in a search for simultaneous X-ray and mm observations of DQ Tau near periastron phase. Getman et al. [524] caught DQ Tau in a flare state in both mm and X-rays. The inferred morphology, duration, and plasma temperature of the X-ray flare are found to be typical of those of large flares from ONC stars. Accounting for the probability of catching such a flare at periastron passage in connection with its consistency of a coronal origin, proper loop size, and peak X-ray luminosity it is argued that the power dissipated is consistent with magnetic reconnection processes associated with colliding magnetospheres. Such events are still rare and the mechanism of interacting magnetospheres has been proposed for the generation of the giant flares in the massive θ^2 Ori A system [525].

7.2.3 AK Sco

This PMS binary is in many instances similar to DQ Tau, except it is more massive with two near-identical stars of 1.35 $M_\odot$ and radii of 1.6 $R_\odot$ likely of F5 type. Its orbit also has not yet been circularized with an eccentricity of 0.47 leading to a periastron separation of about 11 stellar radii [526]. The system does not exhibit signs of enhanced accretion near periastron, in fact the Hα equivalent width displays smooth variations and flux peaks appear near apastron. There are also no inferred flares expected due to reconnection events in interacting magnetospheres as possibly in DQ Tau, as the periastron separation is much larger than the sum of the magnetospheric radii of the two stars assuming standard PMS field strengths. [527] fit the SED from 350 nm to 1,100 μm with a passive disk model suggesting that the accretion rate is barely of the order of 10^{-8} $M_\odot$ yr^{-1}. Tidal forces pushing the disk outward into a ring and the stellar parameters allow for an inner rim temperature much smaller than the dust evaporation temperature (~1,500 K). Combined with an age of at least 10 Myr, AK Sco and its dust disk are quite similar to the outer ring case of Hen3-600.

7.2.4 *θ^1 Ori E*

The fifth star in the Orion Trapezium for a long time eluded observers of its proper identification, mainly due to interference of scattered light from the other bright trapezium stars. Herbig & Griffin [531] finally discovered it to be a double-lined spectroscopic binary. The system is now identified as an intermediate-mass binary with a period of 9.89 days and almost identical mid-G-type giants with masses of about 3–4 $M_\odot$. The mean age of the Orion Trapezium is <1 Myr and to date this system resembles the youngest of its kind in this mass range. In essence this system eventually will cross the HAEBE region and arrive on the ZAMS as late-type B-stars. Interestingly enough, no eccentricity is detected and its orbit appears already circularized. Figure 7.5 shows that the system lies between the tracks from the models of [528] between the 3–4 $M_\odot$ range. It is also remarkable that there is no excess IR detected, i.e. there are no signs of residual disk emissions. θ^1 Ori E is also the second brightest X-ray source in the Orion Trapezium with an X-ray luminosity of $1.2{\times}10^{32}$ erg s^{-1} which with a bolometric luminosity of 29 $L_\odot$ produces an L_x/L_{bol} of 10^{-3} for the pair [530, 532]. Figure 7.5 also shows

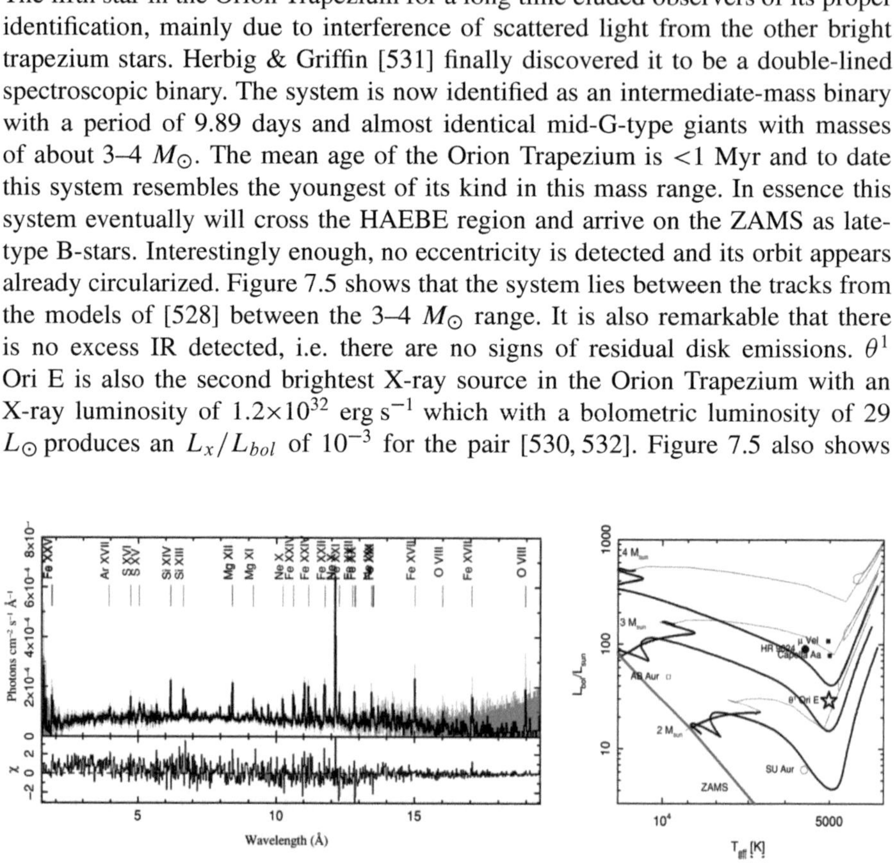

Fig. 7.5 (*Left*) The *Chandra* high-resolution X-ray spectrum of θ^1 Ori E. Some of the most significant inner K/L shell lines are marked. The statistical uncertainty is shown in gray. The lower panel shows the residuals against the hot emission measure model. (*Right*) Stars listed in Table 7.1 are shown on a temperature–luminosity diagram. PMS stars are shown with open symbols and post-MS stars are shown as solid symbols. The shapes encode L_X/L_{bol} as a "star" for $>10^{-3}$, circle for $>10^{-4}$ to $<10^{-3}$, and a square for $<10^{-5}$. The PMS evolutionary tracks (thick gray lines) and ZAMS are from [528]. The post-main-sequence tracks (thin gray lines) are from [529]. Both plots are from Huenemoerder et al. [530]

Table 7.1 Stellar comparisons (see [530] and references therein)

Quantity	θ^1 Ori E[a]	SU Aur	AB Aur	μ Vel	HR 9024[b]	Capella Aa[c]
Spec.Type.	G8 III+G8 III	G2 IV	A0	G5 III	G1 III	G8 III
$T_{\rm eff}$ [K]	5,012	5,550	9,750	5,030	5,530	5,000
Age [My]	0.5	4	4	360	post-ms	525
$M/M_\odot$	3.5	2.0	2.7	3	2.9	2.7
$R/R_\odot$	7	2.6	2.3	14.4	13.6	12.2
$g/g_\odot$	0.07	0.3	0.5	0.01	0.02	0.02
P [day]	9.9	1.8	1.4	117	23.2	104
$v \sin i$ [km s^{-1}]	37	66	80	6.2	22	5
$L_{\rm bol}/L_\odot$	29	6.3	49	108	90	78.5
$L_x/10^{30}$ [erg s^{-1}]	121	8	0.4	2.2	63,125	1.6
$(L_x/L_{\rm bol})/10^{-6}$	1,100	300	2	5.4	181,360	5.3
$VEM/10^{52}$ [cm^{-3}]	1,200	50	5	8	200,800	4
$\overline{T}_x$ [MK][d]	20	20	4.7	7.9	30,80	6.3
$A(Fe)$[e]	0.1	0.6	0.3	1.8	0.2,0.7	0.8
$A(O)$[e]	0.1	0.3	0.2	0.7	0.6,1.1	0.4
$A(Ne)$[e]	0.4	1.3	0.6	1.4	1.1,1.2	0.7

[a] X-ray properties are for the binary system; if each stellar component contributes equally, values should be divided by two.
[b] When two quantities are given they are for quiescent and flare states, respectively.
[c] X-ray properties assume that component Aa dominates.
[d] A qualitative temperature of the high-energy emission measure distribution, adopted from visual inspection of published curves or few-T fits.
[e] Abundances are relative to Solar.

the X-ray spectrum of the system which appears to feature extremely hot coronal plasmas at a very young age. For the X-ray properties it is concluded that as moderate mass stars collapse toward the main sequence, they go through a phase of strong magnetic dynamo generation, very similar or identical to that of coronally active late-type stars which sustain a convective zone and shear-generated magnetic dynamo. Since A-stars are dark or at most quite faint in X-rays [533], at some point, the dynamo vanishes, and the magnetic fields dissipate. AB Aur, which is near the main sequence and relatively faint in X-rays, is a possible future state of θ^1 Ori E. It is also clear from the post-MS objects that a dynamo can be generated as stars of this mass evolve into giants. Table 7.1 and Fig. 7.5 illustrate this point. Of the sample, θ^1 Ori E has the highest relative X-ray luminosity, being as high in L_x/L_{bol} as "saturated" short period active binaries, even though its period is longer than those systems. For θ^1 Ori E's near future step in evolution AB Aur might be considered with smaller L_x/L_{bol}. The presence of a very hot corona and the low probability of disinct flares is common to several evolved giants, such as HR 9024, μ Vel, 31 Com, or IM Peg. While this is not understood, it could be a significant trait of the coronal heating mechanisms.

7.3 Formation of Binaries

7.3.1 Fragmentation and Multiplicity

Formation concepts generally do not directly account for solutions which include bound binary and multiples. Early concepts such as fission of a rapidly rotating protostar have ultmately gravitational failed and it has been shown that gravitional torques ensuing from a bar instability and developing spiral arms can transport excess angular momentum instead. Fragmentation of collapsing cores can lead to the formation of wide binaries but not the close systems where the majority of PMS binaries are actually observed (Sect. 7.1.2).

From the observations described above it seems sensible to assume that binaries form prior to the PMS stage and that there is no random pairing of field stars. Statistical assessments of various formation scenarios concluded that there may be many ways to form binary stars, but ring or disk fragmention should be dominant [478]. Properties like coevality or the higher binary frequency in large regions hardly allow of a different conclusion and a most likely scenario is that of a correlated formation following fragmentation. Physical processes involved in binary fragmentation are not much different to what has been discussed so far, and there are again radiative, gravitational, and magnetic effects that matter. The isothermal phase is most favorable for fragmentation. Specifically the transition between isothermal and adiabiatic phases, with regard to the fragment masses produced during the collapse, proves important as here a critical density is reached [534] (see below). Once non-spherical structures develop and collapse is slowed or maybe even halted, fragmentation can occur [535]. The exact process is still unclear. But it has been pointed out that in the case that rotation becomes important at the final stages of the isothermal phase, the central density distribution flattens into a disk which ultimately fragments [536] (see [481] and references therein).

Fragmentation in conjunction with the formation of massive stars is discussed in Chap. 9 specifically in Sect. 9.4.2. New concepts such as the competitive accretion cluster models implicitely include fragmentation of low-mass cores and subsequent multiplicity. During the very early phases of star formation fragmentation into binaries can happen at various occasions, during any stages of the collapse phase [534,537], but also during the accretion phase in an accretion disk. During the collapse, for example, prolate clouds are statistically more likely to fragment into binaries due to their inherent bar-like structure [538]. Here fragmentation occurs preferentially towards the major axis such that the two ends of the prolate cloud collapse separately but bound resulting in highly eccentric and wide binaries with a variety of mass ratios. Generally it is now thought unlikely that fragmentation during the isothermal collapse phase can form close binaries without some dramatic orbital decay process [539].

Once the heat rate due to compression and the increasing column densities and dust optical depths quench radiative cooling, the gas eventually becomes adiabatic and the core pressure goes from flat to polytropic and heats above 300 K when

H_2 becones rotationally excited. This happens at a critical density of 10^{-13} g cm^{-3} (see Chap. 5). At this critical density the minimum Jeans mass is of the order of 10^{-2} $M_{\odot}$, also called the opacity limit for fragmentation. There is an even lower minimum mass during the opacity phase when H_2 dissociates at a few 10^3 K and which allows further fragmentation (see Sect. 7.3.1). For the formation of multiple systems the rise above the critical density is very sensitive to the polytropic index in the equation of state. On can define a length scale for multiple fragmentation by matching the spherical volume of the core to the critical density [493], leading to

$$R_{scale} < \left(\frac{3M_{core}}{4\pi\rho_{crit}}\right)^{1/3} \sim 125\left(\frac{M_{core}}{[M_{\odot}]}\right)^{1/3} \text{AU}. \tag{7.2}$$

This leads to a minimum separation for the fragmentation limit of 27 AU and the formation of closer binaries needs some dramatic changes in the equation of state or the introduction of a second fragmentation phase.

7.3.2 *Fragmentation Support*

There are many mechanismns to support or suppress fragmentation discussed in the literature and a good review can be found in Goodwin et al. [493] and references therein (see also Chap. 4, Sect. 4.2.4 and 4.2.5). Several processes are discussed involving rotation, turbulence, magnetic fields, and disks with relevance to binary formation.

Rotational fragmentation is based on the argument that rotating spherical cloud with solid-body rotation should generally fragment. Numerical simulations showed that this happens when

$$\alpha_{therm}\beta_{rot} \leq 0.12\text{–}0.15, \tag{7.3}$$

where $\alpha_{therm} = E_{therm}/|E_{pot}|$ is the initial thermal virial energy fraction and $\beta_{rot} = E_{rot}/|E_{pot}|$ the kinetic rotational virial energy fraction. Modes might be added by introducing more prolate geometries, different adiabatic or barotropic equations of state, or choose modified initial conditions such as different denity profiles, differential rotation, variations in external pressures. Most of these modes provide more sensitivity to fragmentation.

Turbulent fragmentation is a rapid and highly dynamic process and has been found to highly support fragmentation (Sect. 4.2.4). However, it is found that turbulent cores generally fragment in several cores where the number of stars that actually form increases with the level of turbulence and it remains unclear how binaries and triples form at sufficiently low turbulence levels.

Fragmentation under the influence of *magnetic fields* is generally thought to be dynamically not important. It has much to do with the fact that at critical densities ionization levels in the gas drop and the field remains detached from the gas and is mostly ineffective. Furthermore it has been shown that even with strong enough fields fragmentation is suppressed rather than supported.

7.3.3 Secondary Fragmentation

Boss [540] studied such a second fragmentation at a later collapse phase when molecular hydrogen starts to dissociate, which could lead to a second collapse phase. However, even though this happened initially, at the end the two seed stars were forced to merge again due to the growth of spiral arms removing angular momentum from the proto-binary.

The general idea is that rotational or disk fragmentation should be able to form binary systems with separations up to the size of accretion disks producing close binaries with separations <500–1000 AU. In an extension to the attempts by [540] to create a close binary during a second collapse phase, Bonnell & Bate [541] investigated the possibility that fragmentation can also occur at the end of this second collapse phase. Two additional constraints were introduced: the initial core conditions at this late phase are that of a pressure- and rotationally supported core with some central condensation and that the equation of state in the second collapse phase reflects the dissociation of H_2 rather than an optically thin gas.

These simulations yield two fragmentation modes. Fig. 7.6 shows the first mode, where fragmentation of a protostellar disk occurs due to rotational instabilities in a

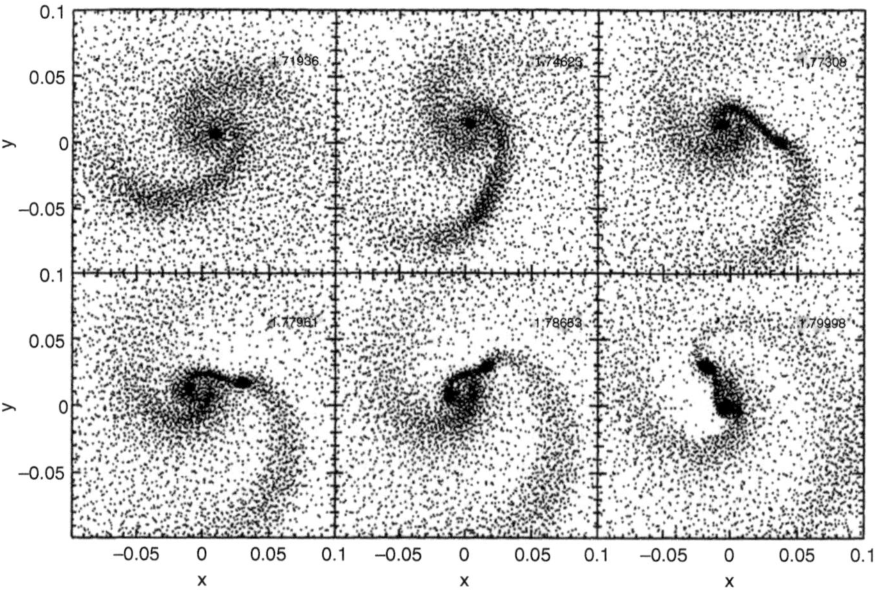

Fig. 7.6 Protostellar disk fragmention due to a rotational instability (from [542])

protostellar core. After about 5×10^4 yr a second core forms which quickly settles and forms a binary of cores with a separation <1 AU. Fragmentation here appears to be relative insensitive to the initial density distribution. A second fragmentation mode involves the the formation of a rapidly rotating core at the end of the collapse phase which is unstable to axisymmetric perturbations. Eventually the core dissolves into a ring which fragments into several components.

It has to be stressed that the process of secondary fragmenation might not always lead to a second seed star as spiral instabilities can always remove enough angular momentum. It is also doubtful that it will provide a mechanism for binary formation much larger than 1 AU separation. It is, however, a rather viable process for very close binaries and brown dwarf systems.

7.4 Mass Transfer in Compact Binaries

Mass transfer has been studied in compact binaries, some of which include symbiotic binaries, dwarf novae and cataclysmic variables in general, soft X-ray transients, low-mass X-ray binaries, and high mass X-ray binaries (HMXBs). Interesting from the star formation standpoint are the latter because not only are they young when observed, their starting point practically is the endpoint of massive binary formation in dense stellar clusters. Furthermore, there are mass transfer processes involved in their formation which have direct relevance to models of massive star formation, specifically with respect to competitive accretion and stellar mergers (see Chap. 9).

HMXBs generally contain a compact object, usually a neutron star or a black hole accreting matter from a massive companion via Roche-lobe overflow or through accretion from a stellar wind or both. Massive stars accreting from a white dwarf companion are very rare as its components evolve on completely contradictory timescales and it is not clear how they form at all. HMXBs containing neutron stars (and black holes), on the other hand, are ubiquitous and here there are plausible mechanisms at hand for how they form. The mere existence of so many of these systems shows that the survival rate, once one or both components engage in a supernova explosion, is rather high. This survival is a consequence of large-scale mass transfer which precedes the explosion and it has been shown that explosive mass ejection from a less massive binary component should not lead to disruption (see Verbunt & van den Heuvel [543] and references therein. Most of the following is drawn from these references (see also Paczynski [544]).

Conservative mass transfer usually happens through the process of Roche-lobe overflow. Rochelobes are equal potential surfaces in a binary and are determined by its main physical properties, the gravitational attraction and the orbital motion of the two stars. The potential then can be written as

$$\Phi = -\frac{GM_1}{r_1} - \frac{GM_2}{r_2} - \frac{\Omega^2 r_3^2}{2}, \tag{7.4}$$

where r_1 and r_2 are the distances from any point in the system to the centers of stars, r_3 is the distance to the rotation axis of the system. Close to each star these surfaces are simply concentric spheres around each star. Further out these surfaces deform and merge into closed Roche lobes. The shape of these lobes only depends on the mass ratio $q = M_1/M_2$. While the exact shape of the lobe has to be calculated numerically, the Roche-lobe radius R_L with respect to the center of mass can be approximated by [545] to

$$\frac{R_L}{[R_\odot]} = \frac{0.49q^{2/3}}{0.6q^{2/3} + \log(1 + q^{1/3})} \frac{a}{[R_\odot]}, \tag{7.5}$$

where a is the orbital separation of the two stars. The mass transfer rate in the case of Roche-lobe overflow has been modeled by [546] to

$$\frac{dM}{dt} = \left(\frac{R}{R_L} - 1\right)^3 \times 10^3 M_\odot \mathrm{yr}^{-1}, \tag{7.6}$$

where $R > R_L$, i.e. the stellar radius R has evolved past the Roche radius, is a necessary condition. Note, that in general, $R/R_L \leq 0.001$ is very small and in standard situations transfer rates are usually of the order of 10^{-6} $M_\odot$ yr^{-1} or below.

There are two modes of mass transfer, conservative and non-conservative transfer. Under the assumption that the binary orbit is circular and in the case that both mass and momentum are conserved, Kepler's laws apply and there is even some small amount of mass loss by some stellar wind, any system will lead to a wide final binary with long orbital periods. For the application of competitive accretion, this is not interesting.

In evolved binaries but also in turbulent scenarios such as competitive accretion systems are subject to substantial amounts mass and angular momentum losses and conservative mass transfer will not work. The mechanism which account for such losses has been decribed by [544] in form of a spiral-in process. Even though much of the details are not fully understood, one might consider the following. The onset of mass transfer in a binary with an initially large mass ratio will cause the separation of the stars to decrease rapidly leading to a common envelope phase. The transfer process ends once the released binding energy through the shrinking orbit is large enough to eject the envelope. Once then can relate the binding energy of the stellar envelope to the difference in binding energies of the binary at the onset and the end of the spiral-in process, i.e.

$$\frac{GM_1 M_e}{\lambda R_1} = \alpha\left(\frac{GM_c M_2}{2a_f} - \frac{GM_1 M_2}{2a_i}\right), \tag{7.7}$$

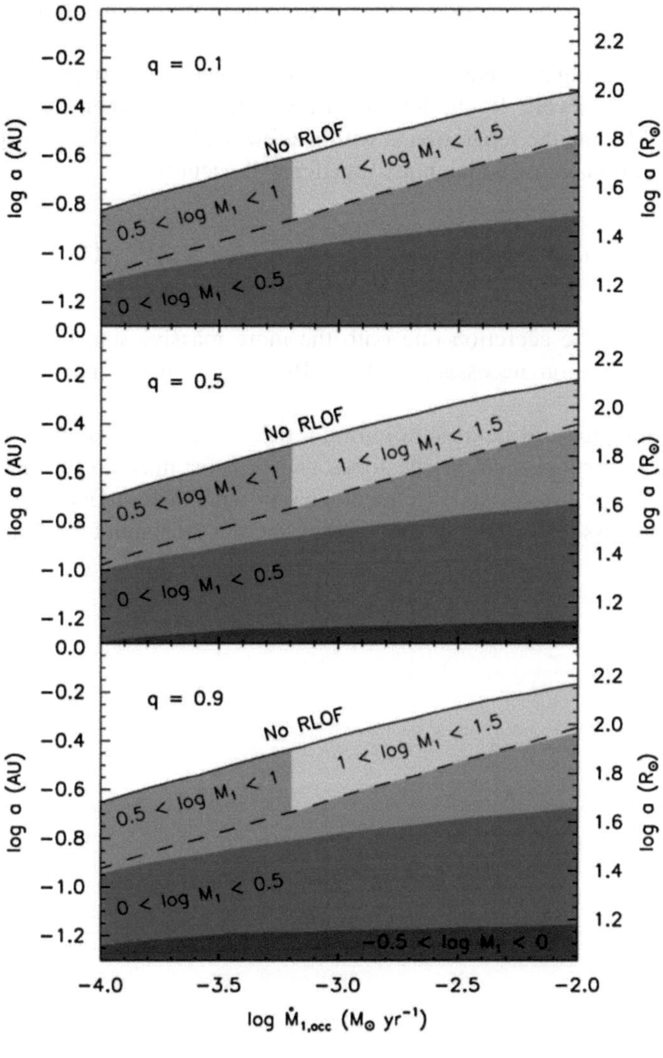

Fig. 7.7 The minimum separation a for Roche-lobe overflow (solid line) as a function of accretion rate ($\dot{M}_{1,acc}$) onto the more massive primary star for three values of q. Below the overflow line in the various shaded regions the range of overflow masses. Binaries with a separation below the dashed line experience overflow before the primary starts deuterium shell burning, while those with larger separations experience overflow after the onset of shell burning (from Krumholz & Thompson [547])

for a donor mass M_1 with a core mass of M_c, and a radius R_1 at the onset of mass transfer, an envelope mass M_e, and a companion mass M_2 also at the onset of mass transfer and initial a_i and final a_f separations. While λ is a weighting factor for the gravitational binding of the envelope to the core. The important factor for competitive accretion concepts is the efficiency factor α which regulates the efficiency by which the released energy is used to in the expulsion of the envelope and equivalent to the tidal factor C_{tidal} in Eqn. 9.12.

Rapidly accreting massive protostars undergo a phase of deuterium shell burning causing them to swell. Roche-lobe overflow will occur when the primary star exceeds 10 $M_\odot$ and the separation is within a few tenths of an AU. [547] estimate a critical separation below which mass transfer will occur to

$$a_{crit} \approx 0.2 \left(\frac{\dot{M}_{1,acc}}{[10^{-4} M_\odot \text{ yr}^{-1}]} \right)^{1/4} (1+q)^{1/3} \text{ AU}, \tag{7.8}$$

where $\dot{M}_{1,acc}$ is the accretion rate onto the more massive star. Figure 7.7 shows the binary separation necessary to have Roche-lobe mass transfer for three q parameters.

During this phase of deuterium burning, those with close enough binary companions will overflow their Roche lobes and transfer mass to the more massive companion. Since massive stars frequently have companions at distances well under 1 AU, this process may affect the early evolution of a substantial fraction of massive stars [547].

Chapter 8
Accretion Phenomena and Magnetic Activity in YSOs

The decades were full of new developments and concepts leading to the present understanding of physical processes in YSOs. While the early stages prior to collapse were governed by large-scale turbulence, magnetic fields and the presence of dust it seems more and more evident that the processes in the aftermath are once again dominated by these constituents. Dust, magnetic fields, and turbulent processes are within the ingredients that characterize the vicinity of protostars. The main contributing processes relate to *accretion*, which is a turbulent process, as well as to *stellar* and *disk magnetic fields*. Omnipresent again is dust, which is constantly modified by radiative heating and grain–grain collisions [548]. Grain properties change dramatically during the later stages of PMS stellar evolution due to a manifold of physical and chemical processes. For now the discussion of some of these processes will be postponed to appear within the context of protoplanetary disks in the final chapter.

The previous chapters guided the reader from the birth sites of stars out of the darkness of molecular clouds through a sequence of processes that were initiated by self-gravitational collapse and led to a radiating protostellar system. This evolution gave way from dusty molecular cores radiating at temperatures of 10 K to young stellar systems with emission across the entire electromagnetic spectrum and temperatures ranging from 10 K to 10 million K. Here investigates in further detail the key phenomena of accretion processes of protostars and PMS stars. Considered interactions focus on stellar and disk magnetic fields and emissivities at various wavelengths. Special emphasis again is placed on processes producing high-energy emissions.

8.1 Accretion Disks

The physics of accretion disks is exhaustively discussed in the literature [290, 549] and many details may therefore not be repeated here. It has been shown in Chap. 5 that the collapse of rotating and magnetized clouds will inevitably result

N.S. Schulz, *The Formation and Early Evolution of Stars*, Astronomy and Astrophysics Library, DOI 10.1007/978-3-642-23926-7_8, © Springer-Verlag Berlin Heidelberg 2012

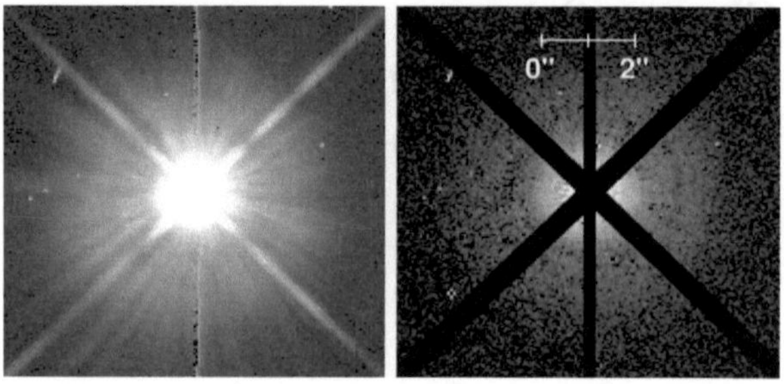

Fig. 8.1 Hubble WFPC2 images of the close CTTS TW Hya. The *left image* shows the raw image, the *right image* the same but with instrumental distortions subtracted. The residual image shows the face-on disk of the stars. From Krist et al. [552]

in the formation of accretion disks. Disk accretion thus dominates the evolution of protostars and PMS stars for a substantial fraction of their evolution (see Chap. 6). Observationally the presence of disks around T Tauri stars is well established and has been proven most recently in stunning optical images of protoplanetary disks (see Chap. 12). Interferometric images of disks around well-known T Tauri stars, including TW Hya (see Fig. 8.1) [550–554], have now emerged as well. The following sections highlight a few physical aspects relevant for the understanding of protostellar accretion disks.

8.1.1 Mass Flow, Surface Temperature, and SEDs

A key function of an accretion disk is its ability to store and transport large amounts of matter and to regulate its release onto the central stellar object. Masses of accretion disks can vary by large amounts depending on the mass of the collapsing cloud and on how much time has elapsed during the evolution of the YSO. From observations, disk masses of Class II sources are best known with typical values of $\sim$0.02 $M_\odot$ [59, 62, 555]. Masses for Class 0 and I are less known simply because they cannot be separated from envelope masses. A recent survey by [556, 557] shows that the majority of the mass in these objects is still in the envelopes and the associated disk mass accounts for 10 percent of the envelope mass at most (see also [558] and references therein). For massive systems disk masses may even be 30 percent of the stellar mass and as much as 15 $M_\odot$ [559], though calculations confirming such a number are still outstanding [352].

Most analytical descriptions assume geometrically thin disks. Thin is defined by the vertical height of the disk H at all radial positions to be always very much smaller than the radial extent of the disk (i.e., $H \ll R_{disk}$). This condition allows a decoupling of the vertical scale from the radial-azimuthal plane in the treatment of the disk structure. Also, any matter in the disk moves in almost circular (*Keplerian*) orbits within one single plane. Such a configuration is shown in Fig. 8.2. Velocities in azimuthal and radial direction, v_ϕ and v_r respectively, characterize the matter

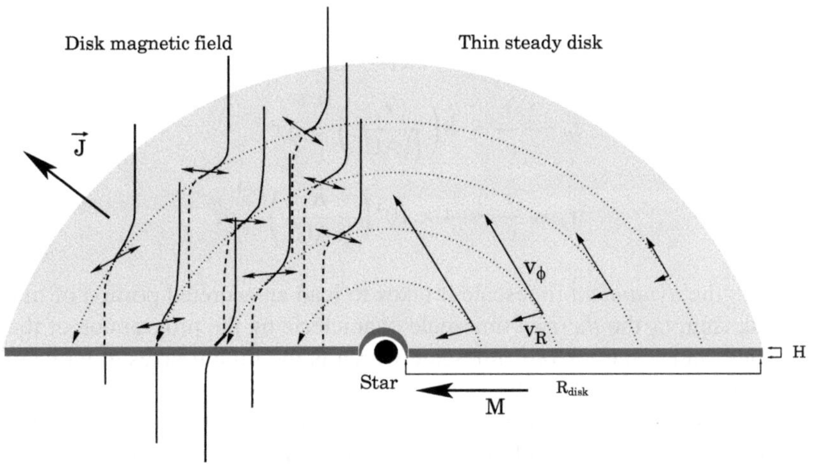

Fig. 8.2 Schematic drawing of some properties of a steady geometrically thin, optically thick inner accretion disk. The right-hand side of the drawing illustrates the geometry and matter flow. Matter follows almost circular orbits with $v_\phi \sim v_K$ and v_R highly subsonic ensuring a slow (in comparison to free-fall) progression of matter towards the star. This can only work if angular momentum is transferred outwards. In the standard picture of accretion this can be achieved through viscous stresses. In protostellar disks the necessary viscosity may be generated through the *Balbus–Hawley* (MRI) instability. The left-hand side of the drawing illustrates how this instability is generated with the help of the disk magnetic field. Magnetic field lines normal to the disk plane get distorted by the velocity gradient between two adjacent disk orbits causing a net drag effect. J here is the transported angular momentum

flow. v_ϕ is almost equal to the local *Kepler velocity* making the circular flow highly *supersonic*:

$$v_\phi \sim v_K = \sqrt{\frac{GM}{r}} \gg c_s \tag{8.1}$$

On the other hand, the radial flow v_r toward the central star has to be highly *subsonic* [549] indicating the presence of *viscosity* ν in the matter flow. Using the α parametrization [560] with $\alpha \sim 0.01$, the radial velocity amounts to:

$$v_r = \frac{\nu}{r} = \alpha c_s \frac{H}{r} \tag{8.2}$$

Typical values from these relations give azimuthal velocities between 1 and some 100 km s^{-1} and radial velocities of <0.01 km s^{-1} for sonic speeds below 1 km s^{-1}. Observations of disk kinematics are difficult [561], but measurements so far confirm fast azimuthal velocities exceeding 2 km s^{-1} [558, 562, 563]. It may also be useful to keep track of specific timescales in the evolution of disks based on these assumptions, which are [564]:

$$
\begin{aligned}
t_{dyn} &\sim \frac{r}{v_\phi} \sim 0.2\left(\frac{r}{[1\mathrm{AU}]}\right)^{3/2} \mathrm{yr} \\
t_{th} &\sim \frac{r}{c_s} \sim 8\left(\frac{r}{[1\mathrm{AU}]}\right)^{11/8} \mathrm{yr} \\
t_\nu &\sim \frac{r}{v_r} \sim 4\times 10^4\left(\frac{R}{[1\mathrm{AU}]}\right)^{5/4} \mathrm{yr}
\end{aligned}
\tag{8.3}
$$

Here t_{dyn} is the *dynamical* timescale it takes to lead an accreted portion of material into a disk ring, t_{th} the *thermal* timescale characterizing the progression of thermal instabilities, and t_ν the *viscous* timescale which is the time it takes to move matter from a radius r to the central star.

Viscous transport is one of the least understood physical processes in accretion disk theory (see discussion in [290]), and for the purposes here can be avoided by assuming a steady optically thick disk. Steady in this respect means that the disk evolves through a persistent flow where the accretion rate is approximately constant in the inner regions of the disk. The luminosity of the disk within a radius r from the central star of mass M and radius R can be formulated using (6.7) and the virial theorem to:

$$
L_{grav} = \frac{GM\dot{M}}{2R} = \frac{1}{2}L_{acc} \tag{8.4}
$$

Assuming that the optically thick disk radiates in LTE, the surface temperature of the inner regions of the disk can then be estimated to:

$$
\sigma T_{surf}^4 = \frac{3GM\dot{M}}{8\pi r^3}\left(1-\left(\frac{R}{r}\right)^{1/2}\right) \tag{8.5}
$$

which has a maximum near 1.36 R at a temperature of:

$$
\begin{aligned}
T_{max} &= 0.488\left(\frac{3GM\dot{M}}{8\pi\sigma R^3}\right)^{1/4} \\
&\sim 6800\left(\frac{M}{[1M_\odot]}\right)^{1/4}\left(\frac{\dot{M}}{[10^{-6}M_\odot\mathrm{yr}^{-1}]}\right)^{1/4}\left(\frac{R}{[2R_\odot]}\right)^{-3/4} \mathrm{K}
\end{aligned}
\tag{8.6}
$$

Figures 8.3 and 8.4 illustrate what inner disk temperatures may be expected for various stellar radii R, stellar masses M, and accretion rates $\dot{M}$. The upper regime assumes a radius of 2 $R_\odot$ for a contracting T Tauri star and that the disk reaches the stellar surface through a narrow *boundary layer*. The inner disk temperature only depends on the energy a particle must lose when brought from the outer disk radius to the innermost circular orbit. Energy release in a boundary layer is discussed in [290]. Thus for the top plot regime relatively high temperatures can be reached even in the case of typical accretion rates for T Tauri stars. Accreting protostars have higher disk temperatures according to their higher accretion rates. As described

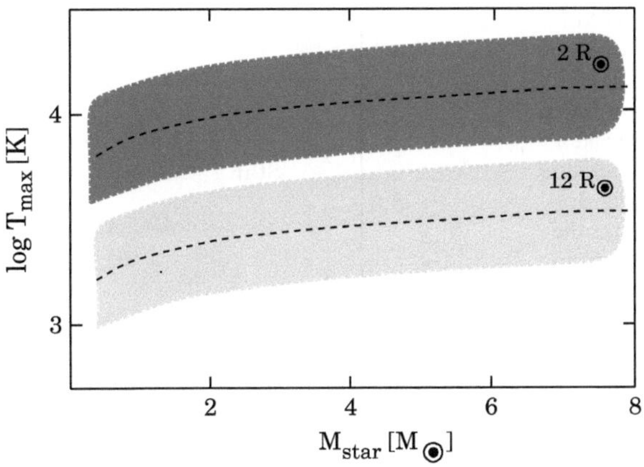

Fig. 8.3 Maximum temperature regimes of the *inner* surface (near the protostar) of a steady geometrically thin, optically thick accretion disk based on (8.6). The top bar refers to an inner disk radius of 2 $R_{\odot}$, the bottom bar to a radius of 12 $R_{\odot}$. The top boundary corresponds to a mass accretion rate of 10^{-4} $M_{\odot}$ yr^{-1} and marks very early protostellar phases, the bottom boundary corresponds to to 10^{-6} $M_{\odot}$ yr^{-1} and likely represents Class II objects. The dotted line in between marks a rate of 10^{-5} $M_{\odot}$ yr^{-1}. The top case should be rather unlikely as it is assumed that the disk reaches the stellar surface. The bottom case assumes that the disk is terminated by a magnetosphere of a few stellar radii (see Fig. 8.4 for SEDs)

earlier, it is not expected that the disk actually reaches the stellar surface because there will be a strong enough magnetic field to disrupt the disk flow. The case has some relevance for estimating surface temperatures at the bottom of *accretion columns* which terminate on the stellar surface.

The second regime in Fig. 8.3 considers an inner disk radius of 12 $R_{\odot}$, which seems more realistic for T Tauri stars. Here the disk surface temperatures are expected to be much more moderate. The diagram also extends these predictions toward higher masses. Though not incorrect such a projection has to be viewed cautiously, since little is known about the optical properties of massive accretion disks at the protostellar stage.

Once the temperatures are known one can predict the energy spectrum of a PMS stellar system. The SED of the star is equal to blackbody emission at the stellar surface temperature. The SED of the inner disk [62] can be calculated by integrating over all disk annuli radiating a blackbody flux at a temperature given by (8.5), i.e.:

$$L_{\nu} = \cos\theta \int_{R_{in}}^{R_{disk}} \pi B_{\nu} T_{surf} (1 - e^{-\tau_{\nu} cos\theta}) 2\pi r \, \mathrm{d}r \tag{8.7}$$

where R_{in} is the inner disk radius, R_{disk} the disk radius, θ the angle of incidence towards the disk ($\theta = 0$ for a face-on disk), and τ_{ν} the optical thickness a defined by:

$$\tau_{\nu} = \kappa_{\nu} \Sigma(r) \tag{8.8}$$

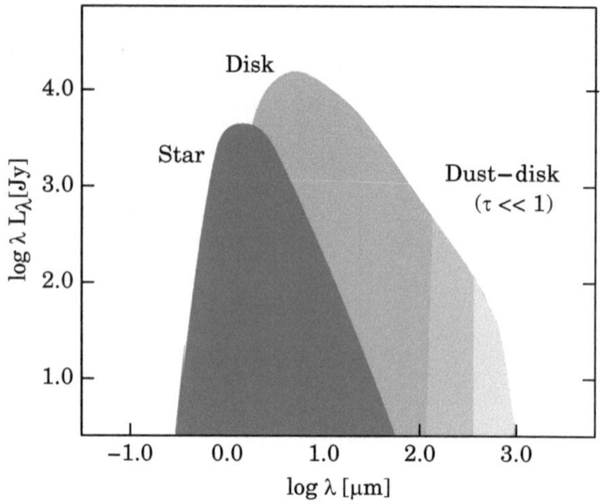

Fig. 8.4 SEDs of a PMS stellar system consisting of the stellar SED for a solar-type star of ~4,000 K surface temperature and a maximum disk temperature of ~1,000 K (see Fig. 8.3). At longer wavelengths the disk emission comes from optically thin dust and opacity and hence emission drops sharply

The radial dependence of the disk surface density $\Sigma(r)$ is generally assumed to be a power law, values for κ_ν are discussed in Appendix F. As long as the disk remains optically thick the SED at long wavelengths will have a dependence $L_\nu \propto \nu^{-1/3}$ ($\lambda L_\nu \propto \lambda^{-4/3}$) [290]. For longer wavelengths the dust emission becomes increasingly optically thin and the SED cuts off somewhere near millimeter wavelengths. In general, at larger disk radii the IR flux is entirely dominated by reprocessed radiation from dust. The dust temperature itself has its maximum near its sublimation temperature which lies between 1,500 and 2,000 K. This places an inner disk radius beyond which dust cannot exist.

8.1.2 Disk Instabilities

One of the biggest obstacles in the understanding of astrophysical accretion disks always has been the problem of angular momentum transport. An extensive review of the detailed physics of disk instabilities has recently be compiled by S. A. Balbus [565]. The common credo is that the transport has to be realized through some kind of disk viscosity. Here is where the problem starts as it is not a straightforward matter to invoke such a viscous flow in protostellar accretion disks. *Ordinary viscosity* on the molecular level can be ruled out as associated timescales are much too long for the anticipated disk sizes. Quite popular for high-gravity compact accretion disks in X-ray binaries, which are semidetached binary systems containing a main sequence star and either a neutron star or a black hole, was the α disk model developed by N. Shakura and R. Sunyaev in 1973 [560]; in absence of a detailed theory they characterized viscous stress through a dimensionless parameter α. The train of thought here was that differential rotation would create enough *shear-turbulence* to transport momentum. Though this is probably true, the problem is that one has to provide for local instabilities that provide a Keplerian rotation profile [565]. Shear-turbulence is highly correlated with radial and azimuthal velocity fluctuations [549]. Other causes for *turbulent viscosity* have been considered. One example is viscosity invoked by disk convection [566], though the effects specifically for T Tauri disks are not clear (see [290] for details).

Key for all viscosity considerations is the *Reynolds number (Re)*, which is defined through the ratio of inertial and viscous forces and thus also the ratio between the first and second term on the left-hand side in (A.24) [549]:

$$Re = \frac{F_{inert}}{F_{viscous}} \tag{8.9}$$

If $Re \ll 1$ viscosity dominates the flow, if $Re \gg 1$ it is unimportant. In the case of ordinary molecular viscosity this number is of the order of 10^{14}. There is a critical Re in laboratory fluids below which turbulence sets in and which is of the order of 10 to 10^3. Thus, one has to look for a process that drops Re by over ten orders of magnitude while retaining Keplerian rotation.

Disks around protostars are less compact and very much larger in size than their X-ray binary cousins with disk radii scaling by factors 10^{2-3}. They are also likely magnetized as they are threaded by the original field of the collapsing cloud (see Chap. 5). Thus protostellar disks may not be so much affected by gravitational but rather by magneto-hydrodynamic instabilities unless one considers disks still embedded in the collapsing envelope fed by high accretion rates [567]. Even weak magnetic fields can lead to a *magneto-rotational instability (MRI)* [13, 369, 568], sometimes also called *Balbus–Hawley instability* [290]. Though its theoretical roots have been in the literature for a long time [13, 569], its impact on protostellar accretion disks has been acknowledged only recently. The instability creates viscosity through unstable rotation profiles in disks where the angular velocity decreases outward. Several studies have shown in this respect that a laminar Keplerian flow can turn turbulent [570–573]. As Balbus [565] states:

The MRI is the only instability to be capable of producing and sustaining the enhanced stress needed for accretion to proceed on viable time scales in non-selfgravitating disks.

One of the decisive differences in the MRI treatment of accretion disks here is that it is not considered as a rotating fluid with a magnetic field in the classical sense but as a magnetized fluid under rotation (see also Appendix D). A full formal description of the transport equations goes way beyond the scope of this book. The interested reader will find a full set of equations to describe the MHD of disks in [565]. They are basically the ones described in Sect. 5.1.2 adding terms for the magnetic energy density $\left(\frac{B^2}{4\pi}\right)$ and adding the induction equation (B.10). The azimuthal component of the equation of motion is then a direct expression of angular momentum conservation [565]. The MRI effect is a relatively simple one. Consider two adjacent Keplerian orbits in the disk. A field line threaded between the two orbits gets stretched towards the inner orbit by the faster azimuthal velocity (see Fig. 8.2). The field line will resist the shear resulting in a net drag in the outer orbit subsequently increasing the outer azimuthal velocity and thus transporting angular momentum outward. Here is the crucial element of the process as it is the angular momentum transport itself that causes the instability and its growth. Thus, however weak the disk magnetic field may be, as long as there is some ionized fraction in the disk threaded by the magnetic field, an instability will grow in a Keplerian disk.

8.1.3 *Ionization of Disks*

The consequences for disks in evolving protostars and PMS stars are likely manifold and the details are still not entirely clear. As collapse calculations assert that disks grow rapidly and angular momentum is effectively removed from the inner disk, it is now evident that disk magnetic fields should be present throughout the later phases of PMS evolution. One train of thought, for example, is that eventually there are areas in the disk in which the ionization fraction drops below a limit where the magnetic field does not effectively connect with the disk flow. Similar to (8.9) one can define a *magnetic Reynolds number Rm* [549], which [574] define as:

$$Rm = \frac{c_s H}{2\eta_o} \tag{8.10}$$

where H is the disk thickness (see Fig. 8.2) and η_o the ohmic resisitivity as defined by [575]. For a vertical field the critical magnetic Reynolds number is about 100 corresponding to ionization fractions of the order of 10^{-12}–10^{-13}. Once the ionization fraction drops significantly below such a limit magnetic turbulence cannot be sustained, viscosity decreases and matter transport stops. These effects are of specific interest in the later, possibly protoplanetary, stages.

One way to ionize disk matter is through internal thermal processes. Using (8.5) one can see that the surface temperature of the disk within 1 AU is larger than 10^3 K. It was shown using the *Saha equation* (A.52) that an ionization fraction of the order of 10^{-12} can be maintained for temperatures larger than 1,400 K (i.e., close to 1 AU) assuming that the ionization due to thermal electrons is provided by alkali atoms with low-ionization potential [576]. Another way to ionize the disk is through external radiation fields. Here the conventional mechanisms considered for molecular clouds, which are cosmic rays and the UV field of nearby massive stars may be effective [577]. For example, one study finds that under certain circumstances cosmic rays can support disk accretion rates of the order of 10^{-7} $M_\odot$ yr^{-1} [578]. However the dominant source is the field from the central PMS star. The impact of this irradiation is evident in the IR/sub-mm dust excess emission in PMS stars. Another likely contributor to the ionization fraction is the irradiation through high energy X-rays (see Chap. 10). The impact of X-rays from the central star has been realized and [579] recently argued that X-ray irradiation alone may

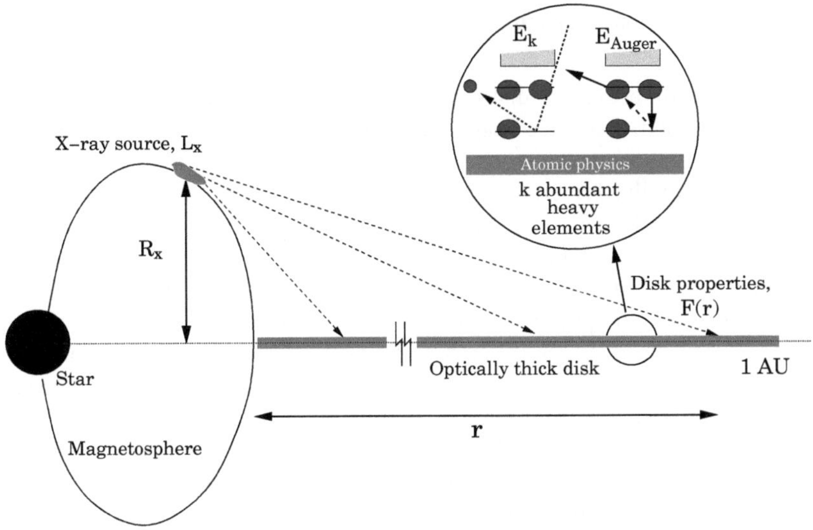

Fig. 8.5 Geometric set-up to calculate the X-ray ionization rate in protostellar disks. The assumption is that X-ray sources originate somewhere near the stellar magnetosphere $\sim 10\ R_\odot$ above the disk plane. The X-rays are characterized by the shape of the spectrum $g(e)$ and the integrated X-ray luminosity L_x. Radiation hits somewhere in the middle of the optically thick disk at a distance of about 1 AU. The response then depends on the irradiated flux, atomic physics, and geometry. Adapted from Glassgold et al. [579]

provide sufficient ionization to sustain the Balbus–Hawley instability. In fact, it is argued that cosmic rays and extra-stellar fields may not contribute as they may be swept away by particle winds from the stellar magnetosphere quite similar to the solar wind (see also below).

The disk ionization fraction produced by X-rays involves ionization and recombination in abundant heavy elements. Thus depending on the evolutionary state of the stellar system many heavy elements may be more or less locked away in grains and effects like grain growth, mantle formation and sedimentation become important. Considering the setting illustrated in Fig. 8.5 and separating the geometrical dependencies from the energy dependence, one can formulate a complex product for the ionization rate [579,580]. For the early stages, where most of the disk material is in gas form, the X-ray ionization rate $\mathcal{N}_{XR}$ can be formulated as:

$$\mathcal{N}_{XR} = F(r) \times \Sigma_{XR}(E) \tag{8.11}$$

where $F(r)$ is the radiation flux hitting the disk as illustrated in Fig. 8.5 and $\Sigma_{XR}(E)$ contains the detailed physics of the X-ray source and irradiated site. This function then contains all the atomic physics and geometric conditions.

8.1.4 Layered Accretion

The ionization of accretion disks primarily happens near the disk surface and it has been argued that once magnetic transport is assumed to remove angular momentum only those layers with sufficient ionization fraction couple to the field and proceed radially. Layered accretion in T Tauri stars has been modeled by Gammie [578]. Disks around T Tauri stars are mostly dense and, cold and for the lack of free electrons, rather poor conductors with respect to electrical currents. At low conductivity the disk gas decouples from the MRI turbulence and these parts of the disk should remain quiescent. Inside some critical radius the field is everywhere well coupled to the disk gas through collisional ionization. Outside this radius a layer of surface density Σ_d is ionized by cosmic rays and radiation from the central star at a certain distance from the disk midplane. In between resides a dead layer which is decoupled from the field. For a typical solar-type T Tauri system, the model predicts a critical radius of about <0.1 AU and surface densities of $\sim$100 g cm^{-2} outside this radius. A stellar accretion rate of 10^{-8} $M_{\odot}$ yr^{-1} results independent of the rate of infall onto the disk. There is matter buildup in the inner few AU which is not accreted onto the star giving rise to the possibility that accretion is not steady. As an interesting effect, the effective temperature profile of $T \propto r^{-1/2}$ outside the critical radius significantly differs from the derived dependence in Eqn. 8.6.

Heating and cooling in protostellar disks with layered accretion has recently been simulated using 3-D radiation-MHD calculations [581]. The accretion stresses in T Tauri disks are assumed to originate from MRI turbulence and reflect the local heating, cooling, and magnetic field coupling properties. As in the model by Gammie [578], Fig. 8.6 shows the simulated accretion stress versus time in Kepler orbits and its dependence on scale height from the disk midplane. Stellar illumination dominates the atmospheric heating as described by [582] and its absorption is shown by the white lines. The dark solid and dotted lines resemble the Planck and Rosseland mean disk photospheres (see Appendix C, C.4). The accretion flow occurs in the disk atmosphere, not the interior. The simulation shows that relocating the dissipation to the atmosphere leaves the interior cooler and almost

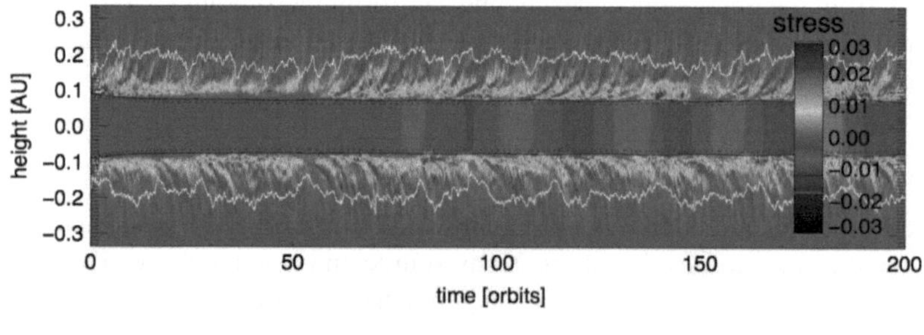

Fig. 8.6 Accretion stress vs. height and time in a simulation of layered accretion of a solar-type protonebula at 1 AU, The stress is horizontally-averaged and the grayscales are plotted in units of dyn cm^{-2} (from Hirose & Turner [581])

isothermal. The magnetic fields are then generated in an optically thin turbulent and stressed layer which rise buoyantly as they dissipate. These hot, opticallythin current sheets might be detectable through their line emission. While the total mass column is 1,000 g cm^{-2}, over half of the accretion heat is deposited at columns within 1 g cm^{-2} of the boundaries, and about 25% within 0.1 g cm^{-2}.

8.1.5 Flared Disks and Atmospheres

The considerations above assumed that the disk is geometrically thin for all radii. In many cases, probably also depending on evolutionary state, this assumption is not entirely valid. It has been emphasized that many, even low-mass, accretion disks can appear *flared* at large radial distances [583], in contrast to the suggestion that specifically flat IR spectra in some YSOs are indicative of massive accretion disks [356]. It has also been shown that intermediate-mass stars like Herbig Ae/Be stars are surrounded by massive accretion disks with geometrically thin inner disks of up to a few ten stellar radii and optically thin outer portions [469]. Aside from mass transport in optically thin disks it is quite likely that disks possess optically thin components and layers which may be attributed to remnants of infalling envelopes, heated disk flaring and atmospheres, disk winds, and evaporation activity.

Another aspect that introduces vertical disk structure stems from the very assumption that MRI indeed is responsible for the redistribution of angular momentum. The disk is then divided into an *active layer* next to the surface of the disk and a *dead zone* centered on the midplane. The active layer is penetrated by high-energy irradiation generating the necessary ionization fraction. Coupling of the magnetic field withing this layer causes the MRI and thus allows accretion of matter [578, 584]. The boundary between these two zones is likely sensitive to physical and chemical properties of the disk [574]. A recent study, in this respect, introduced the concept of three distinct layers by adding a chemically active intermediate layer between the surface and dead layer [585].

In the steady disk model the scale H/r already increases with radius as $(H/r)^2 \propto c_s^2 r \propto r^{-3/4}$ and given the temperature dependence in (8.5) leads to $H/r \propto r^{1/8}$. For large disks such a small dependence is nonetheless significant and the difference in scale height between an inner and outer disk radius can be a factor 5. Disk flaring primarily depends on the temperature distributions on the disk surface as well as its emerging vertical structure. If a disk flares through reprocessing of radiation from the central star it is required that there be a sufficient amount of dust intermixed with the disk gas. The surface height of a moderately flared disk [583] is expressed through:

$$h_d(r) = 0.1\left(\frac{r}{R_{star}}\right)^z, \tag{8.12}$$

which for moderate flaring ($z = 9/8$) results in typical disk heights between 0.2 R_{star} at the inner disk rim to a few 10 AU at the outer disk rim. Numerical calculations of SEDs of flaring disks show that IR excesses sensitively depend on z, which can be as high as $z = 5/4$.

Much of what is known in terms of the internal mechanisms and interactions between the central star and the disk is still in the realm of speculation. In the context of surface properties of protostellar disks there are a few facts to build upon, much of which has been reviewed by [586]. PMS stars have strong stellar winds resulting from the interaction of a rotating magnetosphere and the inner disk (see [587] for a review). Though the outflow is more collimated towards the rotational poles in protostars, observed momentum losses in outflows at later evolutionary stages suggest that the outflow may be less beamed [421]. T Tauri emit a significant EUV, FUV and X-ray continua likely stemming from accretion shocks at the base of accretion columns (see illustration in Fig. 8.7 below and Fig. 6.14).

The persistent irradiation and subsequent absorption of stellar luminosity at the disk surface is the dominant mechanism to heat the disk. Specifically, EUV photons tend to photoionize neutral hydrogen. As a result an ionized atmosphere would form with a scale height of [559]:

$$h_a(r) = R_g \left(\frac{r}{R_g} \right)^{3/2} \tag{8.13}$$

where R_g marks the radius where the atmosphere is still bound by the star. Beyond that radius ionized hydrogen may diffuse into an extended unbound atmosphere or surface layer (Fig. 8.7). This escape radius can be expressed as [588]:

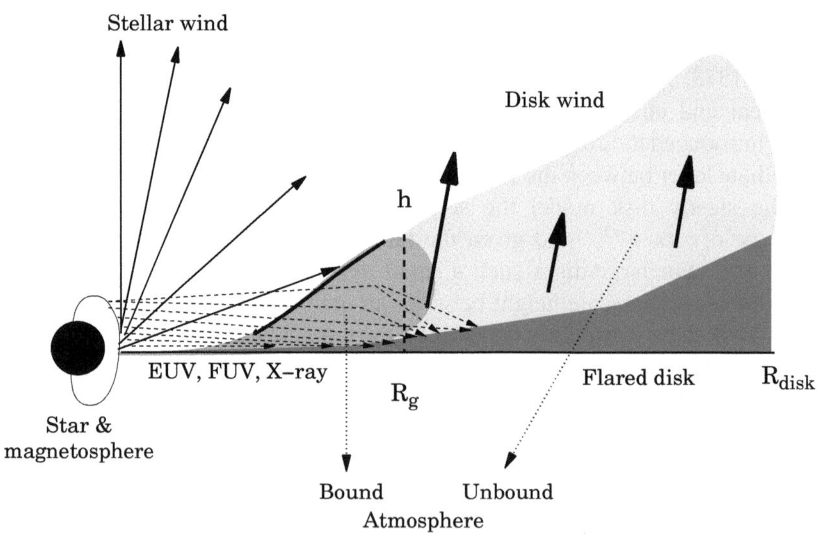

Fig. 8.7 Schematic representation of an irradiated flared disk. Below the radius R_g where matter remains bound by the gravity of the central star an optically thin atmosphere develops. Above this radius flared matter may escape and form some kind of slow wind. Adapted from Hollenbach et al. [559]

$$R_g = 60\left(\frac{M}{[M_\odot]}\right)\left(\frac{T}{[10^3\ \mathrm{K}]}\right)^{-1}\ \mathrm{AU} \tag{8.14}$$

It should be noted that specifically in younger systems there may be an inner disk atmoshere which would not be penetrable by EUV photons. Here irradiation by FUV and X-ray photons need to be considered [7, 588]. The bound atmosphere is optically thin and its number density satisfies the barometric law with a base density of [559]:

$$n_0(r) = 4 \times 10^7 \left(\frac{\mathcal{N}_{UV}}{[10^{49}\ \mathrm{s}^{-1}]}\right)^{1/2} \left(\frac{M_{star}}{[10\ M_\odot]}\right)^{-3/2} \left(\frac{r_g}{r}\right)^{3/2}\ \mathrm{cm}^{-3} \tag{8.15}$$

where $\mathcal{N}_{UV}$ is the EUV photon field in Lyman continuum photons per second from the central star. For a solar mass star this means that densities at r_g can be as high as $10^9\ \mathrm{cm}^{-3}$, which is what is found in the solar corona. Beyond the bound atmosphere mass will diffuse off in a slow disk wind with a velocity close to c_s and a mass loss rate close to r_g of:

$$\dot{M}_{ph} = 3 \times 10^{-5} \left(\frac{\mathcal{N}_{UV}}{[10^{49}\ \mathrm{s}^{-1}]}\right)^{1/2} \left(\frac{M_{star}}{[10 M_\odot]}\right)^{1/2} M_\odot\ \mathrm{yr}^{-1}. \tag{8.16}$$

Disk mass loss rates thus scale with the square root of the EUV radiation field and can be substantial at times.

8.1.6 *Dispersal of Disks*

It has been shown above that the termination of disk accretion is a matter of halting viscous transport of angular momentum and that this is independent of whatever mechanism was providing this viscosity. The ultimate fate of such a disk may then be the formation of planets. Recent data provide and estimate for the lifetime of disks around low-mass stars which is estimated to $\sim 6 \times 10^6$ yr [589]. Hollenbach [586] points out there are several basic processes to terminate an accretion disk besides reaching an ultimate fate of planet formation:

- complete accretion of disk material onto the central star;
- removal of enough disk material due to the effects of winds;
- stripping off disk material due to close stellar encounters; and
- photoevaporation due to UV radiation from the central source or a close companion.

The first item may not happen too often as the viscous timescale in (A.9) for an active accretion disk of the size of a few 100 AU is comparable to or larger than the time a solar mass star requires to reach the ZAMS. The second item has been discussed above and though it has been postulated as a possible mechnism [590–592], the disk dispersal timescale through wind stripping is more than an order of magnitude too large [586]. The third item, which addresses the impact of stellar encounters, though generally ineffective as well, has an intriguing notion with respect to dense young stellar clusters as well as binary or multiple stellar evolution. Tidal disruption of an accretion disk though 3-body interactions have been addressed in some detail [593, 594]. Assuming that the timescale to truncate the disk radius to about 30 percent of its original size is inversely proportional to the cluster density and velocity dispersion of stars, a stellar density of 10^4 stars pc^{-3} would produce an encounter every 20 million years. The stellar density within the central 0.1 pc of the Orion Trapezium cluster reaches about half of this density [331].

8.1.7 Photoevaporation of Disks

The discussion of the *photoevaporation* of disks through EUV irradiation involves several different aspects. These involve the possibility of strong winds from low-mass stars or winds from massive central stars evaporating their circumstellar disk. Many young massive stars are actually embedded into ultracompact H II regions and evaporation can extend to their circumstellar envelopes. For quite some time it was assumed that it is mostly EUV radiation that is relevant for irradiative and evaporative action in disks. However, it is now understood that depending on the location of the disk (i.e., within, outside or independent of r_g) one has to consider EUV, FUV and X-ray wavelengths [7, 588, 595, 596]. Besides the destruction of their own disk and envelopes, massive stars have winds strong enough to destroy star–disk systems in their neighborhood as well. This aspect will be of interest in Chap. 12 and here only mechanisms where the protostar evaporates its own disk are discussed.

Photoevaporation has specific relevance toward the accretion history of massive stars, though some of it may also apply for low-mass stars with exceptionally strong winds. However, it should be noted that the mechanisms for photoevaporative disk winds within low-mass stars are highly uncertain and evidence is at best circumstantial. It also has been pointed out that photo-evaporation is effective in removing disk material at radii larger than ~10 AU, inner material has to be removed by viscous accretion [443, 595, 597, 598].

Information on the evolution of accretion disks in massive stars is still entirely based on theoretical considerations. Collapse calculations show that also in the case of massive stars strong accretion disks may evolve through some kind of angular momentum transport [352] and magnetic fields [358, 360]. How these massive disks transport angular momentum is highly uncertain. Magnetic instabilities, at least for the thick outer parts, are less feasible as there may not be sufficient penetration of radiation to produce enough ionization. Some more momentum may be carried off through a more powerful wind as has been observed specifically for massive YSOs (see Sect. 8.1.8). Alternatively some simulations [352] (Sect. 5.4.3) apply tidally

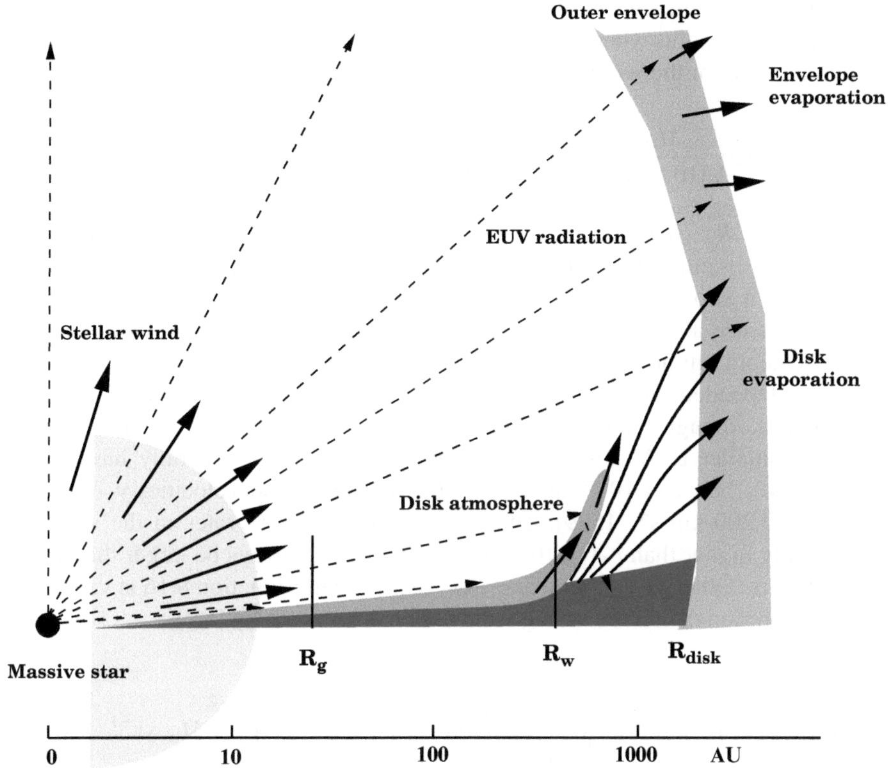

Fig. 8.8 Illustration of a system where the central massive star evaporates its immediate environment consisting a massive disk and maybe a residual outer envelope. In disk evaporation the ram pressure of the stellar wind pushes ionized atmospheric matter beyond the limit R_g where it is gravitationally bound until it is counterbalanced by thermal pressure. At this point matter is removed from the disk. Beyond R_w diffuse irradiation also causes disk evaporation. Similarly the EUV field and the ram pressure of the stellar wind will eventually disrupt any residual outer envelope

induced gravitational torques. Similarly, the disk geometry in massive disks may not be as thin as has been discussed in the context of less massive stars. For starters, it can be assumed that the disk is geometrically thin at least near the central star. That this is the case for stars up to $\sim$10 $M_{\odot}$ has been shown [469], though from Fig. 5.9 it can be deduced that thickness grows quite rapidly with radius.

Very important in this respect is the fact that the evolution of massive stars proceeds so much faster than in low-mass stars and thus timescales are most relevant. Massive stars above 6 $M_{\odot}$ can reach the ZAMS as early as between 4×10^4 yr and 4×10^5 yr (see Table A.9). The time of growth of an accretion disk under the assumption that the stellar wind sets in at full power when the star hits the ZAMS [407] is then limited to radii between 100 and 2,000 AU, depending on what calculation is used (see Chap. 5). The wind's ram pressure prevents any atmosphere from achieving much scale height as long as the wind radius r_w (see Fig. 8.8), which is defined as the radius where the thermal pressure of the ionized hydrogen balances the ram pressure of the wind [559]:

$$r_w \simeq 13\left(\frac{\dot{M}_{wind}}{[10^{-7} M_\odot]}\right)^2 \left(\frac{v_w}{[1000 \text{ km s}^{-1}]}\right)^2 \left(\frac{\mathcal{N}_{UV}}{[10^{49} \text{ s}^{-1}]}\right)^{-1} \text{AU} \tag{8.17}$$

is larger than R_g (see Equation 8.13). Equation 8.17 shows that a strong wind does not simply imply high mass-loss rates and high wind velocities. Considering the fact that a strong EUV field creates more ionized hydrogen this will increase thermal pressure and ultimately counterbalance the ram pressure of the wind. It should be kept in mind that mass-loss rates in stellar winds and EUV fields are not entirely independent and smaller mass-loss rates imply also smaller EUV fields. Most B-stars, though having only moderate mass-loss rates and wind velocities, also have smaller EUV fields than O-stars. Massive stars generally have mass-loss rates $\dot{M}_w$ between 10^{-5} and 10^{-7} $M_\odot$ yr^{-1} and wind velocities v_w between 1,000 and 3,200 km s^{-1} [599]. Thus as long as mass-loss rates in the wind are substantially higher than $\sim 1 \times 10^{-7}$ $M_\odot$ yr^{-1}, r_w will always be larger than R_g in O and B stars. Otherwise r_w is near or less than R_g and Equation 8.16 is valid. The photoevaporation rate in the strong wind case is then given by [559]:

$$\dot{M}_{ph} \simeq 5 \times 10^{-6} \frac{\dot{M}_w}{[10^{-7} M_\odot]} \frac{v_w}{[1000 \text{ km s}^{-1}]} \left(\frac{\mathcal{N}_{UV}}{[10^{49} \text{ s}^{-1}]}\right)^{-1/2} M_\odot \text{ yr}^{-1} \tag{8.18}$$

This rather approximate expression implies two things. One is that the evaporation rate is larger or of the order of the mass loss of the stellar wind. The other one, more importantly, is that it may take of the order of 10^{5-6} yr to evaporate the entire disk in young massive stars. Recent numerical simulations also indicate that dust scattering of ionizing photons in the outer circumstellar disk increases the photoevaporation rate significantly [600].

8.1.8 MHD Disk Winds and Jets

UPDATE with [601,602] There are many ways for protostellar disks to actually lose mass rather than accrete onto the protostar. Specifically in younger evolutionary stages mass is carried out of the system by *bipolar outflows* and *jets* [417, 603], which are both now known to be very common in deeply embedded protostars as well as CTTS. In later stages, specifically past the CTTS stage, mass loss may be achieved through disk evaporation mechanisms (see below). High-velocity winds in T Tauri stars were first detected about 40 years ago [48] and even today their origins are still quite poorly understood. More than 200 sources with bipolar outflows are known from CO observations [604, 605] and their morphology as well as dynamical properties offer a wide range of properties. A comprehensive review of magnetized winds and the origins of bipolar outflows was given by M. Camenzind [606]. Here it was argued that magnetized winds ejected from the surface of disks will be collimated by magnetic effects on scales typically larger than the light cylinder radius for these objects. It was predicted that protostellar jets have moderate velocities and jet radii of a few hundred AU. Furthermore, simulations indicate that the mass flux is an indicator for the degree of jet collimation [607].

For a long time, the general trend indicated that there was a clear distinction between not well collimated cold neutral outflows with relatively moderate outflow velocities of the order of <30 km s^{-1} and highly collimated ionized jets with flow velocities exceeding 100 km s^{-1}. In recent years, though, it has become more evident that molecular outflows are in fact jet-dominated as they also show high-velocity components (∼30–40 km s^{-1}) tracing jet-like activity inside cavities of lower velocity gas (∼5 km s^{-1}) [604].

Three empirical correlations can be summarized from many observations:

- Outflow activity declines with progressing age.
- The mass outflow rate $\dot{M}_{wind}$ as well as the outflow velocity increase with the bolometric luminosity of the central source L_{bol}.
- The degree of collimation decreases with outflow rate.

The first correlation repeats what has already been stated in Chap. 6. Molecular outflows are present throughout the embedded phases of protostars and have a high degree of collimation [421] and speeds. Though a tight correlation between outflow energetics and the driving source is found for Class I sources, Class 0 sources produce almost an order of magnitude more powerful outflows. Many of these outflows appear as large bipolar lobes and are not as fast and highly collimated. These lobes are now understood to be ambient cool material that has been swept up by an intrinsic fast jet or jet fragments. These lobes appear elongated in the direction of the stars magnetic and rotational axis assuming that they are approximately the same. The wind momentum flux is largest in these directions and the density of the ambient material is lowest as most matter is concentrated in the accretion disk. For more in-depth studies of the detailed topology of bipolar lobes reviews by [608] and [609] are recommended. There are outflows observed in Class II sources though even less collimated and weaker in power. To summarize, the first correlation also reflects the fact that the mass outflow rate $\dot{M}_{wind}$ is related to the mass accretion rate through [610]:

$$\dot{M}_{acc} = \left(\frac{r_A}{r_w}\right)^2 \dot{M}_{wind} \tag{8.19}$$

Though this equation is not exact it is close to observed values at least for low luminosity sources. The ratio r_A/r_w between the Alfvén radius and the radius to the innermost field lines where the wind emanates (see also Sect. 8.16) is about 3, and thus the outflow rate is about one-tenth of the accretion rate. Equation 8.19 is a result of the basic MHD wind theory describing centrifugally driven (CD) winds in magnetized stars [611–613] and it applies to magnetized accretion disks [614].

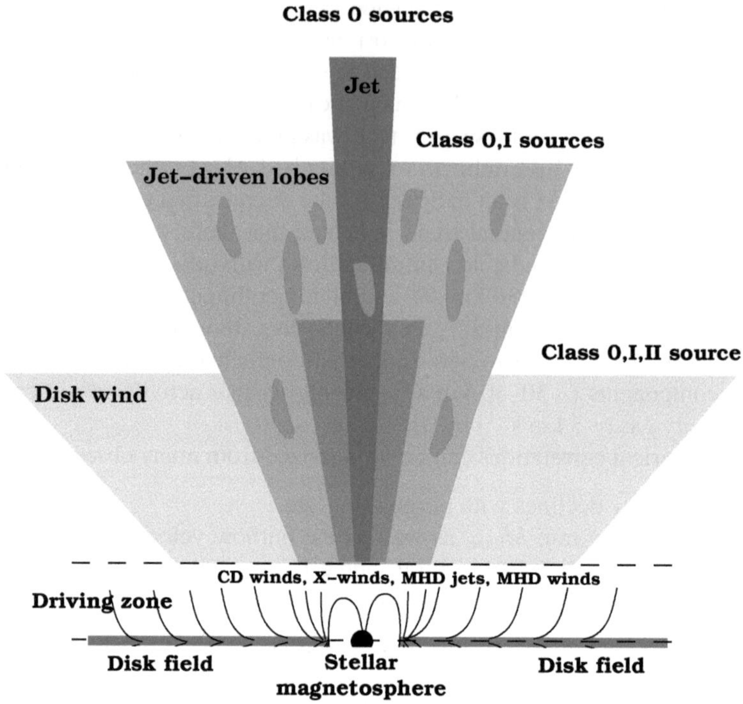

Fig. 8.9 Various types of outflows observed in low-mass YSOs based on observational accounts. High velocity and very collimated jets are predominantly observed in Class 0 sources. The schematic picture indicates the trend that outflow collimation decreases with progressing age. Similarly, as indicated by the lighter color in the outflow, the flow velocity decreases from the jet cores (100–400 km s^{-1}) to the outer edges of a disk wind ($\sim$5 km s^{-1}). A full understanding of the mechanisms that drive these outflows is still outstanding, but the general consensus is that it is likely of magnetic origin possibly in connection with rotation. The main processes discussed in this context are centrifugally driven (CD) winds, the X-wind model, and MHD jets and winds as discussed in the text

The driving mechanism is sometimes also referred to as the *Weber–Davis Model.* A detailed review is given by [610].

The second correlation is entirely based on observations and applies for objects as luminous as $10^6\ L_\odot$ whereas it seems to flatten out for objects with less then 1 $L_\odot$. Jets associated with low-luminosity YSOs have velocities of up to 400 km s^{-1} and very small opening angles (dark jet in Fig. 8.9). Several outflows have been observed in Herbig Ae/Be stars and various high-luminosity objects with jet velocities up to three times higher and outflow rates up to 100 times higher [615, 616]. This result, however, signals that MHD theory for low mass systems as applied in Eqn. 8.19 may not be sufficient. Evolutionary projections for massive stars (see Chap. 9) show that accretion rates in these systems not only need to be much higher than for low-mass or intermediate-mass systems, but accretion in massive stars is much more inefficient and outflow rates contain a much larger fraction of the accretion rate.

The third correlation reflects a suggestion that an apparent decrease in outflow energetics with progressing evolution reflects a corresponding decay in the mass accretion and infall rate [421]. Though details in this context are very complex and not entirely understood one can look at it in a simple way. With (8.19) a lower accretion rate subsequently leads to lower outflow rates. The reduction in outflow rate leads to an increase of the Alfvén radius since $r_A \propto \dot{M}_{acc}^{-2/7}$ [549]. In order to keep the observed relation between accretion and outflow rate intact, r_w has to move outwards. Under the assumption that the injection speed scales with the azimuthal disk velocity, this means a slow down in the outflow. The fact that this could result in a de-collimation is not so straight forward [607, 617–620]. However, since the launch site of the wind is pushed further out into the disk where the magnetic field strength rapidly decreases (i.e., $B \propto r^{-3}$), and is less directed towards the stellar axes, one could intuitively speculate that this also results in a less collimated outflow.

More vigorous theoretical treatments of collimated stellar winds reach back to the late 1970s [621]. In Sect. 5.8 one of the more contemporary calculations produced outflows through poloidal fields and centrifugal forces with an entire collapse calculation. Other numerical calculations deal specifically with mechanisms generating outflows [618, 619, 622, 623]. At the base of the jet, gas is injected into a poloidal magnetic field and centrifugally accelerated. Collimation is achieved through pinch forces exerted by the toroidal field generated by the outflow. The velocities achieved for protostellar systems remain yet short of the observed velocities. The nature of such collimated, jet-like outflows can be steady or episodic and is basically determined by what is called the *mass load* of the wind [624, 625]:

$$k = \frac{\mathrm{d}\dot{M}}{\mathrm{d}\Phi} = \frac{\rho v}{B} \tag{8.20}$$

where v is the speed of the mass flow and Φ the field flux.

Jets and outflows from YSOs have been proposed as candidates to drive some of the supersonic turbulence in molecular clouds. Multidimensional jet simulations, however, showed that the energy and momentum deposition from collimated supersonic jets is not an efficient process. This includes jet-clump interactions, transient jets, and magnetized jets. Supersonic fluctuations are damped quickly and do not spread into the parent cloud [602]. Jets are able to leave strong imprints in the cloud structure and can disrupt dense clumps.

8.2 Stellar Rotation in YSOs

The evolution of stellar angular momentum is one of the most problematic issues in the early formation and accretion history. It was predictable from the beginning that the young star must carry angular momentum and engage in spin-ups and spin-downs during its young life. By studying rapidly rotating massive stars in the mass range 3–12 $M_\odot$ it was found that increased angular momentum for a given mass has a substantial effect on evolutionary tracks with a mass–luminosity relation that is considerably different from that of non-rotating stars [626]. Models of rapidly rotating low-mass PMS stars that contract homologously to the main sequence also showed discrepances in the projected age between rotating and non-rotating stars [627]. Historically, a real interest in rotational properties of PMS stars was sparked with the discovery that many T Tauri stars rotate much slower than their break-up speed [628]. This came somewhat as a surprise since theoretical considerations called for fast rotators rather than the observed slow rotators at that evolutionary stage. Suggestions right away ranged from winds (see above) as an effective braking mechanism to rotational instabilities in the radiative interior of the stars [448, 629, 630]. Many observations have established a rather confusing picture of rotational behavior of PMS stars [628,631–633] and a detailed theoretical treatment going beyond plausibility arguments is still outstanding.

8.2.1 *Fast or Slow Rotators?*

The issue whether protostars and PMS stars should be fast or slow rotators links the angular momentum history directly to calculations from stellar collapse and the PMF phase. Fast or slow rotation here is measured relative to what would be the surface velocity of a star rotating near its break-up velocity after it has accreted most of its mass somewhere at the end of the accretion phase. Why should protostars during their accretion phase become fast rotators? The answer comes with the design of accretion disks discussed above as well as the specific angular momentum available in the collapsing molecular cloud. One should realize that with respect to the break-up rotational velocity the available angular momentum is almost five orders of magnitude larger. Again, accretion disks are the key to the initial angular momentum budget of the young star. The thin disk paradigm has Keplerian behavior, and assumes that the inner disk radius is approximately equivalent to the radius of the stellar surface at 2 to 5 solar radii. A solar mass star should be spinning up to velocities of $v_K \sim$ 200–300 km s^{-1} unless its surface already rotates faster than this Keplerian velocity [587]. The latter is unlikely as this velocity is, of course, the protostars break-up speed. Fragmentation has been considered to be a process that would divert enough angular momentum into a binary orbit and allow for slower rotating constituents. It has been shown even in this case that spin angular momentum deposited from a rotating cloud into a binary system can hardly produce protostars rotating below the break-up speed [631, 634]. It is thus possible that accreting protostars somewhere in their accretion lifetime may reach break-up speed and are *fast rotators*. The observational account so far is sparse as rotational properties of Class 0 and I objects are difficult to determine.

However, recent observations of deeply embedded protostars indicate many slow-rotating stars as well and a few showed break-up speeds [383, 635].

On the other hand, it is well established that PMS stars, after emerging from the birthline are slow rotators with only about 10 to 20 percent of their break-up speed. There is a significant dependence with protostellar mass, which will be discussed separately below. Moreover, none of the observations so far show CTTS rotating close to break-up [346, 636]. This means that a solar mass star has to lose 80 to 90 percent of its angular momentum in less than its accretion time of $\sim 10^5$ yr. How this is achieved is still entirely unclear and so far none of the proposed mechanisms, such as fragmentation, MHD winds or jets, or some other form of magnetic braking, can satisfactorily explain the effect. Specifically the traditionally assumed disk locking, which assumes that the angular momentum deposited on the protostar is largely removed by a connecting magnetic field seems flawed. Differential rotation tends to significantly reduce the spin-down torque in such a case and it remains questionable if the effect could remove the necessary amount of angular momentum [637].

8.2.2 *Contracting towards the MS*

There are now many studies in the literature that contain information on rotational velocities in PMS stars following the first account by [628]. From these studies it follows that rotational properties of PMS stars contracting toward the MS are complex and various contributions from spin-ups and spin-downs have to be considered. PMS stars are expected to spin-up toward the main sequence as a consequence of contraction. This cannot happen by conserving angular momentum as T Tauri stars in the solar-mass range show between 30 and 200 times the present solar value [631]. Many studies also find that WTTS may rotate generally faster than CTTS. Furthermore, there is a trend in which post-TTS (PTTS), low-mass stars with ages larger than 10 Myr, show a larger spread in rotational velocities than in WTTS [642]. It may be speculated that a widening of the velocity distribution from the T Tauri phase to the ZAMS is indeed a general trend [642, 643] (Fig. 8.10).

Angular momentum loss due to wind ejection cannot account for fast losses as projected during the protostellar phase as here timescales of the order of the KH timescale and higher are required [645]; though it may do so during the PMS phase [636]. The required momentum loss can be achieved once a few assumptions are upheld, specifically concerning mass-loss rates and magnetic geometries [613, 646, 647]. These are mass-loss rates of at least $\sim 10^{-8}\ M_\odot\ \mathrm{yr}^{-1}$ and a situation where the magnetic field lines at the origin of the wind are attached to the star [290]. Other mechanisms involve magnetic star–disk interactions as proposed by [648–650] (see also Sect. 8.3) where the stellar magnetosphere truncates the

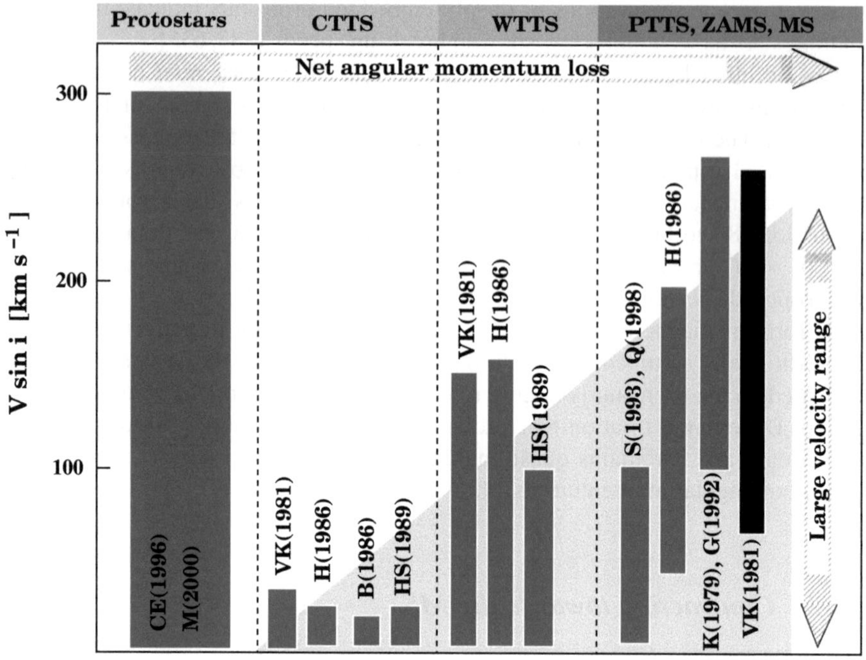

Fig. 8.10 Overview of the domains of various studies of rotational velocities of young stars (K(1979) [638], VK(1981) [628], H(1986) [631], B(1986) [639], HS(1989) [636], G(1992) [54], S(1993) [640], Q(1998) [641]). The general perception is that although PMS stars contract towards the MS they lose net angular momentum on the way. Though the true rotational picture from CTTS to the MS is still fairly unclear, there may be a trend in that the rotational velocity distribution widens

circumstellar disk at a characteristic radius forcing accretion along the field lines towards the magnetic poles of the star. Magnetic torques then transfer angular momentum into the disk. Most recently [642, 644, 645] presented a model for the evolution of surface rotation of solar type stars from the birthline up to the age of the Sun, which is well on the main sequence. The model includes the previous assumptions but in addition requires that, at least for the evolution of moderate and fast rotators, mechanisms exist that efficiently redistribute angular momentum in the stellar interior as well. The evolution towards slow rotators still remains uncertain. Fig. 8.11 shows calculated rotational tracks based on this model for various disk lifetimes showing the spread of rotational velocities from CTTS and WTTS beyond PTTS stages and on the MS toward the Sun's rotational period. Clearly, the upshot is that modeling rotational evolution into the MS requires a better understanding of stellar wind dynamics, interactions with circumstellar disks, as well as angular momentum transport, specifically in convective stars.

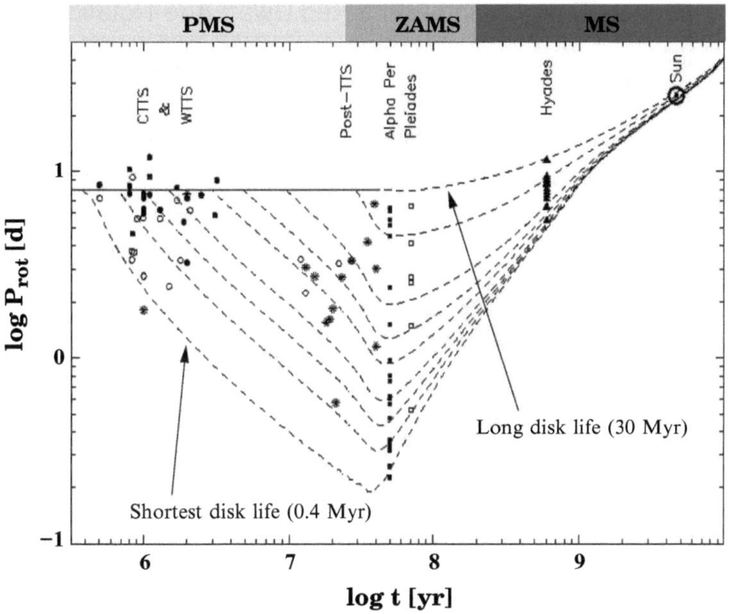

Fig. 8.11 The evolution of rotational periods of young solar-mass stars from the CTTS stage towards the Sun's period on the MS. The lowest dashed curve relates to short disk lifetimes, the top curve to very long disk lifetimes. A description of the data and models is found in [642, 644]. Adapted from Bouvier et al. [642]

A long-term study of the last several decades called the *Sun in Time* program investigates the rotational history of Solar analogs, i.e. G0V to G5V stars, at ages from late PMS stages (50 Myr) to their late MS stages (~9 Gyr) (see Guinan & Engle [651] and references therein). From these studies it is inferred that the young (ZAMS) Sun was rotating more than an order of magnitude faster than today. In a multi-wavelength approach the magnetohydrodynamic (MHD) dynamo activity is monitored through strong coronal X-ray and chromospheric FUV-UV emissions. At increasing age rotational modulated, low-amplitude light variations indicate the presence of large starspot regions covering up to 5–30% of the stellar surface. Figure 8.12 shows a plot of angular velocity $\Omega \propto 1/P_{rot}$ of solar-type stars in units of d^{-1} plotted against stellar age. In comparison below the same for a sample of M0–5 dwarfs is shown. Both the G- and dM-stars decrease in angular velocity with age. For a given age the dM stars have longer rotation periods indicating that they spin down faster than the more massive G stars. It is concluded that rapid magnetic braking effects in dM stars is most likely the reason. More efficient dynamos might exist due to a likely deeper convection zone in these stars combined with their lower masses.

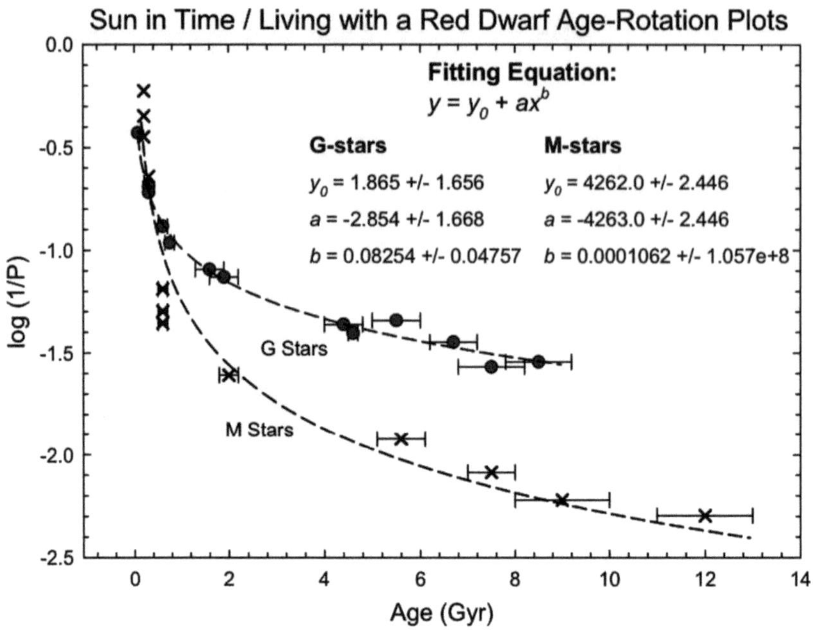

Fig. 8.12 The angular velocity of several G and M dwarfs as a function of age from the *Sun in Time* project (from Guinan & Engle [651])

8.3 Magnetic Activity in PMS Stars

The major role that magnetic fields play in the early evolution of stars has been described so far in this work in connection with almost every aspect of protostellar and PMS properties. However, the stellar interior is of importance as well. Young stellar environments are highly magnetic. Observational evidence for strong activity is ubiquitous and aside what has been presented so far, much will follow in the next chapters. Optical and IR spectroscopic and photometric variability studies of CTTS stars suggest magnetospherically controlled accretion [652–656], whereas X-ray studies testify of magnetic activity throughout all T Tauri phases specifically WTTS [71,657,658]. The following sections will lay the groundwork to understand these observations, which, specifically in the high-energy domain, rely heavily on the presence of magnetic activity.

8.3.1 *Magnetic Fields in PMS Stars*

Magnetic field predictions in T Tauri stars have existed since the early 1990s [649, 650, 659]. They lead to typically strong magnetic fields of 2–3 kG. Based on the principle of *star–disk locking*, also referred to as the *Ghosh and Lamb (GL) model* [648]. Here the stellar magnetic field couples effectively to the gas in the inner region of the accretion disk and one can estimate the magnetic field strength under the further assumption that T Tauri stars are slowly rotating with only a fraction $\epsilon < 1$ of the Kepler velocity. At the inner disk radius R_{in} magnetic stress is large enough to remove excess angular momentum. In the GL model this radius is proportional to the stellar dipole moment as $\mu^{4/7}$ and the mass accretion rate as $\dot{M}^{-2/7}$. The stellar surface magnetic field $B_{surf} = \mu/R^3$ can then be expressed as:

$$
\begin{aligned}
B_{surf} = 1000 &\left(\frac{\epsilon}{[0.35]}\right)^{7/6} \left(\frac{\beta}{[0.5]}\right)^{-7/4} \\
&\left(\frac{M}{[0.8M_\odot]}\right)^{5/6} \left(\frac{\dot{M}}{[10^{-7}M_\odot\ \mathrm{yr}^{-1}]}\right)^{1/2} \\
&\left(\frac{R}{[2.5R_\odot]}\right)^{-3} \left(\frac{\Omega_{surf}}{[1.7\times10^{-5}\ \mathrm{s}^{-1}]}\right)^{7/6} \mathrm{G}
\end{aligned}
\tag{8.21}
$$

The β parameter, like ϵ, is a fiducial one within the GL model corresponding to $R_{in} = 2R$. For typical values of CTTS as used in the equation above, the magnetic field is of the order of 1 kG [649]. As a side note, the GL model was originally developed for the interaction of accretion flows onto magnetospheres of heavily magnetized neutron stars. The adaptation for protostellar fields thus represents more relaxed conditions with longer periods, moderate magnetic fields, and low accretion rates. Similar expressions for the stellar magnetic field strengths taking into account magnetic diffusivity and pinching by the accretion ram pressure can be found in [650, 659]. The latter predict field strengths that are systematically higher than the other predictions.

Indications for magnetic activity are ubiquitous and invoke models of cool star spots or high-energy emission. There are also a variety of studies that directly measure magnetic fields in young stars based on absorption line spectroscopy in optical and near-IR wavelengths. Early measurements date back to the 1980s [660–662]. In most cases these measurements are based on *Zeeman displacement* of lines, which in the IR is large compared to the <25 km s^{-1} rotational line broadening in CTTS. The Zeeman displacement varies with λ^2, whereas the Doppler effect depends on λ only. Other measurements use estimates based on possible starspot activity or spectral fits constraining accretion parameters. Most recent measurements have been performed by [663] measuring the Zeeman displacement in IR ($\sim$2 μm) line absorption. Figure 8.13 shows an example of such a measurement for the CTTS BP Tau yielding a field strength of 2.8 kG.

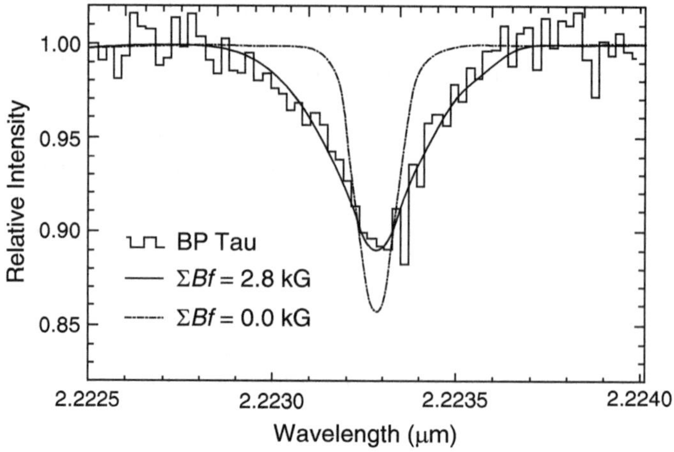

Fig. 8.13 Modeling of the Titanium I line at 2.2233 μ m in the CTTS BP Tau. The dash-dotted curve shows the predicted line profile using a non-magnetic atmospheric model. The solid line includes magnetic fields. From Johns-Krull et al. [663]

8.3.2 Field Configurations

Detailed observational probing of magnetic field configurations is difficult and has yet to be undertaken. A test of predicted correlations of basic measurable quantities such as mass, radius, accretion rates and rotational periods using standard models by [649, 650, 659] showed, for example, that they were not met unless non-dipole field topologies are invoked [656]. Another feature in the models by [650, 664] tend to reconcile some of the correlations once the effect of flux trapping between field lines is included.

Magnetospheric models evolved throughout the 1990s and currently there are various star–disk field configurations under consideration which all address various aspects of the observed IR to X-ray emission. All of these configurations are similar, though the MHD models applied to these configurations differ in important physical features with specific emphasis on how the stellar magnetic field couples to the inner accretion disk and its field. Most controversial in these models are questions about the matter flow between the disk and the star and how energy is deposited. Primarily discussed in literature are four configurations. These are the *X Wind Model* [284], *Magnetically Funneled Accretion Streams* [290], the *Reconnection-driven Jet Model* [665] and the *Slingshot Prominences* [666, 667]. All of these configurations have been introduced fairly recently and it should be kept in mind that none of them has yet been established.

The similarities of these configurations can be summarized as follows. First, all configurations are rotational symmetric as well as symmetric with respect to the accretion disk plane. Second, all of them invoke in the first order a stellar dipole field. Third, the disk magnetic field has poloidal structure. All three aspects evolved out of the grown perception of how disk fields develop from threaded molecular clouds before collapse and from the evolved limit of a dynamo-generated stellar magnetic field as observed in the Sun. Rotational symmetry is simply inherited from the huge initial angular momentum budget of the collapsing cloud. All configurations allow accretion streams onto the star, though in some this is not emphasized. The truncation of the inner disk by the magnetosphere is usually assumed. The truncation radius can be inferred from the GL model:

$$\frac{r_{trunc}}{R} \sim 2.2\left(\frac{B_{surf}}{[1000\ \mathrm{G}]}\right)^{4/7}\left(\frac{M}{[M_\odot]}\right)^{-1/7}\left(\frac{\dot{M}}{[10^{-8} M_\odot\ \mathrm{yr}^{-1}]}\right)^{-2/7}\left(\frac{R}{[R_\odot]}\right)^{5/7} \tag{8.22}$$

R and M are again radius and mass of the central star. The first two configurations have been discussed specifically in the context of the *accretion* of matter and high-energy emission in PMS stars and will be discussed separately in Sects. 8.3.3 and 8.3.4. The other two models deal more with the *outflow* of matter and are briefly introduced here.

The *Reconnection-driven Jet Model* [665], shown in Fig. 8.14, is a numerical MHD model, in which at an initial time setting a stellar magnetosphere truncates an accretion disk which carries a poloidal interstellar magnetic field. The directions of the fields are assumed to be the same. With progressing time, accretion and rotation drive a reconnection scenario where field lines twist through rotation while accretion pushes inward. Eventually, poloidal lines reconnect creating a neutral ring between the polar twisted fields, the stellar magnetosphere, and the disk field. Though not emphasized in the model, accretion via close magnetospheric loops persists during

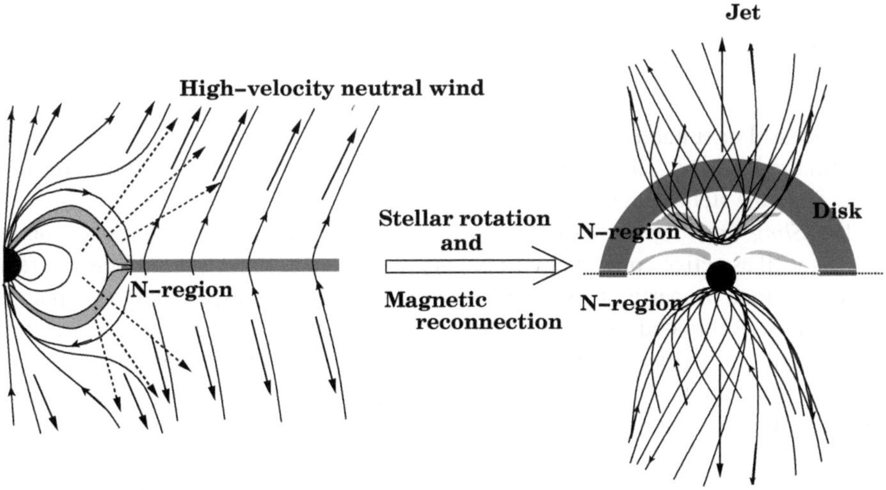

Fig. 8.14 Configuration where accretion drives a jet through rotational twisting and magnetic reconnection (for details see [665]). The model illustrates the possible magnetic connection between the inflow and outflow of matter in YSOs

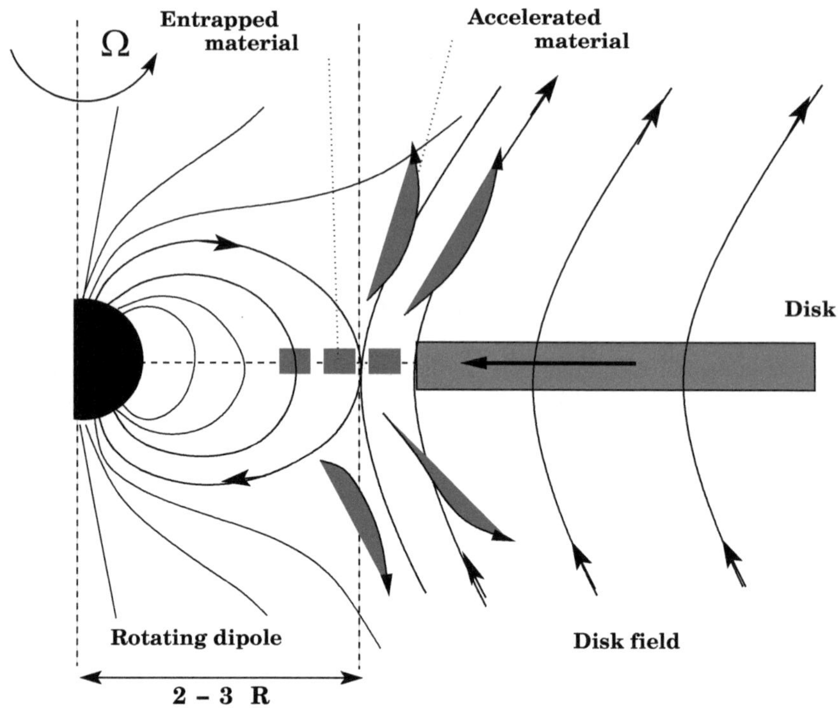

Fig. 8.15 Flares and prominences even occur in our Sun, though possibly on a smaller scale. Slingshot models like the one developed by [666,667] attempt to characterize the stability of matter trapped in magnetic loops with respect to perturbating forces. In the configuration depicted the matter breaks off of an accretion disk, gets embedded in the field and accelerated outward by magnetic perturbations. At older stages the embedded material comes from the stellar surface or may be provided by binary companions

the process. For suitable Alfvén radii the model produces outflows of the order of a few hundred km s^{-1}. The twist effect may also allow swept-up bipolar lobes further away from the stellar environment.

The *Slingshot Prominences* [666, 667] model (see Fig. 8.15) applies to young stars as well as active coronal environments of cool stars. In fact, prominences are a rather common property of rotating later-type stars. In this respect magnetospheric accretion or collimated outflows is not the issue at all, but phenomena associated with magnetically driven flares, prominences, and coronal loops. Prominence-type loops are not unknown to T Tauri stars [668] and may play a more prominent role during giant X-ray flares as well as in coronal emission during later stages. Prominences are not directly associated with the stellar surface and form at heights of a few stellar radii. The model investigates equilibria and stability of trapped matter in magnetic field lines and is pursued analytically [666]. In essence, an imbalance in the equilibrium of rotation, gravity and magnetic forces between the stellar field configuration and an outside poloidal field, mainly caused by the star's rotation, accelerates material out of the system.

8.3.3 *The X-Wind Model*

Most young stars emit large amounts of X-rays throughout almost all evolutionary stages. Not only are X-ray fluxes orders of magnitude larger than those observed from the Sun, young stars also engage in gigantic flares generating hard emission from plasmas as hot as 100 MK (see Chap. 10 for a full review). The fact that X-ray emission can be immediately related to enhanced magnetic activity had been established in the mid-1970s during the *Skylab* mission through the first X-ray imaging of the Sun. Hence, developing a magnetospheric model based on X-ray signatures seems sensible. X-ray observations of young stars throughout the 1980s and 1990s using various X-ray observatories (see Table 2.1) prompted F. Shu and collaborators to develop an empirical model that includes all configurational features expected from stellar formation scenarios and which can account for protostellar X-ray signatures. The X-wind model in this respect expresses a first attempt to fully incorporate all observed phenomena with respect to matter flow and emissivity in YSOs. As Shu writes [669]:
X-ray astronomers have long held the conviction that X-rays provide the crucial link that might unify seemingly disparate aspects of the rich phenomenology that involves YSOs: magnetic activity, protostellar winds and jets, and extinct radioactivity in meteorites.

Fig. 8.16 shows a schematic drawing of the magnetic field geometry as adapted from [669]. The style and some features in Figs. 8.14, 8.15, and 8.16 are deliberately held uniform to highlight the fundamental similarities in all these configurations. The *X-wind* model primarily applies to young stars where accretion is still a driving force in their magnetospheric dynamics and it thus relates mostly to protostars and CTTS. Disk accretion approximately at the truncation radius R_x divides the accretion flow $\dot{M}_{acc}$ into a funnel inflow or accretion stream onto the star $\dot{M}_{stream}$ and into an X-wind outflow $\dot{M}_{xwind}$:

$$\begin{aligned} \dot{M}_{xwind} &= f\dot{M}_{acc} \\ \dot{M}_{stream} &= (1-f)\dot{M}_{acc} \end{aligned} \tag{8.23}$$

where f parameterizes the angular momentum balance of the system. In the case of a centrifugal MHD wind, f is about 0.1 (see (8.19)). It is an important parameter in the model as it reflects the relation between the star's angular velocity Ω_s to the Keplerian velocity at the inner edge of the disk Ω_x. The system co-rotates in its equilibrium state (i.e., $\Omega_x = \Omega_s$), which is supported by magnetic torques. The configuration also needs zero field surfaces separating the stellar dipole and surrounding poloidal fields. In Fig. 8.16 these regions are labeled as H and R.

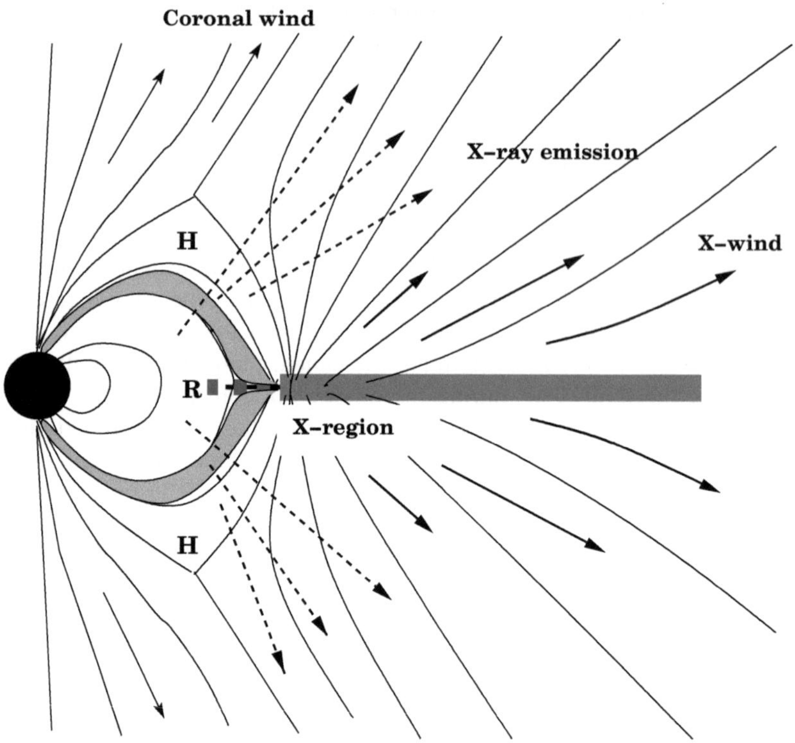

Fig. 8.16 The X-wind model as developed by Shu et al. [669]. This schematic shows the magnetic field configuration with the funnel accretion flow along the stellar dipole field. Key locations are the X-region near the inner disk from which the X-wind is actually launched, the *helmet dome* (H) between the stream and the coronal wind as well as a *reconnection ring* (R) with the possibility of chondrule production

Ampére's law requires that large electrical currents flow within these null field planes. Dissipation of these currents leads to reconnection events of opposite field polarities reducing the amount of magnetic flux trapped in the X-region and pushing the system out of equilibrium. The magnetic energy $\mathcal{M}$ released can then be expressed by:

$$\frac{\mathrm{d}\mathcal{M}}{\mathrm{d}t} = \eta |\Omega_{surf} - \Omega_x| \frac{\mu^2}{R_x^3} \tag{8.24}$$

Here η is a geometrical scaling parameter likely less than unity as not all magnetic field lines participate in a specific release event. For η equal to unity (8.24) allows one to estimate the pool of energy available to be irradiated. For typical angular velocities (i.e., $|\Omega_{surf} - \Omega_x| \sim 10^{-6}$) and field conditions at the truncation radius (see (8.22) and (8.22): $B_{surf} \sim 3$ kG and $R_x = R_{trunc} = 6R$), (8.24) yields between 10^{32} and 10^{33} erg s^{-1}. The true X-ray luminosity $L_x = \epsilon\, \mathrm{d}\mathcal{M}/\mathrm{d}t$ is much lower because of the finite efficiency ϵ for X-ray generation, and some of the energy is diverted into the acceleration of fast particles leading to what are known from the Sun as *coronal mass ejections (CMEs)* [669].

8.3.4 Funneled Accretion Streams

The actual accretion of mass onto the central star mediated by the stellar field happens through accretion streams along field lines. This form of accretion is inherent to all the configurations presented so far, but was specifically emphasized by L. Hartmann [290]. Such a configuration is illustrated in Fig. 6.14. The funneled magnetospheric material emits broad emission lines as it falls along field lines into accretion columns. Eventually the material crashes onto the stellar surface producing hot plasma emission through accretion shocks. The maximum total energy released by such a process can be estimated by:

$$L_{shock} \simeq \frac{GM\dot{M}}{R}\left(1 - \frac{R}{R_{trunc}}\right) \rightarrow \frac{5GM\dot{M}}{6R} \tag{8.25}$$

assuming that the magnetospheric radius is at $\sim$6 stellar radii. How much of this energy is eventually converted into radiation is highly uncertain and depends on the amount of energy that is actually dissipated in the disk (and thus the inner disk radius) and to some extend also on how much energy contributes to the angular dynamics of the system [290]. However since the energy dissipated in the disk at the magnetospheric radius (see (8.4)) can only be a small fraction of L_{shock}, most energy must be released in the accretion shock.

Most of the dissipated energy will be released in the form of UV continuum and X-ray emission. A study of optical excess emission in T Tauri stars [670] finds that the blue portion of the optical emission is notoriously underestimated as a result of the overpowering extinction in the UV band (see Chap. 3). Modeling yields photospheric temperatures of $\sim 10^4$ K and thus large fractions of emissivity is expected to occur below $\sim$3,100 Å [670]. For an in-depth discussion of the UV continuum emission from accretion shocks see [290]. A substantial amount of emissivity can be expected in X-rays as well, which is an issue that has been known for some time, but which remained fairly exotic until recently as it lacked observational support [447]. For large magnetospheric radii, the infall velocity approaches the surface escape velocity, i.e.:

$$v_{in} \simeq \left(\frac{2GM}{R}\left(1 - \frac{R}{R_{trunc}}\right)\right)^{1/2} \rightarrow \left(\frac{5GM}{3R}\right)^{1/2} \tag{8.26}$$

which for a solar mass star with a stellar radius of about 2.5 $R_\odot$ yields velocities of up to 350 km s^{-1}. Velocities of this order of magnitude have been inferred from observations of broad H_α lines in CTTS. Shocks at such velocities reach plasma temperatures of up to 2 MK (see (3.13)), which would be observed as thermal X-ray emission.

8.3.5 Magnetic Reconnection and Flares

One of the integral physical processes many MHD models refer to as a release mechanism for magnetic energy is *reconnection*. Magnetic reconnection events are well known from solar coronal research, specifically as a driving mechanism of CMEs. Models invoking reconnection to explain powerful X-ray flares in YSOs have been considered since the 1990s [664, 669, 671–674]. The modeled timescales and temperatures are of specific interest for comparison with observations. One of the first treatments of X-ray flares from the active binary *Algol* has been performed by [671]. From a balance between conduction cooling and reconnection heating [675, 676] derived a scaling relation between the reconnection plasma temperature and magnetic field strength:

$$T_{rec} = 7.3 \times 10^7 \left(\frac{B}{[100\ \mathrm{G}]}\right)^{6/7} \left(\frac{n_0}{[10^{10}\ \mathrm{cm}^{-3}]}\right)^{-1/7} \left(\frac{l_{loop}}{[10^{10}\ \mathrm{cm}]}\right)^{2/7} \mathrm{K} \quad (8.27)$$

where B is the equipartition field strength at the reconnecting site defined by the equality of magnetic and thermal pressure,

$$\frac{B^2}{8\pi} = 2 n_e k T \quad (8.28)$$

n_0 is the plasma density in magnetic loops before the energy release and l_{loop} is the characteristic length of the reconnecting loop. Simulations, however, show that the actual temperature of the plasma after energy release is only one third of T_{rec} [676].

The reconnection timescale is proportional to l_{loop}/v_A, where v_A is the Alfvén velocity [677]. The range of proportionality (λ_r) is still rather arbitrary, recent observations of solar flares indicate values between 10 and 1,000 [678] regardless of the flare size. With the definition of the Alfvén velocity the reconnection heating timescale can be written as:

$$\tau_{rh} \simeq 1.5 \times 10^3 \left(\frac{\lambda_r}{100}\right)^{-1} \left(\frac{n_0}{[10^{10}\ \mathrm{cm}^{-3}]}\right)^{1/2} \left(\frac{l_{loop}}{[10^{10}\ \mathrm{cm}]}\right) \left(\frac{B}{[100\ \mathrm{G}]}\right)^{-1} \mathrm{s} \quad (8.29)$$

The radiative cooling timescale of a plasma can be estimated from the energy balance of the cooling plasma; assuming radiative losses under LTE conditions [679, 680] to be:

$$\tau_{rc} \simeq 1.4 \times 10^4 \left(\frac{n_0}{[10^{10}\ \mathrm{cm}^{-3}]}\right)^{-1/4} \left(\frac{l_{loop}}{[10^{10}\ \mathrm{cm}]}\right)^{1/2} \left(\frac{B}{[100\ \mathrm{G}]}\right)^{-1/2} \mathrm{s} \quad (8.30)$$

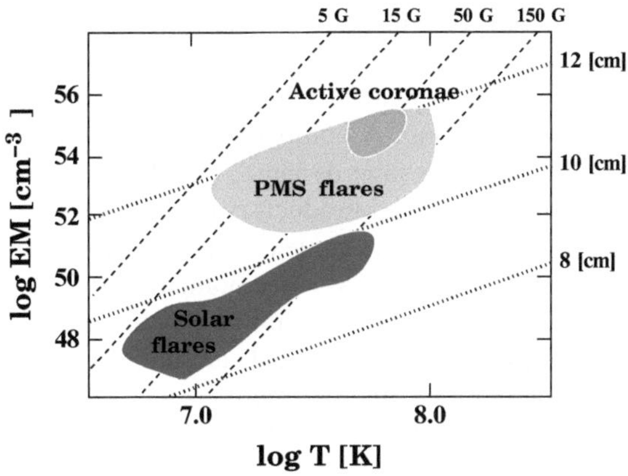

Fig. 8.17 Emission measure versus temperature diagram of solar and stellar X-ray flares. The emission measure is the integral of the emissivity over the entire emitting volume. The data for the region covering solar flares were taken from [681], for the region of PMS flares from [682], and for flares in stars with active coronae from [681,683,684]. The dashed and dotted lines represent scaling laws derived by [675,676,682] for B and log l_{loop}

When applied to radiative flares, the rise time of such a flare is of the order of 10^3 s with the decay time an order of magnitude longer.

Most young PMS stars exhibit field strengths as high as 3 kG. Such levels are only reached in Sunspots, whereas the Sun's surface field is orders of magnitude lower. The situation is slightly different in prominences, which in the Sun may reach field densities similar to loops in T Tauri stars [124]. Stellar flares as compared to solar flares, specifically from T Tauri and ZAMS stars, can be far more energetic. Flares in PMS stars reach temperatures of up to 10^8 K and emissivities are in the same category as observed in stars with active coronae (see Figure 8.17).

8.3.6 Origins of the Stellar Field

Last, but not least, there should be a few remarks about possible origins of stellar magnetic fields in protostars and PMS stars. There may be two ways to look at this issue. The magnetic field could be of fossil origin and inherited from the pre-stellar collapse [685]. Such a scenario is interesting only for young protostars as once the star becomes fully convective and appears on the Hayashi track these fields should diffuse quickly [686], although some relic field strength may persist even after a radiative core develops; arguing that concentrated flux in rope-like structures can resist turbulent diffusion [686]. Furthermore, magnetic flux captured from the convective core could remain trapped for a long time, even well into main-sequence life [687,688]. The latter is specifically interesting in the context of magnetically active intermediate mass and massive stars (see Sect. 10.3.7). Another appealing aspect of relic fields is its likely dipole structure, preferring the magnetic configurations discussed above.

The origins of the magnetic field in young stars is to date unknown. A likely mechanism is the *stellar dynamo*. A comparison of the Sun's coronal properties, which are generated through the solar dynamo, with PMS stars shows similarities, but also compelling differences. An introductory review of the processes involved is given by [689] and specifically in the context of PMS stars by [690]. Some aspects will also be discussed in Sect. 10.1.6. At some point during the PMS evolution such dynamo activity has to initiate. However, up to today the precise timing is entirely unclear. In the dynamo picture, magnetic fields are initiated by differential rotation at the interface between radiative and convective zones. Thus for fully convective PMS stars on the Hayshi track, for fully radiative stars, and, in general, stars with an internal structure different than the Sun, the dynamo picture may have severe shortcomings and theoretical concepts are in need.

Chapter 9
Massive Star Formation

Since the generation of the first stars (see Sect. 9.6) massive star-formation has been the major driver in the evolution of the universe as we know it today. Massive stars are the factories that produce elements up to iron during their lifetimes and enrich the ISM with heavy trace elements during their supernova phases. Furthermore their UV radiation, stellar winds, and explosive deaths deposit much of the energy and turbulence into the ISM we see today. Even though massive stars are comparatively rare at a given time according to the IMF (see Fig. 3.27) their average lifetime of a few Myr (see Eqn. 2.2) makes them very efficient element manufacturers. During the Sun's lifetime of 8 Gyr, massive stars will have enriched a few 10^4 $M_\odot$ by many orders of magnitude if we assume an average mass of 20 $M_\odot$ for each star. In that respect massive stars dominate the total energy output of galaxies as well as the energy balance in the ISM (see Chap. 3).

A massive star is defined by its mass needed to provide the amount necessary to exceed the Chandrasekhar limit during core collapse, which either results in a neutron star or a black hole. The Chandrasekhar limit is the mass above which electron degeneracy in the stellar core is insufficient to resist gravitational collapse at the end of a star's life. This limit is $\sim$1.4 $M_\odot$ and numerical studies show that at least a star of 8 $M_\odot$ is necessary to provide enough core mass. In order to break neutron degeneracy in the stellar core to form a black hole, a mass of about 25 $M_\odot$ is required.

Our understanding of the formation and early evolution is still quite rudimentary. This has to do with the simple fact that very young massive stars are not only rare but on average spend less than 10^5 yr or about 15% of their entire short lifetime [691] in their PMS phase and only a fraction of that time in possible collapse and active accretion phases. Furthermore, high dust extinction and only sketchy guidelines of how to identify cores containing forming massive stars make observations and their interpretation very difficult. Many times one takes advantage of searching in the vicinity of known active star-forming regions as the bulk of massive stars are born in cluster environments. Last but not least, the theoretical treatment is much more complex than what has been described in Chap. 5. It is dangerous to assume that high-mass star-formation is simply a scaled-up version of low-mass formation. This

N.S. Schulz, *The Formation and Early Evolution of Stars*, Astronomy and Astrophysics Library, DOI 10.1007/978-3-642-23926-7_9, © Springer-Verlag Berlin Heidelberg 2012

has to do with many issues which include that there is hardly any contraction time before the star appears on the ZAMS and radiative forces have a much different impact on the early native environment. The most problematic difference perhaps is that high mass cores must accrete large amounts of protocluster gas in almost record times.

Low-mass star-formation concepts developed several mechanisms to regulate collapse and fragmentation in molecular clouds through gravity and magnetostatics and through gravity and turbulence. Figure 4.17 shows that compressible supersonic turbulence can regulate the star-formation efficiency in molecular clouds [294, 302] and it is acknowledged that the presence of supersonic turbulence in particular has important consequences for the evolution of MCs towards massive star-formation. There are three major concepts which are being pursued and all three may actually happen depending on initial and environmental conditions [477]. It should also be emphasized that formation and evolution of seed stars in all these concepts are dominated by accretion processes. Yorke & Sonnhalter [692] and Krumholz et al. [693] consider the collapse of isolated massive molecular cores of masses up to 200 $M_{\odot}$ to produce significant accretion via disk onto a accreting protostar (see also [694–696]). Bonnell et al. [697, 698] take advantage of the fact the most stars form in cluster environments and introduce the concept of competitive accretion where depending on the location in the cluster-forming cloud some stars grow faster than others engaging in runaway growth. A third concept features stellar collisions and mergers in very high density environments of $>10^4$ stars pc^{-3} [699–701]. For some time the latter concept was considered most promising, in part because of radiation pressure problems, and the lack of available gas in highly packed locations was of some concern in accretion scenarios. However, observational evidence of massive disks and disk remnants as well as the fact that not all massive stars form in ultra-dense environments makes the merger hypothesis a more remote possibility. Mergers can happen in some extreme cases but are not considered major drivers in massive star-formation.

9.1 Properties of Young Massive Stars

9.1.1 Basic Parameters

The basic properties of young massive stars are defined by their appearance onto the ZAMS. Once massive stars appear, most of their natal environment has to have evaporated and observationally most young massive stars are likely post-ZAMS. From observations of main sequence stars with masses larger than about a tenth of the Sun an empirical formula can approximate the mass–luminosity relation to [19]

$$\log\left(\frac{L}{[L_{\odot}]}\right) \approx 3.8 \log\left(\frac{M}{[M_{\odot}]}\right) + 0.08, \tag{9.1}$$

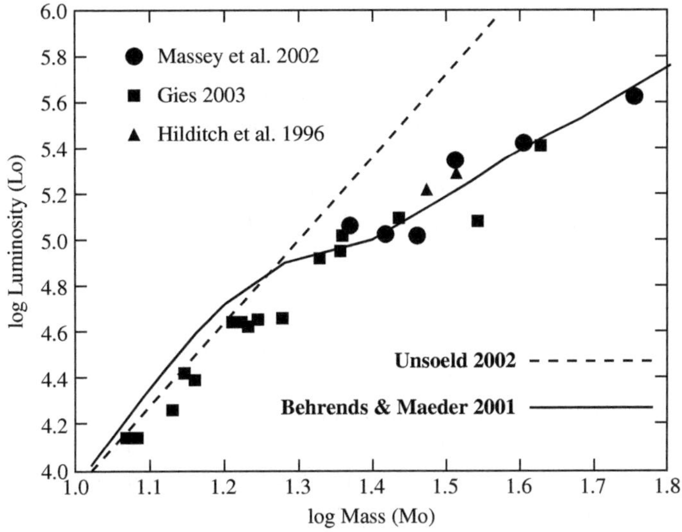

Fig. 9.1 Mass–luminosity relationship for young OB stars from binary orbits with data from [703–705] (from Crowther 2004 [702])

Table 9.1 Massive star parameters on the ZAMS (approx. from [695] and Eqn 2.2)

Spectral types	L [10^3 $L_\odot$]	M [$M_\odot$]	T_{eff} [10^4 K]	t_{life} [Myr]	General designation
B3V–B0V	3–58	8–16	12–31	29–20	Early B-type
O9V–O6V	58–130	16–32	31–43	20–1.8	Late O-type
O5V–O2V	145–500	33–64	45–51	1.8–0.9	Early O-type
WNL–H	500–2700	66–150	51–56	0.9–0.4	O/WR-type

where L is the bolometric luminosity of the star. However, this relation breaks down above about 20 $M_\odot$. Figure 9.1 shows an observational account of masses and luminosities up to 65 $M_\odot$ [702]. The hatched line represents Eqn. 9.1. Its shows that for stellar masses larger than 20 $M_\odot$, the luminosity relation becomes much shallower. A semi-empirical approach using some constant fraction assuming twice the accretion rate with respect to the mass outflow rate (see Sect. 9.4.5 for more details) calculated basic properties of young massive stars from 1 to 85 $M_\odot$ at the actual birthline [695]. This approach produced a fair match to the observed data in Fig. 9.1 in terms of the mass–luminosity relation. Table 9.1 shows some of these basic parameters of designated young massive star groups from these calculations with some extrapolation to higher masses. The lifetimes are from Eqn. 2.2. It defines parameter limits for designations of early B-type, late O-type, early O-type, and WR stars in the text.

9.1.2 Stellar Wind Properties

OB-stars generally show strong outflows generated by UV line-driven radiation pressure. While during pre-ZAMS accretion phases massive outflows dominate the evolution, it is mostly the stellar UV radiation field which drives the evolution of the star after accretion ceases. First through evaporation of its natal environment (see Chap. 8), then through mass-loss by the line-driven wind on the main sequence. Its presence has a major impact on ISMs in galaxies (see Chap. 3). Stars with luminosities larger than $10^4\ L_\odot$ show evidence for the presence of such a wind [706]. The theoretical framework for the mass-loss though line-driven winds was developed by Castor, Abbott & Klein (CAK) [707]. There are many reviews describing massive stellar winds, notably Lamers & Cassinelli [406] and Kudritzki & Puls [708]. However, all these studies deal with winds from "normal" hot stars, OB-stars and A-supergiants, rather than young massive stars near the ZAMS.

The mechanism of hot star winds is based on the transfer of photospheric photon momentum to the stellar atmosphere through line absorptions. Matter will be accelerated until it reaches a terminal velocity v_∞. The mass-loss rate at a given radius is defined by the velocity field $v(r)$ and the equation of continuity,

$$\dot{M} = 4\pi r^2 \rho(r) v(r), \tag{9.2}$$

with

$$v(r) = v_\infty \left(1 - \frac{R}{r}\right)^\beta \tag{9.3}$$

where v_∞ is the final or terminal velocity of the wind, R is the stellar radius, and β a field parameter somewhere between 0.8 and 1.5 in most winds. The absorption of photo-momentum is key and Kudritzki et al. [709, 710] could show that the outflow can be characterized by three parameters describing the fraction of optically thick/thin lines (α), the number of strong lines (k), and δ the ionization balance. It was shown [711] that $k \propto Z^{(1-\alpha)}$ and that

$$\dot{M} \propto Z^{(1-\alpha)/(\alpha-\delta)}. \tag{9.4}$$

Furthermore, wind momentum and stellar luminosity are connected wia the wind-momentum–luminosity relation

$$\dot{M} v_\infty \left(\frac{R}{[R_\odot]}\right)^{1/2} \propto L^{1/(\alpha-\delta)}, \tag{9.5}$$

where L is the stellar bolometric luminosity. The relation provides a powerful tool to diagnose stellar winds. These relations also show the dependence of the wind momentum and mass loss rate on the stellar luminosity and metallicity. It has been shown that in the LMC and SMC, which have lower metallicity than the Milky Way, terminal velocities, wind momenta, and thus outflow rates are smaller on average [708].

9.2 Distribution of Massive Stars

9.2.1 Clusters and Associations

Morphology, size, and distribution of clusters and associations will be the subject of Chap. 11. Here the focus is on the massive stellar content and their formation history. The bulk of massive stars are found in clusters and associations (see also Sect. 11.3.3) and more than half ($\sim$70%) of all massive stars still reside within these structures. There are about 70 OB associations known in the Galaxy but the true number is more like a few hundred spread throughout the entire Galaxy.

Massive stars in associations appear in various environments and have a large variety of kinematic properties. The first to monitor the proper motions of O-star in associations was A. Blaauw in the early 1950s, who determined the average velocities to be a few km s^{-1}, though some can reach over 40 km s^{-1}. The motion is one of expansion emanating from a core region. Most of the time, one finds some giant H II region harboring a young open cluster with massive stars locked into their centers either at or near the association core. Specifically, O-star members exhibit a large variety of kinematic properties (see [712]). Observations of OB associations outside the Galaxy are also helpful in this respect, specifically within the lower metallicity galaxies *LMC* and *SMC* (see, e.g., [713]).

OB star clusters and more loosely bound OB associations are now believed to be the end products of most massive star-formation activity today (see [197,477]). The relation of very massive stars, i.e. late O-type and earlier, in molecular clouds of a certain mass has been addressed in Chap. 4 (see Fig. 4.5), where [210] calculate a probablity of about 50% for a cloud mass of $10^5\ M_\odot$. In that respect smaller clusters are less likely to carry very massive stars and its massive stellar content is restricted to various B-star populations. The ONC (see Sect. 11.2.1) with 414 pc is the closest cluster containing O-stars, most massive clusters containing one or more O-stars are beyond 1.5–2 kpc. Here the Carina star-forming complex at a distance of 2.3 kpc is famous for its large number of OB stars (see Sect. 11.2.4) and for some of the nearest examples of the most massive O-stars including the first identified O3 star [714]. With up to 200 known O- and early B-stars, which include three well known WR stars, and η Car, Carina is the most massive ensemble of nearby O-star clusters. Cyg OB2 is another close massive star-formation field, almost as close as the ONC, which now has been recognized to contain large number of massive stars up to O3 types with an estimated total mass content of $5 \times 10^4\ M_\odot$.

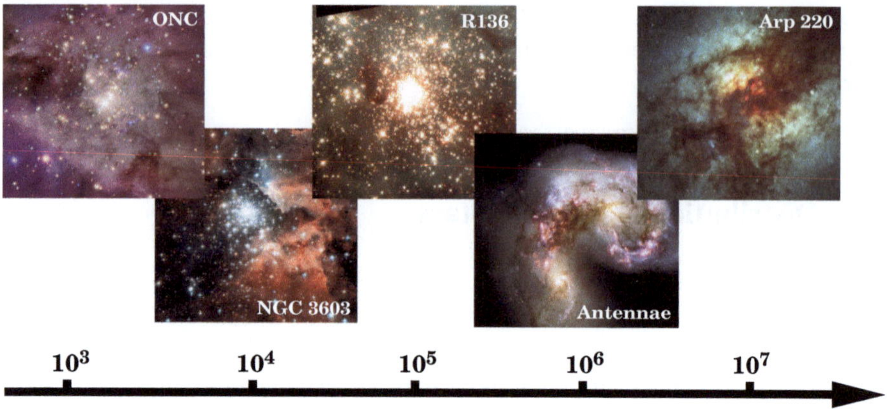

Fig. 9.2 Scaling cluster masses in various environments. The values are in $M_{\odot}$, the ONC (CREDIT: Chandra/CXC) of the order of 10^3 $M_{\odot}$, NGC 3603 (CREDIT: HST) of the order of 10^4 $M_{\odot}$, R136 (CREDIT: HST) of the order of 10^5 $M_{\odot}$, The Antennae galaxies (CREDIT: HST) host clusters of 10^6 $M_{\odot}$, the IR galaxy Arp 220 hosts clusters of 10^7 $M_{\odot}$

Clearly, the heavyweights of clusters are much further away (see also Sect. 11.1.9). While the ONC has a stellar mass of a few 10^3 $M_{\odot}$, even the individual clusters in the Carina region are of rather moderate sizes, the dense compact cluster associated with the giant H II region NGC 3603 with over 10^4 $M_{\odot}$ and R136 in the 30 Dor region in the LMC with over 10^5 $M_{\odot}$ produce the richest bounty of the most massive stars with 10s and 100s of O- and WR-stars, respectively (see mass scaling of clusters in Figure 9.2).

The Carina Region with its richness in massive stars is also structured in such a way that it looks like a precursor of an OB association. These structures on large scales contain remnants from a multitude of hierarchical massive star-formation structures and the OB-star populations extend across the entire volume of the parent molecular cloud or cloud complex. Thus OB-star locations can spread over several tens of pc. Classical examples include the Scorpius–Centaurus (Sco-Cen) association with its subgroups of the Upper Sco, Upper Centaurus Lupus, and Lower Centaurus Crux association. The added stellar mass content is significant with several hundreds of $M_{\odot}$ in OB-stars. These associations are also observed in other galaxies, perhaps the best studied extragalactic OB association is NGC 604 in M33 [715] with over 200 O- and WR-stars.

The cluster IMF will be discussed in detail in Sect. 11.1.4. Here a few remarks are made with respect to the behavior of very massive stellar clusters. There are much more massive clusters in other galaxies than in the MW. This is even true for the LMC which, despite magnitudes of lower mass content, hosts the giant cluster R136. And still, the IMFs of both entities does not differ beyond the expected statistical variation (see Sect. 3.5.2, [159]). This argument still has to be put to the test in extreme cases such as starburst and interacting galaxies. The interacting Antennae galaxies NGC 4038/4039 are specifically rich in very massive star clusters [716] which include high end superclusters with estimated mass contents of $>10^6\ M_\odot$. The determination of the massive end of the cluster function is very difficult for the various reasons outlined in Sect. 9.2.3. The massive clusters ONC, NGC 3603, and R136 exhibit some interesting details on the massive content with respect to the stellar IMF. While only R136 seem to be in good agreement with the Salpeter slope of 2.35, which also appears to be true for massive stars in the wider field of 30 Dor [717]. NGC 3603 appears to show a very flat slope and harbors excess massive stars [718], while the ONC seems to show a deficit of very massive stars [719]. It is not sure what that really means. However [720, 721] have suggested that the cluster IMF is indeed slightly flatter (i.e. 2.0) than Salpeter (see also Sect. 9.3.3), even though in many cases this is attributed to limited statistics or selection biases [165].

9.2.2 Field Stars

The remaining fraction of massive stars not found in clusters and associations are considered field stars. However, about 50% of these stars can be directly associated with runaways from nearby clusters or associations [722, 723]. Only about 2–6% of known field stars are considered to have formed outside any evident cluster environments (see also [724]. The question remains whether O-stars can form in isolation and what this means with respect to formation mechanisms. There are various implications for the formation of field stars:

- There may be a large enough probability that O-stars can form in low mass clouds (see Fig. 4.5)
- Field stars also form in larger clusters which disperse unusually quickly
- Field stars form by a different mechanism than cluster stars

The first item is based on the probability calculated by [210] of a molecular cloud to contain at least one O-star which assesses that even for a cloud mass of $10^5\ M_\odot$ that probability is very low. In fact from a typical IMF in Fig. 3.27 one expects that a minimum mass of about $3\times10^2\ M_\odot$is required to produce at least one O-star [725]. Parker and Goodwin [726] revisited the question if O-stars can form in isolation through random slope sampling of the stellar [159] and cluster [727] initial mass functions. This study suggests that the observed fraction of field stars can be reproduced within the uncertainties of the current cluster mass functions on the high end and that isolated O-stars form as very low-mass clusters. In that case, however, the initial mass function might not be tightly universal in its entirety as suggested by [159], allowing for slight slope variations. This result has the implication that no separate mechanism to form these stars is needed.

The second item seems to be an intriguing possibility as, for example, in the case where the original cluster was dispersed in a more catastrophic event, or the observed field stars could have obtained much larger kicks originally as is now being observed. A recent study by [728] investigates the latter case of the combined effect

Fig. 9.3 Wide scale 2.2 m WFI image of the giant star-forming region 30 Doradus in the LMC, which hosts the dense and very masssive cluster R136. The inset shows a blow-up HST WFPC image a runaway star and its bow shock. The distance between the runaway star and R136 is about 50 pc. (Credit: ESO; HST)

of massive binary ejection from star clusters and a second acceleration during a subsequent supernova explosion. In this two-step-ejection scenario, massive field stars cannot in the vast majority of cases be traced back to their parent star clusters. The O-star fraction affected by this two-step-ejection process is of the order of 1–4%, close to the fraction of isolated field stars in [723]. However, the stars in the latter study do not show unusually high proper motions. However, [728] pointed to the existence of a bow shock in one of these isolated stars [729] indicating that actual stellar velocities might be much higher than those measured. Figure 9.3 shows a massive runaway star with a bow shock that might have traveled over 130 pc from its suspected birth cluster in just about 1 Myr.

The third item must be considered if the initial cluster mass function is indeed strictly universal. There are some peculiar examples of extreme field stars. The high latitude 09.5V star HD935212 is more than 1 kpc above the galactic plane and must have been formed in the Galactic halo [730]. An isolated O6.5V star is ionizing a compact H II region N33 in the SMC with no association with any cluster [731], in analogy to many isolated massive stars fould in the LMC [705] unlikely to have high kick velocities. As an alternative mechanism, stellar mergers in unusually dense natal clusters might be considered.

9.2.3 *Mass Biases and Limits*

There are of course very plausible reasons why the mass function in Fig. 3.27 is so uncertain on the high mass end. Measurements of the initial mass of the stars are diluted by age and dynamical effects such as:

- as outlined in Sect 9.1.2, massive stars lose mass rapidly through their stellar winds. Mass loss rates of the order of 10^{-5} $M_\odot$ yr^{-1} imply that a 25 $M_\odot$ star loses over 35% of its mass during the first Myr of its lifetime;
- many massive stars have unresolved companions, which in many cases involve massive secondaries;
- massive stars born off center in clusters tend to sink towards the cluster center losing their birth siblings which remain in the outskirts.

These effects introduce a substantial bias in the determination of the high mass IMF. However, as [477] point out, it is the IMF for stars below 32 $M_\odot$ where these biases are effective; stars more massive in the range >64 $M_\odot$ are so rare that they are hardly affected. However, it is the first item which is majorly responsible for the difficulty to observe these most massive stars as their youth is critical.

While there is yet no accepted upper limit for the mass of stars through theory, observations place limits of the order of 150 $M_\odot$ based on the absence of higher mass stars in the Arches cluster [168], which has been re-iterated in some other studies [732, 733]. In a statistical study [733] placed a broad limit of 120–200 $M_\odot$. There are some sites in the Milky Way and the LMC, however, which still provide evidence for even larger masses. Crowther et al. [734] find that the Galactic cluster NGC 3603 hosts stars with initial masses in the range 105–170 $M_\odot$ and R136 in the LMC in the range of 165–320 $M_\odot$ suggesting a limit over twice as high. It is suggested that given the very short lifetimes of these extreme stars and potential effective mass-loss rates of 2–5 $\times$ 10^{-5} $M_\odot$ yr^{-1} the Arches cluster may already be too old to diagnose an upper mass limit. Higher limits can exist specifically in metal-poor environments (see Sect. 9.6).

9.3 Observations of Early Stages

In Chap. 5 many of the general physical properties have been described with focus on very basic relations. The observation of cores associated with massive star-formation has always been difficult as not only are early embedded stages in a massive star's lifetime very short [691], identifiers and signposts are still highly debated. It is now generally accepted that massive star-formation occurs inside dense, compact clumps in giant molecular clouds with peak column densities of larger than 10^{23} cm^{-2} corresponding to several hundred A_Vs. These clumps have over 10^3 $M_\odot$, radii $\leq$1 pc, volume densities of several 10^5 cm^{-3}, velocity dispersions of a few km s^{-1}. To observe these clumps one has to look longward of

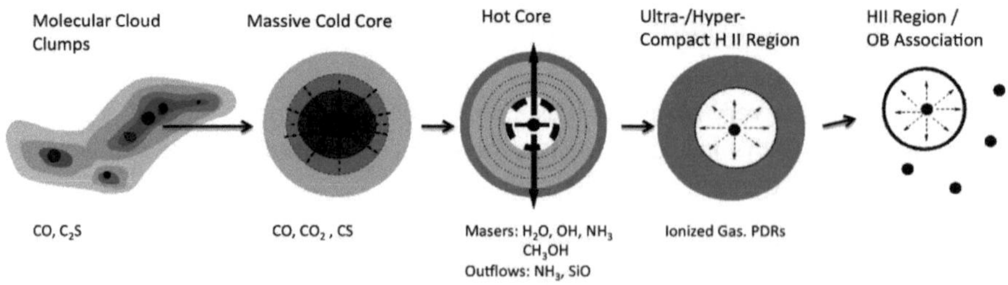

Fig. 9.4 Schematic illustration of the possible sequence from a molecular cloud toward massive stars in H II regions and OB associations. Some clumps in molecular clouds surveys might host dense cold collapsing cores, which develop into hot cores hosting an accreting massive YSO generating massive outflows, which at the end of accretion matures to hyper- and/or compact H II regions during which the young massive star ionizes its environment and eventually appears on the ZAMS in H II regions

the mid-IR into the radio band [735]. Figure 9.4 illustrates the major steps believed to lead from molecular clouds to H II regions and OB associations with molecules contributing major line emissions to identify and diagnose these phases and their structures. Figure 9.5 shows an extraordinary example of the high mass star-formation region W11 at various mid-IR, sub-mm, and cm wavelength indicating the co-existence of many of these phases, maybe the making of a future OB association.

There are a multitude of excellent reviews available and here it is referred to the review by Zinnecker & Yorke [477] for many more references. Chapters 3 and 4 already introduce to potentially relevant molecular cloud surveys. However, several specific types of surveys are predominantly useful for massive star-formation. Many surveys such as CS-surveys for dense molecular gas or methanol maser emission surveys for shock-excited compact regions as signposts for massive star-formation have been carried out. It is now also believed that newly formed massive stars are associated with compact regions of ionized gas and within those embedded in compact cores of warm dust and molecular gas. Searches today also include mm dust continuum surveys for massive cold cores and extreme early stages,

9.3.1 Massive Dark Clouds

Searches for the earliest phases of high-mass star-formation are performed on IRDCs with their high densities and low temperatures representing the initial conditions (see Chap. 4, specifically Sect. 4.1.6). The earliest phases are then expected to be massive, cold and quiescent clouds, emitting primarily at sub-mm and mm wavelengths. With masses of a few $\approx$1,000 $M_{\odot}$ and low temperatures of 10–20 K they contain no obvious IR sources or star-formation tracers. One approach to identify these earliest and cold phases of massive star-formation is to search for objects which appear in absorption at mid-IR wavelengths [735] (see Sect. 4.1.6).

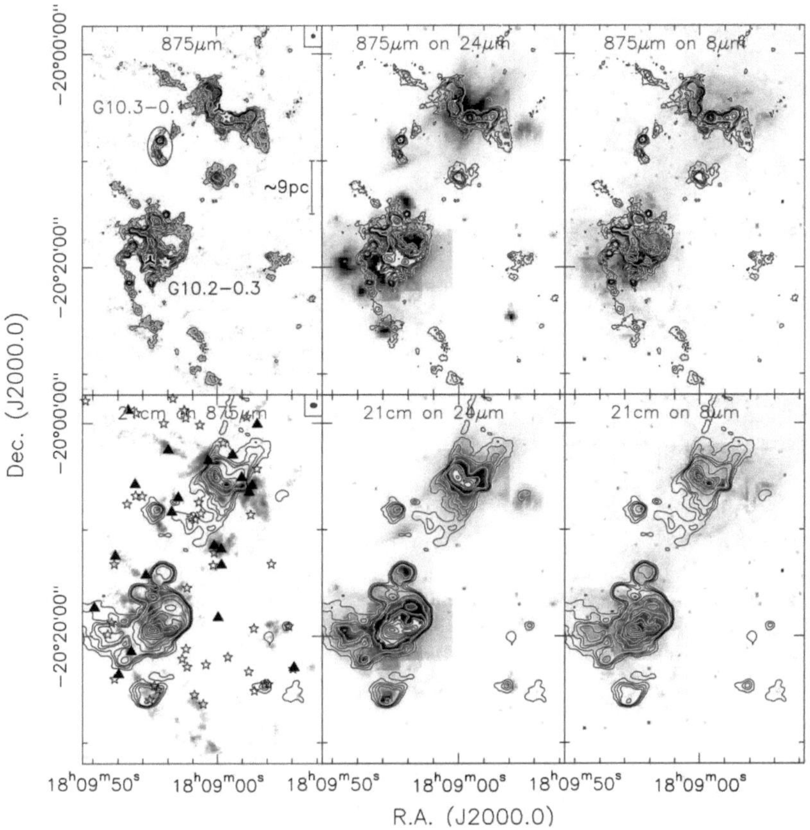

Fig. 9.5 Continuum images of the W31 region. In the *top-left* panel the 3-pointed stars mark the position of an O-star [736] and the approximate center of the southern infrared-cluster [737]. The two white 5-pointed stars show the positions of UCH II regions [738]. The ellipse marks the emission which velocity-wise is associated with cloud complexes outside our field of view. A scale-bar is shown in the *top-left* panel. The 875 μm and 21 cm beam sizes are shown in the *top-right* corners of the *top-left* and *bottom-left* panels, respectively. In the *bottom-left* panel, the triangles and stars show IRAC-identified protostellar class 0/I (the two classes are combined) and class II candidates, respectivelyi, following [423, 424, 739] (from Beuther et al. [251])

Observations indeed reveal that massive IR dark clouds have gas densities above 10^6 cm^{-3}, temperatures <20 K and sizes of 1–10 pc. However, finding actual evidence of their star-formation content is still a daunting task but many studies of these contents have now surfaced.

The *Spitzer* GLIMPSE and MIPSGAL surveys have been an abundant resource and their data yield the extraction of over 10^4 clouds [741–744]. These surveys are continued with *Herschel* and the Hi-GAL Galactic plane surveys [745, 746]. The latter now shows that very dense molecular clouds are not isothermal and that temperature gradients have most likely an important impact on the fragmentation

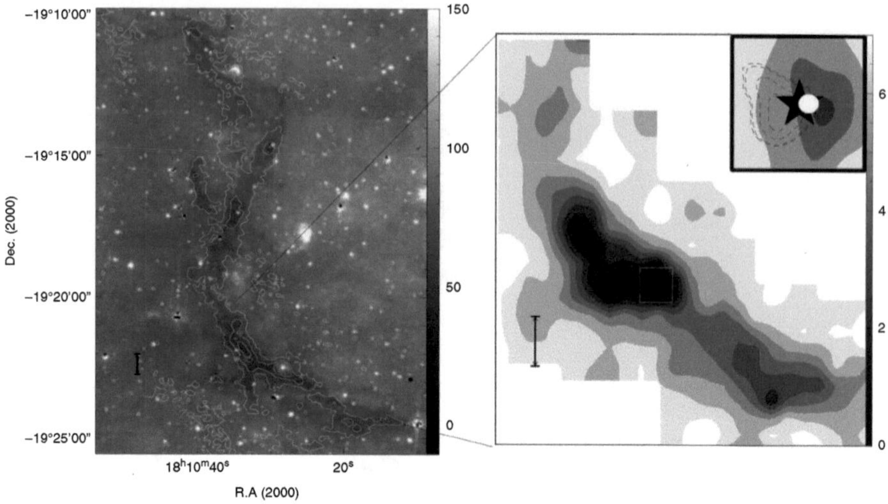

Fig. 9.6 *Left*: An 8-μm image of the IRDC G11.11-0.12 obtained with MXS with the SCUBA 850-μm map overlaid [254]. *Right*: A blow-up of the southern filament in the SCUBA map. The bar indicates a distance of 1 pc. The square inset as shown in the upper right corner shows a BIMA 3 mm continuum image (gray scale) and a 2MASS Ks band image (dashed contours). The star and filled circle denotes H_2O and CH_3OH maser positions, respectively [740] (from Menten et al. [735])

of IRDCs. A group of high mass IRDCs (10^3–10^4 $M_\odot$) show very high mean surface densities similar to values found in the ONC (see Chap. 11) which suggests that these are initial clouds for massive star clusters [747]. Studies also indicate that NH_3 emissions correlate with high mid-IR absorption and with the peaks of dust emission [740] making ammonia emissions another tracer for high mass star-formation in cool cores.

Figure 9.6 shows the case of IRDC G11.11–0.12 [748] which at a distance of 3.6 kpc shows remarkable resemblance to the main filament of the OMC-1 molecular cloud (see Chap. 11) in both dimension and structure, except that it actually is infrared dark [735]. The region appears highly complex but indicates many signposts for ongoing moderate to high-mass star-formation. Strong ammonia emissions indicate a 60 K compact component which is also associated with maser emissions (see below). SED modeling reveals a stellar luminosity of $\sim 10^3$ $L_\odot$, which corresponds to a ZAMS star about 8 $M_\odot$ [740].

9.3.2 Role of Maser Emission

Maser-like stimulated emission is observed from molecules such as water (H_2O), hydroxyl radicals (OH), ammonia (NH_3), methanol (CH_3OH) and produce some

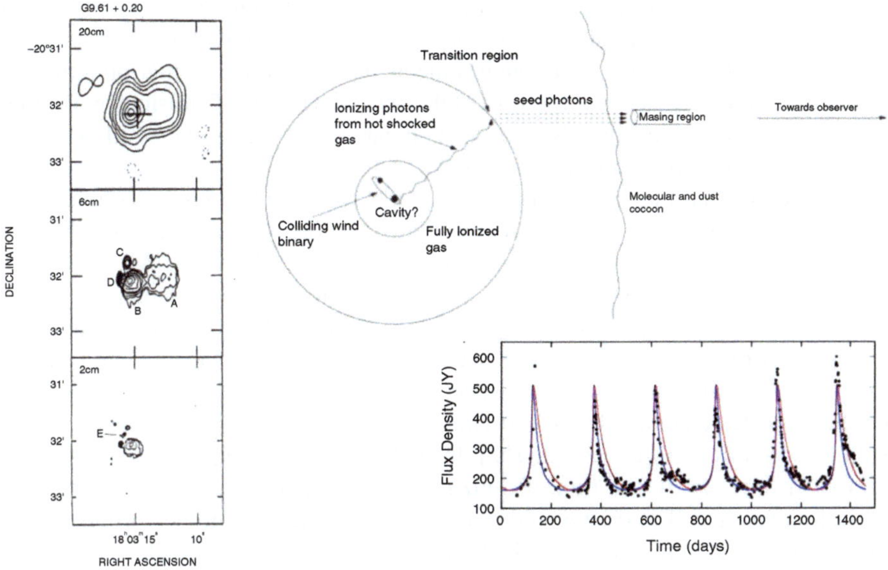

Fig. 9.7 VLT radio continuum maps of G9.62+0.20 (*left*) indicating cores A–E. The bottom panel shows a 1.5 GHz map at 16″ resolution, the middle panel a 4.9 GHz map at 5″ resolution, the top panel a 15 GHz map at 3" resolution (from [757]). The graphical sketch illustrates (*top right*) how the observed periodic maser emissions from G9.62+0.20E [758] (*lower right*) might explain the periodic signal through reprocessed emissions from orbital high-energy flux changes of a young massive colliding wind binary [759]

of the brightest spectral lines in the radio band. All are known to be associated with high-mass star-formation regions. Methanol masers are specifically interesting and have been classified in two categories, Class I and Class II, based on the involved transitions. Class I masers result from collisional excitation followed by spontaneous radiative decay and usually arise at interfaces of H II bubbles and outflows with the interstellar medium. They are therefore often observed offset from H_2O and OH masers, compact HII regions and centers of star-formation, well separated from the source of excitiation and have emissions between 25 and 146 GHz. Class II masers show strong radiatively pumped emissions in the 6 and 12 GHz band of methanol. They are ususally associated with H II regions or near-infrared (near-IR) sources and have now been recognized a reliable tracers of the early stages of high-mass star-formation [749]. Methanol masers are now also been used to trace magnetic fields in star-forming regions through polarization und Zeeman splitting [750–752].

In fact class II methanol masers are somewhat unique as tracers for high-mass star-formation regions, in contrast to OH and H_2O masers which can also trace plasmas in stars and supernova remnants. This is reinforced.by the Arecibo Methanol Maser Galactic Plane Survey, which find a close correspondence with 24 μm point sources [753]. The strongest methanol masers often show emission in multiple transitions and coincide with ultracompact H II regions [754, 755]. Most class II methanol masers show emission from the 6.7 and 12.1 GHz transitions. Modelling of these dense, dusty environments has demonstrated that pumping by infrared radiation can account for most of the observed masers. Collisions with other molecules in the ambient gas also play a significant role. Models of class II methanol masers together with a list of transitions can be found in [756].

As an exotic but maybe revealing example, class II methanol masers associated with the high-mass star-forming region G9.62+0.20E were the first of the few known periodic methanol masers ever observed [758, 760, 761]. At present no other periodic phenomenon is known to be associated with very young massive-star-forming regions. The underlying mechanism responsible for periodic methanol masers is currently speculative, but it seems that it is primarily the methanol masers that show this phenomenon. In the specific case of G9.62+0.20E (see Fig. 9.7) it has been suggested that G9.62+0.20E might harbor a colliding-wind binary system with an eccentricity of about 0.9 [762]. In such a case the orbital motion will result in a sharp increase in the flux of ionizing radiation from the hot shocked gas at periastron passage. The case would provide direct evidence that these cores harbor massive stellar systems as we see them in young stellar clusters [310].

9.3.3 Massive Core Mass Functions

The core mass functions in Sect 4.1.5 indicate that the power law distribution steepens while clump constituents transit from unbound to bound and thus stellar properties. This could also imply that the mass function flattens from MC to the stellar IMF rendering the formation of more massive stars to be a much more inefficient process than the one for low mass stars (see also Sect. 11.1.4).

However, there are some observations of massive clumps which show core mass distributions which cannot be explained by the above predicament. Clump mass spectra from the very massive and luminous cloud W49 [247] with clump masses up to 3×10^4 $M_\odot$ yielded a fairly flat power law index of 1.52, which even though it is still consistent within uncertainties with the others in the sample, the spectrum in Fig 4.9 appears different in slope. A newer observation of the massive-star-forming clump W31 [251] now also shows such a flat slope ($\alpha = 1.56$, see Fig. 9.8). However, in this case the region is already forming stars, i.e. is already on the bound side and should show a slope close to the stellar IMF [251]. Here more studies are necessary in order to confirm an evolutionary pattern.

9.3.4 Hot Cores and Outflows

Perhaps still the most famous hot core has been identified in Orion when Barrett et al. [763] noticed a blue-shifted component in the ammonia line profiles toward

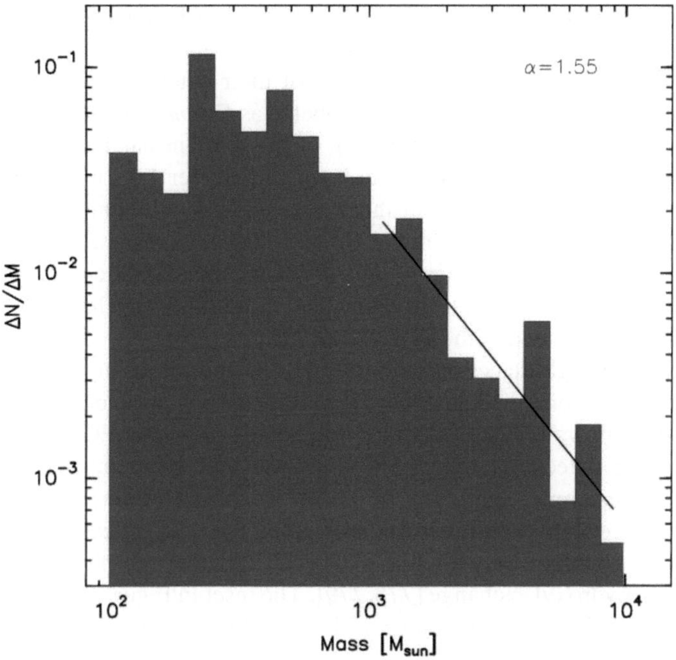

Fig. 9.8 Clump mass function for the W31 complex. Only clumps with masses >1,000 $M_\odot$ were used for the fit (from Beuther et al. [251])

the BN-KL region, which specifically appeared in the highly excited transitions which trace hot gas. The BN-KL region contains the Becklin–Neugebauer (BN) object [764], the first massive YSO discovered in the IR and the Kleinman–Low (KL) IR nebula [765], harboring highly obscured massive stars. The BN/KL region today is known as a much more complex hot core with a rather wide-angle explosive CO and NH_3 outflow [766]. This is unusual as most protostellar outflows are likely to be driven by jets and collimated winds [460, 739, 767]. It now has been proposed that the explosive BN/KL outflow may have been powered by the dynamic decay of a non-hierarchical multiple system ~500 years ago that ejected several massive stars with velocities of about 10 to 30 km s^{-1} [768].

Once young stars form inside the collapsing cold core, their surroundings are heated through radiation, shocks, and outflows driving a rich hot core chemistry [769]. Evaporation of ices produce complex and organic molecules, while shocks and jets propagating through the dense envelope release both refractory and volatile grain material, resulting in prominent SiO, OH, and H_2O emission. Compact ($\ll$0.05 pc) cores with densities of $>10^7$ cm^{-3} and masses up to several 10^4 $M_\odot$ heat up to 90–300 K [770]. Maser surveys, specifically from H_2O, OH, and CH_3OH are most important tools for locating and diagnosing hot cores and their environment.

Large-scale Spitzer surveys of the Galactic Plane have now yielded a new sample of young massive YSOs with active outflows, which may be inferred to be actively accreting. They are identified on the basis of their extended 4.5 μm emission in Spitzer images. These sources have been labeled as *Extended Green Objects* (EGOs) [742, 771–773] from the common coding of the 4.5 μm band as green in 3-color IRAC images. In active protostellar outflows, the Spitzer 4.5 μm broadband flux is likely dominated by emission from shock-excited molecular lines [773–776]. Over 300 EGOs have been cataloged in the *Galactic Legacy Infrared Mid-Plane Survey Extraordinaire (GLIMPSE-I)* survey area [771] and the mid-IR colors are consistent with those of young protostars still embedded in infalling envelopes. A majority of EGOs are also associated with IRDCs [742, 777].

However, the high detection rates of CH_3OH masers in high-resolution VLA surveys provide strong evidence that GLIMPSE EGOs are indeed massive YSOs with active outflows [772]. Molecular line surveys towards EGOs with 6.7 GHz CH_3OH maser detections found SiO emission and HCO tracers of molecular outflows. For example, sputtering of grains in shocks produces SiO molecules, which can trace active outflows because after they are released the molecules remains in the gas phase only briefly ($\sim 10^4$ yr) before they morph back into invisible SiO_2 through hydroxyl exchange [778, 779]. The inset in Figure 9.6 shows positions of a H_2O and CH_3OH maser source within a cold core complex likely indicating the making of hot cores in early phases. Fig. 9.9 shows an example of the hot core G11.92-0.61 likely at some later stages as it is clearly indicates strong outflow activity through ^{12}CO, HCO^+, and SiO emissions [772]. Figure 9.12 shows another example of hot cores traced by class II CH_3OH masers in the neighborhood of a UCH II region (see Sect. 9.3.5) in the G10.3 complex.

In most examples it is observed that these protoclusters host cores of many evolutionary stages making these regions very complex in structure and emissions. Another good example is the protocluster AFGL 5142 [780] which contains two

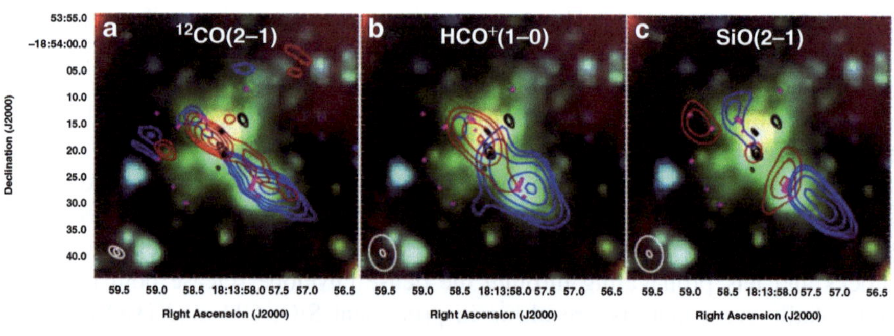

Fig. 9.9 Three-color Spitzer images (3.6, 4.5, 8.0 μm: blue, green, red) of the EGO G11.92-0.61 containing hot cores and outflows. The images are overlaid with contours of 1.4 mm continuum emission (black) and high velocity (**a**) 12CO, (**b**) HCO+, and (**c**) SiO emissions. Positions of 6.7 GHz CH_3OH masers are marked with black diamonds; positions of 44 GHz CH_3OH masers are marked with magenta crosses (from Cyganowski et al. [772])

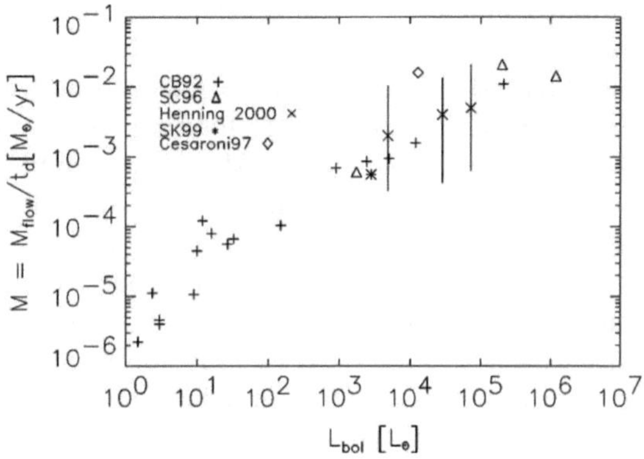

Fig. 9.10 The mass outflow rate of bipolar molecular outflows driven by protostars of bolometric luminosity (from Churchwell [691])

hot cores of similar core mass, but with 90 K and 250 K significantly different temperatures [780]. Other cores cluster there as well and even though there are multiple outflows, they appear well collimated in contrast to the wide-angle outflow in the Orion BN-KL region. The point can be made, that here since jet-like outflows and disk-mediated accretion process are physically connected (see Sect. 9.4.5) the well-collimated outflows indicate that even in this cluster environment, accretion is responsible for the formation of individual stars in the cluster [780]. However, molecular outflows have been observed toward numerous massive star-formation regions [781, 782] and they are also not as well-collimated as those from low-mass protostars (see Chap. 6). There is a tight correlation between the outflow mass flux and the bolometric luminosity of protostars that holds over at least six orders of magnitude in L_{bol} (see Fig. 9.10 and [691] for references).

9.3.5 *Ultracompact H II Regions*

Ultracompact H II (UCH II) regions [738] are the next step in the sequence of observing young massive stars and clusters. A hot core evolves to produce a massive star or cluster of stars within a dense cloud of gas and dust. The Lyman continuum from the evolving star starts to ionize its environment and free–free radio emission is now detected from the UCH II phase. Wood & Churchwell [738] first defined this phase and concluded that these objects not only are distinct from more evolved H II regions, but their precursors. E. Churchwell [691] also defines a class of hot cores which contain rapidly accreting, massive protostars as the precursors of UC H II regions referring to as PUCHs (precursors of UC H IIs). PUCHs are internally heated by the central massive protostar plus any associated lower-mass cluster members. Because of rapid accretion, the protostar does not yet produce a detectable H II region, even though it has a large UV photon flux. PUCH lifetimes are quite short, typically 10^5 yr.

There is now also a growing number of super compact and dense objects which have been discovered mostly at millimeter wavelengths [783]). Typically about

10 times smaller and 100 times denser than UCH II regions and very weak or not detected at all at centimeter wavelengths. They are called hyper-compact H II (HCH II) and the nature of these objects is not yet understood. These objects reside predominantly in massive-star-forming regions and it has been suggested that this phase is either part of the PUCHs or between PUCHs and UCH II [784, 785].

Massive stars begin burning hydrogen while they are still accreting matter, i.e. likely during the PUCH phase. During this phase they still accrete and probably do not form detectable UCH II regions. Once accretion has largely ceased, the massive star is already beyond the ZAMS on the main sequence and begins effectively ionizing the the gas within UCH II. They now become visible as shells of ionized gas and appear as various structures in radio free–free emissions. E. Churchwell [691] classifies roughly four types of structures as shells (fairly round), irregular or bipolar, cometary, and a core/halo combination. Figure 9.11 shows examples of this morphology classification. These morphologies are determined by many factors such as age, the dynamics of both the ionized and molecular gas, the density structure of the local ambient ISM, and the motion of the H II region relative to the ambient medium.

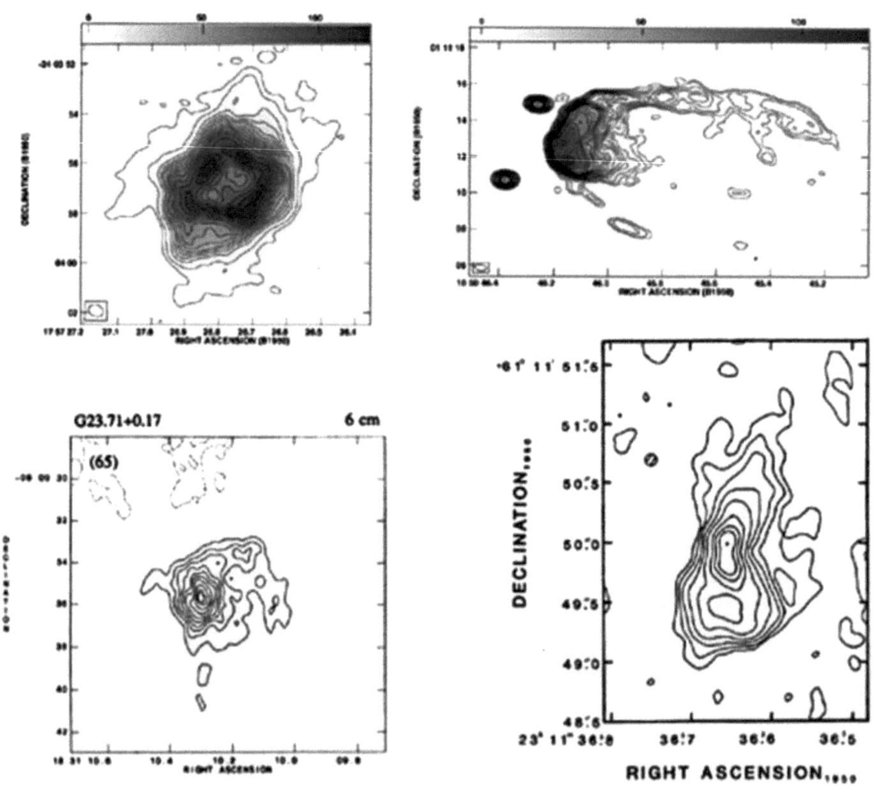

Fig. 9.11 Examples of UCH II morphological types. Top left: G5.89 − 0.39, a shell-like structure. Top right: G34.26 + 0.15, a cometary nebula. Lower left: G23.71 + 0.17 a core/halo morphology. Lower right: NGC7538 IRS1), a bipolar nebula (from E. Churchwell [691])

UCH II regions have distinct properties with respect not only to morpholoy, but also to size, density, dust heating, and molecular environment. For a nebula of uniform density and radius r, the density should scale as $r^{-3/2}$ (see Fig. A.4) as long as the ionizing luminosity remains constant. However, measurements show that the relation is systematically flatter with a slope of approximately unity [788, 789]. The distribution of ionizing luminosities is flatter if are more low than high luminosity sources and the distribution of densities simply reflects the distribution of ionizing luminosities. It also could be attributed to a situation that most UCH IIs are still embedded in a natal molecular cloud and thus tend to be smaller in diameter. The expectation of finding uniform and homogeneous conditions in natal molecular clouds is small in the first place. The $\rho \propto r$ dependence also implies non-standard absorption and extinction properties suggesting that the ionization of these H II halos requires both a clumpy medium and the destruction of dust in the interclump medium [691].

The fact that most UCH II regions are still embedded in large-scale natal MCs also implies that they heat the ambient dust out to very large radii. Thus at far-IR wavelengths UCH IIs are the among the most luminous objects in the Galaxy. Consequently UCH II regions also have diverse molecular properties. [790] has already found that a large fraction also show NH_3 and H_2O maser emissions. Molecular masers are generally very bright and ubiquitous in the vicinity of UCH II regions and include the usual candiates H_2O, OH, and CH_3OH, but also some H_2CO and ammonia. Masers are generally confined to very dense (10^6–10^9 cm^{-3}), small (a few tens of AUs) clumps, which makes them excellent probes of kinematics. In the case of H_2O many velocity components of $\sim$50 to over 100 km s^{-1} are measured. Figure 9.12 exhibits a nice example of the detection of the UCH II region G10.3C (see also Fig. 9.5), which is exhibits clear emission peaks at all observed wavelengths indicating an already quite evolved part of the region UCH II region [251].

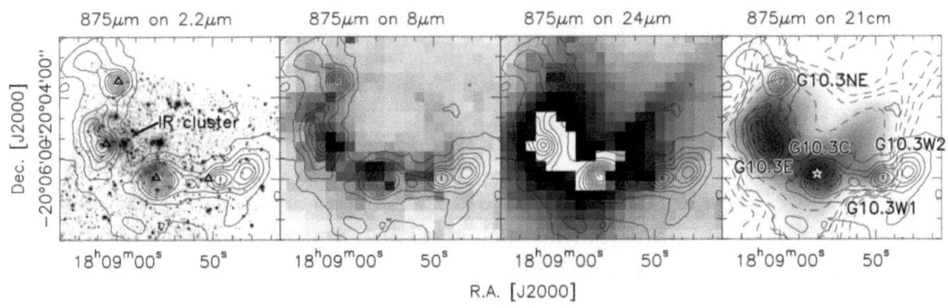

Fig. 9.12 Zoom into the G10.3 complex. In all 4 panels the solid contours present the 875 μm continuum contours, whereas the gray-scale shows other wavelength data as outlined over each panel. The K-band data are taken from [786]. The white central part of the 24 μm image is saturated. The triangles in the left panel mark the Class II CH3OH maser positions from [787]. The white star in the right panel marks the position of an UCH II region [738] (from Beuther et al. [251])

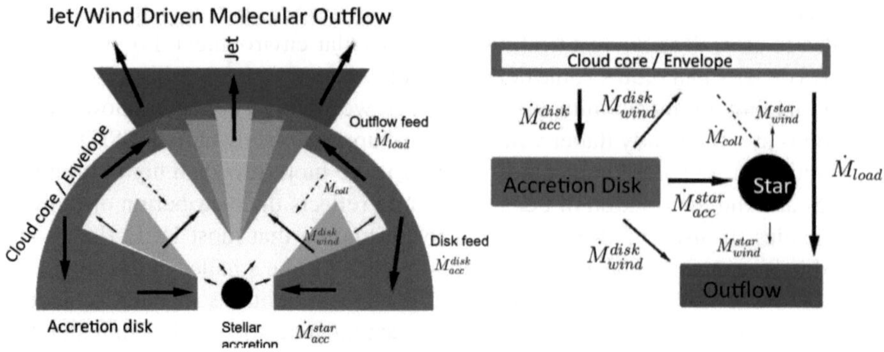

Fig. 9.13 *Left*: Schematic scheme describing various accretion and mass loss rates. The accretion disk provides the transport of matter from the cloud core to the star. The outflow is driven by stellar jets and disk winds and is fed by the cloud core. The various opening angles in the outflow pattern reflects a suggestion by [791] in which the collimating levels of outflows may be part of an evolutionary scenario. *Right*: The same demonstrating the mass exchange components in a flow diagram

9.4 Formation Concepts

This section features various formation and early evolution concepts which are currently under discussion. Massive star-formation mechanism are by far the least understood in the field and concepts are changing on rather short timescales. The situation that there may even be no unique way to form massive stars remains a strong possibility. However, the gap between theory and observations began to close in recent years and, even though much of these concepts will change over time, some of the tools and views provided in the following will prove to be useful. Figure 9.13 provides a scematic overview of the dynamical in- and outflow processes and their rates.

9.4.1 Turbulent Core Model

Even though observational efforts have been stepped up considerable with the availablity of powerful observatories such as *ALMA*, *Spitzer*, and *Herschel*, detailed knowledge about the initial conditions for massive star-formation remains rather vague for the means of modeling. Observations of velocity dispersions of a few km s^{-1} do imply the presence of supersonic motions [767, 792] in fair agreement with the turbulent core model by McKee & Tan [696]. This model is an extension of the classic self-similar paradigm of low-mass star formation (see Chap. 5) for the formation of massive stars. In this approach, core density and velocity dispersion are described as power law functions of the core radius and turbulent motions provide enough ram pressure to support the core against collapse. This accounts well for

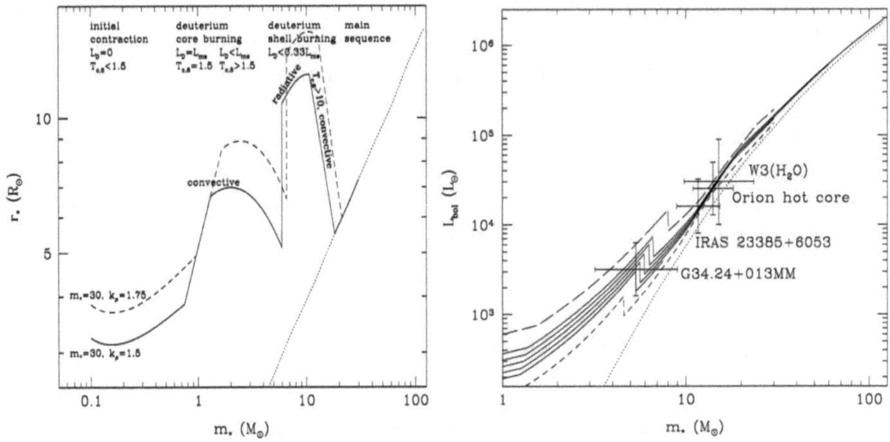

Fig. 9.14 Protostellar properties from the turbulent core model (from McKee & Tan [696]). *Left*: The protostellar radius r vs. mass m for a ZAMS mass of 30 $M_{\odot}$ and a surface density of the core of 1 g cm^{-2} for power law indices 1.5 (solid line) and 1.75 (dashed line). The ZAMS radius from [529] is shown in the dotted line. *Right*: Mass versus luminosity accreting protostars for a power law index of 1.5 and a surface density of 1 g cm^{-2}. The solid lines show ZAMS mass cases of 7.5, 15, 30, 60, and 120 from bottom to top, the top dash-line the case for 10 times, the bottom dashed line for 0.1 times the surface density. The dotted line is again from [529]. The data points are from four observed hot cores

the high pressures found in massive star-forming regions and hints that because of these high pressures massive cores are supported by internal turbulent motions. Corresponding accretion rates reach 10^{-3} $M_{\odot}$ yr^{-1} for most massive stars. As in the low-mass case, a self-similar spherical model is a significant simplification. The self-similarity parameter provides reasonably good fits to observations and enables to calculate quantities such as the timescale for star-formation and the relationship between core mass, column density, pressure, and velocity dispersion.

The model featured by [696] involves a self-similar, spherical medium in hydrostatic equilibrium, where density and pressure each have a powerlaw dependence on radius, i.e.

$$\rho \propto r^{-k_\rho}, \; P \propto r^{-k_p}, \; M = \frac{k_p c_s^2}{G} r, \tag{9.6}$$

and M is the total mass of the star-forming clump. The sphere under these conditions is a polytrope with

$$k_\rho = \frac{2}{2 - \gamma_p}, \; k_p = \gamma_p k_\rho = 2(k_\rho - 1). \tag{9.7}$$

While in the low mass case self-similar parameters adopt $\gamma_p = 1, k_\rho = k_p = 2$ and an isothermal sphere (see Chap. 5), these parameter choice do not seem to form massive stars. Observations of high-mass cores suggest a variety of k_ρ values ranging from 1.45 to 1.8 (see [696] and references therein). Figure 9.14 shows calculated protostellar radii versus accreting seed mass for stars with final core mass of 30 $M_\odot$, a surface density of the core of 1 g cm^{-2} and forming from cores with k_ρ of 1.5 and 1.75 indicating various evolutionary phases from initial contraction towards the main sequence. The smaller accretion rate of the $k_\rho = 1.5$ case leads to 25% smaller protostellar radii during most of the evolution compared to the 1.75 case. The second part of the figure compares the mass–luminosity properties of known accreting high-mass protostars with accretion rates of the order of 10^4 $M_\odot$ yr^{-1} and bolometric luminosities of a few 10^3 to 10^5 $L_\odot$.

9.4.2 Fragmentation in Massive Cores

One of the questions relating cores to final stars is how the core mass is related to the final star mass in the case of single massive star-formation. In reality, however, cores almost always fragment because they contain multiple thermal Jeans masses in the core gas and the real question is whether the core fragments into only a few objects containing one or a few massive stars at the end or in a multitude of low mass stars, the latter maybe without a single early O-type star. At densities of 10^6 g cm^{-3}, temperatures of 10 K, and 100 $M_\odot$ in the core the Jeans mass is $\sim$1 $M_\odot$ and the latter might just as well happen. Simulations [793] of collapse and fragmentation of a core based on the initial conditions given by the turbulent core model above [696] and using several equation-of-state conditions show that all cases fragment but in no cases form massive stars. In general, the effective equation of state is critical to determining whether or not gas fragments [794, 795]. The presence of turbulence, while supporting the cloud on large scales against contraction, provokes collapse and fragmentation on local scales (see Sect. 4.2.4).

These simulations, however, work well for the formation of lower mass stars as here Krumholz et al. argue that radiative feedback is not as critical at this stage. [796] make the point that isothermal and barotropic approximations completely ignore the effect that rapid heating in the densest inner core part suppresses fragmentation. Traditionally high densities produce high accretion rates leading to immediate high luminosites from a seed protostar. For a typical accretion rate of 10^{-4} $M_\odot$ yr^{-1} and massive core densities ($>10^7$ cm^{-3}), a 0.1 $M_\odot$ star at 1 $R_\odot$ produces about 300 $L_\odot$ to be radiated away. However, the innermost core is optically very thick, trapping some of that radiation, and heats up rapidly. Models by [693] indeed show that by applying higher accuracy radiation transfer and forming slightly higher core masses, core temperatures are orders of magnitude larger. As a consequence the temperature at the core is high enough more rapidly and Eqn 5.1 has some relevance estimating accretion rates even for massive stars. Key in this approach is that not only is the temperature higher, but so also is the Jeans mass, which as a consequence inhibits low-mass fragmentation (see also Sect. 9.4.3).

Other approaches acknowledge the role of global shear motions in galaxies, magnetic fields and ionization feedback on fragmentation. On the basis of a model involving turbulent collapse and sink particles (see Sect. 9.4.4) Weidner et al. [172]

examined the influence of different levels of shear on the fragmentation properties of GMCs as massive as 10^6 $M_\odot$. The simulations show a strong dependence of the amount and central concentration of stars on the strength of the applied shear. At lower levels of shear, more sinks are formed and they are much more highly concentrated toward the center of the cloud. As a consequence, the presence of shear in normal spiral galaxies impedes the formation of very large and dense super star clusters and supports the formation of OB associations with a less massive clusters at the center. However, the study has limitations as its lower mass resolution is only $\sim$100 $M_\odot$ and fragmentation down to stellar scales is not possible, as well as the largest sinks are only a few 10^3 $M_\odot$ making the conclusions highly debatable.

Peters et al. [797] include magnetic fields and ionizing radiation into their simulations, in part motivated by the role of magnetic fields in low-mass star-formation (see Chap. 5) as well as by the disruptive UV radiation field once the star exceeds about 10 $M_\odot$ (see Chap. 8). Magnetic fields support against gravitational collapse and drain angular momentum from the system through magnetic braking, disk viscosity, and outflows. The simulations show that magnetic fields can reduce fragmentation. The additional magnetic support prevents the gas from collapsing into as many secondary fragments as compared to the non-magnetic case, leaving the most massive protostar with a larger accretion reservoir. Further studies confirm that the number of fragments is reduced by a factor of the order of 1.5–2 [798].

The question remains whether there is a scenario in which fragmentation can be entirely prevented and this does not seem to be a viable case. However, [799] could show that only clouds with column densities $>$1 g cm^{-2} ($>6\times10^{23}$ cm^{-2}). can avoid fragmentation *and* form massive stars. This threshold would not only imply some environmental variation of the stellar (IMF) (see Sect. 3.5.2), but also that clusters formed with $\ll$1 g cm^{-2} should be deficient in massive stars. It should be noted, however, that densities of $>$1 g cm^{-2} are quite rare even in terms of very massive clouds. In essence this means that suppression of fragmentation should produce top-heavy IMFs at very high gas densities.

9.4.3 Monolithic Core Collapse

Hydrodynamical 2-D calculations as shown in Fig. 5.9 might lead to intermediate-mass stars as, even though the disk grows significantly, the stellar core cannot grow. Yorke & Sonnhalter [692] improved to a frequency-dependent continuum radiation transfer, specifically for the treatment of dust [800], and demonstrated that massive stars can in principle be formed via disk accretion. Overcoming radiative acceleration is the key issue as gravity at each radial point increases linearly with core mass, radiative acceleration of core material is proportional to the seed star's luminosity L_{seed} and in order to have net accretion the condition

$$\frac{\kappa_{eff} L_{seed}}{4\pi r^2 c} < \frac{G M_{seed}}{r^2} \tag{9.8}$$

has to be fulfilled. This implies an upper limit to the effective opacity

$$\kappa_{eff} < 130\left(\frac{M_{seed}}{[10M_\odot]}\right)\left(\frac{L_{seed}}{[10^3L_\odot]}\right)^{-1}. \tag{9.9}$$

In a nutshell, the mean opacity κ_R and/or the luminosity have to be decreased, or the seed star mass M_{seed} has to be increased. Reduction it the effective opacity might be reduced through density inhomogeneities [801] or dust coagulation and dust destruction during the collapse [800]. The improvement in the Yorke & Sonnhalter calculation, however, came through an enhancement of the nonisotropic character of the radiation field where the accretion disk reduces the effects of radiative acceleration in the radial direction, which was called the "flashlight effect". This effect is further enhanced by the fact that the core's optical and UV radiation field drives out matter in polar direction and thus reducing the backscatter of some of the field toward the disk. More recently Kuiper et al. [802] have repeated these calculations with higher resolution, and find that even in 2-D simulations accretion up to very high masses is possible. In 3-D simulations, furthermore, [803] find that instabilities between the gas and the radiation field provide another means of collimating the radiation and overcoming the radiation pressure barrier. This instability can in fact be analyzed formally and shown to be a generalized version of the classical Rayleigh–Taylor instability [804].

In general it has now been acknowledged that treatments in spherical symmetry are not particularly realistic. Furthermore, Krumholz, McKee, & Klein [805, 806] provide radiative transfer calculations which show that outflows with properties comparable to those observed around massive stars lead to significant anisotropy in the stellar radiation field. As a consequence radiation pressure is a much less significant barrier. Starting up with the conditions outlined in Sect. 9.4.1 Krumholz, Klein, & McKee [693] address the early stages by 3-D simulations of the evolution of turbulent, virialized, high-mass protostellar cores. As a result, for a wide range of initial conditions radiation feedback from accreting protostars inhibits the formation of fragments and the majority of the collapsed mass accretes onto one or a few objects. Most of the fragmentation that does occur takes place in massive, self-shielding disks. These produce rapid mass and angular momentum transport which allows most of the gas to accrete onto the central star rather than forming fragments. Besides mitigating the effects of core luminosity on accretion, massive stars can only form from massive cores as already described in Sect. 9.4.1 and in that respect the monolithic core collapse model is the hydrodynamical solution of the turbulent core model.

In that respect massive stars form cores similar to those in low-mass star-formation. A difference is that fragmentation does not necessarily always happens on Jeans mass size cores, but at larger masses. Krumholz & Bonnell [807] argue that at very early times in the star-formation process, radiation feedback powered by the gravitational potential energy of collapsing gas will modify the effective equation of state of the gas on scales of $\sim$0.01–0.1 pc. Equations of state that are isothermal or

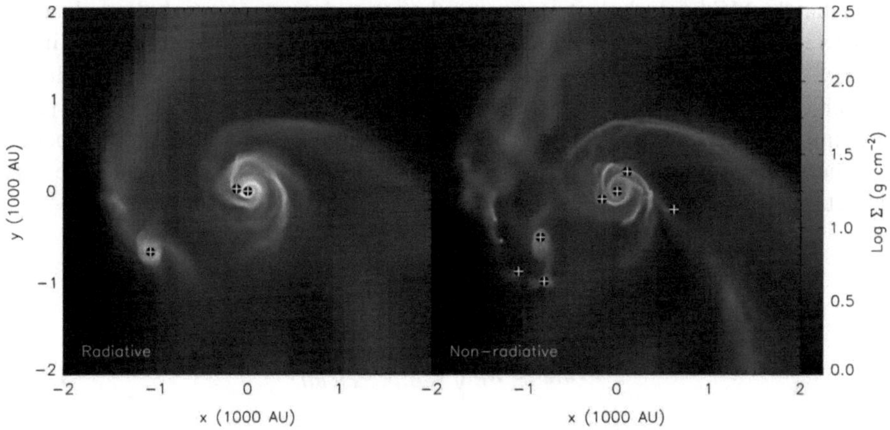

Fig. 9.15 Monolithic core collapse and fragmentation simulated by Krumholz, Klein, & McKee [693]. The simulations show column densities within a 4,000 AU box around the most massive protostars. The one shown on the left includes radiative transfer and feedback; the one on the right does not. Otherwise they are both identical in initial conditions, evolutionary time, and resolution. The plus signs indicate the locations of protostars (from Krumholz & Bonnell [693]

softer favor fragmentation; stiff equations tend to prevent it. In the monolithic core model the accretion luminosity coming from the first $\sim$0.1–1 $M_{\odot}$ seed star provides an equation of state with an effective ratio of specific heats $\gamma = 1.1$–1.2, which is stiffer than isothermal. There is, however, no single defined equation of state, since the gas temperature depends on position relative to the illuminating source and the mass density distribution. Figure 9.15 shows simulated cases with and without radiative feedback. Only a handful of fragments form out of gravitationally unstable protostellar disks that are shielded from radiation by high column densities. The strong suppression of fragmentation in these simulations suggests that monolithic collapse is a very viable process to form massive stars.

9.4.4 Competitive Accretion in Clusters

Competitive accretion relies primarily on the physics of gravity and the fact that massive stars form in clustered environments. That gravity should be the dominant physical process in forming massive stars is not in question as gravity drives formation and evolution on all scales. The primary difference between the core accretion model and competitive accretion is the setup of the initial conditions. In detail it is the physical process invoked to gather the mass of the most massive star.

Models of competitive accretion are derived from the requirement to form a full cluster of stars in addition to forming the few massive stars in the system [808,809]. This is based on the fact that most surveys show that most stars, including massive stars, are born in stellar clusters (see Sect. 9.2.1). The initial cloud conditions are such that the MC has turbulent motions in the gas and supersonic turbulence leads to shocks in the gas producing filamentary structures [808]. Here the turbulent kinetic energy is relatively close to the gravitational energy, which is consistent with MC dynamics (see Chap. 4). By invoking the scaling laws of supersonic turbulence [214] it is implied, however, that there are only small differences in turbulent velocities v_{rel} between protostars and the surrounding gas allowing for Bondi–Hoyle type accretion (see Appendix A, [698]), where

$$R_{BH} = \frac{2GM_{seed}}{v_{rel}^2 + c_s^2}$$

$$\dot{M}_{seed} \approx \pi\rho\sqrt{v_{rel}^2 + c_s^2}R_{BH}^2. \tag{9.10}$$

The initial conditions allow for large-scale fragmentation into many objects to populate a stellar cluster. Fragmentations produce lower-mass stars while the few higher-mass stars form subsequently due to continued accretion in stellar clusters, In this process large masses are funneled into the larger potential well centers and are accreted by seed proto-massive stars [698, 810]. Key here is that there are small relative velocities between the ambient gas and the seed star which is co-moving with the gas into these wells allowing for BH accretion to proceed. The cluster potential then sets another tidal radius owing to multiple gravitational sources (potentials) in the vicinity. This tidal radius and the corresponding accretion rate is then

$$R_{tidal} = C_{tidal}\frac{M_{seed}}{M_{enc}^{1/3}r_{seed}}$$

$$\dot{M}_{seed} \approx \pi\rho v_{rel}R_{tidal}^2. \tag{9.11}$$

This tidal radius is analogous to the Roche-lobe radius in accreting binary systems (see Sect. 7.4). $C_{tidal} \approx 0.5$ in the Roche-lobe approximation [544] and is used for the accretion rate of the seed star $\dot{M}_{seed}$. M_{enc} is the mass enclosed within the cluster at the stellar location r_{seed}.

Figure 9.16 (left) shows a snapshot of a numerical simulation of a 10^4 $M_\odot$ turbulent molecular cloud generating over 2,500 stars [811]. The stars form both in stellar clusters and in a distributed mode which is determined by the local gravitational binding of the cloud. Responsible is a density gradient along the major axis of the cloud. The models not only produce massive stars, but a full IMF and mass segregated stellar clusters [811]. The range of accretion rates and densities produce various star-forming efficiencies, In Fig. 9.16 the overall efficiency in the model is 15%, but with variations from about 1% in distributed regions to about 40% in large clusters, which compares well with recent Spitzer observations (Fig. 9.16, right). More studies investigating competitive accretion with respect to the stellar IMF suggest that the model may be relevant to a wider range of environments, in particular for the upper mass distribution [813].

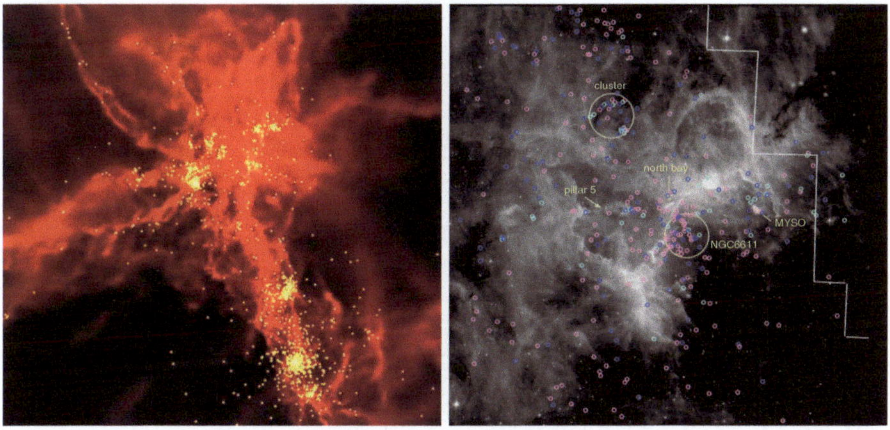

Fig. 9.16 *Left*: Simulation of infalling gas fragments forming small clusters, which then grow through (competitive) accretion of gas and stars as well as merging sub-clusters. The panel shows a 2 × 2 pc region after one free-fall time $t_{ff} \approx 6.6 \times 10^5$yr (from [811]). *Right*: IRAC 8.0 micrometer image of M16 with young stars of various inferred evolutionary states marked in different colors. Those that likely have massive circumstellar envelopes appear in light blue (adapted from [812]). Comparisons of observed clusters with simulation snapshot are limited, because of the vast hierarchical timescales observed in star-forming clusters (see Sect. 11.1.3)

9.4.5 Accretion Rates

Some of the difficulties of simply extending the formation of low-mass stars to high-mass stars are illustrated by the expression for mass accretion rate [696]. In the concept by Shu et al. [284] (see Chap. 5, Eqn. 5.1 and 5.2) low-mass stars form while accreting gas at a constant rate. The rate depends on the observed temperatures of the gas of 10–20 K, which provides reasonable rates for low-mass stars, but not for high-mass stars which also emerge from such cool cores, but it will take simply to long to heat the gas to the necessary ~100 K. Furthermore it generally always has been a problem to form seeds for massive stars because the Kelvin–Helmholz contraction timescale (see Eqn. A.38) is much less than the accretion time (see Eqn. A.39) and the massive seed star would halt the accretion process through radiation pressure and ionization.

While in the monolithic core collapse model and the competitive accretion model accretion rates are implicitly defined, it has been noted even in these treatments that the mass accretion rate is tied to the outflow rate. Empirical fits to data showed that a mass outflow rate $\dot{M}_{out}$ can be related to the luminosity parameterized outflow rate via [814, 815]

$$\log \frac{\dot{M}_{out}}{[M_\odot \ \mathrm{yr}^{-1}]} = -5.28 + \log \frac{L}{[L_\odot]} \left(0.752 - 0.0278 \ \log \frac{L}{[L_\odot]} \right), \qquad (9.12)$$

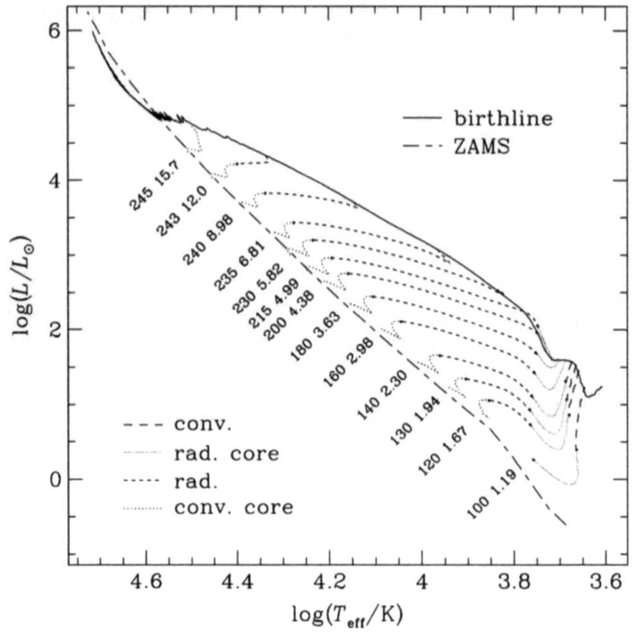

Fig. 9.17 Massive stellar tracks in the $\log T_{eff} - \log L$ diagram for stars which leave the birthline with $\tilde{f} = 0.5$. Couples of numbers represent the age t_{stop} of the star (in units of 10^3 yr) when accretion ceases, and its mass (in $M_\odot$) on the ZAMS (from Behrend & Meader [695])

where L is the bolometric luminosity of the star valid between 10^{-6} and 10^{-2} $M_\odot$ yr^{-1}. So far there is no similar relation known between the mass accretion rate and L and one has to argue whether the ratio of the accretion rate of the seed star $\dot{M}_{seed}$ is related to the outflow rate as an unknown parameter $\tilde{f} = \frac{\dot{M}_{seed}}{\dot{M}_{out}}$ [695].

The latter authors choose $\tilde{f} = f_{acc} = 0.5$ (see Sect. 9.4.1) as this value provides a reasonable fit to the observed data (see Fig 9.1, [696]). However only a fraction $f = \tilde{f}/(1+\tilde{f}) = 1/3$ of the mass transferred from the disk in the central region in accreted by the seed star [366]. In analogy to low-mass star-formation [695] calculate paths from the birthline of the star to the ZAMS. Figure 9.17 shows the result of this empirical approach. Even though this approach does not represent a physical model of star-formation it provides a useful link between observations and expected outflow and accretion rates.

Another study [816] also relies on estimates of observed outflow rates to suggest accretion rates between 10^{-3} and 10^{-4} $M_\odot$ yr^{-1}. The study focuses on accretion disk structure, its stability, and dust sublimation. The most essential finding is a very high temperature of the inner disk which for a 10 $M_\odot$ seed protostar, for example, with an accretion rate of 10^{-4} $M_\odot$ yr^{-1} may reach almost 10^5 K in the disk midplane. The disk luminosity in this case is about 10^4 $L_\odot$and thus potentially higher than that of a massive protostar (see also [817]).

9.4.6 *Stellar Mergers*

It also has been proposed that massive stars can form through coalescence of lower mass protostars or even main sequence stars at the centers of dense clusters. This radical idea gained footage at a time when radiation pressure on dust was outstopping gas accretion in the leading concept of a spherical accretion process, a problem extremely difficult to overcome (see above). Today the concepts abandoning the spherical accretion paradigm overcame this problem. Another concern, which still exists, is that specifically in very dense cluster cores there would not be enough gas available for massive stars even with reduced fragmentation levels (see Sect. 9.4.2) and a solution for this specific case is outstanding.

Even through not thought to be a major mechanism in forming massive stars, it is still possible that a collisional buildup of high-mass stars can occur in the most massive stars in very dense clusters [477]. This specifically relates to the formation of the clusters like NGC 3603 or R136 (see Sect. 9.2.1) . Merger scenarios are not particularly pretty as in the process accretion disks around the lower mass components will be destroyed, outflows disrupted, stellar components ejected. The aftermath may be surrounded by thick tori of expanding debris and perhaps explosive outflows. In this respect Bally & Zinnecker [818] proposed that the outflow emerging from the OMC-1 core in the Orion molecular cloud was produced by a protostellar merger that released between 10^{48} and 10^{49} erg s^{-1} less than a thousand years ago.

Stellar capture and merger scenarios have been invoked for the formation of compact binaries in globular clusters (see Sect. 7.4) and while direct collisions are rare, capture and binary formation are not. Accordingly, simulations by [809] suggest that true mergers of protostars are rare outside of binary systems. However, nearly one-third of all stars and most more massive stars engage in dynamic interactions between stars and suffer disruptive interactions that can truncate circumstellar disks and bring accretion and outflow processes to an abrupt and possibly explosive halt (see above; [818]). It is unlikely that highly-collimated structures could exist in the coalescence scenario and the outflow energetics may be significantly different [819]. For more on the subject, refer to [477].

9.4.7 *Impact of Massive Outflows*

To date there is no real consensus about origin, morphology, content, strength, or collimation of massive molecular outflows. Sect 9.3.4 has already provided an observational account, Sect. 9.4.5 provides a link between accretion and outflow rates, [805] and references therein emphasize the importance of massive outflows towards the star-formation process itself (see also [806]). Massive outflows seems to appear at a large variety of collimation ranges and from observations of certain collimations in specific environments, Beuther & Shepherd [791] propose that collimation decreases from systems containing B5–B2-types in the high-mass protostellar phase, to B1–O8 types in the HCH II phase, to very wide angles for early O-stars in the UCH II phase (see Fig. 9.13). In detail [819], outflows from

- A-type YSOs are similar to those from low-mass YSOs with roughly an order of magnitude higher energetics. Well-collimated jets are common and the associated molecular flows tend to decrease in collimation with age;
- B-type YSOs exhibit a large range of luminosities from 30 (late B-stars) to 10^4 (early B-stars) $L_\odot$. Early B-stars generate UCH II regions and reach the ZAMS while still accreting and generating strong molecular outflows. Intermediate-mass YSOs have mass outflow rates $M_f = 10^{-5}$ to a few 10^{-3} $M_\odot$ yr^{-1}, and mechanical luminosities $L_{mech} = 10^{-1}$ to 10^2 $L_\odot$;
- early B and late O-stars commomly have poorly collimated flows in both young and old sources. However, outflows can be well-collimated when the dynamical timesscale is less than 10^4 yr;
- O-stars with $L_{bol} > 10^4$ $L_\odot$ generate powerful winds 90° wide within 50 AU of the star while accompanying molecular flows can have larger widths. The flow momentum rate is more than an order of magnitude higher than what can be produced by stellar winds with L_{mech} exceeding 10^2 $L_\odot$;

Simulations of the morphology of massive jets and outflows are still rarer than for their low-mass counterparts (see Sect. 8.1.8). Banerjee & Pudritz [602] simulated the 3-D collapse of magnetized, massive molecular clouds comparing three different cases which include an isothermal collapse, a collapse with radiative cooling, and a magnetized collapse. A self-consistent structure of a bipolar outflow merged which produce cavities out of which radiation pressure can be released, thereby reducing the ability to accrete. Figure 9.18 shows a snapshot of this simulation. Two items appear remarkable. First, outflow signatures appear very rapidly, i.e. after collapse

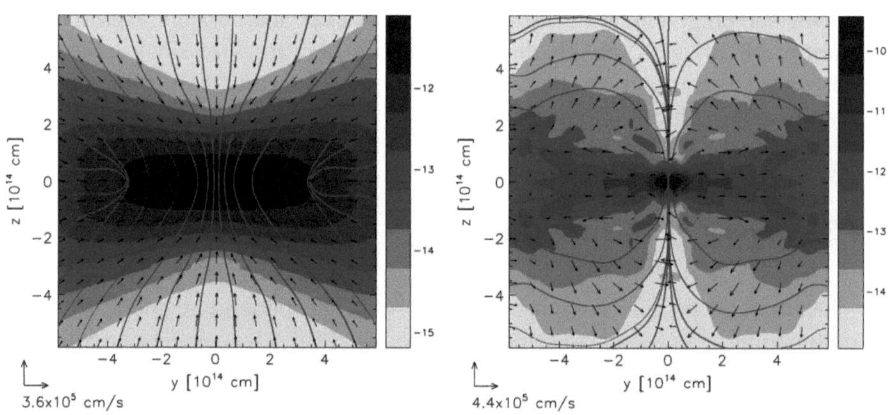

Fig. 9.18 Close-up 2-D snapshots of the disk region normal to the yz-plane in the magnetized case. The left panel shows the situation at $t = 1.45 \times 10^4$ yr ($1.08\ t_{ff}$) into the collapse. The right panel shows the configuration 188 yr later, when the outflow is clearly visible (from Banerjee & Pudritz [602])

for about 1.4×10^4 it only takes 2×10^2 yr to force matter to flow outward, much faster than to assemble the accretion disk. It is also argued, however, that the extraction of angular momentum by a disk-threaded magnetic field significantly enhances the mass accretion rate, thereby helping the massive protostar to assemble even more quickly.

Recent simulations focus on feedback scenarios and also feature the impact of outflows at early times in the massive-star-formation process. Simulations of turbulent, magnetized, large-scale collapsing cores find that outflows not only assemble early but increase the mass accretion rates onto the massive seed star by breaking up dense filaments that feed the star. A first attempt to include feedback effects of protostellar outflows into the process of monolithic core collapse [806] find that that outflows evacuate polar cavities of reduced optical depth through the ambient core. Radiative flux in the polar direction is enhanced by factors of up to 15 with respect to the midplane direction. As a result radiative heating and radiation pressure onto the protostellar disk and infalling cloud gas is largely mitigated.

9.5 Multiplicity

Taking a closer look into the cores of embedded massive steller clusters reveals that a large number of newly formed massive stars appear to be in binary, triple, or even multiple systems [477]. Massive stars very often come in hierarchical triple systems with an almost equal mass close massive or intermediate mass binary and a third more distant companion. A prominent example is θ^2 Ori A [820] , the second massive star in the Orion Trapezium Cluster hosting a pair of intermediate mass stars. The Orion Trapezium itself has now come under even more scrutiny as more and more companions to its primary members are found through ever-increasing observing sensitivities. Figure 9.19 shows the current account of primaries and and secondaries, even tertiaries [820–822]. The most massive star θ^1 Ori C is now known to have a massive companion [822], θ^1 Ori B even is a double binary massive system. θ^1 Ori E used to be an unknown massive type but now has been resolved into a pair of G III PMS stars on their path towards becoming HAEBE stars and to wind up as a late B-type binary on the main sequence [531]. Similar trapezium systems can be found in other clusters such as HD 206267 in IC 1396 or the Trifid nebula. A large study of over 40 of such trapezium systems observed over several decades finds high dynamical potential in these systems, including stellar ejections [823]. It is concluded that these trapezium systems are unstable entities with lifetimes of the order of a few Myr. HST NICMOS imaging of W3 IRS 5 hosts a binary high-mass protostar in addition to three new IR sources, and two point sources to one of the high-mass protostars [824]. This system falls within a region of 5,600 AU in diameter and are coincident with an 10^2 $M_\odot$ dense molecular clump, and might resemble a proto-trapezium deeply embedded in its natal gas.

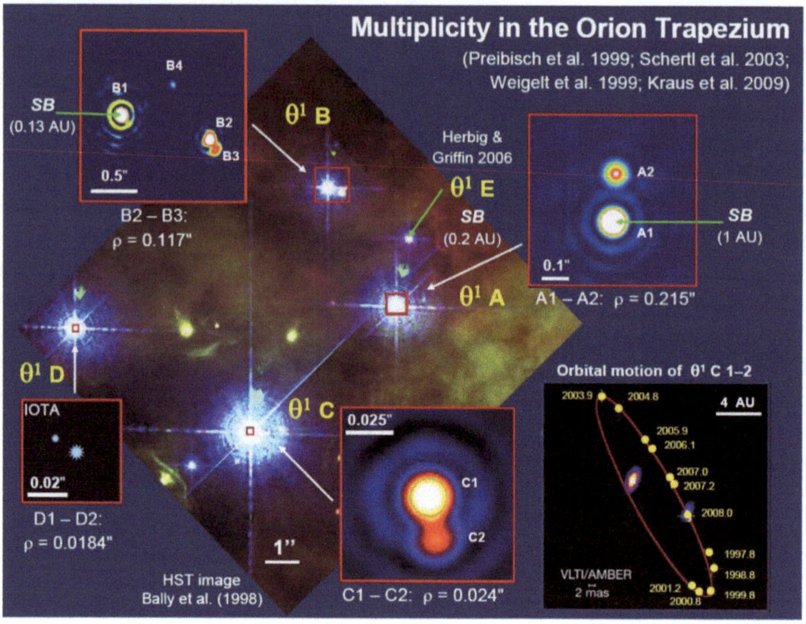

Fig. 9.19 Multiplicity at the core of the Orion Trapezium Cluster. Attached to the HST Image (Credit: J. Bally, CASA) are K-band speckle reconstructions of 13 Trapezium members mostly of O and B type. The inset in the lower right corner shows the orbit of the wide massive binary in θ^1 Ori C (Credit: T. Preibisch, LMU Munich)

It can be speculated how these trapezium systems were created in the first place. However, the aftermath of a competitive accretion scenario apears rather appealing. According to the numerical sink particle simulations (see Sect 9.4.4) gravitational and turbulent fragmentation of a supersonic proto-cluster cloud of a few 10^3 $M_\odot$ develops a few sub-clusters, which sink into the main potential well and merge into a single cluster entity with trapezium systems at the center. The cluster retains its supersonic character and will disperse quickly. Future, more detailed, high-resolution and kinematic studies on emdedded systems are necessary.

Besides fragmentation of gas-rich proto-clusters, there are other mechanisms to consider, which include disk-assisted capture [818], mass tranfer in close accreting proto-binaries [825, 826], and failed mergers [827].

9.6 The First Stars

Stars are very difficult to make in a pure hydrogen gas. Collapse concepts for low-mass stars heavily rely on the presence of dust and trace metals (see Sect. 9.4). The absence of dust at low metallicities requires different cooling agents, of which the only ones available are hydrogen and deuterium molecules. Saslaw & Zipoy [828] found that during the collapse of primordial gas clouds through a density of about 10^4 cm^{-3}, molecular hydrogen is produced and providing subsequent cooling for the formation of proto-galactic objects. Soon after [829] employed this mechanism in Jeans unstable clouds and the formation of primordial globular clusters. In the 1980s Couchman and Rees [830] formulated a first structure model in the context of the standard dark matter model, concluding that the first stellar objects could have reheated and re-ionized the primordial gas and thus influenced subsequent structure formation. Many studies up to the 1990s focused on the chemical evolution and cooling of primordial clouds using idealized collapse models and a complex chemical reaction network. [831] investigated the thermal and chemical evolution of a collapsing spherical cloud composed of pure hydrogen gas over a broad range of initial conditions introducing the effect of shock heating from colliding fragments in a turbulent collapsing cloud. The thermal and gravitational consequences of molecular formation behind fast (20–200 km s^{-1}) intergalactic shocks produced by explosions at large redshift or by cloud collisions during infall into gravitational potential wells were examined by [832] and was shown that significant fractional abundances of H_2/H, roughly 0.001, can be formed behind the shocks in the primordial intergalactic medium (IGM). In the mid-1990s, Tegmark et al. [833] computed the minimum mass that a virialized gas cloud must have in order to be able to cool in a Hubble time. The minimum baryonic mass M_b is found to be strongly redshift dependent, dropping from 10^6 $M_\odot$ at $z \sim 15$ to 5×10^3 $M_\odot$ at $z \sim 100$ as molecular cooling becomes effective. For higher redshifts, M_b rises again, as cosmic microwave background photons inhibit H_2 formation. From a standard cold dark matter power spectrum it is implied that a substantial fraction of baryons may have formed luminous objects by $z = 30$.

Abel et al. [834] presented the first self-consistent 3-D cosmological hydrodynamical simulations of first structure formation in a standard cold dark matter-dominated universe. The simulations featured a treatment of the formation and destruction of molecular hydrogen, which by then had been established as the dominant coolant in low-mass halos ($M = 10^5$–10^8 $M_\odot$) which collapse at high redshifts ($z \sim 30$–50) based on the finding by [833]. The study finds that appreciable cooling only occurs in the cores of the high-density spherical knots located at the intersection of filaments. Furthermore, only a small fraction (<10%) of the bound baryons are able to cool promptly, implying that primordial Population III star clusters may have very low mass. Higher-resolution 3-D simulations that follow the collapse of primordial molecular clouds and their subsequent fragmentation find that primordial molecular clouds with 10^5 $M_\odot$ are assembled by mergers of multiple objects that have formed hydrogen molecules in the gas phase with a fractional abundance of $<10^{-4}$ [835]. As the subclumps merge cooling lowers the temperature to <200 K in a "cold pocket" at the center of the halo. Within this cold pocket resides a quasi-hydrostatically contracting core with a mass of $\sim$200 $M_\odot$ and number densities of $>10^5$ cm^{-3}. Less than 1% of the primordial gas in such small-scale structures cools and collapses to sufficiently high densities to form primordial stars.

Fig. 9.20 Simulation of a highly massive protostar in a zero metallicity environment (Credit: R. Kaehler/T. Abel (Visualization); T. Abel/G. Byron/M. Morman (Simulation))

Figure 9.20 shows the simulation of a highly massive protostar in an almost zero metallicity environment. Today it is believed that first stars formed very much in isolation and they had to be very massive. Many simulations have converged to first star masses between 30 and 300 $M_{\odot}$. The large uncertainties in mass reflect our current imperfect understanding of early stellar evolution. Some interesting aspects also come from observations. Chiappini et al. [836, 837] analyzed eight ancient stars in the 1.2×10^4 Myr old globular cluster NGC 6522 and found unusual abundance ratios with respect to Fe of elements including europium, barium, yittrium, strontium, and lanthanum. These unusual elements can be created through the slow neutron capture process (s-process) even in low-mass stars. The conclusion is that the eight globular cluster stars were formed from gas and dust enriched by supernovae from an earlier generation of metal-poor but fast-rotating, massive stars. The idea is that metal-poor but fast-rotating, massive stars may be able to increase s-process yields of such elements by up to four orders of magnitude.

The notion first stars had to be exclusively massive has now also come under scrutiny as recent results indicate that that turbulence-induced fragmentation produces a wider range of masses than had been predicted before. Clark et al. [839] investigated the role turbulence in primordial gas focusing on two main cases involving two stellar population. *Population III.1* stars form in halos entirely unaffected by any previous star-formation activity. *Population III.2* stars form in halos that already have an increased ionization fraction. Fragment masses cover over two orders of magnitude with the former group exhibiting smaller masses and an IMF

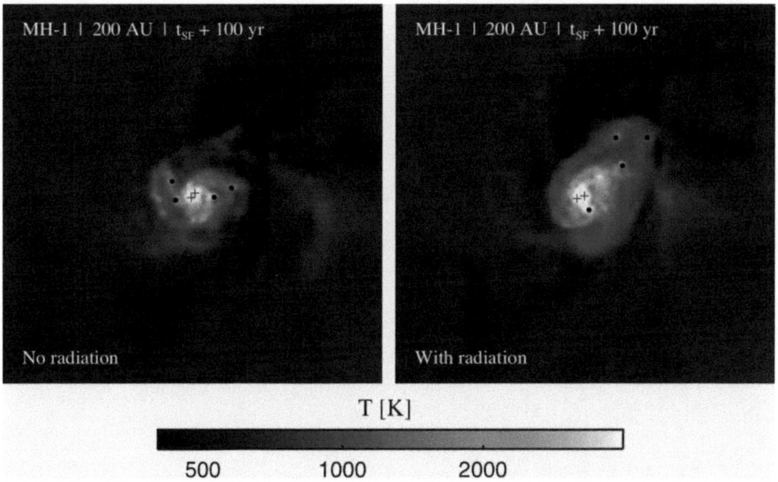

Fig. 9.21 Fragmentation of a primordial gas with and without infrared radiation emitted by accreting protostars. The gray scale is a density-squared weighted temperature projection along the line of sight. The radiation slightly heats and puffs up the disk and fragments form at larger distances from the center of the halo. The qualitative nature of radiation feedback is very similar to present-day formation studies (compare with Fig. 9.15). Due to the high temperature of the gas and its very efficient cooling by molecular hydrogen lines the radiative effect is significantly reduced. Dots, crosses, and stars denote protostars with masses below 1 $M_\odot$, between 1 and 3 $M_\odot$, and above 3 $M_\odot$, respectively (from Greif et al. [838])

much broader than previously thought. In other words, the first stars might already have already formed in clusters rather than isolation. Greif et al. [838] go a step further and model the collapse of minihalos from cosmological initial conditions (see Fig. 9.21). The simulations follow through the runaway condensation of their central gas clouds, to the formation of the first protostar. During the following accretion phase gas accumulates rapidly in the circumstellar disk around the first protostar and heavily fragments successively into small groups of protostars with masses ranging from $\sim$0.1 to 10 $M_\odot$ confirming that Population III stars likely already form in clusters.

Chapter 10
High-energy Signatures in YSOs

The discovery of bright X-ray emission from young low-mass stars in the wake of the launch of the *EINSTEIN* X-ray observatory (see Chap. 2) [66, 67] truly took researchers by surprise. X-rays from Orion had been detected with *Uhuru*, the first X-ray satellite launched in the early 1970s, but due to the lack of spatial resolving power a connection to young low-mass stars seemed far from likely. At the time, these stars were investigated for their complex emission patterns at long wavelengths. Models for gravitational collapse do not provide many clues about mechanisms for emissions at wavelengths much shorter than a few microns (see Chaps. 4 and 5). High energy emission requires special circumstances and the responsible physical processes are often different from the ones at the origins of the optical and IR emissions. In order to emit at short wavelengths, one either needs very high temperatures or very high magnetic field strengths or some mechanism to produce high-velocity electrons. The previous chapters already demonstrated that gravitational potentials in protostellar systems are by far too small to free enough energy (see Sect. 6.7). Thus the energy has to come from a different pool, which is most likely rotational and magnetic energy inherited from pre-collapse cloud dynamics. Synchrotron radiation from magnetic fields as high as 10^{11} G can be ruled out on the grounds that superdense degenerate matter, as found in neutron stars, is needed to carry such high field densities [840]. Chapters 6 and 8 also showed that magnetic field strengths in protostellar environments do not exceed 1 or a few kG.

The existence of high-temperature plasmas in the vicinity of T Tauri stars had been suggested only a year before the detection of X-rays through spectra from the *International Ultraviolet Explorer (IUE)*. Temperatures of up to 2×10^5 K and line emission from N V, C IV, and Si IV were observed in several stars [841–844]. X-ray emission requires temperatures of larger than $\sim 10^5$ K (i.e., wavelengths $<$150 Å). The only place in MS stars where such temperatures are reached outside the stellar interior are strong winds and the stellar corona. In fact, strong stellar winds as they exist in massive stars [406] can produce quite luminous X-rays from plasma heating in shock instabilities [70] and the detection of X-rays from these stars [68, 69] at first nourished speculations of strong winds in T Tauri stars [66]. Massive stars generally do not engage in coronal activities, though some observations suggest

N.S. Schulz, *The Formation and Early Evolution of Stars*, Astronomy and Astrophysics Library, DOI 10.1007/978-3-642-23926-7_10, © Springer-Verlag Berlin Heidelberg 2012

magnetic interactions in the younger proponents [407]. For low-mass stars a coronal interpretation by analogy to the Sun seemed more natural. Massive stars have fully radiative interiors whereas low-mass stars also have convection zones.

It was pointed out that the existence of both UV and X-rays could indicate a coronal connection [66]. A youthful extrapolation of solar-type coronal activity for young stars in connection with optical and near-UV excesses had already been suggested in the early 1970s. The realization that WTTS are actually not accretors (see Chap. 6), but still emit strong X-rays [845] in combination with significant radio fluxes [846] then seemed like a smoking gun. However, such qualitative comparisons with the Sun's corona, though essentially helpful, do not really account for the vast diversity and differences in the characterization of X-ray signatures in YSOs and today many observed features are far from being explained. The bottom line is that coronae play a central role in the emission at high energies, but it is also undisputed that they are only part of the puzzle. Other contributions from special magnetic configurations (see Chap. 8) as well as accretion and outflow activities are now reality as well [71, 447, 847, 848].

The mechanisms which produce X-rays in hot plasmas are manifold, though free–free emission in form of bremsstrahlung as well as free–bound emission due to collisional ionization are most likely. The order of magnitude of the expected total loss rate in the case of bremsstrahlung can be estimated from Eqn. C.28. A typical emission volume using a characteristic magnetic loop length of 10^9 cm (see Sect. 8.3.5), plasma densities of 10^{10} cm^{-3}, Z between 8 and 26 for O and Fe ions, and plasma temperatures typically 1×10^7 K would produce luminosities of the order of 10^{26} to 10^{27} erg s^{-1}, which is close to the Sun's X-ray luminosity. PMS stars are more than 10^4 times more luminous in X-rays (see below) indicating either plasmas of higher densities or larger magnetic loop sizes. As a rule of thumb one can assume a domination of discrete line emission from collisional ionization below 5×10^7 K and thermal bremsstrahlung above this temperature [680, 849].

10.1 The X-ray Account of YSOs

The first extraterrestrial X-rays were detected from the Sun by H. Friedman in 1949 in a V2 rocket flight. It took another 25 years to take the first X-ray image during a *SKYLAB* mission in 1973. Figure 10.1 shows two more modern FUV and soft X-ray images of the Sun as seen with the *YOHKOH* (left) soft X-ray telescope [4] giving a stunning view of a stellar corona. Loops as seen in the *SDO* FUV/soft X-ray image give an impression of loop sizes in MS solar type stars. In PMS stars these loops are significantly larger in size and emissivity. Modern telecopes are not yet powerful enough to image coronae from other stars, but Fig. 10.1 gives a small-scale impression of coronal activity in young stars.

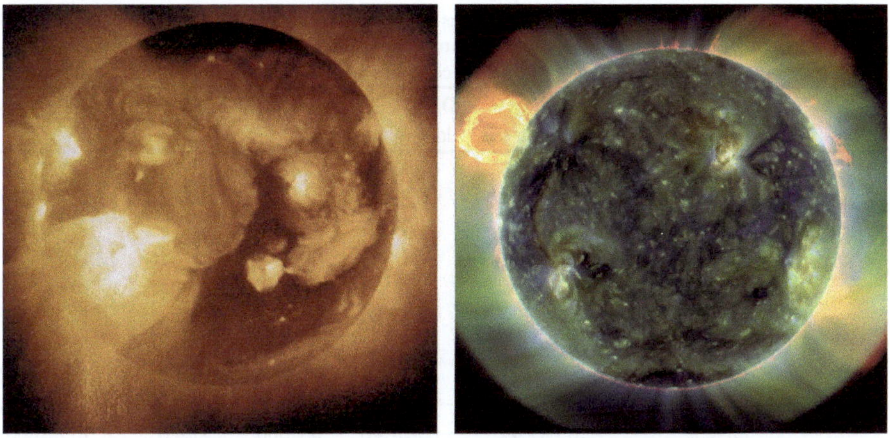

Fig. 10.1 *Left*: The Sun in X-ray light as observed with the *YOHKOH Soft X-ray Telescope* in 1991. Seen is the circumference of the solar surface at a radius of $\sim 7 \times 10^{10}$ cm. Characteristic loop sizes are of the order of a few tenths of the solar radius (credit: from Batchelor [4]; NASA/ISAS). *Right*: A multiwavelength FUV/X-ray image of the Sun taken by the *Solar Dynamics Observatory (SDO)* on March 30, 2010. The colors trace temperatures from $\sim 6 \times 10^4$ K (red) to $>10^6$ K (blue). On the left a giant solar coronal loop emerged during the exposure. In PMS stars these loops are orders of magnitude larger and more powerful (Credit: NASA/SDO/AIA)

10.1.1 Detection of Young Stars

X-ray observations of young stars and star-forming regions intensified with the increased availability of observatories. An early X-ray account of all observations of relevant star-forming regions is given in [71] together with a list of references. Table 10.1 lists results of recent observations of major star-forming regions and prototype YSOs. About 80 percent of detected young stars in IR surveys are found in clusters with at least 100 members, supporting the fact that the majority of young stars form in large clusters [197, 850, 851]. There is no unbiased way to find young stars solely on the basis of their X-ray emission, and several criteria have to be fulfilled to make a positive identification. The majority of stars are found by association with a major star-forming region and ultimately through identification with cataloged stars at optical, IR, sub-mm and radio wavelengths. In this respect X-ray studies of YSOs are never independent of studies at longer wavelengths (see Sect. 10.1.2). Young stars from long-wavelength surveys are usually identified as such if they appear embedded in, or are associated with molecular gas and their positions in the HR-diagram indicate ages of less than a few Myr [197, 850].

To date, several hundred X-ray observations and studies have been performed with various space observatories (see listing in Table 2.1 for a selection). The largest account of X-ray detected young stars comes from the *RASS*. Figure 10.2 shows a distribution of stars as detected by this survey. It shows that the majority of these stars concentrate not only close to the Gould Belt of recent star-formation activity

Table 10.1 Various major star-forming regions with stellar clusters in the solar neighborhood harboring over 100 detected IR stars observed in X-rays since the launch of *Chandra* and *XMM-Newton*. X-ray observations before 1999 are listed in [71]

Region name	Cluster name	RA hh mm ss	Dec dd mm ss	*X-ray*	*IR*	Refs.
Cassiopeia	W3	02 26 51	62 15 52	1,300		[852]
	AFGL 4029	03 01 32	60 29 12		173	[197,853]
Perseus	NGC 1333	03 28 54	31 19 19	95	143	[854,855] [197] [850,856]
	IC 348	03 44 30	32 17 00	215	345	[197,857] [850]
	LkHα 101	04 30 14	35 16 25	213	>400	[197,858]
Orion	OMC 2/3	05 35 27	−05 09 39	400	119	[197,859]
	ONC	05 37 47	−05 21 46	>1,600	2,520	[532,860] [690] [197]
	ONC (COUP)					[861]
	σ Ori	05 38 45	−02 36 00	107		[862]
	NGC 2024	05 41 43	−00 53 46	283	309	[197,863]
Monoceros	Mon R2	06 07 47	−06 22 59	368	371	[197,864]
	NGC 2264	06 41 03	09 53 07	169	360	[197,865] [850,866]
Vela	RCW38	08 59 05	−47 30 42	>200	1300	[197,867]
Carina	CCCP (Tr 14 - 16, Bo10,11, Col228,234)	10 45 04	−59 41 04	>14,000	>50,000	[868,869]
Ophiuchus	ρ Oph	16 26 52	−24 31 42	570	337	[859,870] [871]
	Trifid	18 02 23	−23 01 48	304	85	[197,872]
	M 8	18 03 45	−24 22 05			[873,874]
	NGC 6530	18 07 52	−24 19 32	900	100	[197,875]
	NGC 6611	18 18 48	−13 47 00	>1,000	>1,000	[197,876]
Scorpius	NGC 6334	17 29 54	−35 47 00	1607	>10,000	[877]
Ara	RCW 108	16 40 00	−48 51 45	420	>1,000	[878]
Sagittarius	M17	18 20 26	−16 10 36	900	100	[197,310]
Cygnus	IRAS 20050+2720	20 07 06	27 28 59		100	[850]
	S 106	20 27 25	37 21 40	000	160	[197,850]
Cepheus	L1211	22 47 17	62 01 58		245	[850]
	Cep A	22 56 19	62 01 58		580	[850]
	Tr 37	21 38 58	57 29 11	28		[272]

in the vicinity of the Sun (see Sect. 3.1.3), but also shows that the majority of the stars are concentrated in known major star-forming regions and young stellar clusters. More detailed studies with X-ray observatories with enhanced spatial resolving power such as *ROSAT*, *XMM-Newton*, and specifically *Chandra* reveal large numbers of stars concentrated in stellar clusters.

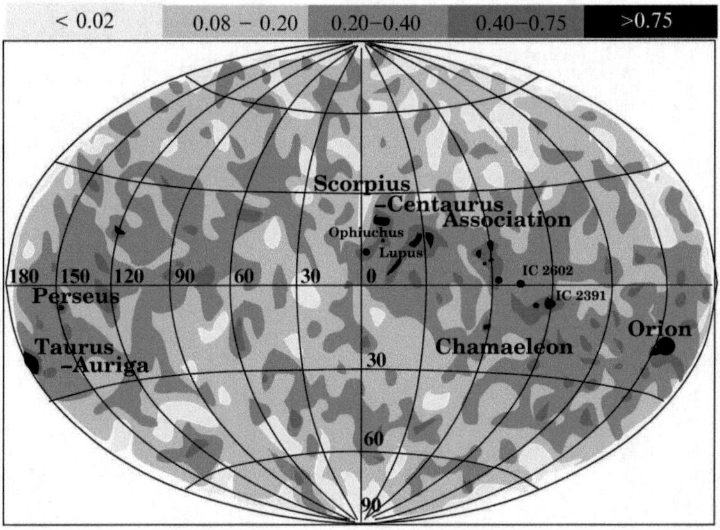

Fig. 10.2 Distribution of X-ray emitting stars as determined from a correlation of over 8,500 identified X-ray sources from the *RASS* (see Fig. 2.10) with the *Tycho* star catalogue [880]. The grayscale indicates the source density in units of sources deg^{-2}. Clearly seen is a concentration of X-ray sources near the central plane of the sky distribution. The darkest spots correlate with well-known star-forming regions, indicating the majority of detected X-ray emitting stars are young. The most concentrated regions of X-rays are distributed along a belt of propagating star-formation, triggered about 60 Myr ago, called the *Gould Belt*. Adapted from Guillot et al. [880]

Distance is a major obstacle for X-ray studies of star-forming regions, a fact which is related to the limited sensitivity of existing instruments. *Chandra*'s sensitivity holds the current standard with its 0.5$''$ spatial resolving power, which is orders of magnitude better than all other X-ray observatories listed in Table 2.1 and all planned observatories for the foreseeable future. It provides a flux limit of $\sim 10^{-16}$ erg cm^{-2} s^{-1}. The density of stars in young stellar clusters can reach up to 2×10^4 pc^{-3} in Orion [331] whereas in regions of average stellar density, as in the vicinity of the Sun or the Taurus–Auriga complex, the average density is more like 1–3 pc^{-3}. Even *Chandra* cannot resolve such extreme stellar densities as in Orion at distances further than $\sim$1.3 kpc, whereas *HST* is able resolve that in the optical band up to a distance of $\sim$6.5 kpc. Such extreme densities are not found in every case as many regions are older than the core of the *ONC* and appear more dispersed. High spatial resolution usually comes at the expense of field of view. Here *ROSAT* proved to be of immense value, because, while not able to resolve the dense cores of clusters, deep observations of the entire cluster were possible. Figure 10.4 shows an example of the young stellar cluster *Tr 37* at the core of the giant H II region IC 1396 in Cepheus at a distance of 0.8 kpc, about twice the distance of Orion. While the core of the cluster remains unresolved many X-ray sources are detected through the full extent of the cluster [272]. So far (see Table 10.1), the *ONC* has the youngest and closest active star-forming region, which outshines others in terms of number and density with over 1,600 detected X-ray sources [861, 879]. The *Chandra Orion*

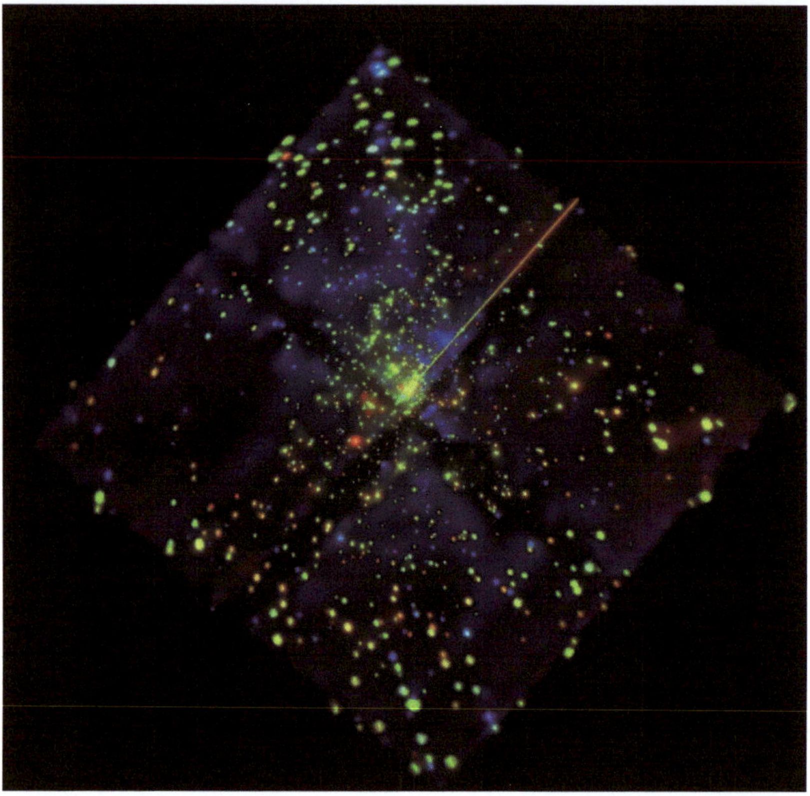

Fig. 10.3 Ultralong X-ray exposure (~10 days) of the *ONC* with *Chandra*. The image contains over 1,600 X-ray sources with several of them now known to be brown dwarfs. The color is encoded as Red = 0.2–1.0 keV, Green = 0–2.0 keV, Blue = 2.0–8.0 keV. Credit: T. Preibisch and the COUP consortium

Ultradeep Project (COUP) under the lead of E. Feigelson performed a 10-day deep exposure of the *ONC* in early 2003 (see Fig. 10.3) which will set future standards of X-ray population studies in star-forming regions. At a more remote location at 2.3 kpc lies the massive-star-forming Carina and in 38 observations the *Chandra Carina Complex Project (CCCP)* lead by L. Townsley detected over 14000 X-ray sources, the majority identified as young PMS stars.

10.1.2 Correlations and Identifications

The study of young stars performed at short wavelengths benefits from correlations with long-wavelength observations as many emission characteristics (see Figs. 6.14 and 6.15), though their underlying physical mechanism may not be related, happen at the same physical site or at least occur simultaneously depending on the

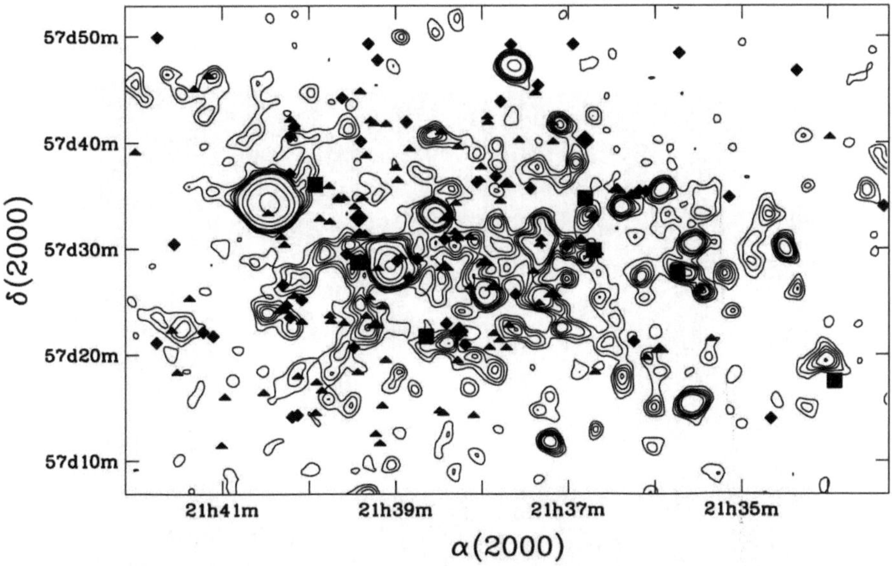

Fig. 10.4 Detection of young X-ray stars in the young stellar cluster *Trumpler 37* at the core of the giant H II region *IC 1396*. While the core of the cluster remains unresolved, 85 X-ray sources were detected within the extent of the cluster. The contours are X-ray flux. The overplotted symbols (small triangles and diamonds) are detected CTTS and WTTS from Hα survey catalogs. From Schulz et al. [272]

evolutionary state. In essence, it is the properties at long wavelengths that in many cases provide the identification as young stars. Examples of correlated emission between X-rays, optical and IR emission can be seen in Fig. 10.4, Fig. 10.3, and Fig. 10.5, respectively. Optical identifications such as those for IC 1396 mostly correlate with Hα surveys and catalogs, IR identification mostly use J (1.25 μm), H (1.65 μm), K (2.2 μm) band surveys [532, 854, 879, 881]. Properties of over 10^4 young stars have been correlated with various source catalogs and are accessible through various online databases with *SIMBAD*, operated by the *Centre de Données astronomiques de Strasbourg (CDS)*.

Many times there are no identifications of X-ray sources with detections at other wavelengths. This can happen for several reasons. One reason is related to the evolutionary stage of a young star. More advanced stages, such as WTTS, do not accrete from active accretions disks and certainly are devoid of a thick circumstellar envelope. Low-mass stars may not be luminous enough to be detected in the optical/IR unless nearby. These stars are still powerful X-ray sources and throughout the 1990s X-ray observations solely contributed to the number of known WTTS [882].

Another reason is absorption. Besides the fact that very young protostars are enshrouded in their infalling envelope, later evolutionary stages can indicate high absorption as well, because they are still embedded in their parent molecular cloud or are located behind a foreground cloud. Resolving the various situations depends on detailed case-to-case studies. In the IR these studies are mostly performed using

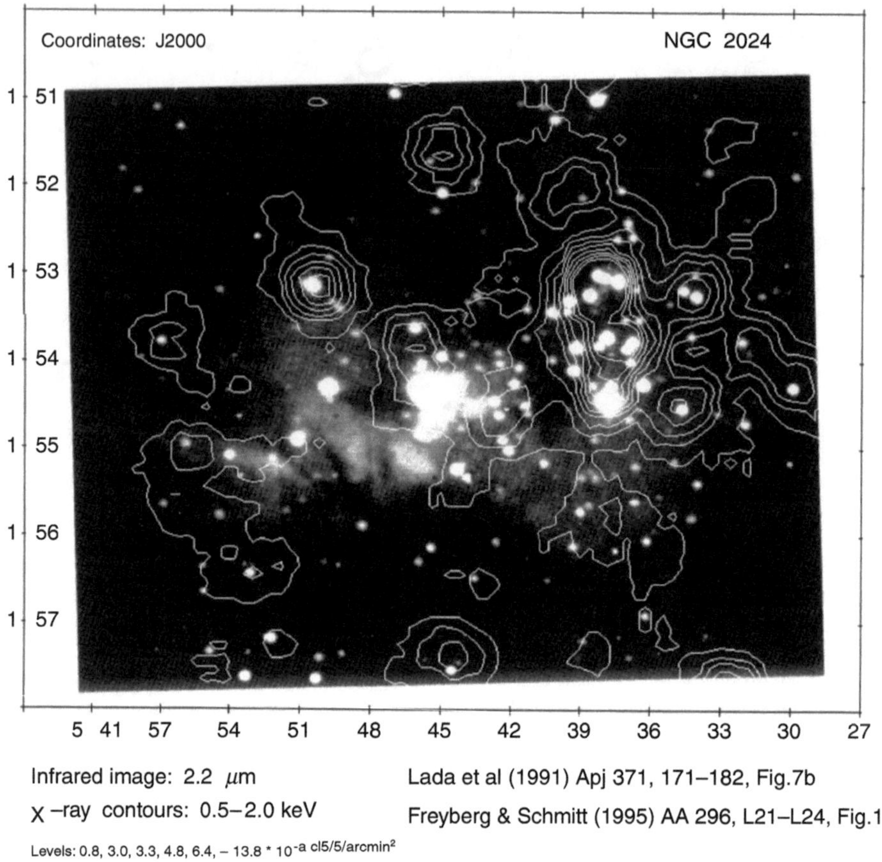

Fig. 10.5 X-ray image of the *Horsehead Nebula (NGC 2024)* with overlaid IR-contours. The X-ray data in the form of contours were taken with *ROSAT*. The image was one of many confirmations during the *ROSAT* period that X-rays from young stars correlate well with IR wavelengths. Credit: M. Freyberg, MPE

the JHK bands, sometimes including the 3.5 μm band. The degree to which a source appears embedded depends on the flux in each band. Most embedded sources are only seen in the 3.5 μm wavelength band. Increases in X-ray absorption (see Sect. 3.3.3) removes photons towards lower wavelength and highly absorbed spectra lack soft X-ray emission in this respect. Observed column densities lie between 10^{21} and 5×10^{24} cm^{-2} [272, 532, 860, 879, 881, 883–885]. Using the relation expressed in Eqn. 3.25 this spans an equivalent range of optical extinction between $A_V = 0.8$–2,800. Above an extinction of a few A_V there may not be an optical counterpart.

During the operation of *Spitzer* between 2003 and 2008 many X-ray observations benefited from simultaneous or follow-up observations such as for the Trifid Nebula [872, 886] or LkHα101 [858] or W108 [878], many observations of star-forming regions also rely on near-IR ground observations such as the *2MASS* survey [858, 887–889] and many other studies. Since 2008 the near-IR imager *HAWK-I* [890] is available at the *ESO* 8m VLT, which with a typical seeing of 0.5″ to 0.8″ matches well to the resolving power of *Chandra*. This has been shown during the large Carina survey CCCP (see Sect. 11.2.4) in the fields covering Tr 14–16 [891] (see Fig 11.12).

Correlations with radio observations are rare, but may bear critical information with respect to the underlying magnetic activity [47, 71, 532, 892]. Non-thermal radio continua have been observed in many WTTS at levels that are many orders of magnitude higher than in most powerful solar flares [893–896]. Feigelson and Montmerle [71] conclude that this radio emission closely resembles characteristics known from magnetically active binaries (i.e., stars with active coronae). The general problem is that even these radio fluxes are low with respect to currently existing instrument sensitivities and thus the stars have to be nearby to be detected. However, radio emission, when available, can be a critical indicator of evolutionary stage. In regions where the direct association with a star-forming region is difficult to establish (e.g., if the stars are far away from a cluster core or near less-dense star-forming regions as in the Taurus–Auriga vicinity) the detection of non-thermal radio emission can be the decisive marker between WTTS and near ZAMS stars [846]. Non-thermal radio emission in stars is likely of magnetic origin when the radio flux exhibits circular polarization [895, 897]. Furthermore, [432] argues that under certain conditions non-thermal radio continua are effectively absorbed in CTTS through free–free events in YSO outflows. Thermal radio emission may arise from partially ionized winds [898]. *VLA* observations of the *ONC* offer particular insights into radio emission from young stars [892].

In this respect the case of young stellar cluster *RCW 38* is of some interest. Here *Chandra* observations (see Fig. 10.6) reveal a diffuse cloud of X-rays among the detected cluster stars. Diffuse X-ray emission has been seen in other

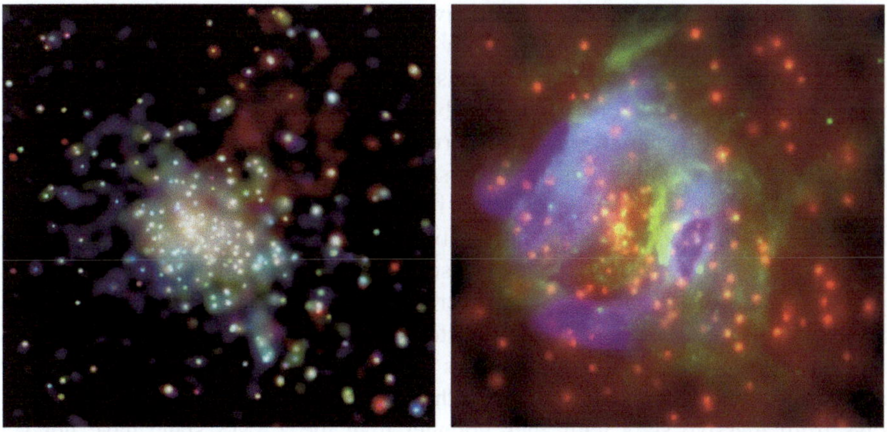

Fig. 10.6 High-energy electrons engulfing the very young and massive ESC RCW 38 observed in the X-ray, IR and radio bands. *Left*: A *Chandra* exposure indicating diffuse high-energy emission between the bright young cluster stars. *Right* Composite image showing the same field of view at X-ray, IR and radio wavelengths. The origins of the synchrotron emission is quite mysterious. One possibility could be a previously undetected supernova event in the cluster. Credit: NASA/CXC/CfA/S.Wolk et al. [867]

star-forming-regions such as the *Rosette Nebula* and *M17* [310] (see Fig. 4.21), the *Arches Cluster* [899], *NGC 3603* [900], Orion [901] and now Carina [868] (see Sect. 11.2.4). The common explanation for this diffuse emission is thermal bremsstrahlung caused by wind-driven processes (see Sect. 4.2.6). In the case of *RCW 38* the situation seems different as the hard X-ray spectrum and the existence of non-thermal radio emission (see Fig. 10.6, bottom) indicate synchrotron radiation from high-energy electrons. There is no direct evidence for any possible cause of the phenomenon. Suggestions range from an old supernova, and some form of shock wave, to winds from rapidly rotating neutron stars [867].

10.1.3 Luminosities and Variability

The measurement of stellar luminosities is usually first on the list of physical quantities to obtain about stellar systems and this paradigm also holds for X-ray stars. Part of the sensation that occurred when X-rays from T Tauri stars were discovered was due to the fact that luminosities were not only high but also showed large excursions on short timescales. It is now well established that X-ray luminosities in these stars exceed those of main-sequence stars by three to four orders of magnitude. There are two luminosity regimes to distinguish: the luminosities of X-ray flares which will be discussed in Sect. 10.1.5, and the luminosity of the persistent emission. The latter is sometimes difficult to determine and in those cases the luminosity, where a source stays longest at its lowest levels thoughout an observed time span, is used. Such a distinction is possible as light curves of young X-ray stars are usually not erratic, but flat with intermittent flaring activity. This behavior is illustrated in Fig. 10.7. The top curve shows a typical flare spectrum of the source COUP891. Massive stars on the other hand have flat X-ray light curves or show some form of slow modulations such as in θ^1 Ori C [848] (middle panel). Some light curves of massive stars might even reveal hidden low-mass companions as demonstrated in the bottom panel in the case of θ^1 Ori A [902].

Persistent X-ray luminosities are correlated with stellar mass. The range of luminosities for young low-mass PMS stars is between 10^{28} and approximately 10^{31} erg s^{-1}, for intermediate mass stars it is between 10^{29} and 10^{32} erg s^{-1} [904], and for high-mass ZAMS stars it is between 10^{31} and 10^{33} erg s^{-1} [407]. The most luminous class I object detected so far measured 2×10^{32} erg s^{-1} in its persistent state [905]. Tt has been determined that the X-ray luminosity scales roughly with mass as $L_x \propto M^2$ for stellar masses below $\sim$2 $M_\odot$ [857, 906]. In connection with higher stellar masses a dependence on X-ray luminosity may also reflect fundamental differences in the X-ray generation process. Though X-rays from massive stars with ages younger than the ZAMS stage have not yet been identified, it is somewhat established that at the ZAMS, X-rays are primarily generated through shock instabilities in radiation-driven winds [70] (see Chap. 9), sometimes modulated by fossil magnetic fields [407]. In low-mass stars, X-rays are thought to be generated by magnetic reconnection processes. Luminosities much

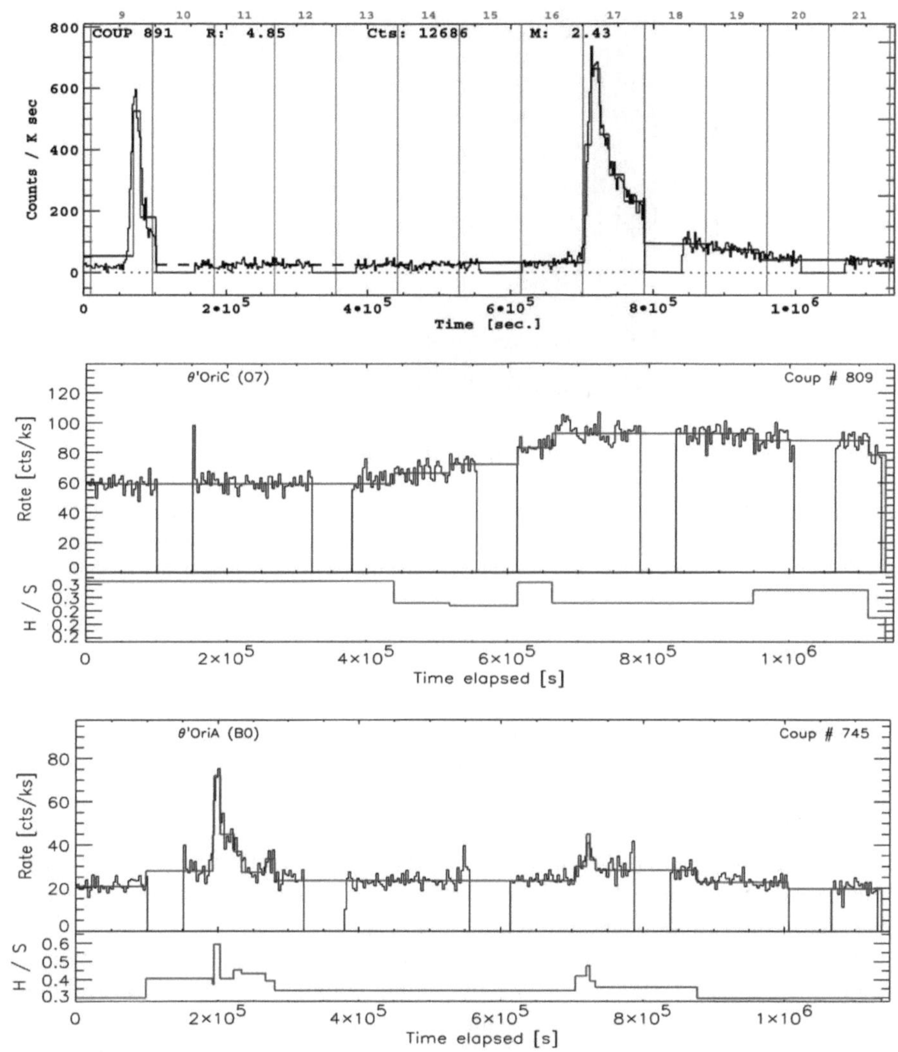

Fig. 10.7 X-ray light curves from various X-ray stars in the vicinity of the *Orion Trapezium* observed during the COUP campaign [902, 903]

larger than 10^{31} erg s^{-1} generated through magnetic reconnection seem unlikely as they would require too large emission volumes. Much of this high luminosity is then produced by wind shocks, if applicable.

X-ray luminosities also depend on evolutionary state. Although such a relationship is difficult to establish, several studies in the 1990s indicated a long-term decline towards the main sequence as shown in Fig. 10.8 [881]. This means that the X-ray generation mechanism seemingly evolves gradually toward the X-ray activity observed in the Sun and mature stars. The difficulty with such studies is that the

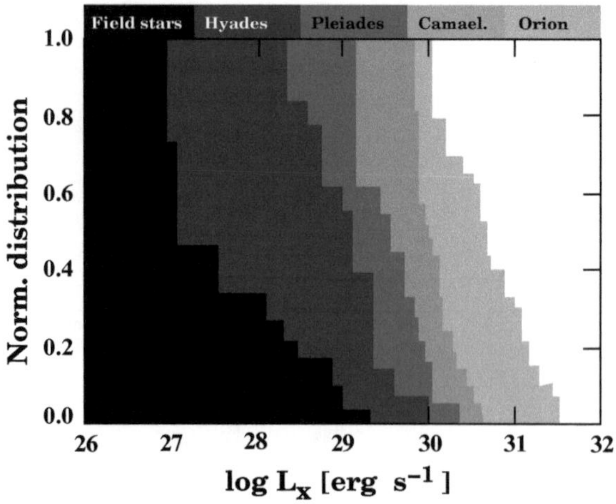

Fig. 10.8 The X-ray luminosity distribution of detected low-mass X-ray sources inside and outside various star-forming regions including the Orion region, the Chamaeleopardis Cloud, and the Pleiades and Hyades Clusters. Orion is the youngest region in the sample with a median stellar age of ~0.3 Myr [331]. Field stars have already reached the MS. X-ray luminosities of T Tauri stars can be as high as a few 10^{31} erg s^{-1}. The majority of the field stars, like the Sun, have X-ray luminosities of $\lesssim 10^{27}$ erg s^{-1}. Adapted from Gagne et al. [881]

luminosity functions as shown in Fig. 10.8 depend on the sampled X-ray population. Here is where observatories with high spatial resolution will make a difference as identifications will improve. For more distant clusters higher average luminosities are more likely determined because X-ray sources are not fully resolved. A recent investigation of close star-forming regions established highly resolved luminosity functions demonstrating that selecting sources with various different masses within one region shows significant differentiation [907].

It also has been established that the X-ray luminosity, at least for low-mass stars (F to early M stars), scales with stellar bolometric luminosity. Although the details of such a behavior remain unclear, it may reflect a dependence on contraction and spin-up [909]. Active regions close to or on the stellar surface occupy a nearly constant area as the stars continue to contract. The surface flux ($F_x \propto R^{-2}$) increases with age as the stars descend on the Hayashi track. Changes on evolutionary timescales should be expected under the premise that X-rays are related to star–disk magnetic interactions, accretion, stellar radius and rotation. All of these properties are subject to change on timescales of 0.1 to 100 Myr. Figure 10.9 shows an account of the ratio of soft X-ray to bolometric luminosity of stellar clusters at various ages from about 0.1 to about 10^4 Myr on the ZAMS showing an increase of the ratio up to about 100 Myr followed by a steady decline afterwards [908, 909].

Many X-ray studies focus on the *ONC* [532, 860, 879, 881, 910] and most of what is known today in terms of variability stems from Orion. The *ONC*, specifically the immediate surroundings of the *Orion Trapezium* at its core, is one of the youngest

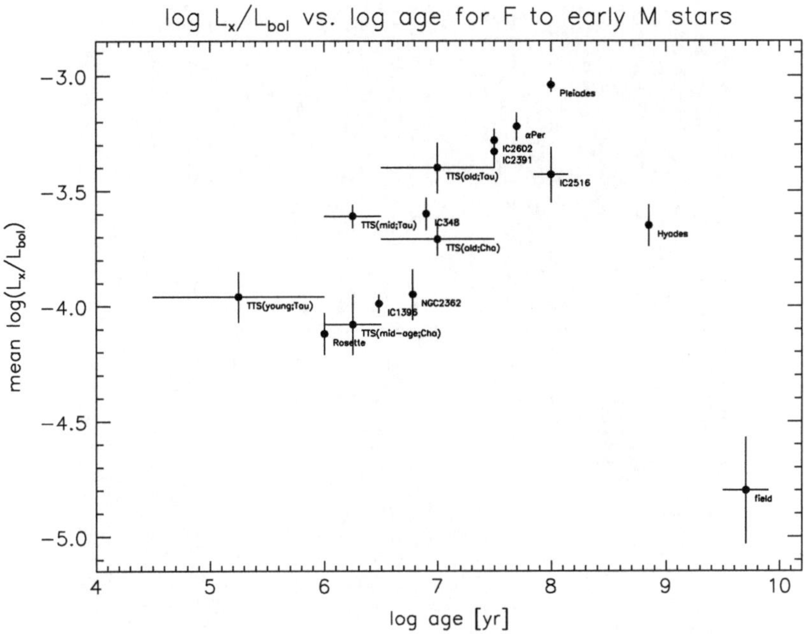

Fig. 10.9 Evolution of the mean of the log of the ratio of X-ray to bolometric luminosity for late type PMS for F-type stars up to M-type stars. From Schulz and Kastner [908]

and densest population of stars in the Galaxy. At a median age of 0.3 Myr [331] this region is populated with likely Class I and II sources, although the stars in the center are massive ZAMS stars [532, 879]. Star formation within the ONC has ceased (i.e., there are no collapsing cores), but it is quite vividly ongoing in the molecular cloud *L1641* behind the *ONC* [911] (see also configuration B in Fig. 4.4). The *ONC* plus the *L1641* cloud offers a prototypical study case for X-ray variability properties throughout the entire stellar mass spectrum ranging from the sub-hydrogen burning limit up to 45 $M_{\odot}$ O-stars. Figure 10.10 shows variable X-ray sources near the Orion Trapezium. While the few massive stars in the region show quite steady fluxes, identified low-mass stars show various levels of variability. A direct comparison of both exposures reveals that a vast majority of the sources are variable. Timescales of days to weeks for changes in persistent emission as well as flux excursions on timescales of hours are observed. Flaring activity is common in YSOs. This activity has been seen before, but not on such a dramatic extent within one single star-forming region. The direct vicinity of the *Orion Trapezium* is a most youthful part of the *ONC*. X-ray observations show persistently bright emission from its most massive members and highly variable and absorbed emission from low-mass stars, some of which are identified with protoplanetary disks (see lower panels in Fig. 10.10).

Details about X-ray variability in young stars is a central issue in high-resolution X-ray studies because so many more sources can be resolved in cluster cores.

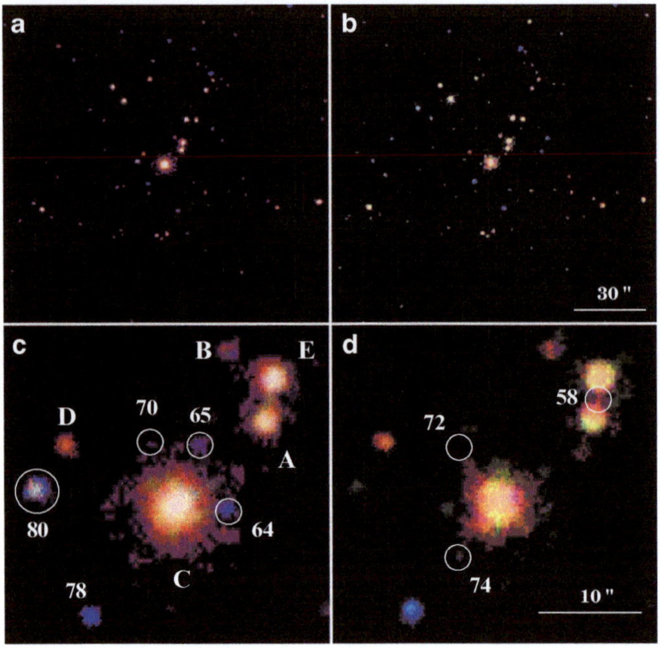

Fig. 10.10 Two *Chandra* observations of the Orion Trapezium in the core of the Orion Nebula Cluster taken three weeks apart (left versus right). The top sequence shows a $3' \times 3'$ field around the Trapezium. The bottom sequence is a close-up of the Trapezium. Most sources in the field are identified YSOs and appear highly variable in both fields. Credit: from Schulz et al. [532]

Not only does this add to the statistics of variable PMS stars, but it also helps identify the culprit of variability. An important issue is the identification of close binary companions, which demands resolutions below 100 milliarcsec and still has to be pursued in the optical through speckle techniques, adaptive optics, and long-baseline interferometry [480]. X-ray variability in PMS stars has many aspects ranging from very long-term, with timescales of years, to seconds. Even variability at the sub-second level has now come under scrutiny [906].

10.1.4 X-ray Temperatures

The fact that X-ray temperatures are as high as several 10 MK should not be surprising once compared to solar coronal temperatures. Figure 8.17 indicates that during large solar flares plasma temperatures may even exceed 30 MK. However, there are distinct differences with respect to young stars. X-ray spectra are the X-ray equivalent to the SEDs at IR wavelengths. They are characteristic of optically thin hot plasmas. *EINSTEIN* and *ROSAT* established a quite impressive account of young stellar temperatures in X-rays. Spectral fits revealed a mostly bimodal behavior in terms of plasma temperatures, with a soft component of 2–5 MK and a hard component of 15–30 MK [71]. In addition, *ASCA* observations showed flares with temperatures of up to 100 MK (see Sect. 10.1.5). High-resolution X-ray spectra confirm these accounts, though with better-resolved temperature distributions (see Sect. 10.3). Compared to the Sun there are basically two major differences to observe. PMS stars show high temperatures also in their persistent emission and much higher temperatures during flares. In fact, it has been recognized many times now that these temperatures are well in line with what is known from stars with active coronae [71, 447, 857].

Equation 8.27 seems to be able to reproduce the temperatures in the correct order of magnitude during flares. The estimated value is fairly accurate under the premise that the emission process is similar to the one in the solar corona since the relation has been scaled from measurement of solar flares. It is also clear that wind shocks cannot supply the high end of these temperatures unless shock velocities surpass 1,500 km s^{-1}, which, according to eqn. 3.13, allows for a maximal plasma temperature of $\sim$30 MK.

10.1.5 X-ray Flares

Statistically, young PMS stars frequently exhibit X-ray flares. A spectacular flare of the CTTS star *LkHα 92* [912] rose in flux by a factor of more than 300 within about 30 min. The first stellar X-ray flares were detected nearly three decades ago in nearby dMe dwarf stars [913]. Today it is acknowledged that all types of late-type stars probably engage in X-ray flaring activity [914], though likely not on the scales exhibited by systems with active coronae such as *RS CVn* and *Algol* type binaries. These stars and PMS stars show exceptionally strong flares (see Fig. 8.17). Today there is now a vast account on PMS flares [429, 673, 857, 903, 912, 915, 916].

The luminosity in these flares can exceed the persistent (or sometimes also referred to as quiescent) luminosity by several factors. In a first study using *EINSTEIN* data it was concluded that during one out of two detected flares, sources are at least twice as luminous as their persistent emission [429]. Recent *Chandra* studies of many variable sources in Orion, IC 348 and many other star-forming regions basically draw a similar picture. Some of these sources have large flares with luminosity excursions that exceeded more than ten times the quiescent luminosity (see, e.g., [857, 917]). Still, most impressive is the giant flare of LkHα 92 with a luminosity increase of a factor 300 [904]. The detection of a hard X-ray flare with *ASCA* from a WTTS [673] was very important. Not only did the luminosity rise by almost a factor of 20 to nearly 10^{33} erg s^{-1}, the peak temperature of the flare exceeded 100 MK. In general, a consistent trend is that the X-ray temperature increases substantially during the rise of the flare.

Such a trend is clearly visible in Fig. 8.17. Solar flares show a similar morphology and PMS flares have to be seen in the general context of stellar flares. This then implies that like the flares in active binaries or the Sun, flares in PMS stars are

Table 10.2 Properties of stellar flares at the flare maximum as observed from the Sun and various different stellar systems including T Tauri stars (from [903,912] and references therein). Although the temperature estimated from the scaling relation in (8.27) appears high, one has to consider that [676] already emphasized that real flares should have about one-third of this value

Stellar system	EM [10^{50} cm^{-3}]	T [10^7 K]	n_e [10^{10} cm^{-3}]	V [10^{30} cm^3]	l_{loop} [10^{10} cm]	T_{rec} [10^7 K]
Sun, compact	0.01	~ 1	10	$\sim 10^{-3}$	0.1	2.7
Class 3	>1	~ 1	10	> 4	>1.0	5.3
Algol	9,800	5.8	24	14	1.5	5.2
UX Ari	80,000	8.0	1	80,000	27	18.7
ρ Oph T Tauris	5,000	2.0	10–20	30	2	5.8
LkHα 92	500,000	5.0	15	2,000	8	9.0
V 773 Tau	540,000	> 10	30	600	40	12.9
ONC	> 10^6	27	260	15	235	53.9

exceptionally strong magnetic reconnection events. Table 10.2 compares properties of various solar and stellar flares with two examples from the Sun, two examples from binary stars with active coronae, and several from T Tauri stars. The parameters from the ρ Oph T Tauri stars from the early detection with *EINSTEIN* already show significantly higher emission measures ($EM = \int n_e^2 \, dV$), temperatures, densities, and emission volumes. The observed parameters match quite well the ones observed for active coronae. These results strengthen the perception that much of the X-ray emission of young low-mass stars, including those with circumstellar disks, is generated by coronal physical processes, which ultimately points to the existence of a stellar dynamo.

The large spread of parameters in Table 10.2 also testifies to the complex magnetospheric physics behind the flare activity. Although a quasistatic model seems to accurately predict temperatures, it has to be acknowledged that loop sizes of more than one or two stellar radii seem unusually large. This becomes more evident in the rise and decay times of flare. Therefore dynamic models with successive heatings and re-heatings may be preferred over static models.

The COUP sample produced a large sample of flares of a variety of remarkable properties [903]. The values in Table 10.2 refer to the top values in each category in the COUP sample and indicate rather extreme conditions. Flare peak temperatures T_{pk} of 32 analysed flares distribute between 40 MK and 270 MK and lightcurves measured the decay times τ_{lc} between 10 ks and 400 ks in one extreme case. COUP loop sizes were determined using an empirical relation which is proportional to $\sqrt{T_{pk}}$ and τ_{lc} and contains a correction factor which depends on the measured slope change of the lightcurve during the decay. Here loop sizes then range from 0.1 to over 5 stellar radii. Much higher values in the sample have also very large uncertainties. However, the study confirms that loop sizes in PMS stars are incredibly large. The PMS flares in Orion are also amongst the most energetic flares observed in stars. Figure 10.11 shows the distribution of the integrated energy of

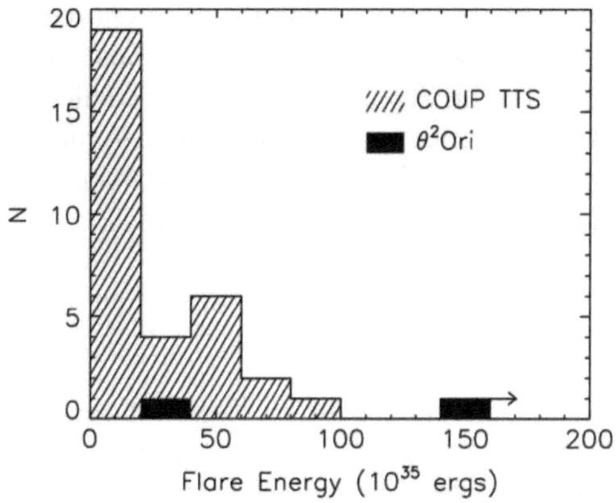

Fig. 10.11 Number distribution of the integrated energy of all bright flares in the COUP study [903] (from Schulz et al. [525])

all flares in the ONC [525]. Flare events marked with a dark square indicate events observed from the massive triple θ^2 Ori A, which at an energy exceeding 10^{37} ergs are unprecedented in stellar systems.

10.1.6 Rotation and Dynamos

Unlike for MS stars, the relationship between X-ray activity and rotation for PMS stars is not well-established [906]. Many observations with *EINSTEIN* and *ROSAT* revealed a strong correlation for late-type stars, independent of the bolometric luminosity, of the form:

$$L_{xs} = 10^{27}(v \sin i)^2 \text{ erg s}^{-1} \tag{10.1}$$

where L_{xs} is the soft X-ray luminosity in the energy band between 0.5 and 2.5 keV and $v \sin i$ is the projected rotation speed in km s^{-1} [918]. More recent studies developed a similarly empirical relationship between stellar period P and L_{xs}. The relation features a strong *correlation* of L_{xs} with P:

$$\log L_{xs} = 31.1 - 2.64 \log(P) \tag{10.2}$$

where P is the period in days and L_{xs} is the luminosity in erg s^{-1} [919, 920]. For a thorough discussion of rotation and X-rays in various MS stellar populations, the review [906] is recommended. X-ray luminosities of early-type stars remain fairly independent of rotation, but show a tight relation to bolometric luminosity of $L_x/L_{bol} \simeq 10^{-7}$. The relation has been verified many times [921] and also seems to hold for early-type ZAMS stars [407]. Again, the empirical difference between early- and late-type stars likely follows from the different physical mechanisms for X-ray generation, generally wind shocks versus coronal processes. In later-type MS stars L_x/L_{bol} may be directly linked to the internal stellar dynamo though its correlation with the theoretically determined *Rossby* number [906, 922]. This number is defined as the ratio of P to the dynamo's convective turnover time near the base of the internal stellar convection zone [923]:

$$R_o = \frac{P}{\tau_c} \tag{10.3}$$

Slower rotating MS stars correlate with R_o between $\log L_x/L_{bol} = [-3, -6]$. Faster-rotating MS stars (>50–70 km s^{-1}) exhibit a saturation level near $\log L_x/L_{bol} = -3$. Recent analyses of low-mass PMS stars shows no particular correlation with Rossby number but a strong correlation of X-ray to bolometric luminosity of around $\log L_x/L_{bol} = -3.7$ [857,906], which is much lower than the MS saturation level. Two main differences in connection with MS and PMS rotation and X-rays become apparent. First, X-ray luminosity in PMS stars seems fairly independent of rotational period, in MS it is not. Second, X-ray luminosities are high in PMS stars despite the fact that these stars are slow rotators (see Sect. 8.2).

These findings suggest that if PMS stars are to have an internal dynamo, it is likely a different mechanism from that of MS stars (see also Appendix B). Currently in the discussion are turbulent dynamos as well as solar dynamos in some super-saturation stage. Sometimes even relic core fields are considered [906]. The most recent X-ray account from the *ONC* does not exclude (but also does not support) strong star–disk magnetic interaction. A similar conclusion was reached in observations of *IC 348* [857,924]. The latter, in the light of magnetic configurations discussed in Chap. 8, will certainly be the subject of controversy for years to come. Some more details on results from various star-forming regions will be presented in Chap. 11.

10.1.7 The Search for Brown Dwarfs

An important topic in stellar X-ray astronomy is the detection of X-rays from *brown dwarfs*. Before the *Chandra* era, X-ray emission from only a few brown dwarfs was known [925,926]. Devoid of any interior hydrogen burning, most of their luminosity comes from gravitational contraction as they cool down with age. Therefore, the window of detection within their lifespans is limited as their bolometric luminosity declines with age. Searches thus focus on ZAMS and PMS associations (see [925] and references therein). X-ray luminosities of identified and candidate brown dwarfs were found to be solar with log L_x of less than 28.5 compared to the Sun's 27.6. A lower limit is difficult if not impossible to establish as it requires X-ray flux sensitivities which will not be available for another decade. *ROSAT* observations yielded upper limits of typically 25. The bulk of the log L_x/L_{bol} values lies significantly below the MS saturation limit of −3, specifically for objects found in the Chamaeleon and Taurus region [925, 927]. Observations with *Chandra* are now adding rapidly to the brown dwarf X-ray account with over 60 detections to date [928]. This became evident very early in the mission when [926] detected a bright X-ray flare from the 550-Myr-old M9 brown dwarf *LP 944-20*. Its quiescent emission remained undetected setting an upper limit of log L_x of 24 with a L_x/L_{bol} close to the Sun's value. Recent observations have now added many more objects to the account [857, 879, 884, 924] and it is now well established that brown dwarfs are significant X-ray sources. Yet, answers to questions about the origins of the X-ray emission and when they appear during their evolution are still outstanding. Once again, the *ONC* proves to be one of the most promising laboratories to study brown dwarf evolution as it harbors one of the largest samples found in a single cluster [929, 930].

From all that has been said about the origins of X-rays in PMS stars, it should not be too surprising to observe X-rays from brown dwarfs. Yet, at the same time they are part of the controversy concerning the X-ray production in PMS stars. Although brown dwarfs cannot ignite hydrogen burning in their interiors, they certainly can go through deuterium and lithium burning cycles before cooling down for the rest of their existence. The only source for their surface luminosity is then gravitational contraction. In fact the limiting mass ^{7}Li-burning is 0.06 $M_\odot$ [346], which is lower than the sub-stellar mass limit of 0.075–0.080 $M_\odot$ [931–933]. Brown dwarfs will not likely develop radiative cores and may initiate convection, which could lead to turbulent dynamos and thus coronal fields similar to their very-low-mass stellar cousins, which are fully convective up to a mass of 0.3 $M_\odot$ [934]. Of course, one has to consider rotation, which in the case of brown dwarfs is an uncertain issue. On the other hand, brown dwarfs are not excluded from the initial mass spectrum of collapsing clouds. Phenomenologically, stars near the sub-stellar mass limit go through a similar pre- and protostellar evolution including the formation of accretion disks. So far the lowest accretion rate in such an object was determined to be $\dot{M} \sim 10^{-12}$ $M_\odot$ yr^{-1} [935]. In some cases, near-IR excesses, which are signatures of accretion disks (see Chap. 8), have been observed [936]. However, the X-ray flare in *LP 944-20*, for example, likely does not relate to disk activity given the high age of the star. Very recently, X-rays from the brown dwarf companion *TWA 5B* in an only 12-Myr-old PMS binary system were detected. The emissions were found to be well in line with coronal characteristics [937].

10.2 X-rays from Protostars

The observation and study of X-rays from PMS stars has become an established field in high-energy astrophysics. With the availability of more sensitive and capable instrumentation, searches into novel domains follow. While the detection of X-rays from brown dwarfs posed a sensitivity challenge for *ROSAT*, another frontier opened up with the possibility of observing star-forming regions at even higher energies.

At high energies one can observe deeply embedded protostars in MC where star-formation is still on-going, where stars are enshrouded in their own circumstellar material. Following the laws of X-ray absorption, this has the consequence that most of the flux in the soft X-ray band is absorbed. A hint that protostars are sources of X-rays came with *ROSAT* by the detection of an incredibly powerful super-flare from *YLW 15 (IRS 43)* in 1997 that lasted for a few hours [938]. In general, the ability to observe X-rays from protostars progressed significantly with *ASCA* and consequently *Chandra* and *XMM-Newton*

10.2.1 The Search for Protostars

The first highlight in this new direction came with the detection of persistently variable hard X-ray emission from two class I protostars in the ρ *Oph F* cloud with *ASCA*: *YLW 15* and *WL6* [383, 939]. At luminosities around 10^{32} erg s^{-1}, they exceed the top quiescent luminosities known for the youngest T Tauri stars. The ρ *Oph* cloud is one of the very few star-forming regions where IR observations have located large accumulations of protostars [62,940,941], with about 17 percent of the young stars of type class I. The properties of the two objects detected with *ASCA* resemble class I sources in many ways.

In the ρ *Oph* cloud core they have ages between 0.75–1.5 $\times$ 10^5 yr. The two protostars, when compared to each other, are somewhat extreme and thus emphasize crucial elements in the evolution of protostars to PMS stars. In the light of X-ray and IR observations [383, 418, 421, 938], *YLW 15* is a 2.0 $M_\odot$ protostar that rotates near break-up speed and exhibits quasi-periodic flaring activity. *WL6* is of very low mass (<0.3 $M_\odot$) but slowly rotating. The protostar could have reached a state closely related to a class II X-ray source. In this respect, it is specifically interesting to investigate how such different properties can coexist within one star-forming region if one assumes that this region represents a common evolutionary environment.

Chandra has observed many more protostars, mostly of the class I type in terms of stellar mass. Variability is as common as in T Tauri stars, and their X-ray luminosities and plasma temperatures are systematically higher than those of PMS stars. Most X-ray studies of class I X-ray sources are still confined to observations of the ρ *Oph* clouds because of their proximity. In a deep observation of ρ *Oph* [884, 942] found over 50 percent of the known class I source candidates. Of this sample, again more than half are classified as class I_c, emphasizing their more evolved evolutionary stage between classes I and II. Other recently identified class I sources appear in the *Monoceros R2* region [857, 864, 943], the *Orion Molecular Clouds (OMC-2,3)* [859, 944] and in *IC 348* [924].

To date, there has not yet been undisputed evidence for X-rays emitted from class 0 protostars. The first detection may have happened with *ASCA* in 1996 [945], although the classification of source *IRS 7* in the *R CrA* cloud core is very uncertain.

In late 2000, Y. Tsuboi [944] reported on hard X-rays from two dust condensations located in the *OMC-2,3* clouds. The sources appear persistent and heavily absorbed with N_H larger than 1–2 $\times$ 10^{23} cm^{-2} which, according to (3.25), results in A_v values between 50 and 150. What suggests an identification with the condensations is that these absorption values are about an order of magnitude higher than one would expect from class I sources. On the other hand, the same argument also casts some doubt on the identification as the column densities may not be high enough [658]. Should the identification be upheld, they may mark a departure from the perception that X-rays from very young stars are highly variable. Last but not least, the discovery of X-rays from a protostellar outflow object *HH2*, may also indicate emission from a very young protostellar object. Although in this case the origin of the X-rays points more in the direction of a jet colliding with its circumstellar environment [946]. Using *COUP* data [947] in conjunction with *HST*, *Spitzer*, and ground *ESO* data found candidate Class 0-I objects with X-ray properties of somewhat lower luminosities than Class II sources but otherwise quite similar properties. The conclusion was that the onset of X-ray emission occurs at very early stages, but a definite Class 0 identification could not be made.

Another quite interesting development concerns the detection of X-rays from massive YSOs. The rapid evolutionary timescales make the observability of such events, if they do exist, extremely difficult. From Table A.9 it can be deduced that the time between collapse and full hydrogen burning may be as short as several 10^4 yr. The chances to observe such events are highest in the youngest star-forming regions. One of them is the *Mon R2* cluster at the distance of 830 pc [948]. Within this massive star-forming cloud there are a few massive condensations. *IRS 3* and three others show hard X-rays at the level of 10^{30} erg s^{-1} [943, 949]. Two of these sources show only low variability and $\log L_x/L_{bol}$ of the order of -7, which is what one would expect from massive stars on the MS [921]. X-rays from the *BN* object near the Orion Trapezium have so far remained inconclusive [532, 860].

A new reflection-type nebula has been discovered, in 2004, near *M78* in the *L1630* molecular cloud in the Orion star-forming region (see Sect. 11.2.1). It was named after its discoverer J. McNeil [951] and V1642 Ori was identified. An early study finds a spectrum consistent with an early B spectral type quite similar to the outburst spectrum of the FU Ori star *V1057 Cygni* (called *FUors*) suggesting a FU Ori type event [952]; others conclude that the object is a heavily embedded class I star [953, 954] or suggest the evidence is inconclusive [955]. The object was also observed with *Chandra* [956] and the X-ray data provide strong evidence that a sudden infall of matter from an accretion disk onto the surface of the star is causing the outburst (see Fig. 10.12). In contrast to these early findings the *Chandra* data show that the outburst of V1642 Ori is of Ex Lupi type (called *EXors*), where flux of the hard X-ray emission and the duration and intensity reflect the degree to which star–disk magnetic fields are reorganized before and during major accretion events. Other recent accretion-related outbursts have been observed in V1118 Ori [957], and from EX Lupi [958] itself.

Fig. 10.12 McNeil's Nebula near M78. (*top*) Optical exposure showing M78 and parts of Barnard's Loop. Credit: Jay McNeil. (*bottom*) Images show an optical close-up of of McNeil's Nebula (Credit: NSF/NOAO/KPNO/A.Block) and the *Chandra* X-ray image. The appearance of source 3 is strong evidence that the probable cause of the observed outburst is a sudden infall of matter onto the surface of the stars from an accretion disk. Credit: NASA/CXC/RIT/J. Kastner et al. [950]

10.2.2 *Magnetic Activity in Protostars*

Protostars drive powerful jets into highly magnetic environments. From Eqn. 8.24 it can be seen that high luminosities, as observed in the most powerful X-ray flares, are easily obtainable. Thus the argument that flaring activity, quasiperiodic or not, has magnetic origins is relatively undisputed even in class I protostars. However, it has to be addressed that configurations such as those discussed in the last chapter are even more difficult to adapt than in the case of T Tauri stars simply because the basic assumption of steady-state conditions is probably unrealistic. Thus, numerical models and analytical formulations assume steady-state as an asymptotic projection. Even if non-steady-state elements are included, the results to date remain incomplete (see [383]). Therefore the origins of magnetic activity in protostars are even less understood than in the case of their older PMS counterparts. In a theoretical attempt to evaluate X-ray emission from protostars, Montmerle [383] conceded that:

...again stellar rotation and magnetic activity are intimately related in low-mass stars... Young, fast-rotating (and hence presumably more massive) class I sources will be strong X-ray emitters due to star–disk interactions while evolved, less massive, slow rotators will exhibit only solar-line X-ray activity near the surface of the central star.

Many of the almost two dozen protostars detected so far show X-ray emission very much reminiscent of coronal activity. Most protostars in this category are evolved and close to the age of 10^5 yr. Deuterium burning is well under way. However, $\log(L_x/L_{bol})$ in several large flares exceeds the saturation limit for typical dynamo activity by orders of magnitude. Its likely origin is a *magnetic braking* mechanism in a configuration very similar to the ones discussed in Sect. 8.3.2 with the underlying physics similar to the GL model. The basis for the statement above were observations of *YLW 15* and *WL6* in the ρ *Oph F* cloud, where at least in the case of *YLW 15*, star–disk reconnection events seem the most satisfactory explanation (see Fig. 10.13). For a more recent assessment of the evolutionary state of *YLW 15* see [959]. The realization that stellar mass matters in protostellar flares [383] was important. Thus for masses smaller than about 3 $M_\odot$ and under the assumption that the initial angular velocity of the star is the break-up velocity, one can estimate the magnetic braking time until the star reaches the birthline to be:

$$t_{br,0} \sim 3.3 \times 10^5 \left(\frac{B_{surf}}{[1{,}000\ \mathrm{G}]}\right)^{-2/7} \left(\frac{R}{[R_\odot]}\right)^{-5/14} \left(\frac{M}{[M_\odot]}\right)^{15/14} \left(\frac{\dot{M}_{acc}}{[10^{-6} M_\odot\ \mathrm{yr}^{-1}]}\right)^{-6/7} \mathrm{yr} \qquad (10.4)$$

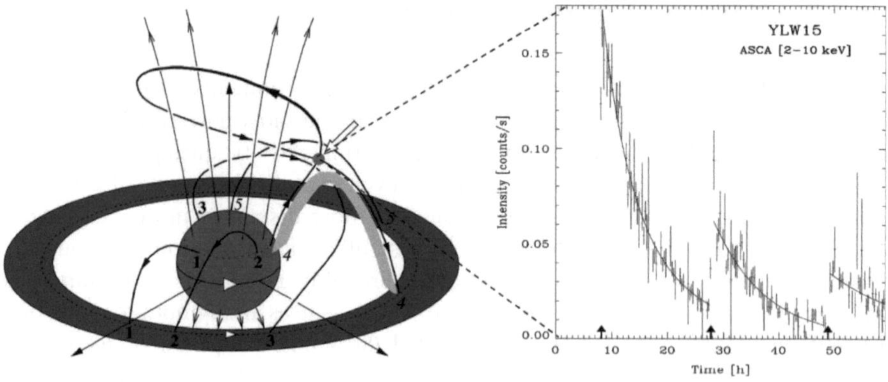

Fig. 10.13 (*left*) The evolution of star–disk magnetic field lines with rotation. Proceeding from (1) to (5), field lines cannot follow the faster-rotating inner star. As a field line winds up (3 and 4), it crosses over and reconnects (5) (from [383], see also Sect. 8.3.2). (*right*) *ASCA* light curve of *YLW 15* showing three successive flares about 20 hr apart. The flares could be modeled using radiative cooling (see Sect. 8.3.5) with loop sizes, densities, and equipartition fields comparable to values obtained from T Tauri and solar flares. From Montmerle et al. [383]

where a disk-like accretion geometry has been assumed (see [383] and references therein for details). Using the mass–radius relation from Fig. 6.4 and recognizing that the relevant mass accretion rate for braking is equivalent to $\dot{M}_{wind}$ in (8.19), one gets a braking time of about 4×10^5 yr for large masses, but much smaller times for low-mass stars. Mass-dependence of magnetic braking can create a distribution of rotational periods within a young star-forming region.

10.3 X-ray Spectra of PMS Stars

First results from *Chandra* observations of nearby star-forming regions have already sparked controversy and infused new life into the question about the origins of the X-ray emission (see Sect. 10.1.6). With new, high-resolution tools, scientists are now able to specifically diagnose the X-ray emitting plasmas. This can be achieved by building a statistical account of luminosities, masses, and rotation periods (see Sect. 10.1). Models then can be verified and tested against specific correlations and diagnostics leading to the determination of densities, abundances, and ionization balances. In X-ray astronomy the problem is a practical one and largely limited by technology. It simply takes more than three times as long to accumulate one high-resolution X-ray spectrum than to perform a wide range of studies on an entire star-forming region. While Sect. 10.1 mainly featured results from low-resolution broadband studies, the next sections introduce X-ray spectra of very young PMS stars at low and high resolution.

10.3.1 Spectral Characteristics

An empirical classification of X-ray spectra like the one with respect to IR SEDs has not evolved. Reasons for this are manifold, including the rather spotty availability of high-quality data. Spectral information has been available since *Uhuru* in the mid-1970s for the Orion region, although observed star-forming regions were not spatially resolved. In retrospect, now that many regions are highly spatially resolved, one recognizes that there were many past X-ray missions recording data from young stars, such as the European *EXOSAT* and the Japanese *GINGA* which otherwise could not be distinguished from diffuse radiation, late stellar stages or possibly coincident sources of extragalactic origin. After the discovery that PMS stars show strong X-rays, *ROSAT* established a substantial account of soft stellar X-ray spectra. Today it is known that spectra of young stars are sometimes strong in either the soft (0.1–2.5 keV), the hard (2.5–10 keV), or a combination of the two bands. *ASCA* established a first account of X-ray spectra throughout the full wavelength band.

Most X-ray spectra are now obtained using X-ray sensitive *Charge Coupled Devices (CCDs)* which offer medium resolution of $E/\Delta E$ between 10 and 60. The application of the energy scale rather than a wavelength scale in low-resolution devices has historical reasons which are related to the electronic properties of the detection devices. At this resolution there is not much possibility of performing detailed line spectroscopy except for very large line equivalent widths with the risk of line blends. There are three basic quantities relevant in the analysis of CCD spectra: the column density N_H as a measure of absorption, the equivalent temperature kT of the X-ray emitting plasma and the X-ray flux f_x in selected energy bands. Figure 10.14 illustrates common spectral shapes of typical CCD spectra from young stars. Although it is true that there is no formal classification of X-ray spectra, there are some global similarities to the IR SEDs due to absorption properties. Hotter absorbed (class 0 and I) protostars emit much harder spectra than cooler and less absorbed T Tauri stars. Expressions like hard or soft spectra are commonly used to describe the bulk emissivity in the X-ray band.

High-resolution X-ray spectra are currently based on dispersive devices such as the *Reflection Grating Spectrometer (RGS)* on board *XMM-Newton* and the *High-Energy Transmission Grating Spectrometer (HETGS)* on board *Chandra* which offer spectral resolving powers of $\lambda/\Delta\lambda$ of 300 and 1,200 at 1 keV, respectively. These spectra are now displayed in wavelengths, which is the linear scale of the detector. Scale conversions are given in (A.3).

10.3.2 Modeling X-ray Spectra

The above shows that modeling medium-resolution X-ray spectra, though highly efficient, has flaws with respect to their interpretation. The energy distributions of coronal plasmas are line dominated unless the plasma temperatures are very high,

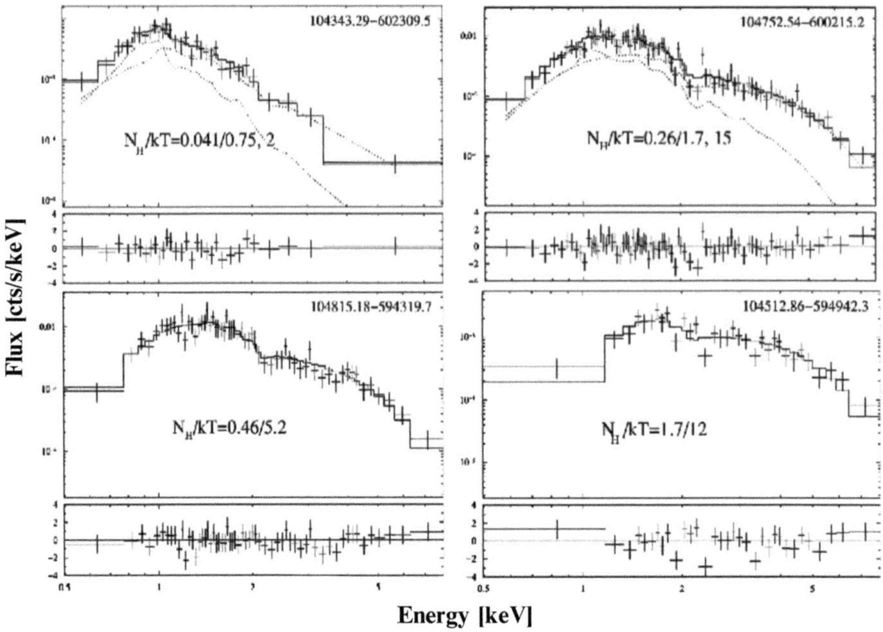

Fig. 10.14 Medium-resolution CCD spectra as obtained with the imaging spectrometer ACIS on board *Chandra*. The shown examples were observed with the Carina CCCP and demonstrate a variety of cases. Top left shows soft and fairly unabsorbed X-rays, The bottom left spectrum shows moderate absorption and temperature, the top tight spectrum moderate and hard X-rays. The bottom right spectrum indicates hard X-rays at high absorption. The units of column density N_H are in 10^{22} cm^{-2}, for the spectral temperature kT in keV (from Broos et al. [869])

in excess of 50 to 60 MK. Consequently, spectral fits to low- and medium-resolution X-ray spectra are not unique and thus the results can be deceiving. In the case of *TW Hya* a fit to the CCD spectrum assumed significant amounts of Fe in the star's corona [909], whereas the line-by-line analysis immediately showed that Fe is extremely underabundant. The bulk of the soft emission stems from Ne ions [447].

Both IR excesses in SEDs as well as absorption in X-ray spectra indicate how much a star is embedded in circumstellar material. However, there exist very distinct differences in the information between X-ray line spectra and IR SEDs. The former describes the hot material and the latter the cool plasma and warm dust. Radiating circumstellar dust occupies much larger volumes than X-ray emitting hot plasmas. Line spectra provide detailed temperature distributions, plasma densities and elemental abundances. If resolved, the lines reveal the dynamics in the emitting plasma, which, when correlated with optical lines, also provide hard geometrical constraints. Matter in stellar coronae exist as an optically thin ionized hot plasma. These coronae have been under study for half a century. Interpretations of X-ray and UV data from the Sun as well as UV data of stars obtained with the *Extreme Ultraviolet Explorer (EUVE)* were fruitful [934, 960]. The modeling of such collisionally ionized plasmas is an ongoing problem in astrophysics as these plasmas are found throughout the entire Universe. Today's models evolved from a large number of codes and analyses [961–963]. In the 1980s to the early 1990s the so-called *Raymond–Smith* model [963] was widely used to fit *ROSAT* spectra from stars. Throughout the 1990s, ever more sophisticated codes evolved, such as *HULLAC* [964], *MEKAL* [680, 965], *CHIANTI* [966], and *SPEX* [967]. Today many of these efforts are combined and developed further into what is called the *Astrophysical Plasma Emission Code (APEC)* along with its database, *APED* [968]. This database currently contains over a million lines from calculations of plasmas within a large range of temperatures and densities. Databases are far from being complete and it has to be acknowledged that updates are made on a continuing basis. The primary goal of the *Emission Line Project (ELP)* within the *Chandra X-ray Center (CXC)* is to produce a catalog of observed spectral lines from stars with active coronae like *Procyon, Capella* and *HR 1099* that can be compared with current plasma spectral models on an ion-to-ion basis. Most models assume an ionization balance from [969] together with standard solar abundances from [970].

The EM is the volume integral over the product of electron (n_e) and hydrogen (n_h) densities at a given temperature. The *Differential Emission Measure (DEM)* is the derivative of the EM with respect to temperature (see [683] for more details). Figure 10.15 shows a variety of such distributions derived from highly resolved X-ray spectra. The distribution for the CTTS *TW Hya* shows that the plasma temperature has a single relatively sharp peak at ~3 MK [447]. The account of such distributions is still very sparse as analyses at this level of accuracy are relatively novel in the X-ray domain. However, such single peaked distributions without at least hot tails as seen in *TW Hya* seem to be the exception but not the rule. Other well-peaked DEMs appear during flaring activity in active stars but at

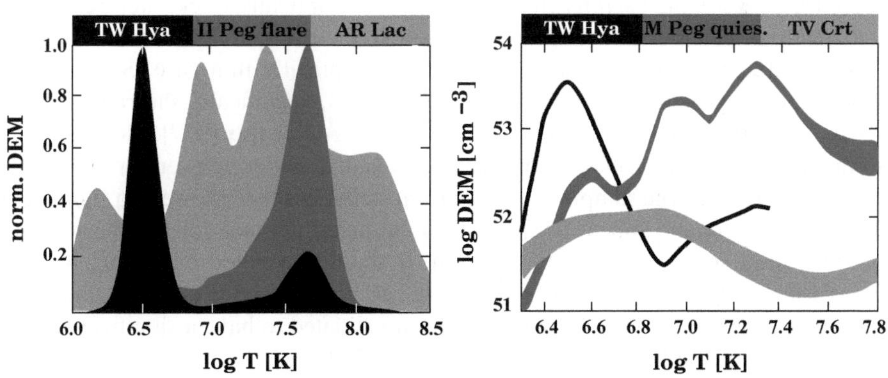

Fig. 10.15 (*left*) Three examples of DEM shapes. The distributions are normalized to their peak values. They show the CTTS *TW Hya*, a flare from the active binary *II Peg*, and the persistent DEM from the active coronal source *AR Lac*. (*right*) The absolute DEMs of the CTTS *TW Hya*, the WTTS *TV Crt* and the persistent phase of *IM Peg*. The width of the curves reflect the measurement uncertainties due to line statistics

much higher temperatures. More complex is the quiescent emission in active coronal sources showing multi-temperature peaks [683, 684]. The DEM analysis allows one to estimate element abundances of the highly ionized material as the X-ray line flux is proportional to the *product* of the total line emissivity and the ion abundance.

10.3.3 Coronal Diagnostics

The properties of X-ray flares stress an analogy with coronally active stars. A rich account of line diagnostics in the extreme UV, specifically from data gathered by *EUVE*, focused on the determination of plasma densities. Densities are a key parameter for inferring the size of X-ray emitting regions since the EM is proportional to the product of the square of the density and the emission volume. High-resolution X-ray line spectra offer an independent diagnostic through the evaluation of specific line ratios. Most of these diagnostics are based on detailed research of the solar corona [971–973] simple for the reason that the solar corona is much easier to study than distant stellar coronae. In this respect *EUVE* offered the first opportunity to study extra-solar spectra of active stars at high resolution. Specifically, observations of *Capella* appear to be at the forefront [960, 974], making the star one of the best studied extra-solar active coronal sources at short wavelengths. As a consequence, it was also the first target observed with the high-resolution spectrometers on board *Chandra* [975, 976].

Though there are many ways to determine the density of a hot emitting plasma through line ratios, they all depend on crucial assumptions about the nature of the emitting plasma. In almost all cases it is assumed that the plasma is optically thin. One of the tests to decide on the optical thickness of a plasma is to search for evidence of *resonance scattering*. If the plasma is optically thick, then resonance line photons can be absorbed and re-emitted in different directions. The line flux would appear changed with respect to the optically thin case. Some of the most frequently used lines are from Fe XVII ions, specifically the ratios of the 15.26 Å and 16.68 Å line fluxes to that of the 15.01 Å line flux [977]. Most analyses so far remain controversial on the issue of resonance scattering even for solar data. Observations of a large sample of spectra from stellar coronae showed no significant damping of resonance lines [978] and thus the paradigm that coronal plasmas are optically thin seems to hold (Fig. 10.16). It should be noted that under various circumstances forms of resonance scattering can be observed even in optically thin plasmas of very large volumes and have been validated to have a significant effect on ionized winds of high-mass X-ray binaries [979, 980].

The other assumption is that the emitting plasma is in *ionization equilibrium*. This simply means that the dominant processes are due to collisions and ionization happens predominantly from the ground state balanced by radiative decay and recombination. There is no significant photo-excitation, which would considerably alter line fluxes. Photo-excitation is expected, for example, in the presence of an external radiation field such as strong UV emission from the hot stellar surface of

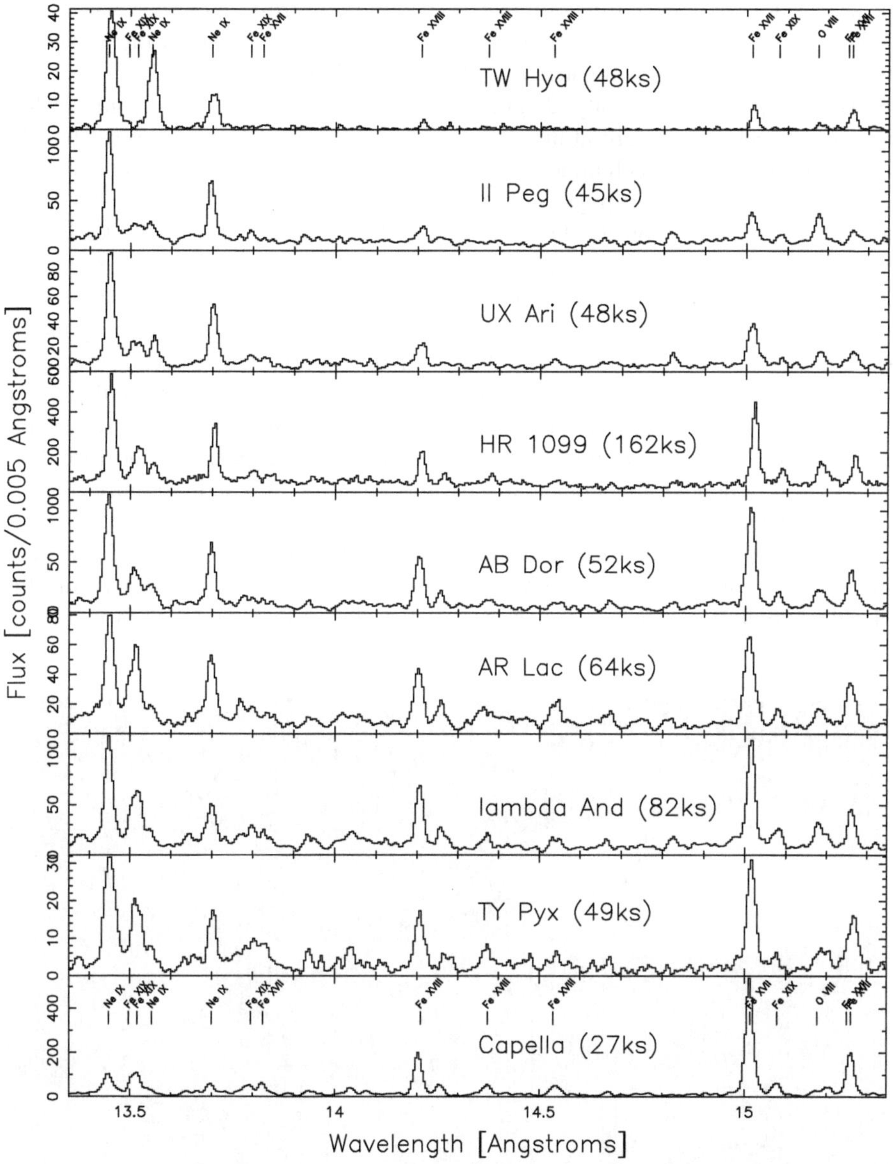

Fig. 10.16 A comparison of high-resolution X-ray spectra between 13.4 Å and 15.5 Å. All stars shown, except for *TW Hya* at the top, are binary stars with active coronae. The panels show at the left the He-like triplet of Ne ions at 13.44, 13.55, and 13.71 Å and, to the right, the Fe XVII doublet at 15.01 and 15.26 Å. The former are used as a sensitive density diagnostic. The latter can test optical thickness properties. From Kastner et al. [447]

massive stars [981, 982]. Line dominated UV emission from stellar coronae are not effective with respect to X-ray photo-excitations, though there may be the possibility of continuum emission from accretion shocks in T Tauri stars.

Most popular for density diagnostics is the use of emission from He-like ions [971, 983], which are atoms where all electrons have been stripped but the innermost two. This technique that was first established by [971] and is now widely used by the astrophysics community [447, 683, 684, 975, 978, 984, 985]. The diagnostic relies on the fact that the K-shell emission from He-like ions is a triplet consisting of closely spaced line components, the recombination line, an intercombination doublet and a forbidden line. The latter is so named as it involves transitions that do not possess a dipole moment and thus violate corresponding quantum selection rules. The fact that it involves metastable states, which are relatively long-lived states, makes them sensitive to collisional depopulation. In the absence of any photo-excitation an enhanced plasma density increases the population of the intercombination line doublet, at the expense of the forbidden line. For an in-depth treatment and discussions see [984, 985] and references therein (see also Appendix D).

Figure 10.17 shows various examples of He-like triplets as they have been observed recently from stars with active coronae. The X-ray line ratios from the He-like triplet of Ne ions, for example, predict electron densities in coronal loops

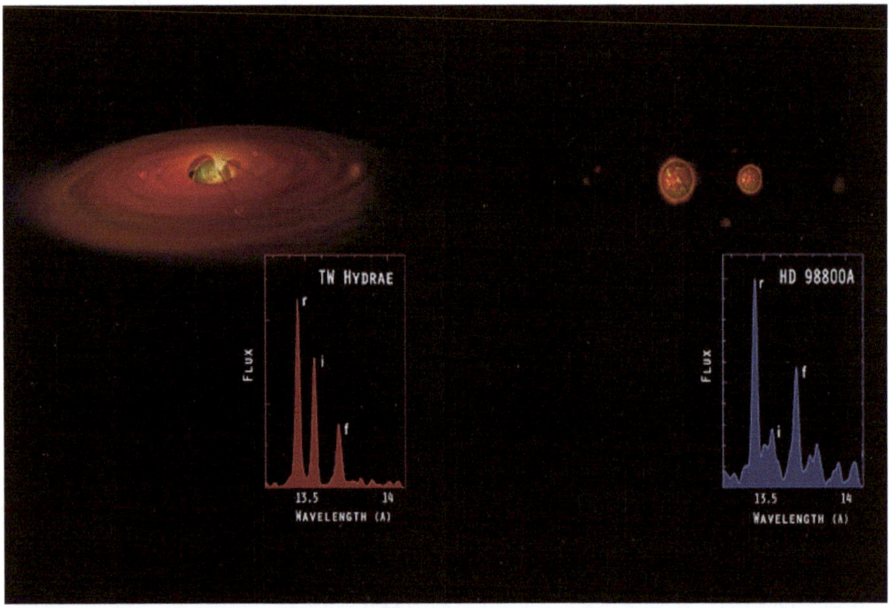

Fig. 10.17 An artist's conception that shows *TW Hydae* on the left, and *HD 98800A* on the right. X-ray spectral line emission from the Ne IX ions show a pattern which for *TW Hydae* indicates disk accretion and the existence of accretion shocks, and coronal emission for *HD 98800A*. Credit: NASA/CXC/RIT/J. Kastner (spectra); CXC/M. Weiss (illustration)

between 10^{10} and 10^{12} cm^{-3}. In many cases, however, a proper diagnostic is a very difficult task as Fe ions are usually quite abundant in stellar coronae and the lines from Fe ions cause significant blends. This is an effect known from spectra in the solar corona. In this respect, the spectrum of the CTTS *TW Hya* is unusual, because it is fairly devoid of Fe lines [847, 986]. Here the density diagnostics from Ne ions is specifically significant leading to values near 10^{13} cm^{-3}.

10.3.4 CTTS versus WTTS

Traditional distinctions of classical T Tauri to weak line T Tauri stars come from the SEDs in the infrared and are based on the evolutionary state of circumstellar dust (see Chap. 6). X-rays generally are thought to reflect magnetic activity and provide a direct measure of coronal activity of the contracting PMS star. While a predominantly coronal origin of X-rays in young PMS stars is still the common paradigm, it is now also established that signatures of accretion can be diagnosed as well. The statistical account in major star-forming clusters, such as the *ONC* and *IC 348*, however, do not support major X-ray differences between WTTS and CTTS other than that the former have enhanced fluxes on average (see Sect. 10.8). To date this manifests the assumption that in both cases X-rays are generated through mostly coronal activity.

A potential way to diagnose coronal activity in young stars is through highly resolved spectroscopy in the far-UV and X-ray domain, where most radiative processes in coronae takes place in a system of complex loops emerging from the photosphere in analogy of the Sun. However, not every hot circumstellar plasma must be of magnetic origin and predictions now also include properties of accretion shocks generated by mass accretion streams from the inner accretion disk onto the star's surface in CTTS (see Fig. 10.17). High-resolution X-ray spectra of several CTTS now show evidence of accretion rather than coronal mechanisms as first seen in *TW Hya* with *Chandra* [447] and *BP Tau* with *XMM-Newton* [989]. In these cases nearly free-falling gas with speeds of the order of a few hundred km s^{-1}, i.e. [990]

$$v_{ff} = \left(\frac{2GM_*}{r}\right)^{1/2} \approx 620\left(\frac{M_*}{[M_\odot]}\right)^{1/2}\left(\frac{r}{[R_\odot]}\right)^{-1/2} \text{km s}^{-1}, \qquad (10.5)$$

will heat up to the maximum temperature of the order of MKs, i.e. (see Sect. 3.1.9)

$$T_s \approx 3.5 \times 10^6 \frac{M_*}{[M_\odot]}\left(\frac{r}{[R_\odot]}\right)^{-1} \text{K}. \qquad (10.6)$$

This gives rise to both optical and UV emissions in the lower-temperature tail, but also soft X-ray lines near the maximum temperature. The high accretion rates predict shock densities of $n_e \approx 10^{12}$–10^{14} cm^{-3} which can be diagnosed by the

Table 10.3 Density measurements in CTTS and Herbig stars based on O VII triplet line fluxes

Star	Spectral type	Mass $[M_\odot]$	$R = f/i$ (see Eqn. D.14)	n_e $[\text{cm}^{-3}]$	Reference
Hen 6-300	M(3+3.5)	–	1.0 (0.5)	6×10^{10}	[518]
TW Hya	K8	0.7	0.054 (0.045)	1.3×10^{12}	[989]
RU Lup	K	0.8	0.26 (0.23)	4.4×10^{11}	[991]
V4046 Sgr	K5+K7	0.75	0.37 (0.16)	3×10^{11}	[992]
CR Cha	K2	–	0.64 (0.44)	1.6×10^{11}	[993]
MP Mus	K1	1.2	0.28 (0.13)	5×10^{11}	[989]
T Tau	K0	2.4	4.0	$<8 \times 10^{10}$	[994]
HD 104237	A(7.5-8)	2.25	1.8 (1.5)	6×10^{11}	[995]
HD 163296	A1	2.3	4.6 (2.3)	$<2.5 \times 10^{10}$	[996]
AB Aur	A0	2.7	4	$<1.3 \times 10^{11}$	[997]

He-like triplets in the UV and soft X-rays. The sketch in Fig. 10.17 illustrates this case on the example of the Ne IX He-like triplet emission. In the case of accretion and low local UV emissions above the H Lyman α break, the forbidden to intercombination line ratio of the He-like triplets becomes sensitive to electron density and a pattern as for the case of *TW Hya* is observed. In the coronal case the low density limit of 10^{11} cm^{-3} applies and a pattern as observed in the WTTS *TV Crt* is observed. Commonly likely both might be expected, accretion and coronal activity. In this case the soft emission from accretion is blended by the relatively hard emission by the coronal emissions. Likely the O VII line triplet is the best measure to determine higher plasma densities depending on whether the hydrogen column in the line of sight is low enough. Table 10.3 summarizes plasma density measurements based on the O VII line triplet in various CTTS and HAEBE stars [990]. Pending which contribution is dominant in the soft range, both cases can be modeled. Accretion signatures beyond the Mg XI triplet at 13.44 Å are not expected due to the limits of the free-fall speed in low-mass PMS stars.

Another way to diagnose the presence of accretion with medium-resolution spectroscopy is the search for X-ray soft excesses in CTTS. In the domain where photoelectric absorption of the gas in the line of sight is sufficiently low, the high-temperature tail of the soft accretion shock emission measure should become detectable mostly in terms of enhanced O line emissions, in practical terms the O VII Lyman α triplet at 21.6 Å, the O VIII Lyman α line at 18.97 Å, and the O VII Lyman β at 18.67 Å. At medium resolution this line emissivity one may then diagnose this

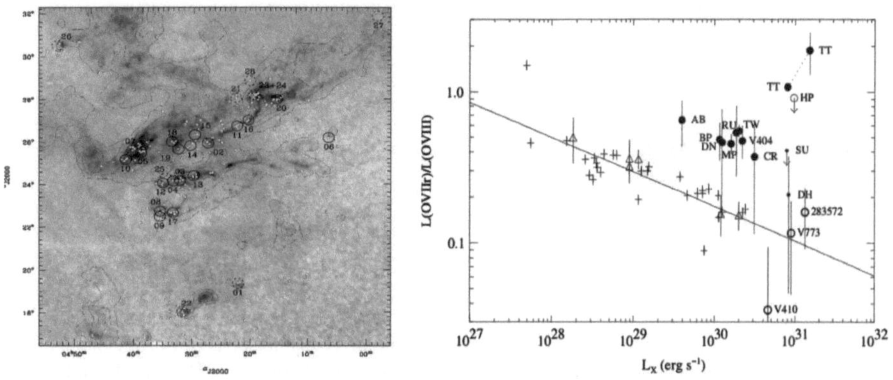

Fig. 10.18 *Left*: Map of the Taurus region (north is up, east to the left) use for the XEST study. The grayscale background map is an extinction (AV) map, contours show the CO emission [107]. Small white dots mark the positions of individual young PMS members. *Right*: The ratio between O VII r and O VIII Lyα luminosities vs. the X-ray luminosity L_X. Crosses mark MS stars, triangles solar analogs, filled circles CTTS, and open circles WTTS. The solid line is a power-law fit to the MS stars from [987] (from Güdel [988])

range by filtering of the O VII and O VIII band separately and view the O VII/O XIII flux ratio with respect to the total X-ray luminosity. Güdel and Telleschi [998] (see also [987, 994] and references therein) utilized data from the *XMM-Newton Extended Survey of the Taurus Molecular Cloud (XEST)* and a few other single CTTS examples to compare the O line flux ratios of CTTS with WTTS and MS stars. Figure 10.18 (left) shows the Taurus fields containing prominent CTTS and WTTS stars such as V773 Tau, V410 Tau, HP Tau, and BN Tau, BP Tau. All CTTS in the sample show an anomalous O line ratio when compared to WTTS and MS stars (see right panel) indicating an excess of cool O VII dominated material. It is concluded that the soft excess depends both on the presence of accretion and on magnetic activity as heating likely comes from both processes.

10.3.5 Observations of X-ray Jets

X-ray images of jets from *Chandra* with its high spatial resolving power of 0.5″ have now emerged as well. Güdel et al. [987] discovered a bipolar jet from the T Tauri star DG Tau consistent with the known optical jet axis. The X-ray jets are detected out to a distance of 5″ from the star and appear extended. The jet spectra are soft, with a best-fit electron temperature of 3.4 MK and shock velocities of a few 100 km s^{-1}. There is also evidence of excess absorption of the counter jet. Similar temperatures and small absorbing gas columns in the jet with the soft spectral component of the star suggest that they are related and resemble the immediate base of the jet. It is speculated that the X-ray emission is either produced either by shocks or by

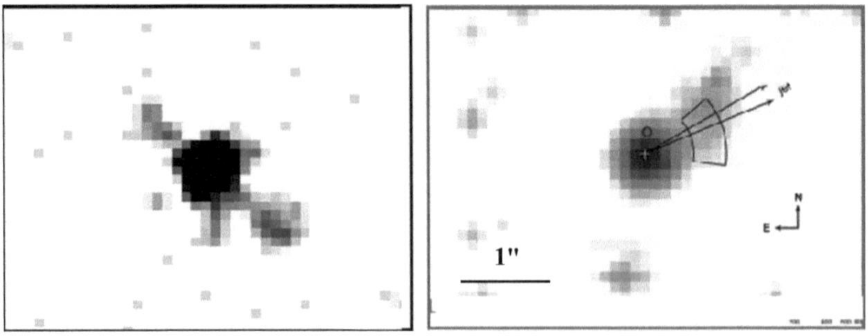

Fig. 10.19 X-ray jets in two T Tauri stars. *Left* A bipolar X-ray jet of DQ Tau on the scale of about 5.″0 (from Güdel et al. [987]). *Right*: An arcsecond scale X-ray jet of RY Tau (from Skinner et al. [999])

magnetic heating in the jets. Cooling estimates hint that the pressure in the hot gas contributes to jet expansion.

Figure 10.19 shows the two X-ray jets known to date. RY Tau is now a second example, but at a much smaller scale. The faint extended structure near the star is visible in deconvolved images and extends out to a separation of 1.7″ resembling the inner parts of the optical jet. As in DQ Tau, the jet emission appears extended and shock heated [999]. In a generalization of these findings [990] introduce a spectral phenomenology in which X-ray spectra of strongly accreting and jet-driving CTTS exhibit two absorber spectra with a cool low-absorbed component associated with the one driving the jet and a hard highly absorbed intrinsic source component.

10.3.6 X-Rays from Disk Fluorescence

Protostars as well as CTTS accrete matter from sizeable accretion disks while being coronally active. Consequently illuminated disks by coronal emission give rise to several instances of feedback. One of these processes is photoelectric absorption of highly energetic X-ray emissions leading to line fluorescence. K-shell line fluorescence is a highly inefficient process in most abundant matter and its yields increase only at high Z materials such as Fe and Ni. At low Z materials the emission of a fluorescent photon is rivaled by internal rearrangement of the K shell though Auger-electrons and possible emissions are deferred to much longer wavelengths. However, at Fe the yields are high enough to observe the effect, which then becomes an important diagnostic of column density. It should also be noted that it takes a rather energetic event to invoke Fe line fluorescence in observational quantitites as enough fluence above the Fe K edge threshold of 7.11 keV is required, which in coronal sources can only be reached in strong and hard flare events. The detection of Fe K fluorescence allows us to diagnose the presence of neutral and near-neutral 1s-2p Fe Kα states of Fe I to X, higher states will show measurable line shifts towards higher energies.

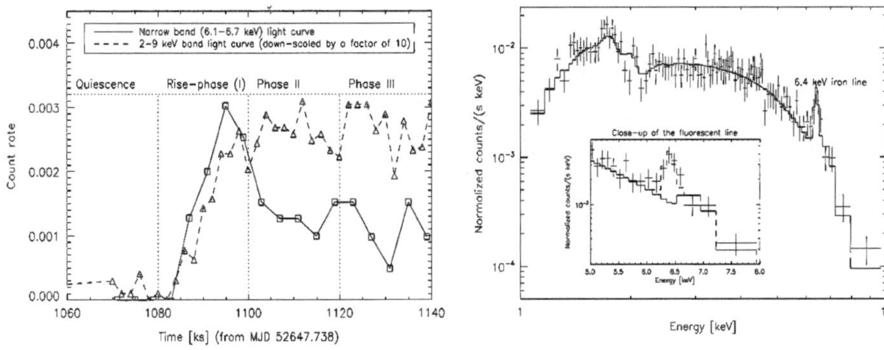

Fig. 10.20 *Left*: Light curves of the flaring protostar V1486 Ori capturing the onset of a giant energy flare during COUP observations. The solid line shows the narrow band light curve around the fluorescence line bandpass of 6.4 ± 0.3 keV. The dashed line shows the integrated light curve in the band pass between 2 – 9 keV. *Right*: Medium-resolution CCD spectrum of V1486 Ori with a strong Fe fluorescence line at 6.4 keV iron line. There are little contributions from ionized lines of Fe XXV (6.7 keV) and Fe XXVI (6.9 keV). Both diagrams are from Czesla and Schmitt [1000]

Such Fe line fluorescence at 6.4 keV was first observed by [1001] in the Class I protostar YLW 16a in the ρ *Oph* cloud. During the COUP campaign Tsujimoto et al. [1002] detected over half a dozen of CTTS exhibiting Fe K fluorescence due to disk illumination following an X-ray flare event. Perhaps the most remarkable example of Fe K fluorescence was detected in the flaring protostar V1486 Ori observed during the onset of a most energetic flare event [1000]. Figure 10.20 shows the lightcurve and spectrum of the observation. The observed flare was one of the most powerful CTTS flare observed in the COUP sample (near 10^{37} ergs in Fig. 10.11). In the flare spectrum of >10 keV even iron is fully ionized and and fluorescence peaks during the rise of the flare and levels off with the continuum flux indicating that some of the initial flare might have been obscured. However, persistent Fe fluorescence is rare but possible, as observed in the ρ *Oph* protostar Elias 29 [1003] and in this case not directly associated with an obvious flare event. Here the persistent fluorescence might hint the presence of non-thermal electrons in the dense magnetic loops. Such an initial burst of non-thermal electrons might also have been present during the fluorescence peak in V1486 Ori.

10.3.7 Massive Stars in Young Stellar Clusters

The study of high-energy signatures from young massive stars has always been difficult because of the short timescales involved in their evolution. Except for some detection of hard X-rays from massive protostellar candidates (see Sect. 10.2.1) there are not many records of X-rays from massive stars in their young, evolutionary phases. Here the term 'massive' may be restricted to very-early-type stars of spectral type B0.5 and earlier (see Table 9.1). In terms of youth, these stars evolve so rapidly that they hardly last the time it takes a low-mass star to pass through the T Tauri stage. The high spatial resolving power of *Chandra* made it possible to separate the X-ray signatures of very young massive stellar cluster cores. In terms of PMS definitions, these stars may not actually be young in evolutionary terms as they all have already reached the MS. In fact, some of these stars could even be further evolved as it may sometimes take about 10^{5-6} yr to disperse the circumstellar disk (see Sect. 8.1.7). The youngest known clusters containing massive cores have ages of less then 7–10 Myr. Most interesting are the ones with ages below 1 Myr, although the age determination using the HR diagram can be highly uncertain (for a more in-depth discussion of embedded young cluster properties, refer to Chap. 11). The youngest regions that have been studied the most in X-rays are the *ONC* [532], *Trifid* [1004], and *RCW 38* [867] with estimated ages between 0.2 and 1 Myr. Other regions, such as the Carina complex hosting several young massive clusters of ages up to 6 Myr (see Sect. 11.2.4) or *M17* with a median age of about 3 Myr [1005] are very important as they contain a large number and variety of massive stars which allow to classify between weak and strong winds, magnetic activity, colliding wind binaries, and maybe more exotic interactions with associated X-ray emissions.

A comparison of resolved X-ray emission from very young clusters (<1 Myr) to young clusters (3–7 Myr) found indications of signatures from the early evolution of young massive stars [407] (Table 10.4). It should be realized, that median cluster ages may not reflect the true age of the core massive stars. It is a reasonable working assumption that the younger the cluster the closer the massive star is to the ZAMS. The first detailed investigations indicate that there is a striking difference

Table 10.4 Young massive stars in stellar clusters at or near the ZAMS

Star	Spectral type	Star-form. region	Age [Myr]	T [MK]	$\log L_x^{mag}$ [erg s^{-1}]	$\log L_x^{wind}$ [erg s^{-1}]
Θ^1 Ori A	B0.5V	Orion	0.3	5–43	31.0	30.7
Θ^1 Ori B	B1V/B3	Orion	0.3	22–35	30.3	
Θ^1 Ori C	O5.5Vp	Orion	0.3	6–66	33.0	32.3
Θ^1 Ori D	B0.5Vp	Orion	0.3	7–8		29.5
Θ^2 Ori	O9.5Vpe	Orion	0.3	5–32	31.1	31.4
τ Sco	B0.2V	Sco-Cen	~1	7–27	31.9	31.4
HD 164492A	O7.5V	Trifid	0.3	<12		31.4
	Be	Trifid	0.3	<40		31.5
IRS 2	O5	RCW 38	0.3	<30		32.2
HD93250	O4 IIIfc	Carina	2.3	6–40	33.2	32.5
Tr16-22	O8,5V	Carina	2.3	20	32.3	31.6
HD 206267	O6.5V	IC 1396	3–7	2–10	-	31.6
τ CMA	O9 Ib	NGC 2362	3–5	3–12	-	32.3
15 Mon	O7V	NGC 2264	3–7	2–10	-	31.7
ι Ori	O9 III	Orion	<12	1–10	-	32.4
ζ Ori	O9.7 Ib	Orion	<12		-	32.5
δ Ori	O9.5II	Orion	<12		-	32.2

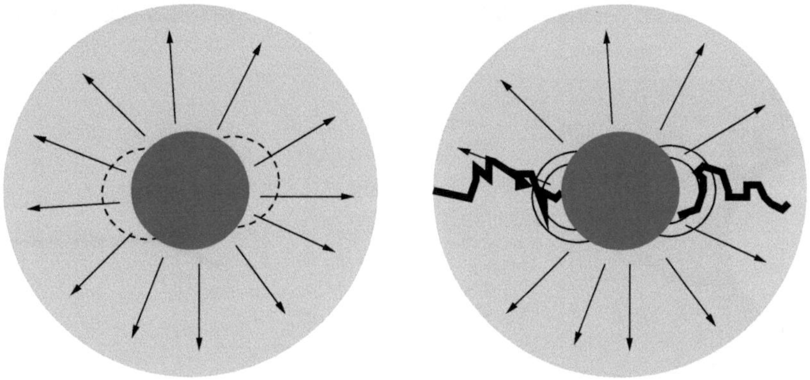

Fig. 10.21 Schematic illustration of stellar wind configurations in massive stars on the ZAMS and beyond. (*left*) A massive star with normal wind activity unperturbed by magnetic fields. Most massive stars seem to follow this scheme. (*right*) The same as left but here the wind is under the strong influence of a stellar field. Most stars near the ZAMS seem to show such a pattern

in the X-rays from various cluster ages. Nearly all of the stars in the older (i.e., >1 Myr) clusters exhibit X-ray properties entirely consistent with shock instabilities in radiatively driven winds. In essence, these properties can be summarized by plasma temperatures not in excess of 15 MK, broad emission lines with velocity widths amounting to roughly 30 to 50 percent of the terminal wind velocity. In addition to the breadth of the X-ray lines, the line shapes are affected by the opacity of the wind, which not only skews the red side of the line but also makes the line appear blue-shifted. Such X-ray properties have been observed in the archetypical O-star ζ Pup and in *HD 206267* in the young cluster core of *Tr 37* [1006,1007]. Most X-ray spectra from cluster cores of less than 1 Myr in age show a more complex pattern, which is very likely caused by the additional influence of magnetic fields.

Detailed models for magnetically channeled winds have recently been developed [1008,1009]. Figure 10.21 illustrates the major differences in X-ray signatures, and their underlying wind configurations. Near ZAMS massive stars confine some of the wind into magnetic field loops which heats the wind plasma to high temperatures. Otherwise it is a normal wind elsewhere as observed in more evolved stars. The magnetic field is then residual from the early protostellar history and is of primordial origin. Temporal and structural properties of such a field are still fairly unknown.

To date, the only convincing case of a very young O-type star showing effective confinement of its wind is θ^1 Ori C (see Gagne et al. [848] and references therein). It is the brightest star in the Trapezium and is only one of very few O-stars with a measured magnetic field of about 1 kG. Even though there are still some uncertainties associated with its spectral type, it is classified as a O5.5V star with a 15.4-day rotation period, an inclination of the rotation axis of about 45°, a mass loss rate of $1.4\times10^{-6}\ M_\odot\ \mathrm{yr}^{-1}$, and a terminal wind velocity of 2,980 km s^{-1}. Some of these stellar and wind parameters are model, dependent and correspond to the hot

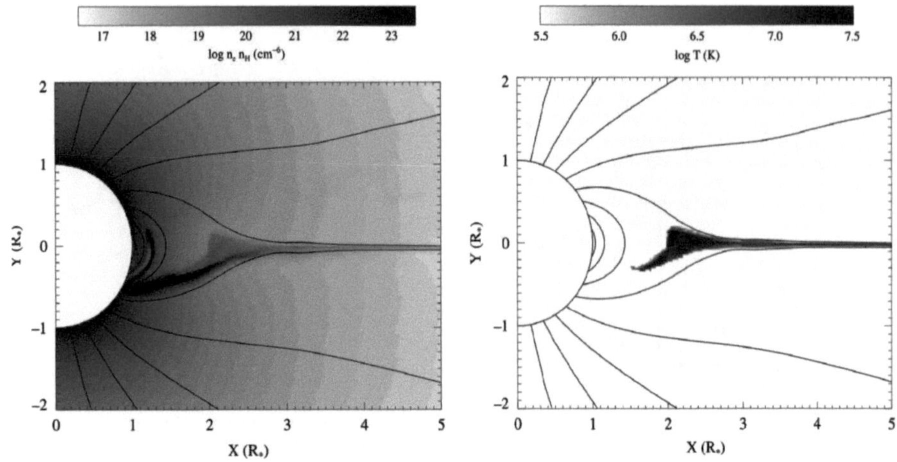

Fig. 10.22 Snapshots from a slow-wind, two-dimensional MHD simulation of θ^1 Ori C with corresponding magnetic field lines superposed. *Left*: The diagram shows the emission measure per unit volume ($\log(n_e n_H)$) for a 180 deg azimuthal cut. *Right*: The appearance of temperature ($\log T$) in the in the range between 0.3–31 MK. In this snapshot, material trapped in closed loops is falling back toward the stellar surface along field lines, forming a complex "snakelike" pattern. The regions of highest emission measure in the assigned temperature range occur above and below the magnetic equator, close to the Alvén radius at $\sim 2R_*$ (from Gagne et al. [848])

model for θ^1 Ori C. The overall degree of wind confinement can be described by a single parameter

$$\eta_* = \frac{B_{eq}^2 R_*^2}{\dot{M} v_\infty}, \tag{10.7}$$

which is the ratio of the magnetic energy density to the wind kinetic energy density. B_{eq} is the equatorial dipole magnetic intensity at the stellar surface. For $\eta_* < 1$ the wind dominates of the magnetic field and effective confinement will not occur. For $\eta_* > 1$ the magnetic field becomes more effective for confinement, but always at small radii, i.e. a very few stellar radii from the stellar surface as the magnetic field energy density drops rapidly at increasing radii. For θ^1 Ori C the model results in a parameter range of $\eta_* \approx 5$–20, a value of 7.5 was used for the model in Fig. 10.22. Based on rather similar spectral X-ray characteristics [1010] make another case for effective wind confinement in the second most massive stars in the ONC, θ^2 Ori A.

10.4 γ-Radiation from YSOs

The quest for even more energetic radiation than X-rays is a recent development. The abundance of accelerated particles and highly energetic gas during solar flare events have been derived from the *Solar Maximum Mission (SMM)* data [1011].

The existence of the acceleration of low-energy (MEV) particles in the solar photosphere has been established. The intriguing consequence of this is that, during protostellar and PMS phases, the stars probably irradiate their circumstellar environment causing nuclear spallation reactions in the disk. Suggestive in this case is enhanced irradiation proportional to the X-ray emission. A high level of particle irradiation is suggested by the observed X-rays and consequences are outlined in the X-wind model [669]. What this means for the history of our Solar System will be investigated further in Chap. 12.

One of the consequences for PMS systems could be enhanced radioactivity. Isotopes may involve beryllium [^{10}Be;^{9}Be], aluminum [^{27}Al;^{26}Al], calcium [^{41}Ca;^{40}Ca], manganese [^{55}Mn;^{53}Mn], and iron [^{60}Fe;^{56}Fe], which have radioactive lifetimes of the order of Myr and are now extinct. The production of such short-lived nuclides in the early Sun is now under increased investigation [1012].

The observational account of γ-radiation from YSOs so far has basically been non-existent. For example, the *Comptel Telescope* on board the *Compton Gamma Ray Observatory (CGRO)* in the early 1990s observed diffuse 1.809 MeV emission from ^{26}Al isotopes from OB associations within the old Gould belt around the Sun (see Sect. 3.1.3) [1013]. The emission is orders of magnitude larger than expected from irradiated young stellar disks [1014] and is likely sparked by nucleosynthesis in supernovas rather than young stellar activity.

Chapter 11
Star-Forming Regions

Satisfactory assessments of star forming regions in the Galaxy are essential for the understanding of early stellar evolution and today a vast amout of astronomical studies are available. Star-forming regions distinguish themselves from other regions in the Galaxy. Many different physical processes are displayed with high levels of variability, a large number of outflow sources, interactions with the circumstellar medium and emissions throughout the entire electromagnetic spectrum. None of the other stages of stellar evolution appear so evidently active in extended, but limited, regions in the sky. Clearly, supernova remnants or planetary nebulae do the same, but they always represent single isolated events. The reason for this may seem obvious. During later stages of stellar evolution, stars with different masses follow different evolutionary paths and only in very rare cases appear clustered. In the early days of stellar formation research these regions were mostly identified as large associations of early-type stars. Even today, OB associations are considered to be synonymous with regions harboring active stellar formation. Although it has been realized that not every star forming region produces massive stars, those are in the minority.

Today, stellar clusters are being used as modern laboratories of stellar evolution research. Very young clusters form in GMCs and remain obscured during the earliest stages of stellar evolution as they are still embedded in molecular gas and dust. Although certainly a challenge to observe, modern observational techniques at sub-mm, IR and X-ray wavelengths now allow one to view into even the deepest embedded stellar clusters. This has been impressively demonstrated in many examples, most prominently in Orion. Figure 11.1 shows the core of the *ONC* in the optical, IR and X-ray bands. While the optical image only shows a few of the brightest and hottest stars, the IR and X-ray images reveal hundreds of newly formed stars in the cluster which may as well are embedded into background molecular clouds. In the near future, it is expected that all sky IR surveys, such as the *Deep Near Infrared Survey (DENIS)* and *2MASS*, will provide the most systematic account of specifically embedded cluster populations in the Galaxy [197]. A multitute of YSO surveys with *Spitzer* are now also available providing an ubiquitous account of detailed IR properties of protostellar systems

N.S. Schulz, *The Formation and Early Evolution of Stars*, Astronomy and Astrophysics Library, DOI 10.1007/978-3-642-23926-7_11, © Springer-Verlag Berlin Heidelberg 2012

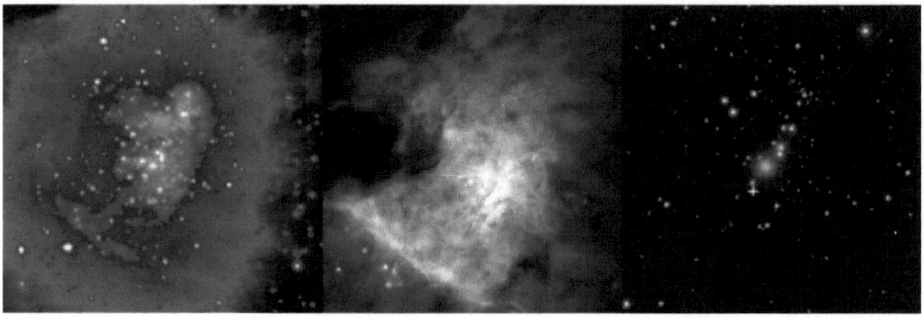

Fig. 11.1 The *ONC* in three different wavelength bands, IR (*left*), optical (*middle*) and X-rays (*right*). The images were generated using NASA's online *SkyView* facility and the *Chandra* data archive

at various ages with quite interesting consequences toward our understanding of the formation and early evolution of planetary systems [1015–1020] (see Chap. 13).

Clusters, in general, cover the entire mass spectrum of stars confined to a limited volume. These stars have more or less formed simulaneously. If one neglects that there are overlapping generations, they can be studied as canonical samples in the HR-diagram. Certain regions are then snapshots of a specific time in the physical evolution of a large range of masses held together by gravity and classical motions. Consequently, they can be described as such.

11.1 Embedded Stellar Clusters

Most stars in the Galactic disk probably originated from stellar clusters embedded in the molecular clouds that formed them. This concerns, specifically, the youngest of the stellar clusters. Clusters older than about 5 Myr are usually not associated with molecular gas any more and appear more dispersed. These *open clusters* will be a subject in the more general Sect. 11.1.5. An in-depth review of the properties of *embedded stellar clusters (ESCs)*, which are clusters with median ages younger than about 3 Myr, has recently been presented by C. and E. Lada [197]. Some of the most comprehensive collections of H II regions and ESCs were accumulation in the *Galactic Legacy Infrared Mid-Plane Survey Extraordinaire (GLIMPSE)* survey (see [877, 1021–1027] for a selection of cluster related papers).

The implementation of advanced IR imaging cameras greatly added to the number of known ESCs. These first IR observations are specifically important as they provide the location, size, and a first characterization of the cluster, which then can be studied further at other wavelengths. The account so far indicates that large ESCs are surprisingly numerous. Projections indicate that they are the birthplaces of the vast majority of the stars in the Galaxy. One of the big challenges ahead is thus to produce a full accountability of the stellar formation process from the ISM to GMCs to ESCs and finally to the dispersal of MS stars. The study of ESCs, in this respect, addresses an integral part of this process. It also means that stellar formation is not an isolated process but a crowded one (see Sect. 11.1.2). Cluster dynamics and timescales influence the evolution of young stars. The material below gives an overview of the basic properties of ECSs.

11.1.1 The Account of ESCs

The number of known young clusters has increased vastly in recent years. It has to be realized, though, that even with the most modern cameras and telescopes, the detection and characterization of ESCs is still severely limited by the large distances of most objects. Much beyond a 2 kpc radius around the Sun, the uncertainties in the determination of observational parameters increase rapidly and only the largest components can be studied. A prime example for how large embedded clusters can be lies in one of the Milky Way's neighbors. Figure 11.2 shows *NGC 604* in the 0.92 Mpc distant galaxy *M33*. It is a massive cluster, harboring as many as 200 massive stars at its core (see Sect. 11.1.9) in comparison to only a handful in the *ONC*. These examples are quite rare and ESCs in our Galaxy are usually less massive. Detailed accounts of ESCs, even within the Galaxy, are still restricted to a radius of about 2.5 kpc around the Sun.

The detection of ESCs is based on various criteria that have been adopted over time as properties of these clusters are better known. In general, these match quite well the properties that characterize major star forming regions and include the detection of molecular gas, outflow activity and the existence of Herbig Ae/Be stars. Many are associated with H II regions within OB associations. For many references of case studies of ESCs see [197]. This compilation of ESCs adopted 76 clusters within a radius of 2.4 kpc of the Sun with 35 or more identified members. The account is impressive as it includes over 1.3×10^4 IR sources and a total estimated mass of about $1.2 \times 10^4\ M_\odot$. This leads to an average of 0.85 $M_\odot$ per detected star, which is, as expected, below unity as the majority of young stars are T Tauri stars, but is still close enough to unity to indicate the existence of a significant number of massive stars. Massive stars reside predominately at the center of ESCs (see Sect. 11.1.3). The numbers directly reflect the mass functions of these clusters, which peak below 1 $M_\odot$. The vast majority of stars form in rich clusters with 100 and more member stars [1028]. Compared to the total mass of GMCs, these stellar masses comprise only a small fraction (see Sect. 4.1.3). This has a profound impact on the morphology of clusters.

The advent of *Spitzer* brought unprecedented advances in the observation of ESCs from space. Gutermuth et al. [1018] presented a survey of 38 regions of candidate nearby young star forming clusters and groups of YSOs with identifications, basic spatial distribution measurements and analyses. Identified were over 2,500 YSOs, of which 80% are classified as Class II sources, and about 20% as Class I protostars. The survey separated 39 cluster cores containing more than 10 YSOs

Fig. 11.2 *NGC 604* in the neighbor galaxy M33 as observed with Hubble. Credit: NASA and the Hubble Heritage Team (AURA/STScI)

each and the majority of the detected survey sources. The mean of the median spacings of YSOs in these cluster cores is 0.072 ± 0.006 pc. The survey infers natal cloud properties from the mean spacing between YSOs which are similar to observed clump properties in IRDCs (see Sect. 4.1.6, [777]).

An enormous account of ESCs at a large variety of ages comes from *GLIMPSE*, one of *Spitzer's* large legacy programs. The survey is used in many ways, largely to find stellar clusters, study their morphology, and correlate known YSOs with regions of various extinction properties. In the following a few examples are mentioned to show the variety of ways the legacy archive is used for cluster research. The presence of a highly obscured young cluster of massive stars coincident with *RCW 121* was revealed in the direction of a previous prominent *IRAS* source [1021]. In a program to search for young massive clusters containing WR stars, [1023] report on a young massive cluster hosting at least two WR stars, but less massive by several factors than some of the better-known clusters containing such stars. [1025] used a correlation of compact radio sources with mid-IR images from *GLIMPSE* to reveal a cluster of young protostars in the region together with multiple collimated outflows.

Here [1027] identify 458 members of the open cluster NGC 6611 with circumstellar disks to study disk evaporation due to its significant population of massive members. [877] reanalysed *Chandra* observations of *NGC 6334* and used *GLIMPSE* maps to correlate the detected X-ray populations with highly obscured regions.

11.1.2 Formation

Most stars in the galactic disk originate one way or another from embedded stellar clusters [197, 1029]. So far, when dealing with collapse conditions and scenarios in the previous chapters, clustered formation was not the center of the focus but rather the physics of stellar formation as single entities. As illustrated in Fig. 4.14, initial conditions for stellar collapse are part of the fragmentation process in GMCs. This implies that clouds collapse into multiple cores leading to clustered stellar formation. With vast advances in the field of computing power studies of cluster formation have recently become more a center of focus. Based on the paradigm that star formation is closely linked to the dynamical evolution of MCs, R. Klessen, for example, computes the formation of model clusters of low and intermediate masses through the interplay of turbulent fragmentation and gravity [301] (see also Sect. 4.2.4). Earlier studies also suggested that without turbulent support clouds would form more dense clusters [303, 1030]. A detailed picture of all physical processes involved has not yet emerged because too many processes can influence the collapse of a cloud into a star cluster [1029]. Besides gravity, magnetic fields, and turbulence, these processes may involve local triggers, such as a blast wave from a supernova or, on a grand scale, spiral density waves across the Galactic plane.

Most GMCs are gravitationally bound and if it was not for magnetic and highly supersonic and turbulent velocity fields, none of these entities would be stable against gravitational collapse. The numerical simulations mentioned above then suggest that only under specific conditions may fragments defy the turbulent mass flow and form clusters [363]. How this happens in detail is still poorly understood. At least in the case of very dense cluster cores the scenario of the collapse of single stellar cores as described in Chap. 5 may not be feasible as solitary cores have sizes very much larger than the density of stars in dense cluster cores would allow. In this respect a formation mode is desired in which large solitary cores break into multi-entities. Such a mode may not be confined to cluster formation but seems necessary to explain the formation of binary and multiple systems as well. Myers [1031] proposed a cooling scenario for the case of fragmentation due to MHD turbulence. Once extinction in a massive core becomes high enough so that the only ionizing source is cosmic rays, the overall ionization fraction in the core is reduced. MHD waves, smaller than a certain wavelength, cannot couple well to the neutral gas leading to fragmentation at this wavelength. What is left is a dense but fragmented collapsing core with a fragmentation distance typical of the distance of stars in cluster cores (see Chap. 5 and 9).

Critically impacted by formation scenarios is the mass function of ESCs and clusters in general. Today it is still entirely debated how an invariable initial distribution of stellar masses (Sect. 3.5.2 and 11.1.4) comes about in clusters. Protostellar cores with different masses grow on different timescales. Specifically in dense cores the formation scenario is likely tainted by competitive evolutionary processes such as competitive accretion from residual molecular gas outside the fragmented cores or from collisions and coalescence of collapsing cores in the very center of massive cores (see Chap. 9 for a review on the subject)

Other aspects of cluster formation are formation efficiency and formation rates which subsequently contribute to the clusters evolution. The star forming efficiency in a cluster is defined as:

$$\epsilon_{SFE} = \frac{M_{stars}}{M_{stars} + M_{gas}} \tag{11.1}$$

which simply is the ratio of the stellar to the total mass of the cluster and embedding cloud. From a collection of various measurements [197] determine a range of ϵ_{SFE} to be 10–30 percent. In detail, efficiencies of star formation in various ESCs give values of 25 percent for *NGC6334* [1032], 23 percent for ρ *Oph* and 19 percent for *NGC 3576* [1033]. Though one may infer from these numbers that star formation in clusters is relatively inefficient, the reader should keep in mind that it is still higher than the overall efficiency in the Galaxy, which is of the order of 5 percent (see Sect. 4.2). The formation rate of ESCs, which in the Galactic plane can be defined as the number of clusters per unit time and unit area, can be deduced from observations of ESCs in the solar neighborhood. Here the rate is determined to be 2–4 Myr^{-1} kpc^{-2} [197] (see Sect. 3.5.3 and 3.5.4), confirming an earlier determination based on only the Orion complex [1034]. These high rates and efficiencies are the basis for the argument that most stars in the Galaxy have, in fact, originated from ESCs (see Sect. 11.1.5 for more discussion).

11.1.3 Morphology

The internal structure of stellar clusters is of great interest as it is a direct consequence of the star forming process and its intitial conditions. Naively one would expect, for example, that the stellar density structure of the cluster reflect the density structure of the initially collapsing cloud. Although this is probably true, this structure will be diluted by various effects, which can include cloud turbulence, massive star formation or external forces, just to name a few.

Embedded clusters exhibit several basic structures, which can be characterized to be *hierarchical*, *turbulent* or *centralized* (see illustration in Fig. 11.3). Hierarchical clusters exhibit multiple clusterings, at least in the form of double clusters. Each of these structures then has its own substructure, which may appear centralized or show sub-clustering. Not every sub-clustering has an *a priori* centralized structure.

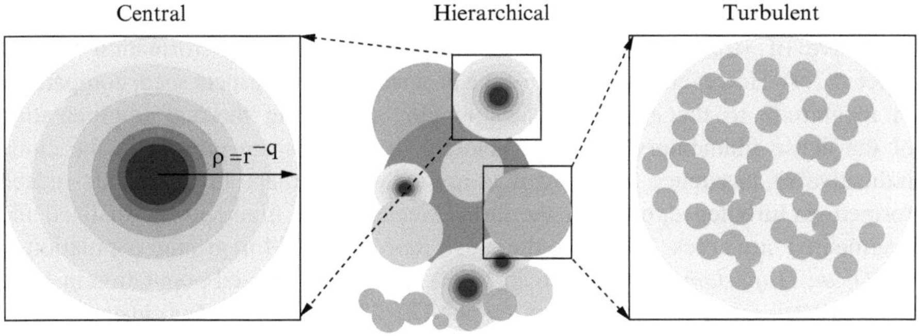

Fig. 11.3 An attempt to illustrate the morphology of most young ESCs based on the three basic elements describing a central, turbulent, and hierarchical structure

More erratic structures persist, reflecting the turbulent properties of either the parent cloud or external influences. The turbulent nature of some clouds may lead to completely unstructured clusters. Examples of typically hierarchical ESCs are *NGC 1333* [854, 1035] and *NGC 2264* [1036]. Large-scale star forming regions are hierarchically structured, though with many sub-clusterings so that they appear centralized. Clusters in the Orion A and B clouds (see below) are in this respect hierarchical, though they contain many centralized clusterings such as *NGCs 2023, 2024* [1037, 1038] and the *ONC*. Even giant isolated and centralized clusters such as *NGC 604* should be part of a hierachical structure, although one should be aware of the fact that starburst activity will likely change formation rates, efficiencies, and timescales dramatically. The hierarchical structure of clusters is likely a result of the fact that many GMCs have complexes with different star forming histories. Centralized clusters indicate a global dominance of gravity in the cluster over turbulent forces. Whether this morphology really reflects the primordial morphology of the parent cloud is not clear. One may be poised to see a congruence between the formation scenarios (see above) and morphology elements. Although this is certainly not coincidental, as observations have determined vital inputs of calculations, one should be careful not to stretch these similarities too far. Many issues on the observation and theoretical sides remain unresolved in this respect.

Complex hierarchical substructure has been observed in deep ground based observations of the *Orion Trapezium* and in the *NGC 2264* star forming region. [1039] find primordial filamentary substructure in the *Orion Trapezium* through deep *VLT* observations reflected by a deeply embedded population of YSOs tracing the molecular ridge behind the main cluster (see Sect. 2.12). *Spitzer* observations have documented similar primordial substructure in great detail. In the ESC *NGC 2264* [1015] used *MIPS* to identify the most highly embedded YSOs. In one particular region of *NGC 2264* bright 24 μm sources are configured in some linear structures with quasi-uniform separations. When the spatial distribution of the majority of the identified Class I sources were compared to the sub-mm and mm structure of the region, a close correlation between the dust filaments and the linear distribution of protostars was recognized suggesting primordial substructures indicative of structures predicted from gravi-turbulent star formation models (Sect. 4.2.4). However, separations of the protostars in this respect seem comparable in magnitude to the expected Jeans length, suggesting thermal fragmentation of the dense filamentary material. Velocity dispersions are present in the cloud indicative of turbuence and here it is argued that perhaps the filaments indeed formed via turbulence, but once the turbulence decayed, filaments thermalized and fragmented into cores. Similar to the older, more evolved foreground population in the *Orion Trapezium*, there is also an older and more dispersed population in *NGC 2264* [1040]. On the othet hand, in emphNGC 2264 in the *Spokes* clusters [1041] detected a dense microcluster of several Class O sources and the mean separation is much smaller than observed otherwise in this region emphasizing the hierarchical structure of the region.

11.1.4 Cluster Mass Functions

One of the fundamental questions surrounding the formation of clusters is what determines the mass of a stellar cluster. The concept of a universal IMF has been introduced in Sect. 3.5.2 which is used to estimate the total mass of a stellar cluster. The idea is that fragmented cloud collapse generates a characteristic distribution of stellar masses. The basis for these functions are the observed luminosity functions. E. E. Salpeter (1955) found that the relation between the original mass function and the observed luminosity function of a sample of main-sequence stars is a fairly smooth one without rapid changes in slope (see Sect. 3.5.2). In other words, the cluster luminosity function has a uniquely corresponding cluster mass function (CMF).

In the past CMFs have been derived from modeling of measured IR *K-band Luminosity Functions (KLFs)* of PMS cluster stars. Salpeter's relation between the luminosity function $f(L) = dN/dm_K$ and mass function $g(M) = dN/d(log\ M)$ is expressed in Eqn. 3.37 (see also [197]). A useful parameterization decribes a tapered power law [171] of the form

$$\frac{dN}{dM} \propto M^{-\alpha}\left(1 - \exp\left(-\frac{M}{M_p}\right)^{\beta}\right), \tag{11.2}$$

where M_p is the peak mass, α the power law index (see Eqn. 3.39), and β the cut-off component to describe the lower end of CMFs. The observations are consistent with a single underlying CMF for masses larger than M_p, the scatter at and below the stellar/sub-stellar boundary requires further study. Figure 11.4 shows a variety of CMFs from young embedded clusters, associations, young open clusters, and old globular clusters shifted by a constant for visibility [165]. A very detailed determination of the CMF from K-band measurements has been performed on the *ONC* [1042] where the stellar mass population was modeled as shown in Figure 11.4, the substellar population though required two more power laws [1042]. The *ONC* CMF peak appears as a broad plateau between about 0.1 and 0.6 $M_{\odot}$.

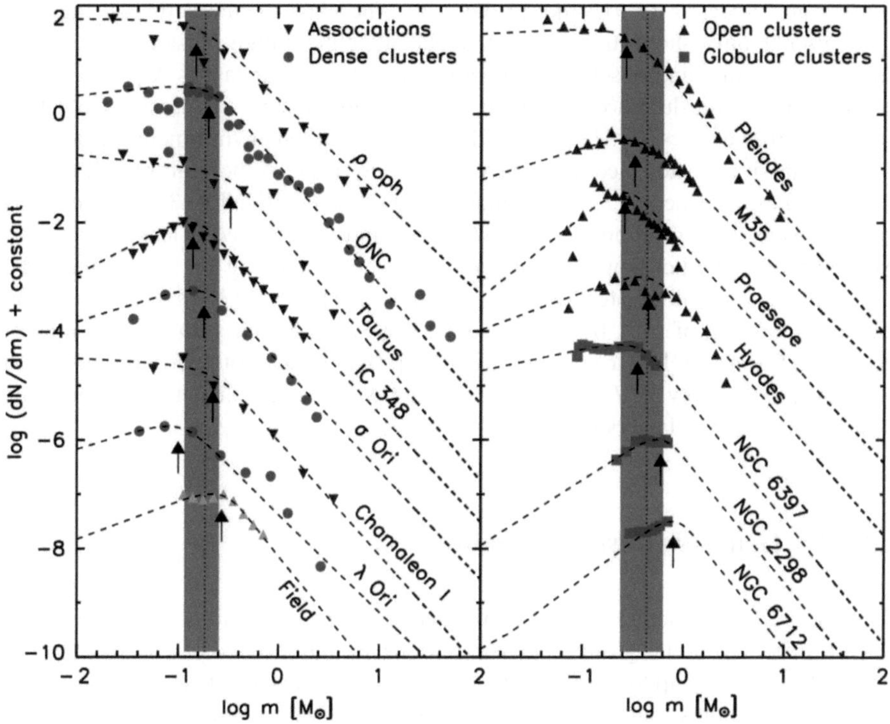

Fig. 11.4 Current-derived mass functions of a sample of star forming regions including embedded clusters, associations, open clusters, and globular clusters. In the left panel at the bottom the inferred field star IMF is shown. The dashed lines represent power-law fits following Eqn. 11.2 (from Bastian et al [165])

Additional apparent variations in the CMF can come from mass segregation in ESCs or associations where mass also depends on a spatial distribution. The issue of mass segregation in ESCs is of great interest. Like the *ONC* and other centralized clusters, many ESCs harbor more massive stars at their cores than in their outskirts [1043–1046]. In ESCs there are two mechanism to consider. A primordial segregation directly related to the fragmenation and massive star formation process (see Chap. 9). A second possibility arises because clusters evolve dynamically and more massive stars tend to sink to the center of clusters. However, ESCs are likely too young to have evolved mass segregation which means that this pattern has been imprinted by the star-formation process itself, although the picture is not unique and the question of imprinted mass segregation remains unclear (see [165] for more discussion).

There are now also many studies of CMFs in other galaxies. Weidner, Bonnell, & Zinnecker [172] compiled an interesting statistic comparing most massive (<50 Myr) star clusters in galaxies of different types. Irregular and dwarf galaxies have very low SFRs (see Sect. 3.5.3) but some host super clusters, regular spirals exhibit factors larger SFRs, barred spirals even larger SFRs, yet their maximum cluster masses are moderate or low like in the MW. The class of interacting or merging galaxies not only show much higher SFRs, their maximum cluster masses very much exceed anything observed in other classes. A peculiar case is Arp 220, which is the nearest ultre-luminous IR galaxy with many clusters of masses $>10^6\ M_\odot$, some possibly even $>10^6\ M_\odot$ [1047]. Thus depending on environment the possible size of massive clusters seems to have hardly any limit of the high mass end (see Fig. 9.2). In this respect [725] argue that from a typical IMF in Fig. 3.27 one expects that a cluster mass of about $3 \times 10^2\ M_\odot$is required to produce at least one O-star. However, other studies now also find sparse clusters containing OB stars in the SMC and simulations with cluster lower-mass limits of $M_{cl} \geq 20\ M_\odot$and $N \geq 40$ and random sampling of stars from a universal CMF match well with observations of SMC and Galactic OB star populations [1048]. Integrated photometry of clusters in the Whirlpool galaxy *M51a* also constrain the CMF in small clusters which appears consistent massive clusters [1049].

11.1.5 General Cluster Properties

The hierarchical structure seen in embedded star forming regions can be extended even further into stellar groups and associations as an imprinted pattern that defines the stellar structure up to the scale of spiral arms in the Galaxy. Under the paradigm that these groupings come in two types, *bound* and *unbound*, ESCs and open clusters represent the former type, whereas loose aggregates or associations of stars the latter (see Sect. 11.3). Today, the big picture of stellar clustering is very well documented. For a thorough review of the matter, refer to [1050].

Stellar clusters within the galactic plane that are not embedded are generally labeled as open clusters. These clusters and stellar associations have been the focus of astronomical studies since the 1930s, when R. J. Trumpler compiled his catalog of cluster HR-diagrams. Similar works were published by A. R. Sandage in the 1960s. These and others are now part of every standard astronomy textbook. They are not to be mixed up with *globular clusters (GCs)*, which reside outside the Galactic plane. GCs are extremely massive and strongly gravitationally bound stellar entities. They are very old, probably as old as our Galaxy, and their origin is still pretty much a mystery. Here, the discussion is restricted to bound stellar clusters at a stage prior to the ZAMS.

Open clusters are less bound, less massive, and much younger than GCs, though the oldest ones are known to be older than a few Gyr. The median stellar density in open clusters in the Galaxy, for example, is on average an order of magnitude smaller than the one in GCs. Galactic open clusters are not as centralized as some ESCs and GCs. The stars in ESCs are more strongly bound by gravity than open clusters. This is not due to the stellar content, but rather for the molecular mass in between the stars. The stellar content of the most massive known galactic ESCs is of the order of 10^3 stars, which is less than the 10^4 stars needed to remain bound by self-gravity. Without the gas, the stars would likely disperse and appear like open clusters. As an interesting side note, the mass cut-off of $\sim$1,000 $M_\odot$ is not only valid for ESCs, but generally for open clusters within the Galaxy. It is suspected that larger clusterings observed in neighboring galaxies (see Sect. 11.1.9) are signposts for starburst activity, where massive star forming activity is caused by exceptionally strong external triggers.

11.1.6 Cluster Age and HR-diagrams

One of the decisive differences in stellar clustering, though, is undoubtedly age. ESCs are the youngest of them all with ages of less than 3 Myr. In these very young clusters, star formation is still ongoing. Older clusters up to 10 Myr, are usually labeled as *young* open clusters. In these clusters most, if not all, parent gas has been relinquished either through star formation or photoevaporation by the hot winds of massive stars. Since many of them are likely to produce massive stars, they appear at the centers of giant H II regions. Even older stellar clusters are called *old* or simply *open* and are clearly past the stellar T Tauri phase.

Assessments of the dynamical state of stellar clusters are determined by testing for various timescales. Typically involved are, of course, the *cluster age* $\tau_{age,c}$ of a single cluster and the *cluster age spread* $\Delta\tau_{age,c}$. They characterize the actual age and the timespan of ongoing star formation within the cluster, respectively. Also relevant are dynamical timescales such as the *crossing time* τ_{cross}, describing turbulent motions (see Sect. 3.1.9 and 4.2.4), and *evaporation time* τ_{ev} when winds from massive stars are involved (see Sect. 8.1.7 and 4.2.6).

Cluster ages τ_{age} are generally determined using HR-diagrams. Although, as outlined in Sect. 6.2.5, this is the most reliable method known to determine the age of a star or a star cluster, the outcome is still very uncertain. This situation is specifically critical for ESCs and young open clusters where uncertainties are sometimes of the order of, or larger than, the cluster age itself. The application of various PMS models can lead to considerably different results as was demonstrated by [1052] in the case of NGC 2264. Fig. 11.5 shows HR-diagrams for three ESCs with median ages between ~0.3 Myr and ~3.0 Myr. The ESCs occupy almost the entire area in the HR-diagram representing early stellar evolution for low-mass stars. Above a temperature of 4,000 K, the stars are mostly of masses above 2 $M_\odot$, which have birthlines too close to the ZAMS (see Chap. 6). Most interesting for low-mass stars is the age spread. Depending on youth, more stars seem to accumulate close to or further away from the birthline with some age spread indicating the period of active star formation. The older the cluster, the closer stellar members move to the ZAMS and the smaller is the age spread relative to the median age of the cluster. Studies show that typical age spreads are of the order of 3 Myr [1053, 1054], though most measurements would show more like 5–10 Myr due to the already mentioned uncertainties. For a 100 Myr old open cluster, such a spread is very narrow, but for ESCs, it is not. The period of ongoing star formation also seems to show its own dynamics across the age spread. A recent study of a few ESCs and active star forming regions indicates that star forming activity is accelerating with time [410, 1051]. On the other hand, these very young stars are formidable

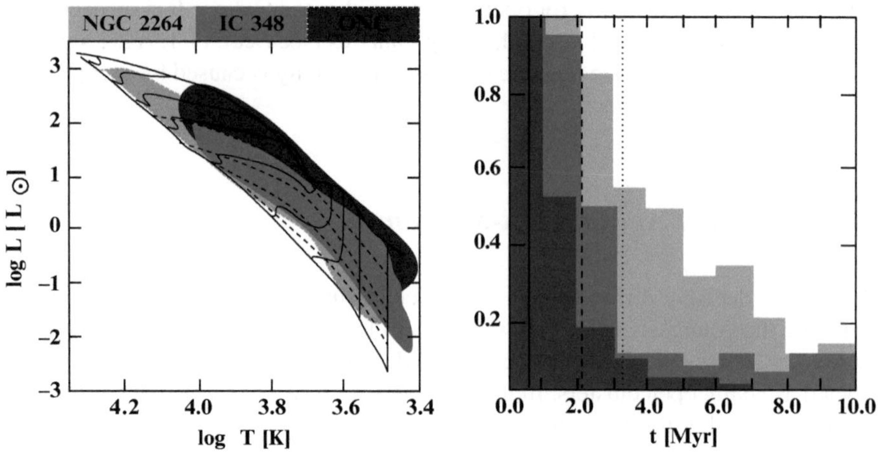

Fig. 11.5 (*left*) HR-diagrams of three very young embedded clusters. The shaded areas engulf the measured data. The data and HR tracks (*solid lines*) are taken from [1051]. The dashed lines are timelines. (*right*) The same data binned by time bins according to the tracks in the left HR-diagram and normalized to the maximum time bin. The solid vertical line marks the median age of the *ONC*. The dashed and dotted lines are for *IC 348* and *NGC 2264*, respectively. Adapted from Palla and Stahler [1051]

X-ray sources and should actually curb star forming activity. In the dense cores of clusters, the ionization fraction should increase significantly (see Sect. 4.2.6), which increases magnetic diffusion times and slows down collapse activity. The notion of accelerating star formation with age spreads is clearly not undisputed. Caution due to the presently poor understanding of theoretical, systematic and observational uncertainties seems advised [1055]. The detailed interplay of these effects is not yet understood. The right-hand part of Fig. 11.5 shows that half of the plotted stars in these ESCs have developed within what is denoted as the median age. In fact, most of the stars in very young clusters are much younger than the median age. It seems that star forming activity has reached its peak within the last 1–2 Myr when the cluster is still embedded into molecular gas.

This issue is quite interesting in the light of dating massive stars in cluster cores. It has been suggested for a long time that massive stars form late in the development of a cluster [51, 1056–1058]. Although [1059] suggests that this is more related to statistical fluctuations.

11.1.7 Cluster Distribution

The hierarchical clustering of stars, specifically in stellar clusters and stellar associations, is quite reminiscent of a similar hierarchy in the interstellar gas content within the Galaxy [1060] (see Chap. 3). Star complexes are well studied within the

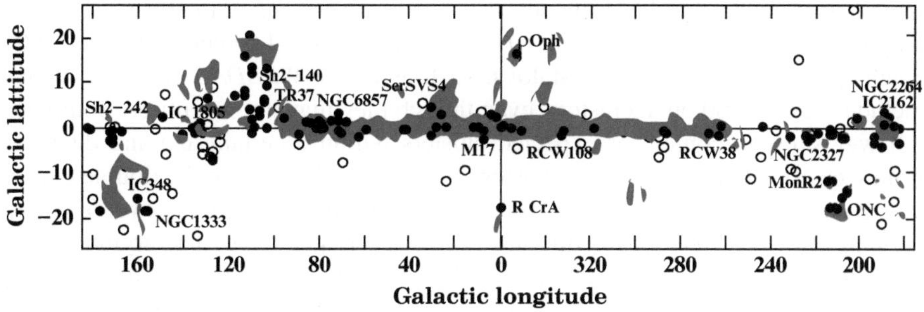

Fig. 11.6 Distribution of stellar clusters with molecular clouds in the galactic plane [107]. The solid points represent ESCs [1064]. The open circles older, open clusters [1065]. The former match well with the molecular cloud distribution. The latter scatter more into low-density areas of the galactic plane

Galaxy as well as in neighboring galaxies such as the *LMC* [1061], *M31* [1062], and *M81* [1063]. The last chapter demonstrated that the distribution of X-ray stars within our Galaxy seems to follow the same irregular and clumpy pattern [880].

ESCs match up very well with the distribution of molecular material (see Fig. 11.6). Here a large number of ESCs and open clusters taken from recent catalogs [1064–1066] are overplotted onto the molecular medium in the Galactic plane. The figure demonstrates that embedded clusters follow closely the distribution of molecular material. Open clusters, even the young ones as long as they are significantly older than 3 Myr, are devoid of their parent gas and distribute more evenly in the Galactic plane. There are probably an estimated 2×10^4 open clusters in the Galaxy, but only about 5 to 10 percent are observed due to the fact that most parts of the Galaxy are too extinct. For recent and more detailed maps, refer to [104].

Stellar clusters disperse – the more massive the more rapid – because of increasing destructive ionizing radiation toward the parent cloud. This becomes specifically obvious in the OB associations, which are generally more random than bound. The distribution of clusters is, to some extent, dominated by galactic tidal forces. This has the effect that the oldest open clusters are more concentrated at large galactic radii where the likelihood of disruptive encounters with GMCs is lower [1067] (see also below).

11.1.8 Cluster Evolution

The fate of a stellar cluster is not only restricted to their distribution in the Galaxy and their association with dense gas. Surely, the latter is very defining for the structure of a cluster. Once the parent intermittent gas is entirely depleted, or in the case of massive star formation also effectively destroyed, the cluster looses its tight grip and begins to disperse. There is convincing evidence that the population of stellar clusters experiences very high accounts of infant mortality [197], meaning that the vast majority of ESCs do not survive beyond ages of 10 to 100 Myr. In fact, maybe only 10 percent live beyond 10 Myr and less than ~4 percent of former ESCs reach ages beyond 100 Myr. Catalogs, such as as the *WEBDA* [14, 1065], which include roughly 300 open clusters, even though they may not be entirely complete, fall way short in content of expected clusters within a certain age and distance. It may be argued that large open clusters such as the *Pleiades* or the double cluster $h + \chi$ *Persei* are survivors of very large ESCs with mass contents of over 500 $M_\odot$. Although such a direct correlation has not yet been quantified. The argument that the percentage of clusters older than 100 Myr is of a similar order of magnitude as the percentage of very large ESCs may be biased by the initial mass content. The mass content of open clusters is significantly reduced compared to ESCs containing a significant number of massive stars due to the very limited lifetime of massive stars.

However, the fact that clusters in excess of 100 Myr are very rare was pointed out in the late 1950s by J. Oort and L. Spitzer. They argued that tidal encounters with GMCs likely disrupt open clusters with mass densities of less than 1 $M_\odot$ pc^{-3} in a few 100 Myr. As has been mentioned above in the context of their galactic distribution, the life expectancy of open clusters is then a function of galactic radius.

Cluster evolution is also impacted by its stellar content and the evolutionary stages of the stars. Figure 11.7 shows a CMD of older open clusters as it typically appears in a standard astronomical textbook. Clusters offer well-defined test beds for stellar evolution studies as they, as a whole, represent a homogeneous sample of stars at approximately the same distance that have a common cluster age and chemical composition. The CMD in Fig. 11.7 then shows the time of the main sequence turn-off, when stars begin to wander away from the main sequence and begin their lives at the end of their evolutionary scale.

11.1.9 Super-Clusters

It was more or less the advent of the HST, which through observations of star forming regions in neighboring galaxies, found something that does not seem to exist in our own Milky Way: *super-clusters*. Of course, given the fact that only a small part of the Galaxy is visible to us, cluster masses exceeding a few 1,000 $M_\odot$ have so far eluded observers in the vicinity of the Sun. Probably the most famous protagonist of this cluster category is *NGC 604* in *M33*. Its total mass exceeds that of the *ONC* by almost an order of magnitude, accounting for its bright appearance (see Fig. 11.2). Among other examples, the large clusterings in interacting galaxies such as the *Antennae* stand out [1068]. Here, a mass function of clusters was determined, over the range $10^4 \lesssim M \lesssim 10^6\,M_\odot$, to be proportional to M^2. This is quite similar to the power-law mass function in molecular clouds and Galactic clusters. Reports on other super-clusters are now emerging, see [1069–1071]. The topic will certainly gain momentum in the future as instrument sensitivities allow more and more studies of stellar properties in other galaxies (for more see Chap. 9).

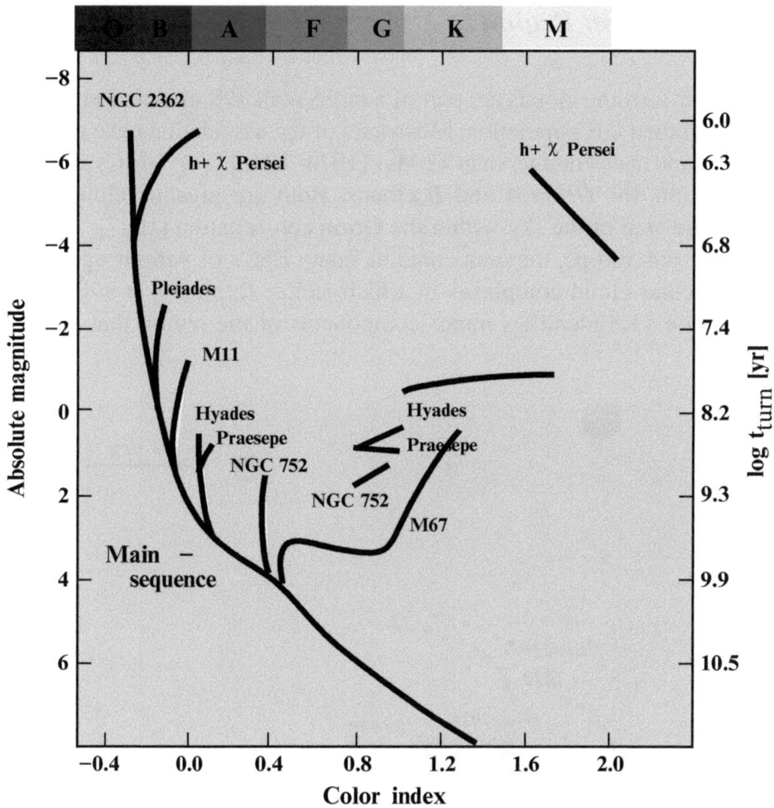

Fig. 11.7 HR-diagram of open clusters as it appears in any standard textbook. The right-hand scale are estimates of the turn-off time for clusters from the main sequence

11.2 Well-studied Star-forming Regions

Star-forming regions within the solar neighborhood have been studied since the 1950s, when it was realized that young stars are born in designated regions. Though the association with MCs and GMCs had not been made yet – the existence of dense molecular gas in the Galaxy was still unknown – the existence of young associations guided most studies to designated areas. Surveys now show that many star forming regions have the configurations outlined in Fig. 4.4 [417]. For example, a survey of a large number of open clusters at the end of the 1980s [1072] already demonstrated that clusters younger than 5 Myr have dense gas more massive than $10^4\ M_\odot$ associated with them. Clusters older than 10 Myr do not have significant molecular gas, but can be associated with clouds of masses up to $10^3\ M_\odot$. The following sections introduce general properties of three of the best studied star forming regions in the Galaxy. Each of them matches one of the configurations in Fig. 4.4. A very comprehensive review of stellar populations in major star forming regions, as known in the early 1990s, is given by H. Zinnecker et al. [1073].

11.2.1 *The Orion Region*

The Orion star forming clouds are part of a large-scale OB association ($\sim$150 deg^2) with a well studied OB population. Most stars of the association have masses larger than $\sim$4 $M_{\odot}$ and ages younger than 12 Myr [1074–1076]. Two GMCs are associated with this region, the *Orion A* and *B* clouds. Both are gigantic cloud complexes that fill a large area of the sky within the Orion constellation [1077]. At a distance between 450 and 500 pc, the area contains many ESCs of various ages as well as several molecular cloud complexes in which stellar formation is still an ongoing process. Figure 11.8 identifies major components of the region that are currently

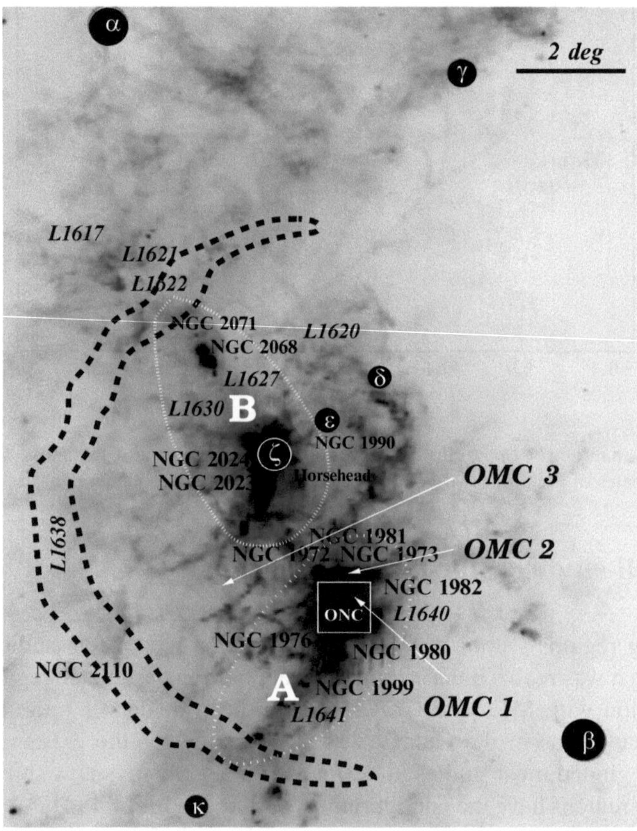

Fig. 11.8 Negative scale IRAS field of the Orion star forming complex zoomed from the field region shown in Fig. 4.2. Highlighted are the main Orion stars flanking the area. The Orion star forming region contains three major molecular giant complexes called *OMC 1–3*. These complexes are broken down into a manifold of subclouds of which the major ones are listed. Most important is the L1641 cloud complex residing behind the famous Orion Nebula *M42*. Also indicated are major embedded clusters of the region

under study. *Orion A* and *B* are marked by dotted white contours engulfing the location of their core regions.

The *Orion A* cloud covers roughly 29 deg^2 of the the sky and contains various dark clouds, most importantly *L1640* and *L1641* which are located behind the H II region of *M42*. This complex shows the classical blister structure as a consequence of massive star activity [1078–1080]. Such a blister structure is illustrated in configuration B in Fig. 4.4. In the case of *M42* the system is viewed with the blister in front and the molecular cloud complex behind. The total mass of *Orion A* is estimated to be about $1.0 \times 10^5\ M_\odot$. The *Orion B* cloud, though slightly smaller (19 deg^2) and about 20 percent less massive, is very similar in structure to the *A* cloud. One of the exciting features of these clouds is that they contain several cold but also hot molecular cores. Most famous are the ones in the A cloud, specifically the *OMC 1–3, OMC 1S* and the *BN-KL* nebula (see Sect. 9.3.4). Several of these clumps contain embedded stars exhibiting far-IR luminosity peaks or are associated with mass outflows and water masers [1081–1083]. The most luminous and hottest core is the *BN-KL* nebula with a luminosity of $10^5\ L_\odot$ and temperature of 70 K. Therefore, it is also the most massive of them all with a mass of 150 $M_\odot$[1077]. Others, like the cores in the *OMC 1* clump, are cold and serve as candidates for collapsing cores probably at the verge of forming new clusters [1084, 1085]. Cluster formation has been quite prolific in these clouds as the many ESCs in the region testify. Until recently sub-mm and mm dust continuum measurements have been the most compelling evidence for very early and on-going star formation in the OMC clumps. Today, evidence for protostellar activity comes from the entire wavelength band from radio to X-rays [859, 944, 1086, 1087].

The origins of the Orion star forming region are founded on a quite interesting speculation. It seems to be a demonstration of how giant star forming regions are born within turbulent interactions of gas in the Galactic plane. According to [1089], the progenitor of the Orion–Monoceros complex (see Fig. 4.2) formed by the collision of a high-velocity cloud, moving from the direction of the southern Galactic hemisphere, with the Galactic disk [1077]. This collision should have taken place about 60 Myr ago and fragmented the progenitor into the Orion and Monoceros complexes. Tidal forces induced a kinematic pattern that forced the fragments to oscillate with respect to the Galactic plane. After one passage, the Orion–Monoceros fragments are now about 150 pc ($\sim$19 deg) below the Galactic plane. Such a disruptive motion with respect to the Galactic plane within 100 and 200 deg longitude is indicated in Fig. 11.6. The kinematics and energetics of the fragmental motion is visible today. There is a substantial large-scale velocity gradient along the *Orion A* cloud which changes from 12 km s^{-1} near the *OMC 2* location to 5 km s^{-1} at the south end of *L1641*. In general, velocity structures are complex and show turbulent motions that could be driven, in part, by outflows from embedded stars, although such a notion is still controversial [1077]. Support could come from magnetic fields.

Right in the middle of all this action resides the *ONC*, which is the largest assembly of young stars in Orion. The *ONC* is way too massive to be dominated by the ongoing turbulence. Consequently, the *ONC* is one of the largest centralized stellar clusters in the Galaxy. It has a diameter of about 100 pc with 80 percent of the stars younger then 1 Myr [331]. The *ONC*, with over 1,000 $M_\odot$, has ~1,600 optically detected sources. Spectral types have been determined for about 60 percent of them. In addition there are over 2,000 additional IR sources [331] and over 1600 X-ray sources [861] of which most have been identified with IR sources. Not yet counted here is the expected substellar population of objects such as brown dwarfs [1090]. Morphologically, the *ONC* is a central cluster and historically, it has been subdivided into three radial zones [1091]: the *Trapezium Cluster* with a radius of ~2 arcmin (0.3 pc), what is usually referred to as the *ONC* with a radius of 20 arcmin, and finally the *Orion Ic* association at least 3 degrees in radius [1074, 1091, 1092].

The top panel of Fig. 11.9 shows the population of the *ONC* at about its 50 percent radius. The size of the symbols denotes stellar mass. As already determined in Sect. 11.1.4 most of the stars are of low mass. θ Ori is a massive star system at the center. The five massive stars of θ^1 Ori form the *Orion Trapezium* and θ^2 Ori is just a bit off-center. The Orion Trapezium was discovered by R. J. Trumpler in 1931 and its main stellar component, θ^1 Ori C, was identified as the main source in excitation of the Orion Nebula. The more detailed stellar properties of this cluster have been under investigation throughout the entire wavelength band for a long time [331, 532, 879, 1090, 1092, 1093]. Still, many issues remain unresolved. Most observations of the central Trapezium have been made at infrared wavelengths in the K and H bands at 2.20 μm and 1.65 μm, respectively. θ^1 Ori A–E are the brightest stars in the cluster at optical, infrared and X-ray wavelengths and are early-type

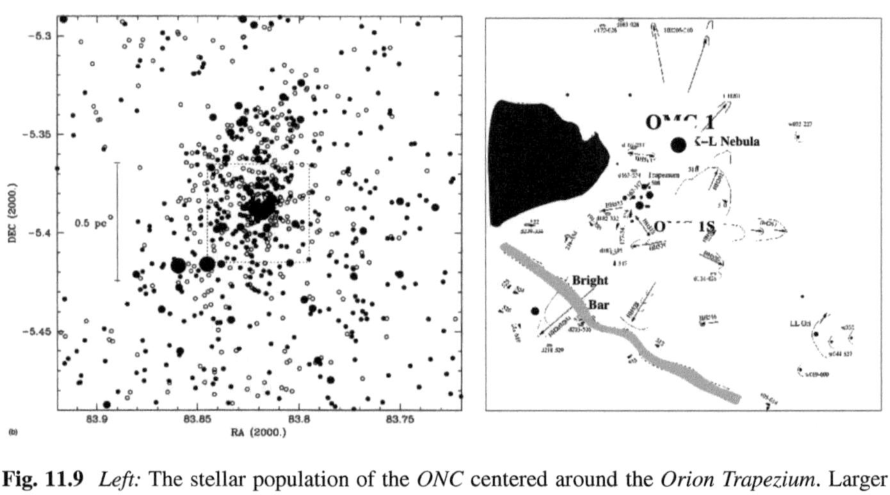

Fig. 11.9 *Left:* The stellar population of the *ONC* centered around the *Orion Trapezium*. Larger point sizes indicate larger stellar masses between 1 and 10 $M_\odot$. (Masses above that limit have the same size.) The field contains over 1,000 stars. From Hillebrand [331]. *Right:* The area around the Trapezium is not only one of the most densely populated stellar areas in the Galaxy, but also features many very special structural features of intermittent gas and interactions of YSOs within the gas. Adapted from Bally et al. [1088]

OB, or intermediate mass stars. Many members of the Trapezium and the *ONC* are multiple systems, some with separations much smaller than 90 AU [497, 1094] (see Sect. 9.5). One big problem in the optical/IR domain is the brightness of the H II cloud, which simply overpowers stellar point sources. Until recently, there had been no study of the stellar mass and age distributions that was not severely extinction limited [331]. The *ONC* also came under high scrutiny in X-rays with the advent of the *Chandra* observatory, which allows to resolve all the individual stars, even down to substellar masses, in X-rays (see Chap. 10).

Besides the bulk stellar population, the vicinity of the *Trapezium Cluster* exhibits many more features quite classic to regions of very early stellar evolution. The bottom panel of Fig. 11.9 shows a sketch by J. Bally and colleagues [1088] of major features in the region. It shows a variety of Herbig–Haro objects, i,e, outflow remnants, *protoplanetary discs* (see Chap. 13), dark dust bays, and bright illuminated ridges of gas. The region includes the famous *OMC-1* clumps in the background as well as the *BN-KL Nebula*.

11.2.2 The Rho Ophiuchus Cloud

The ρ *Oph* cloud is one of the closest star forming regions to the Sun. Although it is well situated within the *Upper Scorpius* OB association, the cloud itself is yet devoid of wind erosion from massive stars. The distance of the association has recently been determined from *Hipparchos* data to be 145 pc. It is an example for configuration A in Fig. 4.4. Although the association is several Myr old (see e.g., [1095]), the ρ *Oph* embedded cluster is probably one of the youngest ESCs known in the vicinity of the Sun [418,940]. Figure 11.10 shows the massive but already fragmented cloud in a large scale within the O and B stars of the surrounding association. There is no erosion from inside, but it is likely under the radiative and tidal spell of the association. The large-scale distribution of dense gas over about 6.4 deg^2 is shown in [1096].

There are three cloud complexes, the main cloud *L1688* and two smaller cloud fragments *L1689N* and *L1689S*. The main cloud *L1688* itself is broken into several cloud fragments A–E, some of which host recently formed sub-clusters containing a large number of Class 0, I, II, and III sources. Mainly, the stellar content of this cluster led to the now widely use SED IR classification of low-mass YSOs (see Chap. 6). Specifically, jet-like outflows from Class 0 sources in ρ *Oph A* helped researchers understand the circumstellar environment of YSOs [421, 1098–1100]. Figure 11.10 shows a more recently determined account of the stellar population in these clouds from *ISOCAM* measurements [1097]. It not only contains many Class I and II sources, but also four young early-type stars. Like in the Orion star forming clouds, X-ray observations of the ρ *Oph* clouds with *EINSTEIN*, *ROSAT*, *ASCA* and *Chandra* have, in major ways, revised researchers perception of young stellar evolution.

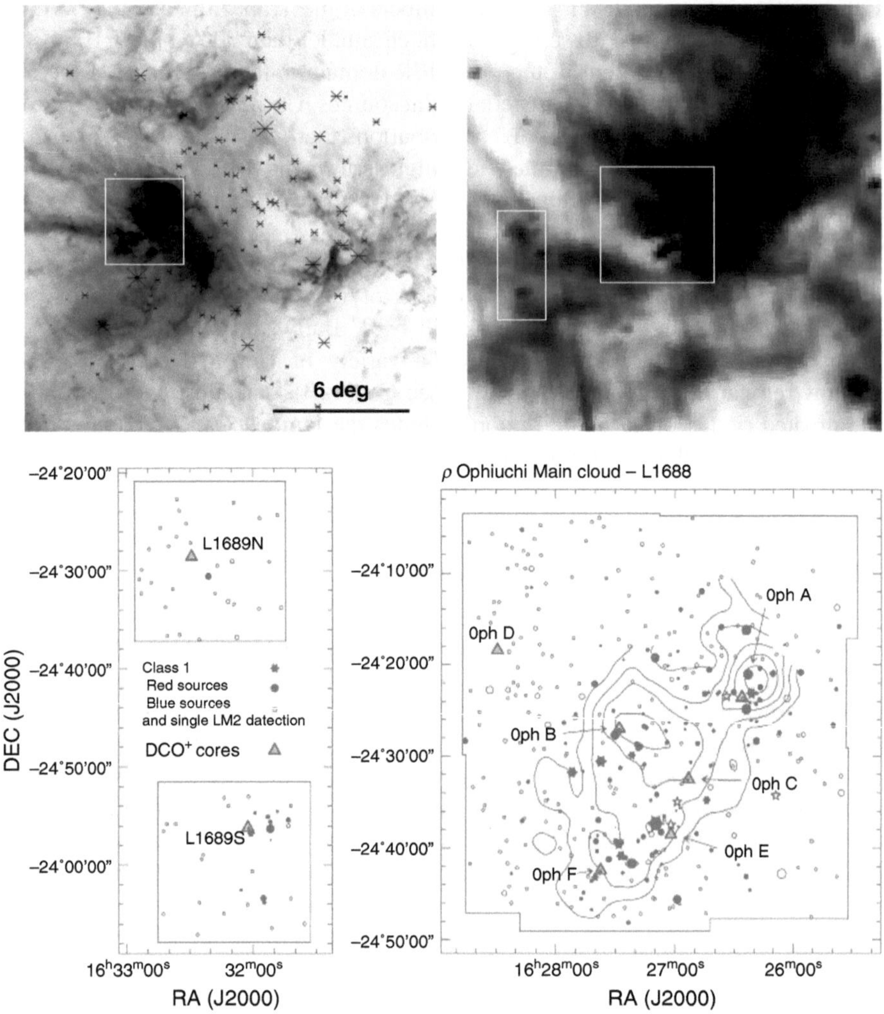

Fig. 11.10 The morphology and stellar content of the *ρ Oph Cloud*. The upper-left portion of the figure shows the *IRAS* large-scale view of the environment of the cloud, which is marked by the white square. The cloud is embedded into the *Upper Scorpius* region with a large association of OB stars (marked by asterisks). The upper-right portion shows the cloud zoomed in from the white square in the upper left image. There are no massive stars associated with the cloud itself. The cloud is fragmented into several subclouds and clumps. The diagram at the bottom shows the stellar and morphological content of the cloud (as marked in the upper-right image) as recently measured by S. Bontemps and colleagues using the *ISOCAM* on board *ISO* at mid-IR wavelengths. Filled stars mark Class I, filled circles Class II, and open circles are likely Class III YSOs. The contours show the extent of *L1688*. Credits: upper-left image by T. Preibisch; bottom diagrams from Bontemps et al. [1097]

As part of a large survey using the *Spitzer Infrared Spectrograph* [1101] a large number of T Tauri stars in the Ophiuchus, Taurus, and Chamaeleon I star forming regions were observed [1017]. The Taurus and Chamaeleon I star forming regions are very similar to Ophiuchus with median ages of <1–2 Myr and relatively unclustered appearance. It was found that the median mid-infrared spectra of most objects in these three regions are similar in shape, suggesting, on average, similar disk structures. The SED shape between 13 and 31 μm in conjunction with the width of the silicate emission feature at 10 μm was used to search for objects with transitional disks. Cases where a near-IR inner disk clearing was suggested by a steep slope remain rare in all three regions, while objects with large silicate emission features appeared much more common. Moreover, it was found that the Ophiuchus core *L1688* as the youngest region had more settled disks containing larger grains indicating protoplanetary evolution. This means that even as early as about 1 Myr the formation of protoplanets is under way.

11.2.3 IC 1396

The last example introduces a star forming region resembling configuration C in Fig. 4.4. *IC 1396* is an extended H II region at a distance of about 800 pc [1102]. Its location coincides with the core of the Cep OB2 association which is divided in two subgroups differing in age. The total mass of the entire region was estimated to $\sim 4 \times 10^5\, M_\odot$ [1103]. An older and more dispersed group of roughly 75 O and B stars is confined within a maximum size of 170 pc [1104] and has an estimated age of 6–7 Myr, and a younger more confined group, the young open star cluster *Trumpler 37 (Tr37)*, with an age of 3 Myr in its core. The nucleus of the cluster bears the star complex *HD 206267*. It resembles the *Orion Trapezium* at the center of the *ONC*, though because of the larger distance its components are much harder to resolve. Its hottest component, an O6.5f star [1105] is the main source of excitation in *IC 1396*. A proper motion study identified over 480 stars within a radius of 7 pc from *HD 206267* with a probable membership of Trumpler 37. A decisive difference from the *ONC* is that these stars are mostly of A and B type, whereas the *ONC* is composed of mostly low-mass stars.

Still, the larger vicinity of *IC 1396* has also been identified as a T association [52, 430] with F-G spectral types and masses below 3 $M_\odot$. These studies also indicated that T Tauri stars should reside in the vicinity of *HD 206267*. Its most prominent proponent is the double T Tauri star *LkHα 349* (see below) at the center of globule *IC 1396A* (see Fig. 4.13 and Fig. 4.12). The vicinity of *Tr37* is very difficult to observe in the optical and IR as it is highly extinct by dust. The existence of an X-ray cluster at the center of *IC 1396* was shown though X-ray observations [272]. These X-ray sources had to be T Tauri stars as most of the B stars of *Tr37* are of a later type, which do not exhibit strong X-rays.

IC 1396 is quite rich in bright rimmed globules as can be seen in Fig. 11.11. In particular the bright rimmed globules *IC 1396A* and *B* [1106], centered to the west

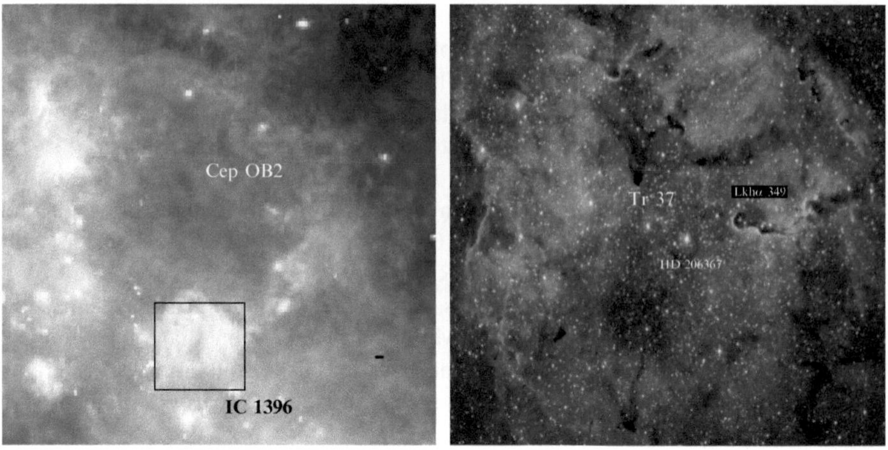

Fig. 11.11 *Left:* The wide scale view of the *IC 1396* H II cloud environment as observed with *IRAS* within the *Cep OB2* association. *Right:* The IC 1396 cloud as marked by the black square in the left image with the young B cluster Tr 37, the trapezium system HD 206267, and the cometary globule hosting Lkhα 349. A color image of this nebula appears on the front cover of this book (Credit: R. Crisp)

of *HD 206267*, have been the focus of previous investigations in the X-ray, IR, sub-mm and radio domain [272, 1107–1110]. Near the center of *IC 1396 A*, two equally young T Tauri stars have been identified *LkHα 349/349c* and are 17 arcsec apart [49, 428]. The spectral types were given as F8 for *LkH 349* and K7 for *LkH 349/c* with bolometric luminosities of 84 $L_{\odot}$ and 9 $M_{\odot}$, respectively. From measurements of the CO structure, [1110] suggested that the central part of this globule resembles a cavity that is formed by an interaction of the stellar wind from the central PMS stars of the globule. Many other globules contain or are associated with *IRAS* point sources [1111] or show signs of active star formation [1112].

Dynamically, the entire H II region belongs to a class of clouds in which star forming activity, on a grand scale, has clearly ceased and most molecular gas has either been accreted or destroyed. What is left is the aftermath, which is mostly gas of the heated H II region and cold fragments now detected as globules (see Sect. 4.1.9) of the old GMS. CO maps show that many bright rimmed globules lie on a ring about 12 pc from the clouds center (see Fig. 11.11). The ring is expanding with a velocity of $\sim$5 km s^{-1} suggesting the dynamical age of these globules is about 2–3 Myr. Activity in these globules seems to represent the most recent wave of triggered star formation [1113]. The entire region is full of turbulent dynamics and is part of the global activities of the ISM. Situated within the *Cep OB2* association, the entire cloud seems to be associated with the expansion of a giant bubble, which may have originated from a supernova explosion 3 Myr ago [1114]. The top part of Fig. 11.11 shows the large-scale environment of *IC 1396* as observed with *IRAS* at 60 μm. Clearly seen is the expanding shell with *IC 1396* residing at the bottom rim.

11.2.4 Carina

The Carina region is a massive star forming complex located in the Sagittarius–Carina spiral arm at a distance of 2.3 kpc (see Townsley et al. [868]) and references therein). The region hosts a variety of young stellar clusters of ages <6 Myr which resemble some of the highest concentrations of massive and low-mass stars known today in the Milky Way. The complex contains more than 3 WR stars, >65 O stars, hundreds of imtermediate mass stars and several tens of thousands of low-mass OMS stars. The only system closer than Carina harboring such a large number of very massive stars is the Cygnus OB2 association. The most massive star in Carina has a mass of 120 $M_{\odot}$and the CMF is well represented in the high mass end as it is in the low end down to 0.1 $M_{\odot}$. In this respect, the *Chandra* Carina Complex Project (CCCP) scanned the entire Carina Nebula in X-rays and detected over 1.4×10^4 X-ray point sources with average luminosities of $\log L_x \sim 30.7$ erg s^{-1} typical of low-mass PMS stellar populations [869] and over 90% confirmed membership of Carina. Figure 11.12 (left) shows a wide-scale mosaic view of the Carina complex

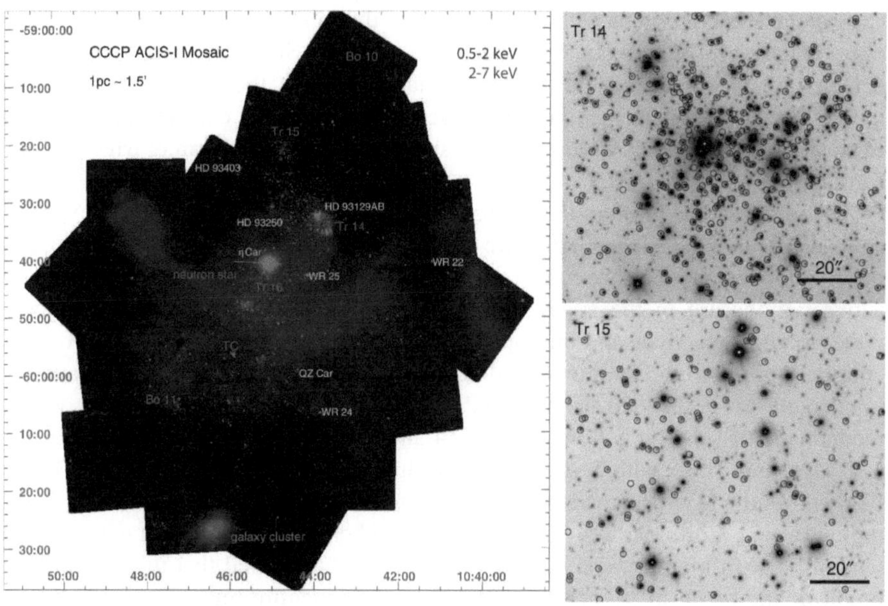

Fig. 11.12 *Left:* The CCCP mosaic in the 0.2–7 keV X-ray band with some of the major components of the regions indicated. Major clusters in the nebula are Trumpler 14, 15, 16, and in the outskirts Bochum 10, 11 and Collinder 228. At the center resides the Homunculus cloud with the highly evolved colliding wind binary η Car. Below Tr 16 resides what is now called the Treasure Chest (TC), a highly embedded young cluster (from Townsley et al. [868]). *Right:* The clusters Tr 14 and 15 as observed in the IR with *HAWK* and with the locations of the *Chandra* X-ray sources overlaid. The difference in stellar density in the cores of the two clusters reflects the state of dispersion with age (from Preibisch et al. [1342])

from CCCP and its constituent elements, such as the hierarchical ensemble of clusters Tr 14–16, BO 10/11, Collinder 228/234, the central homunculus cloud of η Car, or the now detected large area soft X-ray diffuse emission regions. On top infrared and optical images show structures rich in highly-structured arcs, filaments, pillars, and shells on 0.2–10 pc scales. Residual molecular material is scattered throughout the cloud region-likely remnants of the photoevaporating activity of the large OB-star population. The region in general is a prime example of the violent histories in hierarchichal star-forming regions as ionizing radiation fields, potent stellar winds from many very early type stars, and perhaps supernova events left imprints of violence.

Even though the history of supernova activity in Carina is still disputed, there are some tell-tale signs that such events have already taken place. *ISO* observed broad dust emissions around 22 μm highly reminiscent of dust formed in the supernova remnant of Cas A [1115]. The multi-age structure of the region certainly allows for such events to have happened. It has been suggested that at least one supernova might have already occurred, driving Carina's infant superbubble and enriching its interstellar medium (ISM). The recent X-ray discovery of a 10^6-yr-old neutron star in the Carina complex [1116] provides strong evidence for past supernovae in Carina. This is consistent with some evidence that a slightly older, up to 10-Myr-old population of young stars pervades the complex. Six more X-ray sources are detected which might be neutron star remnants. Clearly, the void of massive stars in Tr 15 is likely a sign that these massive stars has already engaged in supernova explosions and feedback of these events should be present in form of triggered star formation.

The many stellar clusters in the region are a prime example of what is expected from a hierarchical cluster structure as illustrated in Fig. 11.3. There are many observational signposts that concurrent star formation activity is ongoing. The most active regions are in the south pillars which host a wide array of HH-objects, dark globules, proplyds and microjets. The south pillars are an eroding set of dust columns similar to the ones observed in the Eagle Nebula (Fig. 4.11). The youngest stellar cluster embedded in the cloud is the Treasure Chest (TC) cluster with a median age of $<$1 Myr followed by Tr 14 and a more loosely clumped Tr 16 with 1–2 Myr, and Tr 15 with 3–4 Myr. Other clusters such as Bochum 10 and 11 are likely around 6 Myr or may even be older. A total of 20 clusters are identified in the region [1117]. Collinder 228 appears to be of rather unusual type as it extends over tens of pc with many sparse compact groups. It is suspected that these groupings might be the result of triggered star formation. The entire region is overlaid by a widely dispersed population of X-ray sources giving testimony of past star forming activity. Embedded in this wider distributed population is Collinder 234, a loose group of massive stars which do not exhibit much clustering with low-mass PMS stars. Figure 11.12 (right) shows the core regions of Tr 14 and Tr 1r56 in the IR taken by *HAWK-I* and with overlaid positions of X-ray sources [891]. The core of the 6-Myr-old Tr 15 has a much lower stellar density as it is already further dispersed.

The population of massive stars deserves some special consideration. In the ONC it seemed that OB-stars appeared rather sparse and the high mass end of the CMF

Fig. 11.13 The *Chandra* Carina data shown in three soft bands (0.50–0.70 keV in red, 0.70–0.86 keV in green, and 0.86–0.96 keV in blue) to highlight the diffuse X-ray emission (Credit: NASA/CXC/PSU/L. Townsley)

is not particularly well populated. In IC 1396, very early type stars were sparse as well, but Tr 37 at least provides a sizeable number of B-stars. However, while in Orion the two most massive stars appear magnetic [525, 848, 1010], only two out of the 65 massive O-stars in Carina qualify to be of a magnetic nature [1118]. Even though the majority of massive stars in Carina are single stars with a standard $\log(L_x/L_{bol}) = -7.26$ and only a shallow trend towards harder emission in brighter objects, there are about 15% of massive binaries in the sample, leaving not many candidates for colliding wind binaries either and most stars likely are classified as weak wind stars [1119].

Perhaps the most interesting part in the X-ray study is the soft diffuse emission. Figure 11.13 shows the distribution of this radiation in Carina. While certainly a significant portion can be attributed to the winds from the ubiquitous massive stellar population, some is likely a testimony of past stellar explosions. A comparison with other sites of massive star formation with pronounced diffuse X-ray emission showed that all models in these cases need at least two diffuse thermal components and in several cases the hot plasmas appear to be in a state of non-equilibrium ionization, indicating recent and current strong shocks [1120].

11.3 Formation on Large Scales

The previous sections demonstrated that most stars do not form in isolation but in rather crowded groups. However, if this is the case then not too many young PMS stars should be detected too far away from ESCs. In this respect, it is quite a twist of fate that the first very young late-type stars were detected in the *Taurus* region, rather far away from any major stellar clusters [39]. Thus there are regions of molecular gas that spread over larger volumes without accumulating large dense cores but still exhibit moderate star-formation rates. Two examples are the *Taurus-Auriga* region and the *Lupus* region. Other sites are OB associations, which, besides dense stellar cluster cores, are also associated with extended molecular gas. Stellar groupings in these regions are generally unbound. They are loose aggregates of stars that formed either in weakly bound cores or in completely unbound structures. That these areas have to exist as well becomes clear by looking at the large-scale structure of interstellar gas. The mass spectrum of this gas includes (see Fig. 4.6) giant cloud complexes with masses near the cut-off at 10^6 $M_{\odot}$and extends hundreds of parsecs. The following two sections introduce the reader to exemplary sites of star formation on large spatial scales.

11.3.1 The Taurus–Auriga Region

This region is an intensely studied star forming region with a research history that reaches back to the early discovery of young stars. It is a prime example for an entirely unbound star forming complex. With 3.5×10^4 $M_{\odot}$it has only a fraction of the mass of the Orion clouds. CO maps show it has an extent of 30 pc which is over ten times more extended [1121] (see Fig. 11.14). Morphologically, the cloud appears branched with many string-like patterns, which were already well documented in E. E. Barnard's photographic atlas of 1927 [1122]. Proper motion measurements by [50] revealed the quite bizarre phenomenon that young stars drift along the filaments, wheras the ambient magnetic field runs roughly perpendicular to the filament's orientation [1123]. The verdict on the effect is not out yet, though promising scenarios have been proposed ranging from magnetic instabilities [1124] to coeval star formation [1125] to star forming bursts in gas collisions of high-speed flows [1126]. Besides the filament phenomenon the stars in *Taurus–Auriga* overall are not really distributed in a random pattern but exhibit some clumpy structure [1125, 1127].

The *Taurus–Auriga* complex has long been one of the fertile grounds where T Tauri stars at optical and long wavelengths are found [39, 46, 49, 414, 428, 915]

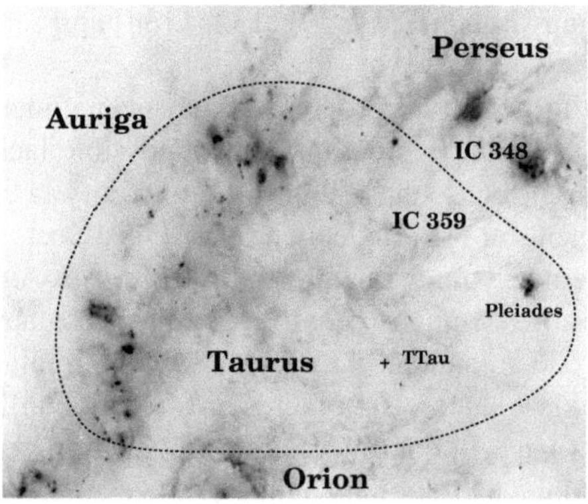

Fig. 11.14 IRAS view of the *Taurus–Auriga* complex which is approximately engulfed by the dashed line. North-west is the *Perseus* cloud complex hosting the ESCs *IC 348* and *NGC 1333* (off the plate to the left of *IC 348*). To the south one finds the Orion region shown in Fig. 11.8. Extended over a much larger area, the region is composed of a lot of diffuse molecular gas with many clumps interspersed. A CO map with identifications of most of these features can be found in [1121]. Also marked is the stellar location of *T Tau*

(see Chap. 4). Still, the region is so vast that many early studies remained patchy in coverage. In addition, since the detection of low-mass stars is directly linked to instrument sensitivity, mass coverage remains incomplete as well. The account of optical stars in the region has recently been extended to objects as faint as 17th magnitude [1128]. X-ray survey studies proved to be very powerful in closing gaps left by long wavelength studies which specifically helped find and identify new WTTS [73, 450–452]. The unique coexistence of T Tauri stars and late-type stars as present in the *Pleiades* and *Hyades* clusters allows one to study long-term relations in young stars with respect to luminosity, rotation, and age [657].

To obtain an objective account of young stellar properties in a vast and sparse region such as *Taurus–Auriga* is very difficult as there is no coherent sample to use. In concentrated regions, such as ESCs, stars have well defined physical conditions and the age spread is confined to several Myr. The *Taurus–Auriga* is relatively mature. Older open clusters, such as the *Pleiades* and *Hyades*, with approximate ages of 100 Myr, tell of active, probably Orion-like star forming activity in the past. Typical estimates for cloud lifetimes are 10 Myr. The fact that star formation within some filaments is still ongoing at the present [1129] as well as the richness of T Tauri stars throughout the region would imply an enormous age spread of the order of 100 Myr. More recent studies, though, caution on the interpretation that all these X-ray stars are really PMS stars [1128, 1130]. Specifically, neither X-ray emission nor the detected Li I absorption provide useful age discriminators, at least for certain spectral types [1128, 1130]. It has even been suggested that many of the sources detected with the *RASS* are not WTTS but closer to stars observed in the older clusters [1131]. Still, a long-standing problem remains in that there appears to be an age gap between $\sim$10 and 100 Myr in stellar age. Such a lack of PTTS has been recognized in many optical surveys [1132–1134] and a satisfactory solution is still pending.

The way star forming activity is progressing has been under study for a long time and the *Taurus–Auriga* stellar population – or its possible lack thereof – presents a challenge. The finding in ESCs that star forming activity is characterized by rapid acceleration within each star forming epoch [1051] (see Sect. 11.1.6). Although this perception is not yet entirely undisputed [1055]. In *Taurus–Auriga*, the star forming picture has become more detailed recently [1122]. These authors find that stellar births occur over a broad area under the cloud's own contraction (i.e., without much influence by internal turbulence). If, in fact, star formation in *Taurus–Auriga* is also an accelerating process, it will continue for several million years. This puzzle is far from being solved and the evolutionary status of *Taurus–Auriga* is still unknown.

11.3.2 Turbulent Filaments

The filament phenomenon in the *Taurus–Auriga* cloud has lately become under more scrutiny. A more recent approach to explain the phenomenon places form and dynamics within the filaments entirely into the domain of turbulence. The filaments are then a consequence of shock compressions and shear flows. Although these structures are normally short-lived and certainly not in hydrostatic equilibrium, it can happen that density fluctuations become gravitationally unstable, which leads to star formation [11, 1135]. Simulations of this model have been performed by R. Klessen (see Fig. 11.15), which successfully reproduce not only the filamentary structure, but also, after the application of self-gravity, lead to stellar collapse. The Jeans length and the free-fall timescale for the filaments require that [535, 1136]:

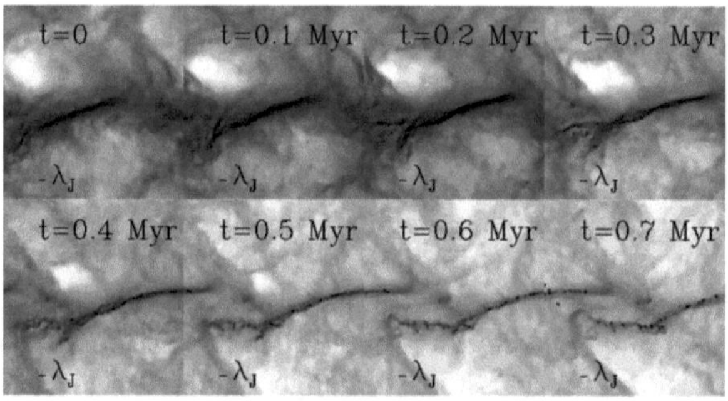

Fig. 11.15 The evolution of a molecular filament structure and subsequent star formation along the filament strings. Each frame shown has a baseline of 6 pc and depicts column density maps of the gravitational fragmentation of a filament generated by supersonic turbulence. The sequence depicts the system at intervals of 0.1 Myr. The small bar at the bottom-left of each frame indicates the Jeans length λ_J, the size of gravitationally unstable regions. Note that protostellar cores are separated by roughly λ_J. With proceeding time, the filament undergoes gravitational fragmentation. A color version of the final result is shown in Fig. 4.18. Credit: R. Klessen, AIP

$$\lambda_J = 1.5\frac{T}{[10\ \mathrm{K}]}A_V^{-1}\ \mathrm{pc}$$

$$\tau_{ff} \sim 3.7\left(\frac{T}{[10\ \mathrm{K}]}\right)^{1/2} A_V^{-1}\ \mathrm{Myr} \qquad (11.3)$$

Starless cores show $A_V \sim 5$ [1137] which leads to $\lambda_J \sim 0.3$ pc and τ_{ff} of 0.7 Myr, well in agreement with observations and simulations (see Fig. 11.15). In general, the influence of turbulence in isolated star forming regions will be a central issue for star formation in the coming years.

11.3.3 Observations of OB Associations

OB associations are generally regarded as fossil star forming regions [1138]. Morphologically, there are two aspects to these associations. One is the formation and kinematics of the high-mass population. The other is the interspersed low-mass population. A legitimate question is certainly how such large assemblies of O and B stars constitute themselves in the first place. Spectral types of these stars were previously thought to be dominated by very early-type stars and include O stars and as well as B-stars of types B0 to B2. Later B types are less massive and appear more in clusters, such as *Tr37* in *IC 1396* (see above), although some small associations embedded into reflection nebula can contain types up to early A type stars. More recent observations now indicate that this perception of the distribution of spectral types in OB associations is misleading and it is now established that late B type stars as well as low-mass stars are quite present as well (see below). These associations have extents of up to 200 pc and, in terms of hierarchy, are the next step beyond open clusters. However, OB associations contain over 50 percent of all massive stars in the Galaxy and subsequently, are the sites where most of the Galaxies' supernovas occur. These catalysmic events inject material and energy into the ISM with the consequence that these fossil star forming sites turn into new sites of star formation. A visible example in the solar neighborhood is the *Gould Belt* OB associations which likely have had their origins through such an event (see Chap. 13).

There are still many open issues concerning star formation in OB associations. One of them is the question about large-scale isolated low-mass star formation. It is yet unclear what the distribution of such a field population would be and how it may differ from the one in clusters. Nevertheless is important to account for all low-mass PMS stars in these associations to determine the *IMF* of the field (see, e.g., [1073]). There are indications that high-mass star forming regions, specifically the ones considered to be starbursts (see also Sect. 11.1.9), contain smaller numbers of low-mass stars than expected from the *IMF* [1139]. Clearly, the comparison of mass functions in the field to the ones in clusters is an important step towards answering questions about a possibly universal *IMF*.

The problem is that the low-mass stellar content is only poorly known, even in nearby OB associations, mainly because it takes an enormous observational effort to identify the widespread population of PMS stars among the many thousands of field stars. During the last few years, the situation started to improve. Specifically, the advent of *Hipparcos*, allowed one to identify the high and intermediate-mass stellar populations [1140] in associations. Very recently a full account of the stellar population of the *Upper Scorpius* association produced an empirical mass function covering the mass range from 0.1 to 20 $M_{\odot}$ that, within errors, remained within the expected field *IMF* [1095]. In fact, these measurements showed an excess of low-mass stars, which raises this issue of stellar populations in possibly triggered star formation scenarios [1095].

Chapter 12
Proto-solar Systems and the Sun

The question about the origins of our Solar System has always been part of human speculation going back at least to Kant's and Laplace's hypotheses in the 18th century. The previous chapters have illustrated how stars emerge from molecular cloud cores and demonstrated that the evolving stars are surrounded by what is called an accretion disk, which evolves with the central star. This disk is synonymous with the *Solar Nebula*, cited in many referrals to the protostellar disk from which the evolution of today's Solar System began. With all the technical advances available at present, the observation of protostellar disks and the detection of evolved planets outside the Solar System is as popular a topic as ever. The following sections attempt to shortly introduce the reader to advances in observations of protostellar disks, their dust properties, as well as some concepts in the discussion of how protoplanets could have formed out of dust disks.

12.1 Protoplanetary Disks

The use of the expressions *circumstellar* and *protoplanetary* in the context of disks is not entirely clear in current literature. It should be realized that the former is a very general expression and represents a large range of phenomena including not only protostellar, PMS stellar, and MS stellar disks but also exotic disks that form in Be-stars as a consequence of equatorial outflows or through other outflow phenomena. Protoplanetary disks are solely restricted to young systems likely from the CTT stage into the early MS as they are believed to be the true precursors of solar systems. They include accretion disks in CTTS, inactive disks in WTTS, and dust disks in more evolved young stars.

The account of circumstellar disks detected in YSOs is growing rapidly. Again, the major star-forming regions are at the forefront of today's investigations. The focus of recent studies of the youngest systems include the *ONC* [1044], *NGC 2024* [1141], *IC 348* [589], and the nearby *TW Hya* association [909] (see also Chap. 6 and 8) to name a few. To possess circumstellar disks is almost mandatory

N.S. Schulz, *The Formation and Early Evolution of Stars*, Astronomy and Astrophysics Library, DOI 10.1007/978-3-642-23926-7_12, © Springer-Verlag Berlin Heidelberg 2012

for YSOs as it is a direct consequence of the star-forming process. The question is, how long in their evolution is such a disk retained. Modes of disk dispersal have been discussed in Sect. 8.1.6. A comprehensive study in the *ONC* found that at least 55 percent, but probably no more than 90 percent, of the stars retain their inner (<0.1 AU radius) disks much beyond T Tauri stage with no obvious correlation to the age or age spread of the cluster [1044]. A slightly lower maximum disk fraction of around 65 percent is found in *IC 348* [589]. What actually determines this fraction in clusters is still unknown. Generally, the fraction of accretion disks should be high given the youth of these clusters. An interesting find in the *ONC* is the disk frequency with stellar mass. Understandably, more massive stars are less likely to have disks (see Sect. 8.1.7). Even if they would still have their disk, such a disk would be hard to detect via IR excess as the stars luminosity is overpowering. In *IC 348* all stars of spectral type earlier than G appear diskless. Peculiar in these clusters is certainly the indication that disk frequency is decreasing toward the very low mass end of very late spectral types, though it is emphasized that stars of all masses can have and retain disks.

The lifetime of T Tauri disks is a central issue in the study of protoplanetary disks. The definition of lifetime in this respect is an important detail. Survey studies such as the one performed above, use near-IR photometric data and are thus bound to the standard SED classification. In these studies, the disk disappearance is linked to the fact that the IR excess is created though the illumination of the disk (see Chap. 8). Disk sizes range from $\sim$10–$\sim$1000 AU (see Chap. 5). A more recent assessment of disk dispersal in T Tauri stars emphasizes that 30 percent of stars should lose their disks within 1 Myr, while the remainder have disk lifetimes between 1 and 10 Myr [1142]. This is quite within the range predicted from meteoric evidence from the Solar Nebula which projects up to $\sim$5 Myr for the formation of chondrules [1143]. The actual disk dispersal process proceeds rapidly on a timescale that is probably an order of magnitude faster than the average disk lifetime (i.e., $\lesssim 10^5$ yr). Although the mechanisms that lead to this rapid timescale are uncertain, observations show only very few stars so far that could actually be identified with dissipating disks' signatures [1144–1146]. On the other hand, some caution is advised when associating IR excesses with dust mass only. There is the possibility that disks may retain their gas mass beyond the dust mass lifetime [1147, 1148].

Disk dispersal in some cases does not mean that all material around the star is entirely relinquished. Besides forming *planets*, it rather transforms into something called a *debris disk* that contains larger forms of dust grains and small planetesimals. The presence of micron-size dust grains in optically thin disks requires that such grains are continuously supplied to the disk (see Sect. 12.1.2). This can only happen if there is a sufficient reservoir of grains or the dynamics within the disk allows the destruction of large grains and planetesimals. As soon as particles in the disk grow to sizes beyond microns, they loose their optical sensitivity. The optical properties of disks are complex and depend on disk mass, disk radius, and, inclination towards the viewer. From observations of IR excess radiation, S. E. Strom and collegues [1149] define three classes of disks: *optically thick* disks ($\tau > 1$) at all IR wavelengths <100 μm, *optically thin* disks ($\tau \ll 1$), and *transition disks* with τ depending in

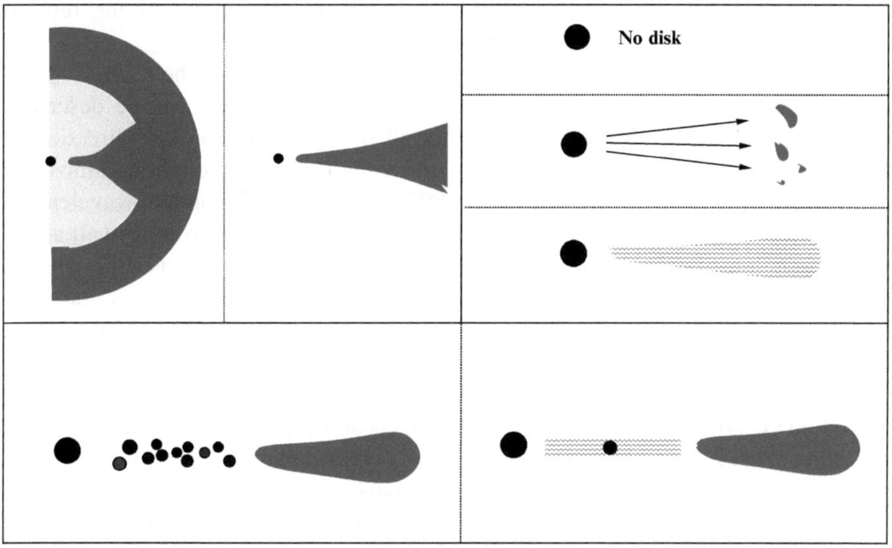

Fig. 12.1 The cartoon shows three classes of circumstellar disks deduced from IR excess observations as suggested by [1149]. The optically thick case (upper-left box) basically describes the situation of an outer optically thick envelope or disk as known for class I to II sources. The optically thin case (upper-right box) throughout the whole disk is a scenario past active accretion where either all the material has already been accreted or the disk has been blown away, disrupted, or is in a state where disk material remains dominated by small dust. Finally the transition case (lower boxes) features a still optically thick outer disk, but indicates two scenarios for in the inner disk. One features rapid conglomeration (lower left), the other the formation of a giant planet accreting from tidally fragmented optically thin material (lower right)

various ways on wavelength. Of interest for the detection of protoplanetary disks are optically thin and transition disks. Suggested scenarios are illustrated in Fig. 12.1.

12.1.1 Proplyds

Some of the most exciting observations of protoplanetary disks were obtained through imaging of the central regions of the *Orion Trapezium Cluster* with *HST*. These highly resolved images revealed breathtaking structures of illuminated YSOs which appear suspended in the intercluster gas [1150, 1151]. The existence of a limited number of such externally illuminated YSOs, was known since the late 1970s [1152, 1153]. Specifically, free–free radio continua were detected from some of the objects with the *VLA* [892, 893, 1154]. In the early *HST* images, about 150 such structures were catalogued and identified as protoplanetary disks under the influence of the ionizing radiation field of the nearby massive Orion Trapezium star [1151, 1155–1158]. The name *proplyd* then simply refers to a short form of *PROtoPLanetarY Disk* [1151]. Numerous more proplyds have now been found [1159, 1160]. Some have been seen in other young star-forming regions containing massive stars such as the *Trifid* cluster [268, 1161].

Fig. 12.2 shows three examples of proplyds as observed near the *Orion Trapezium* with the *HST*. The morphology, though complex in detail, can be described as a simple head–tail structure, where the tail points away from the direction of the illuminating radiation. Proplyds have now been observed thoughout almost the entire wavelength band, from Radio, IR [892, 1159, 1162] and optical wavelengths (see above) to X-rays [532]. Emissions from the ionized shock front as well as the central star have been identified. The basic structure of proplyds is illustrated in Fig. 12.2 showing several ionization fronts caused by the FUV and EUV radiation largely from Θ^1 *Ori C*, the brightest and most massive star of the *Orion Trapezium*. Models investigating the interaction between disks and an external UV radiation field using 2-D hydrodynamical simulations show these disks being gradually destroyed via photoevaporation as UV photons heat the gas in the outer layers of the disk to thermal escape velocities [1158, 1163]. The simulations favor the formation of low-mass stars at the centers of the proplyds, which has also been suggested by X-ray observations. Figure 12.3 shows X-rays from the region surrounding the

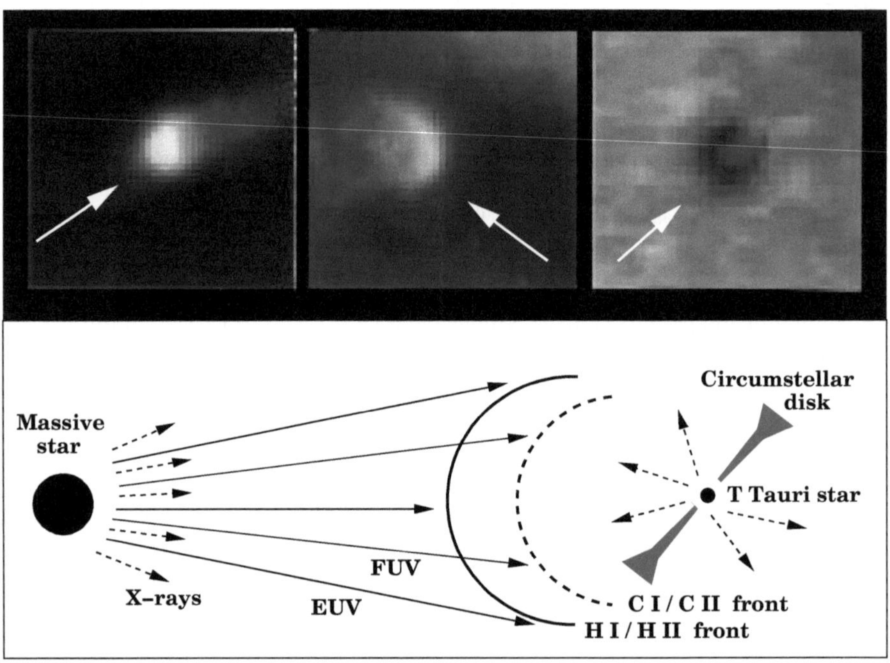

Fig. 12.2 (*top*) The panel shows three Orion proplyds observed with *HST* in 1992. The white arrows indicate the direction of the ionizing radiation from the massive *Trapezium Stars* specifically Θ^1 *Ori C* [1151]. Credit: C.R.O'Dell and NASA. (*bottom*) The sketch illustrates how ionizing radiation from a massive star interacts with nearby protostellar disks primarily through EUV and FUV radiation. Though the massive star is a strong X-ray emitter ionizing effects on the proplyd is unknown. X-rays have been detected from the central T Tauri star (see Fig. 12.3)

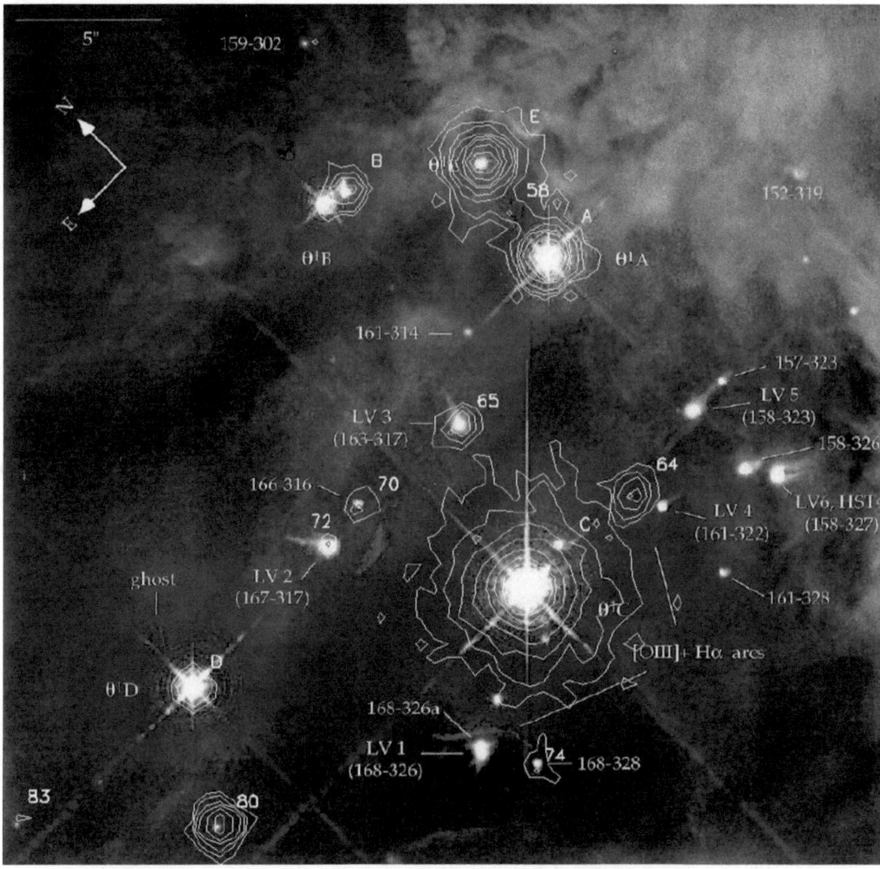

Fig. 12.3 X-ray contours from observations of the *Orion Trapezium* obtained by *Chandra* overlaid onto an optical *HST* observation. The field of view is 30 × 30 arcsec^2 and the contours represent total X-ray flux levels. Credit: HST image by J. Bally [1159]/NASA; Chandra contours by Schulz et al. [532]

Orion Trapezium. Besides the strong emission from its massive stellar members, the image shows X-rays from several proplyds with characteristics that project the emission from T Tauri stars rather than ionization fronts.

12.1.2 Disks of Dust

It has been asserted throughout previous chapters (specifically Chap. 6) that models of irradiated passive disks explain the SEDs of T Tauri stars. A recent review of the disk structure and its emission, including scattering of radiation by disk grains, is given by [1164]. This section now focusses on disk properties around older stars which are already much beyond the T Tauri stage. No optically thick disks have been identified for stars with ages above 30 Myr [1149]. Although it has been established that many young stars (<10 Myr) have optically thick disks. In order to investigate steps that lead to disk appearances in older stars it seems warranted to shortly review some issues concerning the third class of young protoplanetary disks, denominated as *transition disks* [1149].

The cartoon in Fig. 12.1 emphasizes a mode of transition in which the inner disk region is optically thin while the outer regions remain optically thick. The idea behind this is that at inner disk radii, dust agglomerates more rapidly due to the shorter dynamical timescales in disks. The transition mode featured in Fig. 12.1 is also suggested from observations, which seem to favor an optically thin inner disk. A search through IR excess data gathered in the early 1990s [440, 1144] reveals that there are indeed young stars that appear to have holes in their disks. Fig. 12.4 compares two stars of this analysis. Of interest are IR excesses beyond the characteristic photospheric emission of the stars. The observed excess pattern of a few young stars range from $\tau \ll 1$ emissions below 25 μm to $\tau > 1$ emissions

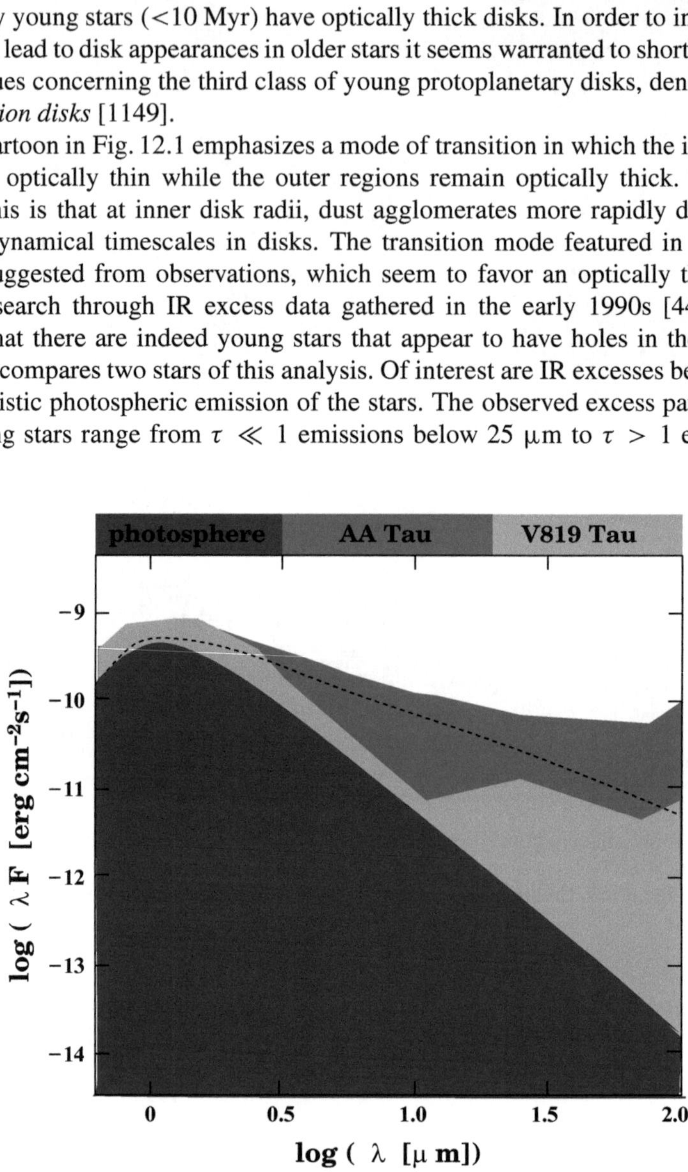

Fig. 12.4 SEDs of stars with disks of various optical thinkness. The dark function shows the optically thick emission from the stars' photosphere. The dotted line indicates the level of IR emission from an optically thick disk (see Chap. 8). The median shade represents data from *AA Tau* indicating an optically thick disk in this young star. The light shade represents data from *V819 Tau* suggesting that for the inner portion of the disk (i.e., towards small IR wavelengths) the emission is optically thin compared to the thick ($\tau > 1$) case. At long wavelengths the emission reconciles with the optically thick case. Adapted from Strom et al. [1149]

above 25 μm and may serve as a probe to identify transition disks [1149]. From the limited number of such disks in the sample, it is also concluded that the transition phase must be short with a specific projection of 0.3 Myr for an optically thick disk lifetime of 3 Myr. What they also show is that protoplanet formation happens during very early evolutionary phases.

The latter suggestion has relevance for the appearance of disks in older stars [1165, 1166]. Protoplanetary disks have now been observed at various evolutionary phases. Stars in the vicinity of the Sun have the best prospects of visibly revealing their disks. There are many ways to infer spatial structure in protoplanetary disks. Most obvious is direct imaging of disks. Here, the nearby *TW Hya* association at a distance of about 60 pc [909], proves fruitful. The prime example is the face-on disk in *TW Hya*, which has been shown in Fig. 8.1. Another one is the A-type member star *HR 4796A* [1167]. A significant fraction of its emission comes from its disk, indicating large amounts of dust. Near-IR images obtained with *NICMOS* on board the *HST* shows a rather well defined ring of dust with a diameter of about 150 pc [1168]. Other examples are *Hen3-600, HD 98800,* and *HD 141569* [1169, 1170]. Many of these young stars are actually binary stars, which render conclusions based on observed asymmetries very speculative. Another way is to deduce disk structure from SEDs, as demonstrated in Fig. 12.4. In the SED from *Hen3-600* this is particularly interesting since it is a triple star system. *SCUBA* detected a cool dust ring of <200 AU radius around the entire system and a warm dust ring originally detected by *IRAS* around the spectroscopic binary only, with a radius of 35 AU. A third way is to extract protoplanet–disk interactions from observed eclipses in conjunction with the central star. A possible example is *KH 15D* [1171], which exhibits deep and periodic eclipses that could be a feature in a nearly edge-on disk [1172]. A review of young binary stars and associated disks is given in [1173].

Knowledge of the path from optically thick disks in YSOs to optically thin disks of dust and debris in MS stars is only patchy. One study showed that observed excess luminosities relative to the stellar photosphere declines steadily with stellar age [1174], which may be interpreted as disk masses decreasing in a similar fashion. It is important to realize that not all disk material agglomerates into solar system bodies, and when it does, the process cannot be particularly efficient. Some of the material may still migrate towards the central star. Some may get heated by the star's UV radiation such that it evaporates away. Only a very small fraction of the dust may grow into *planetesimals* and finally into *planets*. After all, typical circumstellar disk masses are of the order of 0.05 $M_{\odot}$ (see Sect. 6.3.1) whereas the mass of the Solar System, without the Sun, is less than 0.0015 $M_{\odot}$, a mere 3 percent. More interestingly, the process also goes backwards. This is implied by the fact that older stars have dusty disks. In order for the dust to sustain its presence for a long time, it somehow has to be replenished by collisions of larger particles, planetesimals, and protoplanets. Thus, some disks in evolved stars are reduced to mere *debris disks* of dust-like material, quite similar to the material and radius of the *Kuiper Belt* in the Solar System [1175], and may have masses of not more than a few tens of lunar masses up to maybe 1 percent of $M_{\odot}$ (for a review see [1170]). The lower mass limit is driven by the fact that a minimum amount of mass is needed to maintain *collisional cascades* for the 100–500 Myr lifetimes of debris disks [1166, 1175, 1176]. In these cascades, planetesimals are ground into dust through collisions.

These older stars reveal their existence at far-IR and sub-mm wavelengths. Observations of *IRAS* detected far-IR emission in excess of expected photospheric emission from the vicinity of the MS star *Vega* [1178]. These Vega-like IR stars are now quite common and, like in YSOs, the dust of the disks is heated by stellar radiation [1179]. More recently, 850 μm imaging with *SCUBA* provided the first detailed images of dusty disks of nearby stars including ϵ *Eridani*, β *Pictoris* [1180–1182] and Vega [1183, 1184]. For example, the morphology of the disk around the nearby star ϵ *Eridani*, a K-type star at least 500 Myr old, shows a ring structure of dust peaking at 60 AU with some lower emission inside 30 AU and some up to 75 AU. The total mass of this disk is at least 0.01 $M_{\odot}$ [1185]. Figure 12.5 shows images of the disk in β *Pictoris* taken with *HST* at 8–13 μm [1177]. In contrast to *TW Hya* in Fig. 8.1, the disk is viewed edge-on. It shows a warped structure within 20 AU and other asymmetries that are not easily explained by dust but by the influence of

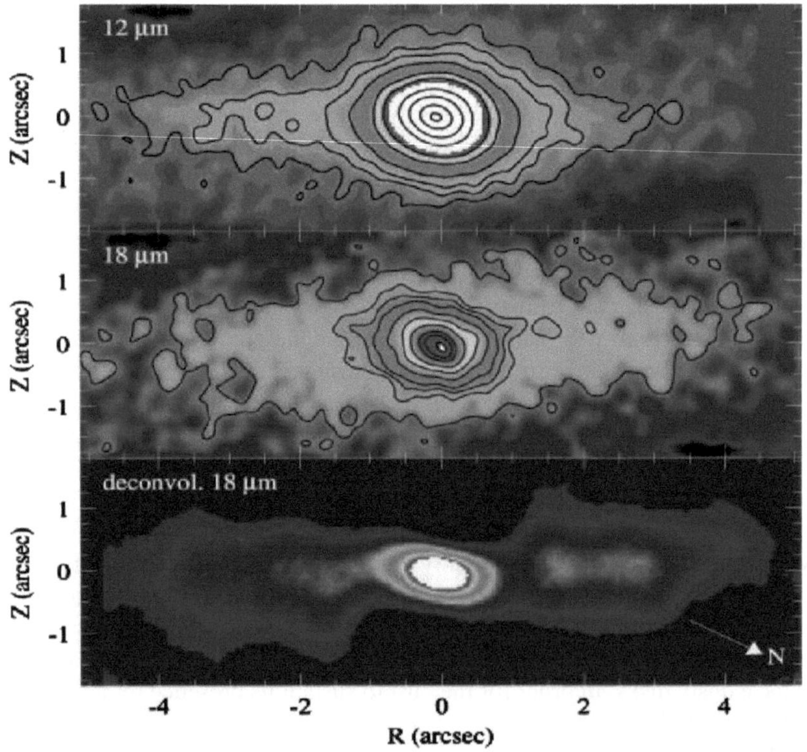

Fig. 12.5 (*top*) Image of the central star protoplanetary disk of β *Pictoris* at 12 μm with flux contours overdrawn. (*middle*) Same view at 18 μm. (*bottom*) The 18 μm image as a result of a deconvolution of the imaging point spread of the telescope. Credit: from Weinberger et al. [1177]

more massive bodies. One interpretation is that there are planets present in the inner disk, which would then have a morphology quite similar to the Solar System.

12.1.3 HAEBE Disks

Some remarks should be spent on the disks of HAEBE stars as they have been subject to much attention in recent years (see [473, 1186] for reviews). Since the studies by *L. Hillenbrand* [469], it has been known that intermediate-mass stars emit a significant fraction of their luminosity in the near-IR specifically with peaks around 2–3 μm. It has been pointed out in the literature many times that the properties observed in HAEBE disks are inconsistent with standard disk models. Various scenarios have been proposed ranging from a more spherical appearance [61] to dust rings [1187]. Accordingly, recent ideas feature optically thin inner disks up to the dust sublimation radius, which is a few tenths of AU, followed by an optically thick *puffed up* rim whose IR signatures allow one to fit the SED [1188, 1189].

It should be realized that HAEBE stars cover a large range of spectral types and surface luminosities, As the interface between low and high mass stars their early evolution happens on many different time spans. In this respect, not much of a protoplanetary disk evolution may be expected from B-stars more massive than 4 to 5 $M_{\odot}$ (*Herbig Be (HBe)* stars) as they reach the ZAMS with disks of very little mass, if any at all. That all HBe stars at some point had quite strong disks follows as a consequence of star formation, though it is also likely that these disks dispersed (see Sects. 8.1.5–8.1.7). In fact, in an analysis of 25 HAEBE disks, all stars of type later than B8 (*Herbig Ae (HAe)* stars) showed dust emission and derived disk masses ranged from 0.02–0.06 $M_{\odot}$. Only one object of earlier type indicated a disk mass of 0.1 $M_{\odot}$. The latter value is near the upper limit of HAEBE disk masses [1186]. The disk frequency in later-type HAEBEs is thus very large [1190]. One of the trademarks of HAe stars seems to be a *silicate* emission feature around 10 μm in the spectra [1191]. The emission is likely related to the hot grains just outside the dust sublimation radius. The temperature of these grains can reach over 500 K and the optical depth with respect to silicate is almost unity.

The question remains, whether disk evolution in HAEBEs allows for the formation of protoplanets. In the case of HBe stars the answer is quite difficult, not only because of the apparently rapid disk dispersal, which is projected to a timescale of less than 1 Myr [1192]. The fact that recent observations show HBe exhibiting so called *unidentified IR bands (UIBs)* often associated with *polycyclic aromatic hydrocarbons (PAHs)*, should not be much concern as these are now observed generally with early-type stars.

Therefore, they are unrelated to protoplanetary disk activity. On the other hand, the lack of the *silicate* feature clearly indicates that there are no significant amounts of hot dust around HBe stars. The ability to form protoplanetary disks thus is reserved to stars later than ∼B8–A0 type.

12.2 The Making of the Sun

The Sun is by all standards an average star. In its young years it certainly was a T Tauri star, though with a mass that puts it already beyond the peak of a typical cluster mass function. The Sun is a G-type star. Such a stellar type is not uncommon in the *Taurus–Auriga* region, though the Sun clearly descended from a much older generation of star formation. The following sections attempt to present a rough overview of likely properties of the early Solar Nebula and the Sun's evolutionary path.

12.2.1 The Sun's Origins

Finding the Sun's original birthsite is simply impossible. At its current age of 4.6 Gyr the Sun is already too advanced to have any aspects of its original environment revealed today. Having traveled over 19 orbits around the Galactic Center since the birth event, all environmental structure has long been dissolved by the dynamics of the interstellar medium. The average lifetime of molecular clouds is only a few 10 Myr and most stellar clusters disperse on a similar timescale. The previous chapter made it clear that most stars are born in more massive clusters and that was likely the case for the Sun as well. However, the present Solar System gives some clues about the Sun's early environment [1014]. For example, from extinct ratios of isotopes in meteorites as well as relatively unperturbed planetary orbits, one can project a maximum size of the parent cluster of maybe 1,000–2,000 stars. On the other hand, the odds of the Solar System forming in such a type of environment are less than 1 percent [1193]. Without constraints from the isotope ratios (see below), it is quite likely that the Sun was born in an intermediate size ESC of maybe 100 members.

Although it is not possible to peek into the Sun's birthplace, one may have a look into the visible history of the Solar neighborhood. The Solar neighborhood is very active and has been so for quite some time. As has been indicated in many studies of the LISM (see Sect. 3.1.3) and by projections of the distribution of X-ray stars (see Fig. 10.2), the Sun is engulfed by an old ring of progressing star formation, called the *Gould Belt*. This assembly of stellar associations and molecular gas (see Fig. 12.6) has been very actively forming stars for the last 10 Myr, though it seems that the star-forming activity was initiated about 60 Myr ago. Appearance and size of the belt indicate that one or more supernova explosions from an ancient association near the center of the ring triggered the activity. Traces of such explosions in the solar neighborhood are plenty, such as the presence of the *Local Bubble* [1194], the *Loop I* [1195], and other candidates [1196]. *Spitzer* surveyed regions of the *Gould Belt* extensively. As part of the *Core to Disk* legacy survey several dark clouds in the *Lupus* region were observed identifying a variety of thick and thin disk objects with a rather high abundance of Class III sources and thin disks on the *Lupus* V and VI clouds indicating active planet formation [1197].

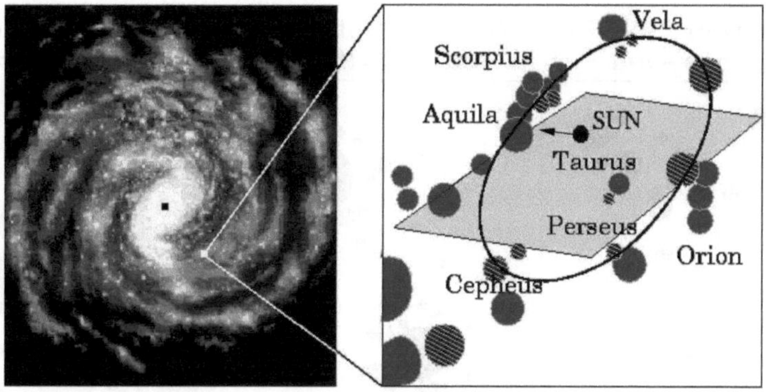

Fig. 12.6 (*left*) Illustration of the relative position of the Sun with respect to the Galactic Center. The distance towards the Galactic Center is about 8 kpc. (*right*) The same schematic illustration of the distribution of molecular clouds and stellar associations as shown in Fig. 3.2, but now without the low-density regions and the warm diffuse gas. The gray plane represents the nominal Galactic plane. The stellar associations as well as associated clouds align roughly with the *Gould Belt* and ancient belt of progressing star formation triggered about <60 Myr ago. The *Gould Belt* has an inclination with the Galactic plane of ~20 degrees

12.2.2 The Solar Nebula

What was matter like in the *Solar Nebula*, from which ultimately the planets formed, and how does this matter relate to materials found in the Solar System today? This fundamental question about the origins of the planets and the Solar System taps into a vast field of material science and kinematics which cannot be covered here in detail. However, this section will deal with some basic ideas of how modern research argues from the point of astronomical observations and basic material chemistry. The observations of ϵ *Eridani* and other Vega-like stars may seem reminiscent of what the early history of the Solar System might have looked like when the Sun was at that age. The similarity of dust observed in β *Pictoris* to cometary and interplanetary dust in the Solar System has been stressed through linear polarization properties [1198]. The disks of these stars are less dusty than the ones in younger stars, but dust grains in these old disks differ from the dust of their younger cousins. Material loss and evolution in the Solar Nebula are subject to radiation pressure, collisions, sublimation, and drag forces. In some systems these materials have evidently either been incorporated into planets or destroyed and are now part of debris disks. One has to be truly careful in the search for primordial material as the presence of any dust in old disks implies that it has to be of at least second generation. From observations of Vega-like disks one can also infer that protoplanetary grains are larger than interstellar grains [1179]

Creation and destruction of small solid bodies have been studied at various levels of depth [1179] [1179, 1199, 1200]. Calculations of growth rates indicate that the build-up of rocks from dust should proceed relatively quickly. Such a growth

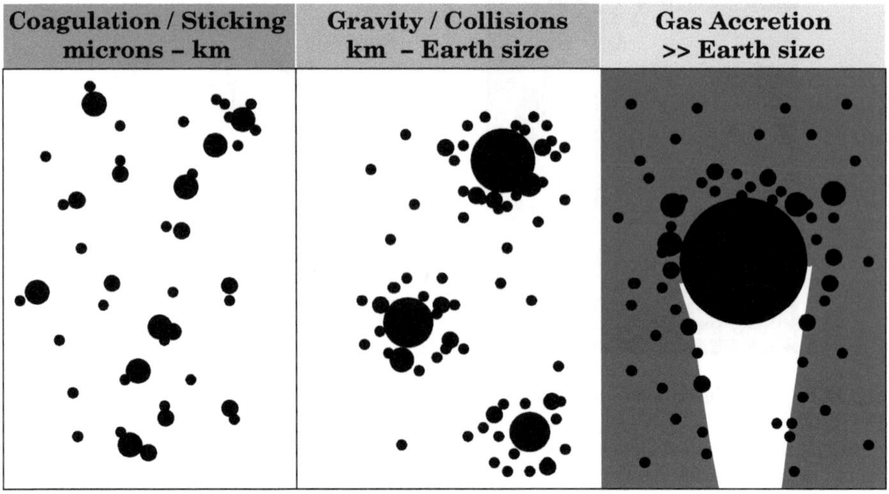

Fig. 12.7 A cartoon to illustrate various stages of particle growth. (*left*) The first stage involves grain growth via dust particle drag forces and sticking and coagulation. (*middle*) The second stage is two-body tidal interactions and collisions via gravity. (*right*) The last stage is past Earth-size growth and gas accretion

process passes through roughly three phases, which is illustrated in Fig. 12.7. The detailed physics that actually leads to grain growth is highly complex and may generally be characterized by sticking and coagulation in the early phases, followed by gravitational attraction of planetesimals and accretion of material during later stages. Particles can grow from micrometer size to meter size within about 10^4 yr, which on a similar timescale grow to bodies of kilometer size. Beyond that size, growth rates are more complex, and planetesimal sizes should be reached within 0.1 Myr [1200–1202]. Models to calculate growth of grains and planetesimals depend on assumptions of the disk structure. In contrast to the accretion disks discussed in Chap. 8, which are heated through viscous dissipation during the accretion process, protoplanetary disks are *passive* and only heated by the central star or in the case of the Solar Nebula by the Sun. The disks have likely vertical structure. Models to calculate the growth of grains and planetesimals depend on assumptions on these disk structures as well as intrinsic velocity gradients caused by convection and turbulence (see [1200, 1202, 1203] for more details). For more updated details see Chap. 13.

Figure 3.21 shows the fluffy shape of planetary dust as it should have been produced in the Solar Nebula. Initially the grain population consists of *olivine* ((Mg, Fe)2SiO_4), *orthopyroxene* ((Mg, Fe)SiO_3), volatile and refractory *organics*, *water ice, quartz, metallic iron* and *trolite* (FeS) [1204]. During disk evolution these materials will change through annealing processes, chemical reactions and radiative destruction, to name a few. The actual results of these processes are highly uncertain and a better understanding is pursued by laboratory experiments (i.e., [1205, 1206]).

Some of the survivors of these processes are *chondrites*, which constitute about 84% of the consistency of meteorites. Chondrites come in various types depending on their element content. Their substructure is complex and consists of a *matrix* in which Si-rich *chondrules* as well as Ca and Al-rich inclusions called *CAIs* are embedded. This matrix contains various substances including organic material. It is a common perception today that the planetoids, the meteorites, and the bodies of the outer Kuiper Belt in the present-day Solar System are remnants of the old population of planetesimals of the Solar Nebula. The mineralogy of chondritic meteorites though including unequilibrated phases is generally consistent with condensations of Solar vapor material except for volatile elements [1207]. Most interesting for the origin of planets are actually the so-called *carbonaceous chondrites*. They differ from normal chondrites in that they contain up to 4 percent carbon [1208] mostly in organic form. How these organic compounds found their way into chondrites is today still unclear [1206, 1208, 1209].

12.2.3 The T Tauri Heritage

In order to date materials as early as the ones in the Solar Nebula one often uses existing and extinct isotopic abundances in Solar System bodies [1210]. Studies of meteorites provided evidence for an early presence of several short-lived nuclides in the Solar Nebula with half-lives $<10^7$ yr through observed excesses in their decay products. Longer-lived, known nuclides are of higher mass, they are products of r-processes in supernovas and in line with galactic abundances. They can hardly be used to date the Solar Nebula. On the other hand, there are about a dozen short-lived nuclides that offer the possibility to stem from a one-time presence in the early solar system as inferred from meteorites (see [1211] for a review). In these cases there are a few more options. One is a stellar origin, which would imply that these nuclides are most likely products of supernova explosions, but for some thermally pulsing asymptotic giant branch stars or Wolf–Rayet stellar winds can be production sites as well. In that case one would opt for a scenario where these nuclides were injected into the collapsing protostellar cloud prior to collapse [1212–1214]. Due to the short lifetime of these nuclides there is the consequence that the injection process that could have triggered the collapse as a standard star-forming scenario as outlined in Chaps. 5 and 6 would likely have to be modified [1215]. In that case the stellar origin for the short-lived nuclides in the Solar Nebula contains the timescale for collapse and Solar Nebula to roughly 1 Myr [1211]. This would be consistent with somewhere within the WTTS phase (see Fig. 6.11).

Another option could be that these nuclides were produced in the Solar Nebula through the bombardment of rocks by cosmic rays from the central protostar [1216, 1217] close to the inner disk boundary. Under conventional conditions this would produce some of the nuclides but would significantly underproduce ^{41}Ca, ^{26}Al, and ^{60}Fe. Fluctuating X-winds (see Sect. 8.3.3) as proposed by [669, 1218] offered progress by invoking powerful reconnection events, which on the one hand generate

Fig. 12.8 Structure of ordinary chondrite originating from a meteorite found at the Hammadah al Hamra Plateau in Libya. The central feature is a cut through a 2.2 mm chondrule visibly showing a well defined fine-grained rim. Credit: Photograph by R. Pelisson, SaharaMet

high X-ray fluxes, but on the other hand start an X-wind which is able to accelerate large numbers of ^{3}He particles to energies above MeV per nucleon. Though ^{26}Al would now be produced in sufficient amounts, ^{41}Ca would be severely overproduced while the amount of ^{60}Fe would still remain insufficient (see also [1211, 1217]). More recent reports on evidence of the decay of ^{10}Be also supports an *in situ* origin since ^{10}Be is not produced by stellar nucleosynthesis but can only be due to nuclear reactions by energetic particles [1219]. In these respects, though the stellar origin of these nuclides is still favored by many, there is a likelihood that at least some of the material is produced in the Solar Nebula.

The X-wind model has in this respect another powerful consequence. The unusual formation history of chondrites and their substructures which are the Si-rich *chondrules* as well as Ca- and Al-rich inclusions called *CAIs* should reflect on their possible formation and origin (Figure 12.8). The fact alone that two seemingly incompatible inclusions appear in chondrites poses constraints. For a review of the details of the formation of these entities the reader is referred to [1209]. The thermal histories of chondrules and CAIs depends not only on the ambient temperature at the time of formation but also on length and peak temperatures during melting processes as well as cooling rates. The presence of volatile elements such as Na and S within chondrules helps determine temperature limits of melting processes these chondrules must have endured. Ranges for estimated peak melting temperatures under certain circumstances can exceed 2,000 K. The coexistence of chondrules and CAIs is thus a consequence of a formation under low- and high-temperature regimes.

The main problem with high melting temperatures is that they cannot be generated by the stellar system beyond maybe 2 AU. Rings of material are observed much further out where such temperatures cannot be reached. A variety of models have been proposed. Models featuring melting on impact between planetesimals are too inefficient, while periodic heating during outbursts of *FU Orionis* type have inappropriate timescales [1208, 1209]. More viable are models involving some form of lightning [1220, 1221] or shock waves unleashed into the Solar

Nebula [1222, 1223]. However, the former raises some concerns about feasibility and the latter lack a viable source for the shock wave (see [1209] for discussion).

Yet the only scenario that generates a plausible thermal environment places the formation of chondrules and CAIs into the very early stages of stellar evolution specifically into the CTTS and even protostellar stage. The Sun (see below) during its early evolution once was an accreting T Tauri star and subject to mass accretion, winds, jets, and enhanced X-ray emission. The X-wind model produces a wind as a result of the interaction between the accretion disk and the magnetic field of the central star. When accreted material reaches the X-region (see Fig. 8.16) accreted gas is launched from the inner edge of the disk (R) and forms an X-wind. Grains are lifted in the gas wind through a strong gas–grain drag and once the grains leave the shade of the disk, these grains are fully exposed to the protosolar radiation field where molten CAIs can form and subsequently cool while outflowing with the wind [669]. More specifically, the model specifies processes by which chondrules and CAIs might have formed by flare heating of precursor rocks [1218]. The X-wind itself is fluctuating due to periodic radial excursions on a timescale of roughly 30 yr. The melting rock sticks during collisions with other melted material. In this scenario irradiated rocks have a chance to grow to mm and cm size. These processes are time-critical. Peak temperatures up to 1,800 K can be reached assuming a mass loss rate of about 1×10^{-7} $M_{\odot}$ yr^{-1} (see [1218] for further details). Observations of YSOs support this scenario (see Sect. 10.1.5).

In this respect Feigelson and colleagues characterized magnetic flaring activity in the *ONC* and discussed the finding with respect to isotopic anomalies in carbonaceous chondritic meteorites of the Solar System [1224]. The sample of flaring PMS stars was selected to best represent the PMS analog of the Sun. It was found that that the PMS Sun exhibited over 30 times more powerful flares and a frequency that exceeds the one of the current Sun by a factor of over 300. It was inferred that the PMS Sun exhibited enhanced proton fluences that were enhanced by many orders of magnitude. From that it was concluded that the observed meteoritic abundances of several short-lived isotopes could be explained. The observation of star-forming activities specifically with X-ray telescopes is thus capable of directly studying the properties of the Solar Nebula. In a more systematic approach, the project *Sun in Time* (see Sect. 8.2.2) [651] studies the long-term evolution of solar type stars in terms of coronal and consequently dynamo activity and find the Suns violent activity as a T Tauri star dropped by orders of magnitude during its life so far. The mean X-ray luminosity is a good indicator of coronal activity (see Chap. 10) and the observation of solar type stars at various ages during PMS phases and throughout the MS provide a detailed history of how a the activity of the Sun might have evolved with time. Figure 12.9 shows the evolution of the mean X-ray luminosity of solar analogs with age, which decreases by a factor 10^3 from the youngest to solar-age stars in the sample. Not only has such dramatic change with time consequences for the internal stellar evolution but also affects the photochemical evolution of paleo-planets and their atmospheres.

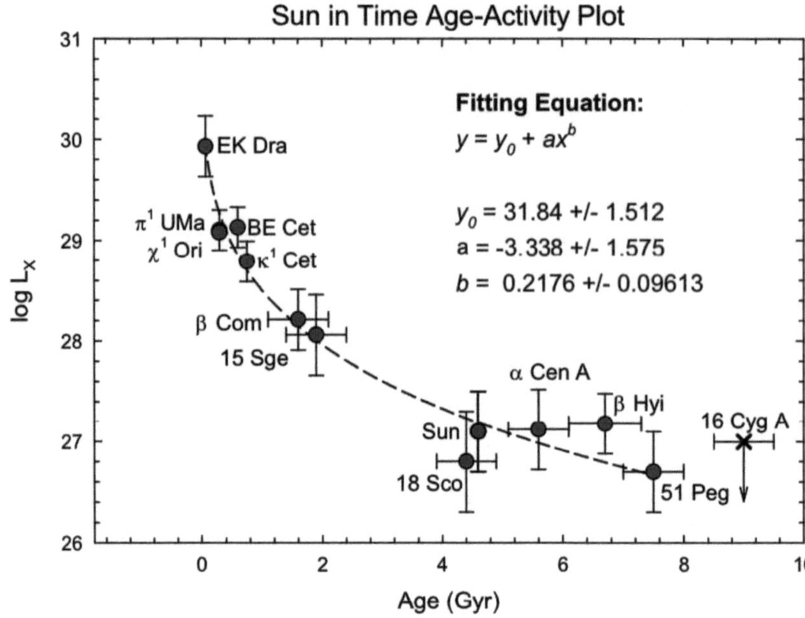

Fig. 12.9 The mean X-ray luminosity of solar analogs over a time span of about 9 Gyr. The Sun is found at about 4.3 Gyr at an X-ray luminosity of $\log L_x[\mathrm{erg\ s^{-1}}] = 27$ (from Guinan & Engle [651])

12.2.4 Evolution of the Sun

What is left is a look at evolutionary scenarios that lead the Sun as it is seen at present times. The Sun today is a mid-aged MS dwarf star of spectral type G2. The previous chapters have outlined how such a star emerges from the collapse of a molecular cloud under well-defined initial conditions and how it grows up passing the T Tauri stages and contraction towards the main sequence. At some point one would like to have a complete description of the Sun's evolution reaching exactly the state it is in today. Luminosity and surface temperature of the Sun have been determined to the best accuracy known for any star and thus the Sun is a specifically good probe to test evolution in the HRD. It is therefore not a coincidence that many collapse calculations are performed for a solar-mass star and results are scaled by solar properties. At the end it is the comparison of different star-formation models that gives an impression of how well the protostellar history of the Sun is known today. Though in a more basic form such a discussion has already been started in Chap. 5 and 6, there are some more recent contributions that add a few more details specifically of interest toward an understanding of the Sun's evolution. The reader should be cautioned at this point and be aware that the following discussion does not invalidate classical PMS calculations and projections presented in Chap. 6 but merely attempts to put theoretical results into perspective with observations.

Wuchterl and Tscharnuter illustrate and compare various evolutionary paths that lead to the ZAMS position of the Sun in the HRD. This includes physical conditions of molecular clouds and cores, collapse treatment including radiative transfer at avarious optical depths as well as a time-dependent treatment of convection [404]. In an earlier study it was shown that there are substantial deviations in the results to classical hydrostatic and initially fully convective models once one applies an integral approach of cloud fragmentation phase, collapse, and early evolution without parameter adjustment during these phases [409]. Yet at the end of most previous collapse models (see Chap. 5) it is assumed at the end that the protostar reaches hydrostatic equilibrium and full convectivity. One of the interesting features of this earlier study is a comparison of evolutionary tracks that assume the Sun either in a dense stellar cluster environment [303, 1030] or the situation where the Sun would have evolved in isolation represented by the collapse of a Bonnor–Ebert sphere [404]. It is argued that mass accretion rates of evolving protostars in clusters differ significantly from rates in isolated stars. This affects predictions drawn from positions in the HRD (see Fig. 12.10). How far such a result can be generalized is unclear. However, since it is quite likely that the Sun evolved in a cluster environment, interactions between cluster stars affect the Sun's evolutionary path. At the end of the protostars accretion phase evolutionary tracks converge as star-forming history becomes more and more irrelevant. For the Sun this would have happened after 0.95 Myr.

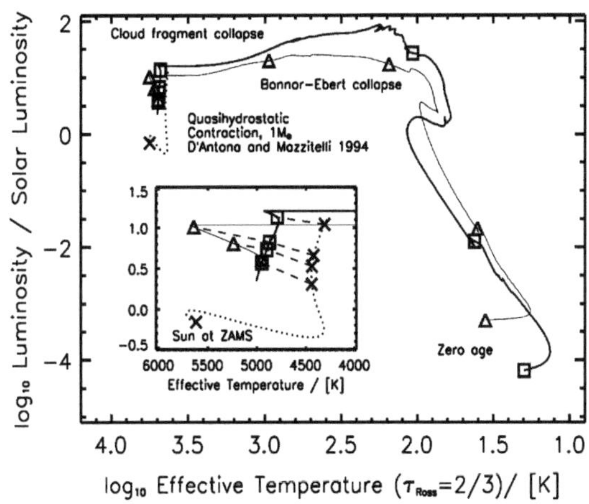

Fig. 12.10 Early stellar evolution in the HR-diagram as calculated by [409]. Three evolutionary tracks for a solar-mass star are compared. The dotted line is a classical stellar structure, hydrostatic equilibrium PMS track, for an initially fully convective gas sphere [411]. The two other lines are obtained by describing the formation of the star as a result of the collapse of an interstellar cloud. The thin line is for a Bonnor–Ebert sphere [404]. The thick line is for a cloud fragment that results from the dynamical fragmentation of a molecular cloud [303, 1030]. The triangles (Bonnor–Ebert), squares [409], and crosses [411] along the respective evolutionary tracks mark ages of 0, 1, 10, 100, and 350 kyr and 0.5 and 1 Myr. The cross at the end of the hydrostatic track denotes the zero-age main sequence. Corresponding age marks for 0.1, 0.35, 0.5, and 1 Myr are connected by dashed lines in the inset. Credit: from Wuchterl and Klessen [409]

A valid point is made in that the determination of accurate initial conditions for stellar evolution in terms of mass, radii, and internal structure is still highly uncertain [404]. This is specifically true for stars that enter the stage of hydrostatic equilibrium. The result of these calculations are surface temperatures that are 500 K hotter and bolometric luminosities that are twice that which classical calculations would predict for the proto-Sun [411] (i.e., 4,950 K and 4 $L_{\odot}$). The reasons for this are not the possibly different formation histories, but the classical predicament that once the protostar ceases to accrete at high rate, the star has to be internally fully convective. Instead it is argued that the proto-Sun hosts a radiative core that could be quite similar to today's Sun. Whether this result holds any truth is clearly a subject for future research. However, the now seemingly open discussion about the initial conditions specifically with respect to full convectivity as the endpoint of star formation has repercussions not only for the theoretical understanding of proto-Suns but even more so for the interpretation of observations of young stars. A factor two in luminosity and 500 K may not have much effect on the placements of protostellar and PMS stellar parameters in the HRD as uncertainties in observed quantities are almost of the same order (see Sect. 6.2.5). It does, though, reflect on the discussion of high-energy results (see Chap. 10), which seem to favor some dynamo activity throughout most evolutionary phases and where observers seek to reconcile radiative signatures between early and late stages of PMS stellar evolution.

Chapter 13
Protoplanets and Exoplanets

The previous chapter touched on the early evolution of the Sun, the protoplanetary nebula, and the formation of planetesimals. This chapter presents an update on the detection of exoplanets, quickly summarizes planet detection techniques, gives an account of planet detection in PMS stellar systems, and introduce some issues related to the formation and early evolution of terrestrial and giant planets. The latter focus on the evolution of planetary systems and on current theories of moon creation and planet migration. For more detailed readings, monographs and courses such as Cassen et al. [1225], Ollivier et al. [1226], Armitage [1227], and Seager & Deming [1228] are recommended.

13.1 The Search for Exoplanets

The first exoplanet orbiting a solar-type star was discovered in 1995 designated 51 Peg b. About a decade later a few hundred of such objects were detected, primarily though radial velocity determinations and stellar transits. By 2006 about 300 exoplanets had been discovered of which which 31 are in multiple systems with as many as 5 planets orbiting a single star [1226, 1229]. The mass distribution of exoplanets until recently was heavily biased towards large masses because of methodical biases that the sensitivity of the radial-velocity method ofdetecting planets increases with the mass of thc exoplanet. Most were found to be a few Jupiter masses with a decreasing tail to the mass limit of 13 Jupiter masses. The planetary mass limit is defined by the minimum mass at which an object becomes dense and sufficiently hot to trigger deuterium burning. Any bodies more massive are likely to be of a stellar nature, such as Y and brown dwarfs.

The advent of space-based observations with the launch of *Kepler* in 2009 radically changed the observational account as now searches for earth-like and even smaller planets become possible. To date over 1,200 exoplanets have been found by *Kepler* with the vast majority of bodies to be of Earth-like masses. This leads to the conclusion that small planets seem much more common than giant planets.

N.S. Schulz, *The Formation and Early Evolution of Stars*, Astronomy and Astrophysics Library, DOI 10.1007/978-3-642-23926-7_13, © Springer-Verlag Berlin Heidelberg 2012

The search for exoplanets, whether directly or indirectly, requires observational separation of two objects in the sky. Even today it is a challenge to, for example, measure binary separations in Orion to better than a few tens of AU (see Chap. 7). However, there are a few advantages to the fact that stars and planets are so vastly different in size, mass, and spectral appearance. There are several critical issues associated with highly sensitive detection methods.

- The *spectral contrast* between a star and its planets is crucial. Different surface emissions from stars of different types vary in contrast with respect to reflectivity and emissions from certain planet types. This contrast is also affected by a planet's distance from the star. A hot Jupiter at a distance of <0.1 AU with respect to a solar-type star has an effective temperature of over an order of magnitude higher than it would exhibit at a distance of 5 AU. At a wavelength of 10 μm the contrast changes from 10^4 for the hot Jupiter to 10^9 for the cool Jupiter.
- The *angular separation* as mentioned above is still a huge challenge as even the closest stars in the solar neighborhood have distances of several to several 10s of parsecs. One AU at a distance of 10 pc subtends an angle of $0.1''$, the resolution limit of the *Hubble* space telesope.
- *Planetary disk interference* is problematic not only in very young systems or debris disk systems where there is still substantial disk material, but even in systems like the solar system, where intermittent dust is heated and emits at IR wavelengths. The presence of these disk emissions will likely affect contrast, e.g., from collisions between planetesimals, comets, and other bodies.

Assuming that both, star and exoplanet radiate as blackbodies one can estimate the contrast at any (IR) wavelength via

$$f = \left(\frac{R_p}{R_*}\right)^2 \frac{B_\nu(T_p)}{B_\nu(T_*)}, \tag{13.1}$$

where R_p, T_p, and R_*, T_* are radius and temperature of the planet and star, respectively. For a solar-type star of 5,500 K and and earthlike planet of 290 K the contrast at 10 μm amounts to about 7×10^6. The peak of $B_\nu(T_p)$ lies in the infrared. For optical light a planet reflects a fraction of the incident starlight, i.e.

$$f \approx 10^{-10}\left(\frac{A}{[0.3]}\right)\left(\frac{R_p}{[\mathrm{R}_\odot]}\right)^2\left(\frac{a}{[1\ \mathrm{AU}]}\right)^{-2}, \tag{13.2}$$

where A is the albedo (see Eqn. 3.27) and a the orbital separation. Table 13.1 illustrates the vast differences in contrast and angular separations for planets orbiting a solar-type star at a distance of 10 pc [1226]. Contrasts for a proto-Earth, assuming Class III type systems with properties described in Chap. 6, are difficult to assess. In the visual range the view toward the star is likely obstructed. Not much is reflected by the protoplanet and the contrast with respect to the star appears high. However, there is severe blockage by the disk, which not only affects visibility of the protoplanet but destroys much of the contrast.

The following summarizes existing search methods for exoplanets and gives a current account of results. By no means is this intended to be complete, owing in part to the ongoing rapid changes in the field.

13.1.1 *Standard Detection Methods.*

The most direct way to detect exoplanets is *imaging* with high resolution in conjunction with coronographs, nulling interferometers in the infrared, or UV/optical telescopes with effective starlight blockage. Imaging is also the most difficult method, not only because sub-arcsecond resolution is required, but also because of the extreme contrasts between star and planet (see Table 13.1). In the IR contrasts for Earth-like planets improve by orders of magnitude; but extreme telescope diameters would be required to achieve sub-arcsecond resolution [1227].

In order to detect *transits* of exoplanets through the stellar disk one relies on the fact that a fraction of planetary systems is aligned to the line of sight in a near edge-on manner. During the transits the lightcurve of the star experiences dipping events caused by the photometric occultation of the transiting planet. The amplitude of the transit signal is independent of the distance between the planet and the star but provides a measure of the relative sizes of the two bodies. The fractional decrement in the stellar flux during the transit, assuming that the stellar disk brightness is uniform, is then

$$df = \left(\frac{R_p}{R_*}\right)^2. \tag{13.3}$$

For a giant planet of Jupiter size it is about 0.01; for Earth-like planets the decrement is accordingly much lower. The probability that a transit can be observed follows from the condition that it be observed nearly edge-on, i.e.

$$\cos i \leq \mathcal{P}_t = \frac{R_* + R_p}{a}, \tag{13.4}$$

Table 13.1 Separation, and contrast values of various planet types orbiting a solar-type star at a distance of 10 pc [1226]

Object	Dist. [AU]	Separation [mas]	Contrast [6,000 Å]	Contrast [10 μm]
Jupiter	5.2	520	5×10^8	5×10^7
Hot Jupiter	0.05	5	4×10^4	1×10^3
Saturn	9.54	954	2×10^9	5×10^8
Neptune	30.06	3,006	1×10^{11}	1×10^{10}
Earth	1	100	5×10^9	7×10^6

with a maximum duration of

$$t_{max} = \frac{P}{\pi} \sin^{-1}\left(\frac{\sqrt{(R_* + R_p)^2 - a^2 \cos^2 i}}{a}\right), \tag{13.5}$$

where P is the orbital period [1230]. For fully edge-on view, $P = 1$ yr, and $a = 1$ AU, such a transit would last about half a day (~13 h). Figure 13.2 shows some configurations. The example transit for HD 189733 is with respect to a near Jupiter-like planet orbiting a slightly sub-solar mass star in 2.2 h at an inclination of 86° [1231]. Based on Eqn. 13.5 the distance of the planet to the star is <0.073 AU.

Strictly, planets do not orbit around stars, they move around the center of mass of the system. This leads to small periodic motions of the star. The Sun at a distance of 10 pc would then show motion variations due to the presence of the giant planets as a function of time of the order of 0.001″. The motion is likely some form of ellipse which *astrometrically* can be decribed as

$$r(\delta) = \frac{a(1 - e^2)}{1 + e\ \cos(\delta)}, \tag{13.6}$$

where a is the semi-major axis, e the eccentricity, and δ the angular position of the star relative to the orbit's periastron [1226].

The determination of *radial velocities* is also based on the fact that a star with orbiting bodies must show some orbital motion around the center of mass of the system. Consequently, one can observe this motion through Doppler shifts in the stellar spectrum. In case of a circular orbit one defines $K \equiv v_* \sin i$ with $M_* v_* = M_p v_p$ giving in terms of simple Keplerian mechanics

$$K = \left(\frac{M_p}{M_*}\right)\sqrt{\frac{GM_*}{a}} \sin i \quad \text{and} \quad P = 2\pi\sqrt{\frac{a^3}{GM_*}}. \tag{13.7}$$

where P is the orbital period. The inclination usually remains unknown and the measurement produces $M_p sin\ i$ which is a lower limit to the planet mass. In reality most orbits are not circular and introducing an eccentricity into Kepler's equations leads to

$$K' = K\frac{1}{\sqrt{1 - e^2}} \tag{13.8}$$

for $M_p \ll M_*$. Examples of measurements of perturbed stellar radial velocities are shown in Fig. 13.2.

Gravitational *microlensing* is another way to sensitively detect low-mass stars but even more than transits is subject of opportunity. A photon which passes a star at an impact parameter b is deflected by an angle

$$\alpha = \frac{1}{b}\frac{4GM_*}{c^2}, \tag{13.9}$$

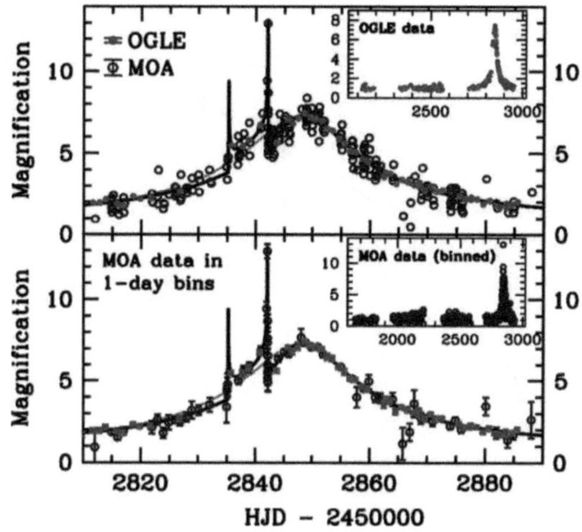

Fig. 13.1 Light curve of *OGLE* and *MOA* measurements of gravitational microlensing by a planet. The top inset shows the *OGLE* data from 2001–2003 and the bottom inset the *MOA* data from 2000–2003 (from Bond et al. [1232]

and for two masses lying at different distances along the same line of sight, the image of the background mass is distorted into what is called an *Einstein ring*. In the case of plane searches this distortion is observed when light from a distant source star is observed as another star intercepts the line of sight while passing through. Several campaigns are under way, such as *OGLE* [1233] and *MAO* [1234] to observe such events. Lensing events are also sensitive to the situation that the passing star has orbiting planets, which will leave imprints on the stellar lens. Several microlensing events have been observed today and Fig. 13.1 shows an example involving the same star in both campaigns.

13.1.2 The Rossiter–McLaughlin Effect

Many exoplanets are observed to have eccentric orbits, and one can speculate that any process which excited their eccentricities may also have caused the inclinations. For close-in planets, there are various migration scenarios for bringing the planets inward from beyond the snow line (see Sect. 13.3.2 and 13.3.6), which can be relevant for predictions about spin-orbit alignment. Stellar obliquity can be assessed by observing a phenomenon known as the Rossiter–McLaughlin (RM) effect, which was predicted in the late 1900s and observed on separate occasions in by Rossiter [1236] and McLaughlin [1237] through line absorption. During a transit, part of the rotating stellar surface is obscured, weakening the corresponding radial velocity components of the stellar absorption lines. When the approaching (blueshifted) part of the star is blocked, the spectrum appears slightly

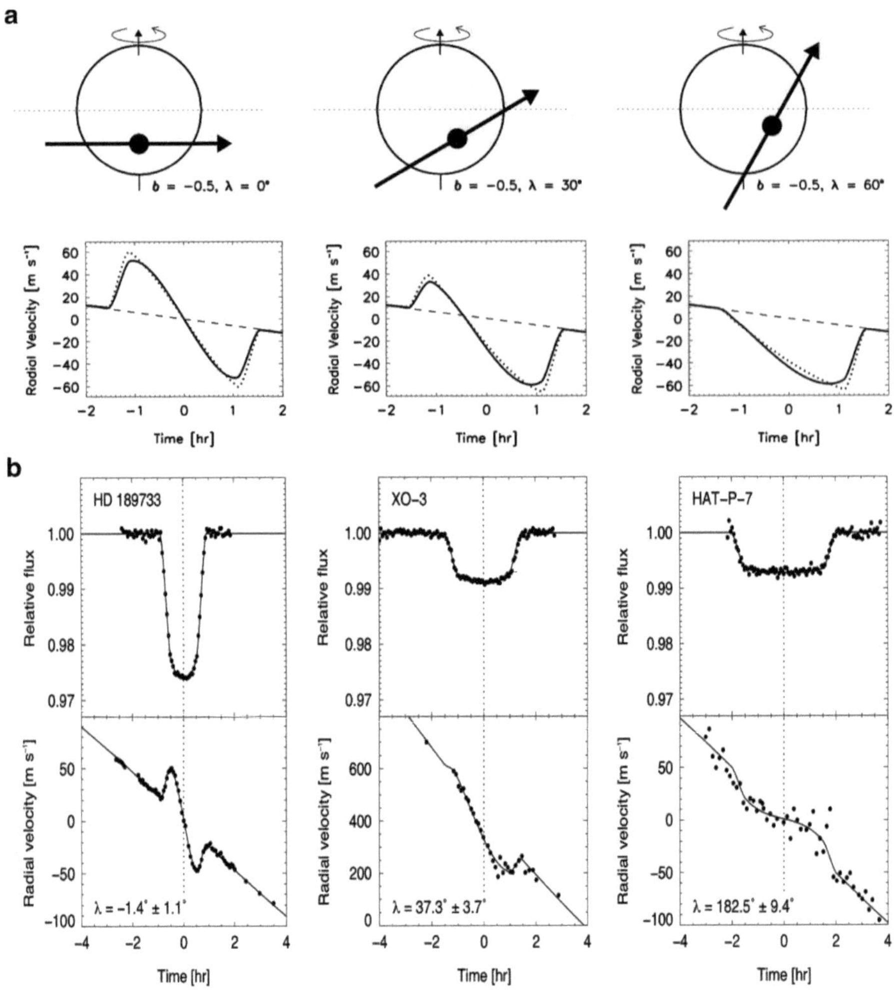

Fig. 13.2 The RM effect as an anomalous Doppler shift combines transit and radial velocity measurements. (**a**) Three transit geometries that produce identical light curves, but differ in spin-orbit alignment. Below the corresponding radial-velocity signals. Good spin-orbit alignment (left) produces a symmetric "redshift-then-blueshift" signal. A 30° tilt (tilt) produces an asymmetric signal. A 60° tilt (right) produces a blueshift throughout the transit. (**b**) Three examples of the RM effect showing the transit photometry (top) and the apparent radial velocity of the star, including both orbital and anomalous Doppler shift. The cases correspond to a well-aligned (left), misaligned (middle), and fully orthogonal case (right) with respect to spin and orbital axes. (Credit: J. Winn, MIT)

redshifted. Likewise, when the receding (redshifted) part of the star is blocked, the spectrum appears blueshifted. The result is an anomalous Doppler shift that varies throughout the transit (see Fig. 13.2). By monitoring the stellar line profiles during transits, the angle between the sky projections of the orbital and spin axes can be determined [1235].

The first RM effect detected in an exoplanet was in HD 209458b [1238] and many measurements since then have revealed a variety of orbits ranging from systems which are well-aligned, to misalignments of larger than 30°, and to systems which are apparently retrograde. Figure 13.2 shows three examples of such measurements, all hot Jupiter candidates. HD 189733 [1231] appears very well aligned with a practically non-existing projection angle of the stellar spin axis to the orbit normal. Other systems seem to be similar [1239, 1240], but others like XO-3 [1241] or HAT-P-7 [1242] seem to have more dramatic histories and in the latter case an estimated angle of 86° with respect to the star's equatorial plane suggests a retrograde motion. Some excess radial velocity variation might imply a third body.

13.1.3 Detection from Ground

The photometric precision of ground-based observations is limited to atmospheric seeing. With very few exceptions the detection of exoplanets is indirect, i.e. relies on the methods as described, mostly the determination of radial velocities, transits, and microlensing. In fact, almost 90% of exoplanet candidates detected from ground are detected through radial velocity measurements. Accuracies to detect sizes of $f_t \sim 10^{-3}$ is sufficient to detect massive extra-solar planets. The first successful exoplanetary *transit* was observed in the star HD209458 [1243, 1244]. Before that the radial velocity technique has been, except a few direct imagings, the exclusive method to detect exoplanets and their periods, orbital axes, eccentricities, and to produce some mass estimates. Radial velocity surveys of nearby F, G, K, and M dwarf stars found a class of close-in extrasolar massive planets that orbit their stars with separations of <0.1 AU (Hot Jupiters) and until the discovery of transits about a dozen candidates were known [1229, 1245, 1246]. From the precise shape of transit curves the stellar and planetary radius, the stellar mass, orbital inclination, and limb darkening parameters are inferred adding to the list of exoplanets. Last but not least, the RM effect (see above) added rotational and alignment information. The caveat in observations from ground is that the photometric precision of ground-based observations is limited due to atmospheric seeing. Accuracies of $f_t \sim 10^{-3}$ are sufficient to detect massive extra-solar planets while they transit through the stellar disk but not a possible low-mass population.

Before the launch of *Kepler* more than 400 planets had been detected with masses from about half of Neptune up to the deuterium burning limit. Statistics from ground based planet detections (see Fig. 13.3) show that minimum masses ($M \sin i$) distribute such that the number of planets increases with a power law of about index -1, with most candidates' masses between Neptune and Jupiter. Orbital periods of a day up to a few thousand days peak around 2–3 days and around 600 days. Separations seem to follow that trend with values between 0.01 and 100 AU with peaks around 0.05 AU and 2–3 AU. The metallicities are defined as the ratio of the abundance of iron in the star to that in the Sun scaled to hydrogen. That distribution seems to peak around 0.1 to 0.3 with a tail to lower values.

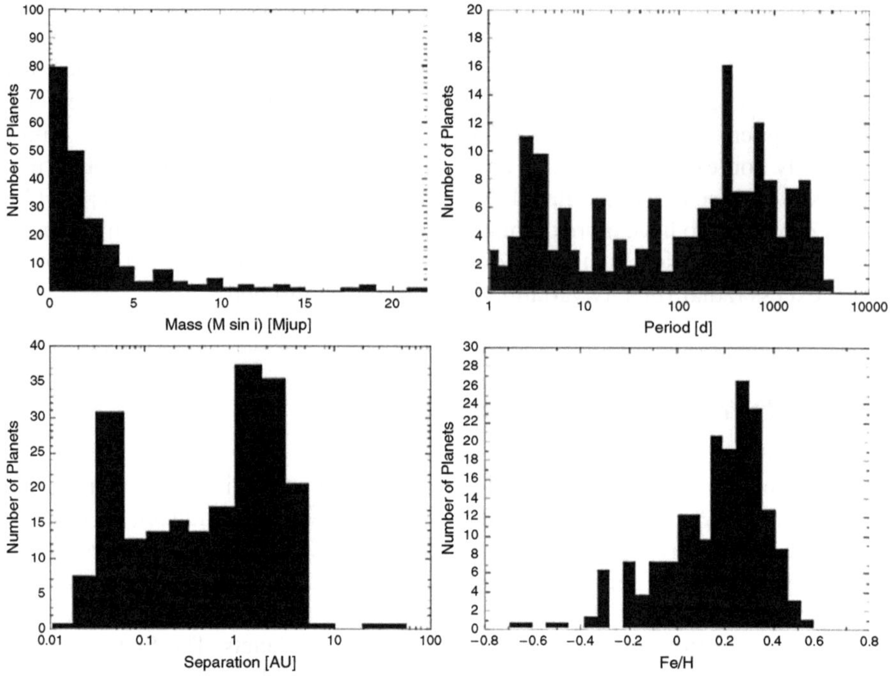

Fig. 13.3 Statistical assessment of ground-based observations of exoplanets. *Top* (*left*): Distribution of $M \sin i$. *Top* (*right*): Distribution of orbital periods. *Bottom* (*left*): Distribution of exoplanet separations. *Bottom* (*right*): Distribution of metallicities

However, with the arrival of *Kepler* and refined selection criteria of ground-based observations through the NASA-UC Eta-Earth program [1247], super-Earths and Neptune-sized planets with periods between 5 and and 75 days have been detected.

13.1.4 Detection from Space

Obervations from ground severely limit observations of exoplanets of masses less than about half the mass of Neptune. The Kepler mission is designed to determine the frequency of Earth-sized planets in and near the habitable zone of Sun-like stars. It was launched in early 2009 into an Earth trailing orbit. Key to the mission is a differential photometer with a 115 square degrees field of view that continuously monitors the brightness of approximately 150,000 MS stars. Its 1 m aperture is sufficient for 4 σ detections of a single transit from an Earth-sized planet passing in front of a 12th-mag G2 dwarf, i.e. a Sun-like star with a 6.5-hour transit [1248].

During the first 6 weeks of observations, the mission detected five new exoplanets with sizes between 0.36 and 1.5 R_{Jup} and masses between 0.08 and 2.1 M_{Jup}.

Table 13.2 Properties of five exoplanets detected by *Kepler* during its first weeks of operations [1248]

ID Kepler-	M_* [$M_\odot$]	R_* [$R_\odot$]	T_{eff} [K]	M_p [M_{Jup}]	R_p [R_{Jup}]	P [d]	a [AU]	i [°]	ρ [g cm^{-3}]	T_{equ} [K]
4b	1.22	1.49	5,860	0.08	0.36	3.21	0.046	89.8	1.91	1,650
5b	1.37	1.79	6,300	2.11	1.43	3.44	0.051	86.3	0.89	1,870
6b	1,21	1.39	5,650	0.67	1.32	3.23	0.046	86.8	0.35	1,500
7b	1.35	1,84	5,930	0.43	1.48	4.89	0.062	86.5	0.17	1,540
8b	1.21	1.49	6,210	0.60	1.42	3.52	0.048	84.1	0.26	1,760

Table 13.2 summarizes these findings. The density of the Neptune-sized Kepler-4b is similar to that of Neptune and GJ 436b, even though separations are in the sub-AU level leading to irradiation levels of several 10^5 times higher than Neptune experiences in the Solar system. Kepler-7b is one of the lowest-density planets (0.17 g cm^{-3}) yet detected. Kepler-5b, -6b, and -8b confirm the existence of planets with densities lower than those predicted for giant gas planets.

Kepler 11 is a specifically spectacular detection. Lissauer et al. [1249] report on multiple planet transits (see Fig. 13.4). Kepler 11 is a single Sun-like star which reveals six transiting planets, five with orbital periods between 10 d and 47 d plus a sixth with a longer period. The five inner planets are among the smallest whose masses and sizes have both been measured. The measurements imply substantial envelopes of light gases (see also Fig. 13.5 below). The degree of coplanarity and proximity of the planetary orbits implies energy dissipation near the end of planet formation.

In addition to ground observations *Kepler* detected 1,248 Earth-size exoplanets, 68 Neptune sizes and 11 trans-Jupiters [1250]. This is consistent with models of the occurrence rate with a power law [1251]

$$\frac{df}{d\ \log M} = k_m M^{\alpha_m}, \tag{13.10}$$

where $k_m = 0.39$ and $\alpha_m = -0.48$, which predict that 23% of stars should harbor a close-in Earth-mass planet in the mass range between 0.5 and 20 $M_\oplus$. From 1,235 *Kepler* planet candidates that include detected planets as small as 2 $R_\oplus$, [1247] determined the detectability of planets as a function of planet radius and period. The study found that planet occurrence varies by more than three orders of magnitude and increases substantially down to the smallest radii near today about 2 $R_\oplus$ and out to the longest orbital period af 50 d within separations of 0.25 AU. The rate is also given by a power law,

$$\frac{df}{d\ \log R} = k_r R^{\alpha_r}, \tag{13.11}$$

where where $k_r = 2.9$ and $\alpha_m = -1.92$ (R is units of $R_\odot$). The projected increase in planet occurence with decreasing planet size is consistent with core accretion models [1247], with the caveat that data below 2 $R_\odot$ are still incomplete.

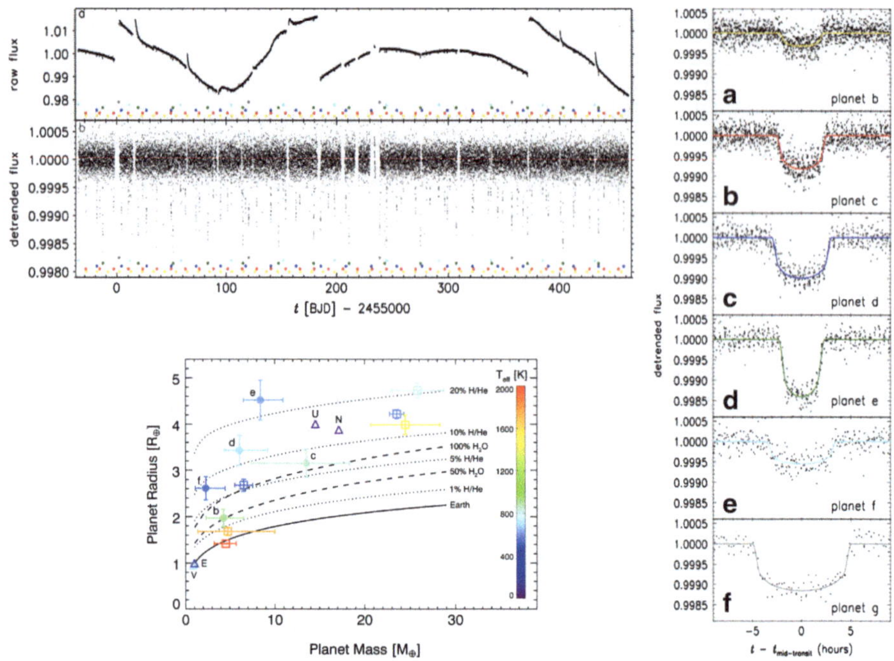

Fig. 13.4 *Top* (*left*): Lightcurves of Kepler 11, raw (top) and detrended (bottom). Kepler-11 is a G dwarf star about 613 pc from Earth. The large gradual flux variations in the raw lightcurves are of instrumental origin. *Right*: Folded data of the lightcurves shown phased at the period of each transit signal and plotted within an 18-hour region around mid-transit. *Bottom* (*left*): Mass–radius relationship of small transiting planets, with Solar System planets shown for comparison. Planets Kepler-11b-f are represented by filled circles. Other transiting extrasolar planets in this size range are shown as open squares. Other transiting exoplanets are shown, namely Kepler-10b, CoRoT-7b, GJ 1214b, Kepler-4b, GJ 436b, and HAT-P- 11b. The triangles (labeled V, E, U and N) correspond to Venus, Earth, Uranus and Neptune, respectively. The colors of the points show planetary temperatures (measured for planets in our Solar System). Mass–radius curves for 8-Gyr-old planets are plotted using contemporary planetary structure models. The solid black curve corresponds to models of planets with Earth-like rock–iron composition. The higher dashed curve corresponds to 100% H_2O. All other curves use a water or an H_2/He envelope atop the rock–iron core. The lower dashed curve is 50% H_2O by mass, while the dotted curves are H2/He envelopes that make up 2, 6, 10, and 20% of the total mass (from Lissauer et al. [1249])

Orbital periods, on the other hand, fall off more quickly and have different parameters for different ranges of radii. The data here can be described with a power law with an exponential cut-off,

$$\frac{df}{d\ \log P} = k_p P^{\alpha_p}\left(1 - \exp\left(\frac{P}{P_0}\right)^{\gamma}\right), \tag{13.12}$$

where P_0 is the cutoff period and γ a sharpness parameter. Fits to the measured planet occurrence with *Kepler* data various radius ranges shows that for smaller

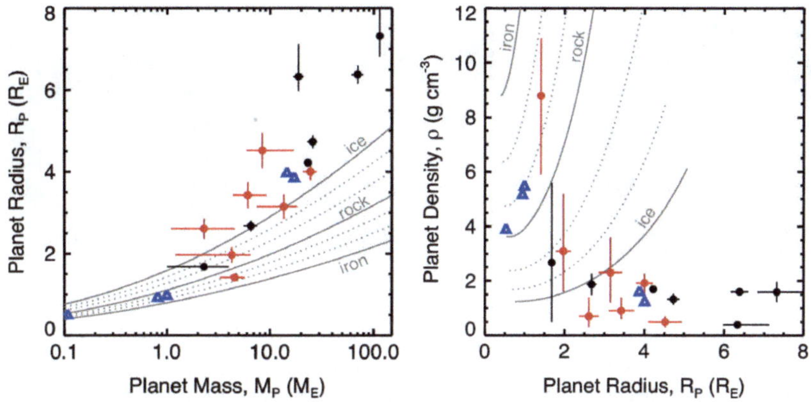

Fig. 13.5 "Equations of state" for planet radii <8 $R_\oplus$ and masses >0.1 $M_\oplus$. The left panel shows radius–mass functions, the right panel shows density as a function of radius. Mars, Venus, Earth, Uranus, and Neptune are depicted as blue triangles. Kepler planets (*filled circles*) are colored red and black for ground-based data. *Solid gray lines* indicate the densities of solid planets composed of pure ice, pure rock, and pure iron from [1252]. Dotted gray lines show the densities of ice/rock/iron admixture compositions (from Howard et al. [1247])

radii (2–4 $R_\odot$) the number of planets with larger periods is an order of magnitude higher than for large radii (8–32 $R_\odot$). The data also show that the cutoff period is larger for smaller planets and it is a suggested parking distance for migrating planets (see Sect. 13.3.6) that move outward with decreasing planet size [1247].

Figure 13.5 shows a first attempt to test "equations of state" for exoplanets in comparison with Solar System planets. They show that mass and radii grow faster than predicted from solid iron, rock, and ice compositions, suggesting that specifically higher mass planets contain a higher fraction of volatile gases including H and He. Most densities of planets with higher radii appear below the ice contour indicating the presence of gaseous components.

13.2 The Search for Protoplanets

There have been several accounts of the detection of planets in PMS stellar systems. Conventional search methods as decribed above are difficult to apply PMS stars (<50 Myr) as the protostellar and in later stages sometimes the debris disk strongly interferes with measurements. Direct imaging have shown that planets have already formed during these early phases. Perhaps the most prominent detection of a protoplanetary system with formed planets in a residual disk is the case of Fomalhaut.

Future high-resolution sub-mm observations with *ALMA* have the potential to resolve a protoplanet's warm dust with a separation of 5 AU from the central star and within 100 pc [1253]. Figure 13.6 shows a model configuration and projected

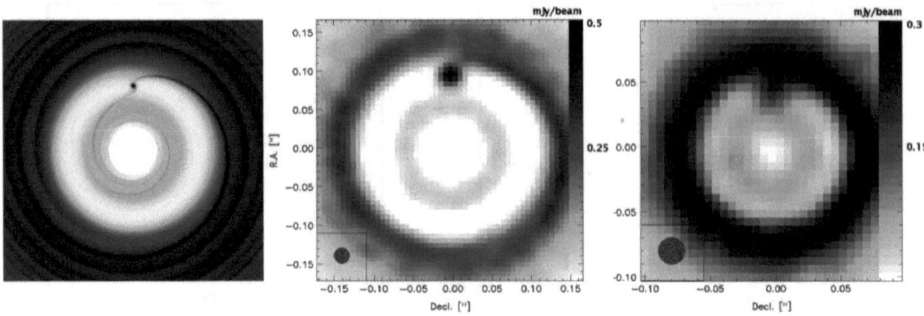

Fig. 13.6 Simulations of *ALMA* observations of disks with an embedded planet of 1 M_{Jup} around a 0.5 $M_{\odot}$ star separated by 5 AU. In the model (right), high- and low-density regions are represented by a dark and bright shades, respectively. *ALMA* images are shown for a distance of 50 pc (middle) and 100 pc (right). Dark and bright shades here represent intensity. The disk mass amounts to $M_{disk} = 10^{-2}\ M_{\odot}$. The insets at the lower left indicate the relative beamsize (from Wolf & D'Angelo [1253])

images for two distance cases. Even though the SEDs cannot be distinguished, dust re-emission images of the disks show that the hot region in the proximity of a young planet, can be resolved alongside with some of the spiral-in structures.

In the following three examples of exoplanet detections through imaging in young (<300 Myr) stellar systems are presented. It is not well defined when a planet is no longer considered a protoplanet. However, it takes a Jupiter, mass planet over 1 Gyr to cool down to below 500 K at large separations.

13.2.1 1RXS J160929.1-210524b

This young protoplanet was detected in *Gemini* observations in the optical and near-IR band [1254]. It belongs to a rare class of very-low-mass stellar objects sometimes classified as Y dwarfs and it may be questionable whether some of them qualify as planets. Zuckerman & Song [1255] released a list of half a dozen lowest-mass companions ($M \leq 20 M_{Jup}$) with ages ranging from 3 to 300 Myr and separations of 46 to 795 AU. In this sample only one is below the deuterium burning mass limit of 13 M_{Jup} and was found at the lowest separation of 46 AU.

The image of 1RXS J160929.1-210524 shows a planetary mass companion of $\sim$8 M_{Jup} at a separation of 330 AU. Its membership of the 5-Myr-old Upper Sco association [1256] was established from near-infrared, H- and K-band spectra and its companionship is confirmed through astrometry [1257]. These confirmations are important because these sub-stellar objects are so unusual in many respects. 1RXS J160929.1-210524b is particularly interesting as its mass, location, and age defies common formation scenarios. Its spectrum and temperature of 1,800 K is consistent with confirmed brown dwarfs in the vicinity. Its location and age rule out standard migration scenarios (see Sect. 13.3.6), its mass the possibility for deuterium ignition.

The question remains whether this object is a product of failed stellar formation or at rapid giant planet formation can occur at larger separations. However, there are quite a number of large systems with debris disks (see Sect. 12.1.2) which might provide some indirect evidence of large planets that might have once existed at wide separations. Some of these disks show morphologies consistent with gravitational perturbations from unseen massive planets.

13.2.2 HR 8799

A recent study of seven sub-giants with the Keck Observatory found radial velocity characteristics consistent with super-Jovian planet-mass characteristics and separations larger than 1 AU and, even though A-stars make up only about 3% of the stellar population, they seem make good targets for finding planets in imaging surveys [1258]. The reason for this might lie in the fact that A-stars are likely accompanied by large debris disks and studies of the nearby young A4V-star Fomalhaut (see below) and the A5V dwarf HR 8799 show just that.

HR 8799 is, like βPic (see Chap. 12), 60-Myr-old A-star located 40 pc from the Sun. In high-contrast observations with the Keck and Gemini telescopes three Jovian-size planets orbiting the central star with projected separations of 24, 38, and 68 AU are observed [1259]. *IRAS* and *ISO* observations of HR 8799 indicate that it has a dominant dust disk, likely residing outside the planet with the largest separation. The planets have masses between 5 and 13 M_{Jup} and it looks like a younger scaled-up version of the outer portion of the solar system. The photometric properties, like the systems descibed in [1255], and in the previous case (see Sect. 13.2.1), are similar to brown dwarfs. Figure 13.7 shows an age vs. luminosity diagram indicating stellar, brown dwarf, and planet regimes [1259] based on evolutionary track from [1260]. Here the HR 8799 planets line up with the previous case just below the deuterium burning limit, an order of magnitude older.

13.2.3 Fomalhaut b

Fomalhaut is an A4V-star 7.69 pc from the Sun and is an excellent example of a young system with a dusty debris disk (see Sect. 12.1.2). With an estimated age of 100 to 300 Myr it the oldest system in the sample and the central star is already on the MS. However, a planet body in the system would still be radiating its formation heat and thus be directly detectable. In this case the high contrast expected can play out as here the disk is already porous and the system is extremely close-by. It harbors a cool outer dust belt with a structure which is expected to host a planet through gravitational attraction. Kalas et al. [1261] detected an exoplanet candidate about 119 AU from the central star with a mass of less than 3 M_{Jup}.

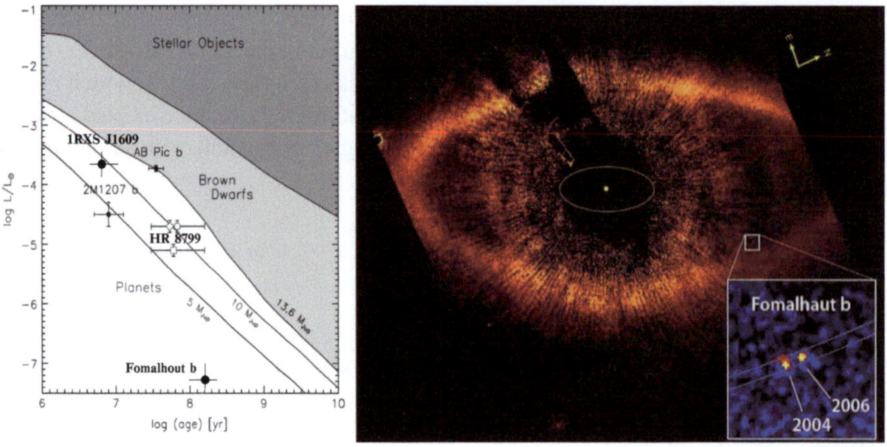

Fig. 13.7 *Left*: An age vs. luminosity diagram for a variety of masses and detected post-Jovian planets residing just below the deuterium burning limit (from Marois et al. [1259]). *Right*: HST coronagraphic image of Fomalhaut at 6,000 Å showing the location of Fomalhaut b (white square) at a 12.7″ from the central star. The yellow ellipse has a semimajor axis of 30 AU which corresponds to the orbit of Neptune in our solar system. The inset is a composite image showing the location of Fomalhaut b in 2004 and 2006; the inner and outer annuli have semimajor axes of 114.2 and 115.9 AU, respectively (from Kalas et al. [1261])

Figure 13.7 shows the HST coronagraphic image at 6,000 Å. It appears that the planet actually orbits about 18 AU inside the outer dust belt. The maximum mass of the planet is contrained by modeling its gravitational influence on the dust belt, reproducing properties of the belt inferred from the HST scattered light images. The model assumes that Fomalhaut b is solely responsible for the observed belt morphology [1261].

13.3 Planet Formation

The formation of planets evolves during the PMS stellar phase when dead zones in the proto-stellar disk appear. Chapter 8 addressed basic phenomena related to the physics and evolution of accretion disks; Chap. 12 outlined later stages such as dust disks and the making of the proto-solar nebula. This section focuses on planet formation, from planetesimals to core and terrestrial planet bodies, and the further development to a giant gas planet, as depicted in the cartoon in Fig. 12.7. Many details are left to the cited references and only some the main concepts are discussed in more detail.

13.3.1 Planetesimals

These are bodies large enough that their orbital evolution is entirely dominated by mutual gravitational interaction rather than aerodynamic coupling to the gas disk. The term planetesimal usually refers to bodies in the size range scale between 1 and 100 km. To from dust and rocks, aerodynamic forces between solid particles and the gas disk are as important as the actual efficiency with which particles coagulate. In order to build up bodies from the gas phase to planetesimal scale, coagulation has to cover over 10 orders of magnitude in size from molecular (nm) to km bodies. For a detailed review the reader is referred to Armitage [1227] and Dominik et al. [1262] and references therein.

On the smallest scales acceleration by drag forces are most important. The *Epstein* drag is effective if the size of a particle is smaller than the mean-free path of the gas molecules with the disk, i.e. dust is suspended in a collisionless and ususally subsonic flow of gas molecules. The drag solid particles feel leads to a collision frequency which is proportional to the gas density and the square of the particle size leading to relatively fast growth. Once particles grow much larger than the mean-free path the interaction with the gas is that of a fluid (*Stokes* drag). The drag force now also depends on the shape of the particles and case of spherical shapes only on the Reynolds number.

At $\sim$1 μm dust particles are tightly coupled to the gas flow. In a protoplanetary disk these particles are distributed within non-zero scale heights above the disk midplane. The z-component of the stellar gravitational force will drag dust towards the midplane. Pressure support above the midplane only acts on the disk gas and dust eventually will settle downwards By equating the z-components of the drag force and stellar gravity one can estimate the velocity at which the particles settle towards the disk midplane to be

$$v_{settle} = \frac{\rho_d}{\rho} \frac{a_d}{v_{th}} \omega_K^2 z, \tag{13.13}$$

where ρ_d is the matter density of the dust particle, ρ the gas density, a_d the size of the dust grain, v_{th} the thermal speed of the gas, ω_K the local Keplerian velocity. Assuming a uniform density, the settling time $t_{settle} = z/v_{settle}$ for a solar mass star would be of the order of 10^5 yr, which would mean that dust settles towards the midplane quite quickly. However, non-uniform densities and the effect of turbulence is expected to dynamically delay the sedimentation process. Turbulence in protoplanetary disks is a basic property (see Sect. 8.1.2) and the vertical profile of the distribution of dust species is very difficult to calculate. In the thin disk approximation the vertical density distribution can be written as a Gaussian function with a width equal to the pressure scale height $H_p = c_s/w_K$. With turbulence and nonlinear density profiles local settling time scales between disk surface and midplane can differ by orders of magnitude. Even for very small dust grains down to a few pressure scale heights, settling can be quite rapid; settling to the midplane can exceed the lifetime of the disk [1263]. More accurate results for dust settling come from numerical approaches. Dullemond & Dominik [1263] also find that a steady state solution between turbulent dust stirring and settling can be reached after a few 10^5 yr and settling times remain below the accretion time for grains up to 100 μm. [1264] simulated dust settling in locally turbulent protoplanetary disks and that dust dynamics is further affected by MHD turbulence. For particles of sizes up to a few cm, dust successfully decouples from the gas and settles; larger bodies, however, seem not to.

Rapid growth beyond the centimeter and towards the meter scale has to happen during sedimentation and certainly before small dust particles reach the radially inner regions of the disk and photoevaporate. Little is understood how at these early stages of PMS evolution the heavy bombardment of high-energy photon and particle radiation influences particle growth (see Sects. 8.3.3 and 12.2.3). High-energy photons significantly alter the thermal structure of the disk with X-ray and FUV photons penetrating to different depths in the disk, producing a complex layered thermal structure with specific chemical reaction environments. Radiation also contributes to gas dispersal and early dust will be lost in the gas dispersal and there will be no build-up of planetesimals. [1265] considered both the effects of vertical sedimentation and external photoevaporation. Due to the evaporation effect the dust to gas ratio f grows anomalously high and triggers gravitational instabilities and the spontaneous formation of kilometer-size planetesimals. Shu et al. [1266] suggested that FUV photons of the early Sun likely evaporated too much gas beyond the Saturn orbit before the cores of Neptune and Uranus were formed, leaving them with less gas to accrete. It is questionable, however, whether the early Sun could provide enough radiation for efficient evaporation. [1267] showed that rapid photoevaporation inside 10 AU leads to implausible timescales.

Drifts and turbulence also affect the radial flow of particles and radial re-distributions lead to local enhancements and depletions. The particle growth timescale ($t_{growth} = m/\dot{m}$) once compared to radial drift times ($t_{drift} = r/|v_r|$) provides that the condition $t_{growth} < t_{drift}$ is valid for particle sizes up to several meters.

In order for particles to grow to km size bodies, they collide frequently and stick effectively. While concepts of coagulation up to m size bodies might be theoretically feasible, the build-up of larger bodies through coagulation appears highly problematic at best. Gravitation does not play much of a role at sizes less than km and the growth dynamics is still governed by coagulation through collisions from drags and drifts. For example, a body of about 1 m has a velocity dominated by the radial drift resulting from the fact that solid bodies in the gas do not experience the same pressure forces as gas. If the coupling to the gas is strong, relative velocities remain relatively low. In the case of a poor coupling as expected for larger bodies it can reach $\Delta v \sim 10^{-3} v_k$ which at 1 AU amounts to almost 30 m/s and it is highly questionable to assume that many bodies would survive collisions at such velocities unless the distribution of relative velocities is skewed to low values.

Two parameters are also important: the collision timescale and the sticking efficiency. For a population of particles the collision time is $t_{coll} = 1/(n\sigma\Delta v)$, where n is the particle density $n = f\rho/m$, $\sigma = 4\pi a_p^2$, f the ratio of the density of solid bodies to the gas density, and a_b the body size. With a body mass $m = (4/3)\pi a_b^3 \rho_m$ the collision time then can be written as

$$t_{coll} = \frac{1}{3}\frac{\rho_m}{f\rho}\frac{a_b}{\Delta v}. \tag{13.14}$$

It is important to realize that that this timescale is proportional to the size of the body, but inversely proportional to the "dust-to-gas" ratio. Standard disk models usually have $f = 10^{-2}$, which for 1 m bodies leads to long collision times of 10^3 yr. Increasing f for shorter collision timescales seems to be the only viable possibility, as increasing the relative velocity appears counterproductive.

The other parameter is the sticking efficiency. Two colliding bodies can adhere for two reasons. Either their surface morphology is such that the two surfaces stick through entanglement, or their internal structure can efficiently absorb the impact energy. While for small dust particles surface sticking is more efficient, large bodies have to utilize the second possibility. In that respect the first larger bodies were not like solid rocks but more like fluffy conglomerates to grow though inelastic collisions. Further solidification might be achieved through ram pressures in collisions, melting and baking in high temperature and radiation environments.

13.3.2 The Snow Line

In a study of surface densities of dust materials and gases in a proto-solar accretion disk which was designed to study magnetic field effects on ionized gas, collisional charging and the resultant electrostatic attraction are important for initial dust aggregration. [1268] generated a minimum mass Solar Nebula model which today is widely used for reference. The surface density of solids is not a continuous function with separation from the central star owing to the fact that at a certain distance water freezes and icy structures add to the surface density, in detail,

$$\Sigma(\text{rock}) = 7.1 a^{-1.5} \text{ g cm}^{-2} \quad \text{for } 0.35 < a < 2.7. \tag{13.15}$$

and

$$\Sigma(\text{rock} + \text{ice}) = 30 a^{-1.5} \text{ g cm}^{-2} \quad \text{for } 2.7 < a < 36. \tag{13.16}$$

The location at 2.7 AU in the minimum Solar Nebula is called the *snow line*. The gas surface density

$$\Sigma(\text{gas}) = 1.7 \times 10^3 a^{-1.5} \text{ g cm}^{-2} \quad \text{for } 0.35 < a < 36 \tag{13.17}$$

is unaffected by this effect.

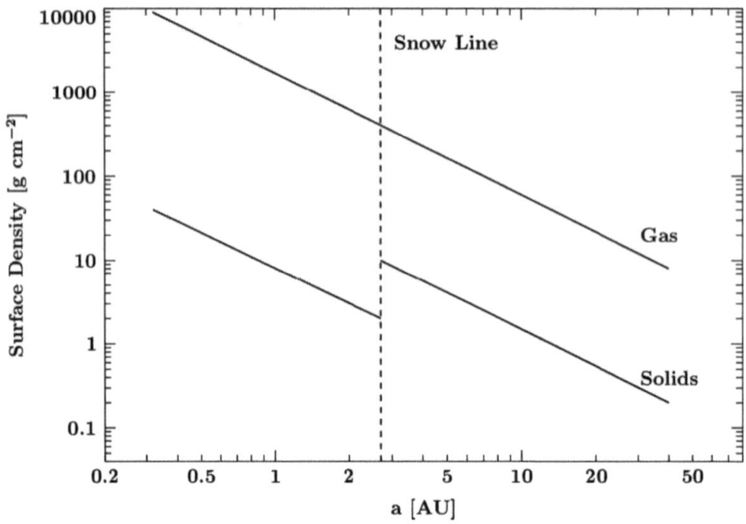

Fig. 13.8 The surface density of gas and solids as a function of radius from a minimum mass Solar Nebula model. The dashed line marks the location of the snow line. The surface density of solid increases significantly at the snow line (adapted from Hayashi [1268])

Figure 13.8 plots the surface densities as a function of orbital separation illustrating the large jump at the snow line. Kennedy and Kenyon [1269] further investigated the location of the snow line in evolutionary terms by locating the snow line where the mid-plane temperature of the disk is ~170 K. They find that for the solar case the snow line moves inward over time, crossing the canonical distance of 2.7 AU at 5×10^5 yr. A comparison of the snow line distance with typical disk lifetimes of several Myr leads to an increasing distance with stellar mass.

13.3.3 The Condensation Sequence

The snowline concept is only part of an entire series of condensation limits which is determined by many molecular substances in the protoplanetary disk. These tend to be rather volatile with low melting or boiling temperatures. Under low pressure conditions as they exist in protoplanetary disks, molecules condense directly into solid grains. And, as such, different molecules have vastly different condensation temperatures, assuming common pressure and volume conditions. For example while *methane* (CH_4) still condenses at a very low temperature (~40 K), oxides like *aluminum oxide* (Al_2O_3) require rather high temperatures (~1600 K). Water ice under proto-planetary mid-plane conditions is stable below about 150–170 K.

The elemental abundance in the disk should reasonably reflect the abundance of the central star. However, knowing the abundance does not easily translate into knowing how abundant elements are chemically distributed. In the Solar Nebula this discussion comes back to the study of *chondrites* with their substructures and

Table 13.3 Condensation temperature sequence for major elements and compounds for Solar System compositions and a pressure of 10^{-4} bar (from Lodders [1270])

Mineral species	Composition	Condensation temperature [K]	a [AU]
Corundum	Al_3O_3	1,677	0.03
Grossite	$CaAl_4O_7$	1,542	0.03
Ca titanate	$Ca_4Ti_3O_{10}$	1,512	0.03
Perovskite	$CaTiO_3$	1,441	0.04
Spinel	$MgAl_2O_4$	1,397	0.04
Fe alloy	Fe	1,357	0.04
Fosterite	Mg_2SiO_4	1,354	0.04
Enstatite	$MgSiO_3$	1,316	0.05
Schreibersite	Fe_3P	1,248	0.05
Trolite	FeS	704	0.16
Magnetite	Fe_3O_4	371	0.57
Water ice	H_2O	182	2.4
Ammonia hydrate	$NH_3.H_2O$	131	4.6
Methane hydrate	$CH_4.H_2O$	78	13
Argon hydrate	$Ar.6H_2O$	48	34
Methane	CH_4	41	47
Ne ice	Ne	9.3	906
He ice	He	<3	8,711

inclusions (see Sect. 12.1). Lodders [1270] provides an extensive study of Solar System abundances and condensation temperatures of a large variety of elements and compounds under various conditions, deduced from photospheric abundances and meteoritic CI chondrite abundances.

Table 13.3 shows some of the values for a number of more prominent compounds as expected in condensates and a few persistent gases. The values show that for heavier metallic compounds, many oxides and silicates constituting rocky matter, condensation temperatures are high, i.e. these can condense under the right circumstances well within 1 AU. For an estimate of the location of the condensation lines for the Solar Nebula one can use the assumption for the minimum nebula [1268]

$$T = 280\left(\frac{a}{[\mathrm{AU}]}\right)^{-1/2}\left(\frac{L}{[L_\odot]}\right)^{1/4} \mathrm{K}. \tag{13.18}$$

At early stages dust grains are still highly depleted in the nebula and the gas is fairly transparent to visible, but not to UV radiation. The values for a in Table 13.3 assume $L = 1L_\odot$ for simplicity. The pressure value used for the temperatures is higher than what is expected at a protoplanetary midplane and T values thus appear a bit on the high side. The snow line therefore is located slightly inward at 2.4 AU. There is a transition between the purely rocky condensates in the hot inner disk <0.1 AU and a combination of rocks and ices in the outer disk parts. While some the very volatile gases can condense as hydrates at separations of a few tens of AU, pure methane and ammonia remain gaseous for most the protoplanetary disk (<25 AU) [1268].

13.3.4 Core (Terrestrial) Planets

Planetesimals are bodies of size of about 1 to 100 km with large enough masses to further to grow through mutual gravitational attraction. In this new picture solid bodies grow, disassemble, reassemble, or get distroyed. Figure 13.9 illustrates three major modes of possible outcomes of collisions between two (or more) planetesimals. The outcome of a collision qualitatively depends on the masses of the impacting bodies and their direct velocity vector components. For two bodies on a direct collision course the specific kinetic energy

$$\mathcal{Q} = \frac{m\Delta v^2}{2M}, \tag{13.19}$$

where m and M are the two masses with $M > m$ and Δv is the relative speed, becomes relevant. Qualitatively one can separate three cases:

- The mass m merges with mass M and forms a higher-mass planetesimal of mass slightly less than $M + m$ as small shrapnel fragments eject.
- The mass m breaks up the mass M, but fragments remain within gravitational reach of the most massive remnant and eventually fall back and reassmble. The final result is a more loosely bound higher-mass planetesimal of mass slightly less than $M + m$ as small shrapnel fragments eject.
- The mass m breaks up the mass M and fragments disperse. In this case several outcomes are possible, ranging from complete destruction to a range of shrapnels all $\ll M$.

Even though these regimes appear clearly defined, the amounts necessary for low, medium, and high impact depend on the composition and material strength of the bodies. In that respect icy bodies can be weakly or strongly bound, basalt-rich bodies are usually strong, and reassembled bodies have ususally lower binding strengths. Larger bodies are less susceptible to this predicament and two regimes can be defined,

- a *strength-dominated regime*, where material strength dominates the impact properties. In this regime the material strength declines with increasing body size and the specific kinetic energy needed to destroy the body actually decreases with increasing body size.
- a *gravity-dominated regime*, where large bodies are held together by gravitational forces dominating all others. This is more or less the classic regime where specific kinetic energy to destroy the body exceed the gravitational binding energy of the target, i.e. GM/a_b.

Destructive properties of small targets can be investigated experimentally, larger bodies require numerical hydrodynamics of rigid bodies. Simple parameterized solutions for impacts on icy and rocky targets can be found in [1271, 1272]. These projections show that some target materials are also sensitive to impact velocities on top of the specific energy parameter value. The transition between strength- and

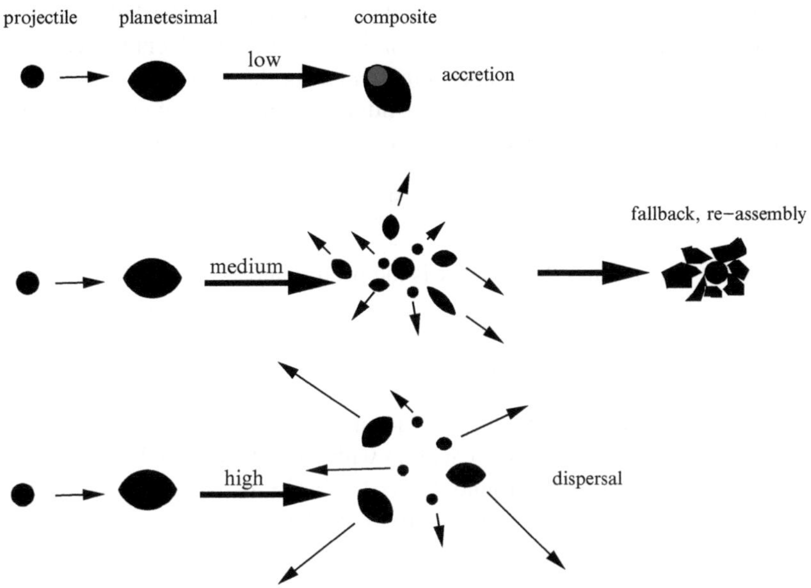

Fig. 13.9 Impact of physical collisions of planetesimals leading to either direct accretion (low ...), accretion through shatter and back-fall (medium ...), or destruction and dispersal (high specific kinetic impact energy)

gravity-dominated regimes, however, appears very clear. Only bodies well above the 1 km size range are clearly dominated by gravity, the 100 m to 1 km range largely depends on material strengths and specific energies that lead to dispersal can vary by over an order of magnitude between weak icy bodies and solid rocks. The results by [1271, 1272] show that in the gravity regime there are also "weak" and "strong" targets and specific destruction energies vary from 10^5 to 10^9 erg g^{-1}. The latter is sufficient to destroy a 100-km-sized asteroid.

The growth of a protoplanet depends on its feedback on the properties of its surrounding planetesimals. One simple concept is the *isolation mass*, the maximum mass a protoplanet can achieve through runaway growth clearing its environment of all planetesimals and with no threat of a destruction event. The *Hill sphere* is important as it defines whether an encounter is a simple two-body interaction and thus *dispersion dominated* or if three-body effects, i.e. effects that include the stellar tidal field, have to be included. By equating the orbital frequency of the protoplanet around the central star with a ($w_a^2 = GM_*/a^3$) to that of the planetesimal orbiting the protoplanet at radius r ($w_r^2 = GM_p/r^3$) and accounting for the limit $M_p \ll M_*$ in the Roche volume, one yields the Hill radius and velocity

$$r_H = \left(\frac{M_p}{3M_*}\right)^{1/3} a; \quad v_H = \sqrt{\frac{GM_p}{r_H}}. \tag{13.20}$$

In the shear-dominated regime where a planetesimal enters the Hill sphere from outside, tidal interactions are important. For example by introducing an eccentricity $e \sim \Delta a/a$ to the orbit of the planetesimal entering the Hill sphere, where Δa is the perturbation of the orbit by the stellar tidal field. In the frame rotating with the protoplanet, the average relative velocity v_{shear} of a planetesimal entering the Hill sphere is

$$v_{shear} = \frac{9}{8}\omega_a \Delta a, \tag{13.21}$$

and the rate at which mass flows towards the protoplanet is

$$\frac{dM_H}{dt} = v_{shear}\Sigma_p \Delta a. \tag{13.22}$$

Σ_p is the surface density of the planet. One can now also define the radial extent of a planet's feeding zone $\Delta a_{max} \propto r_H$ scaling with the radius of the Hill sphere. Scale factors have been given to be around 2.3 to 2.5 [1273, 1274]. Using the expression in [1275] the *isolation radius* can be approximated to

$$R_{iso} \approx \left(\frac{15}{4}\frac{\Sigma_p a}{\rho\alpha}\right)^{1/2} \quad \text{with} \quad \alpha \equiv \left(\frac{9}{4\pi}\frac{M_\odot}{\rho a^3})\right)^{1/3}, \tag{13.23}$$

or in terms of

$$R_{iso} \approx 1.606 \times 10^3 \left(\frac{a}{[\mathrm{AU}]}\right)^2 \left(\frac{\Sigma_p}{[\mathrm{g\ cm^{-2}}]}\right)\left(\frac{\rho}{[\mathrm{g\ cm^{-2}}]}\right)^{-2/3}. \tag{13.24}$$

The isolation mass then follows as

$$M_{iso} \approx \frac{4\pi}{3}\rho R_{iso}^3 = 7.028 \times 10^{-4}\left(\frac{\rho}{[\mathrm{g\ cm^{-2}}]}\right)\left(\frac{R_{iso}}{[1{,}000\ \mathrm{km}]}\right)^3. \tag{13.25}$$

For typical surface densities [1275] use 5.5 g cm^{-2} and find an isolation mass of about 0.01 $M_\oplus$. Others estimate a more rapid growth and reach a higher value of 0.07 $M_\oplus$ by using a slightly higher feeding zone and higher surface densities [1227]. Simulations by [1276] reach even 0.16 $M_\oplus$beyond the snowline where water condenses as ice from disk gas at about 170 K.

In general a first runaway growth period leads to body sizes up to 0.1% of $M_\oplus$ succeeded by what is called *oligarchic growth* reaching isolation within 10^5 yr. Further oligarchic growth continues by stirring up planetesimals, losing energy to small bodies through gravitational scattering, damping, and further growth through collisions. The number of oligarchs around 1 AU in a solar mass system after about 10^6 yr is about 100 to 1,000 with masses between 10^{-2} and 0.1 $M_\oplus$, still far from an Earth. It will take several 10^7 yr of collisions of large bodies to finally assemble an Earth-mass planet. There are now high-resolution numererical N-body simulations available [1276–1278]. In the one by [1277] it takes up to 200 Myr to finally clear

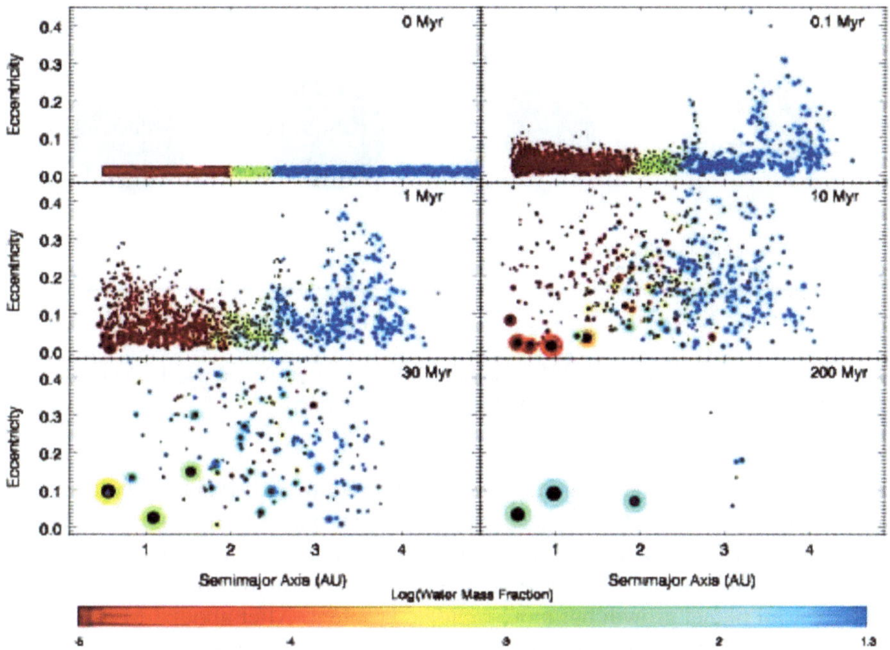

Fig. 13.10 Results from a high-resolution simulation of rapid and oligarchic growth to the final assembly of terrestrial planets. The simulation includes about 2,000 planetary seeds. The shading is according to the water content. A Jupiter planet is beyond the scale at 5.5 AU (from Raymond [1277])

of gas and planetesimals to completely clear out, though after 30 Myr most final bodies have already been assembled. It should also be mentioned that at these stages the Solar Nebula is already entirely devoid of volatiles and any water, bearing the implication that these substances have to be provided by bodies from the outer parts of the Nebula at later stages (see Fig. 13.10).

13.3.5 Giant Planets

In the formation of terrestrial planets, bodies even at Earth masses are too small to trigger gas accretion. Two modes of formation have been proposed for massive planet formation, each backed up by obervations. In the *disk instability* model, giant planets form via fragmentation of an unstable proto-planetary disk. Fragmentation follows the gaseous analog of the Goldreich–Ward mechanism where a protoplanetary disk sediments, cools quickly, and freeze-fragments. However, there is the class of substellar protoplanets described in Sect. 13.2 which observationally show properties close to brown dwarfs and alternately might represent failed fragmentation into a stellar binary (see Chap. 7). This section focuses mostly an

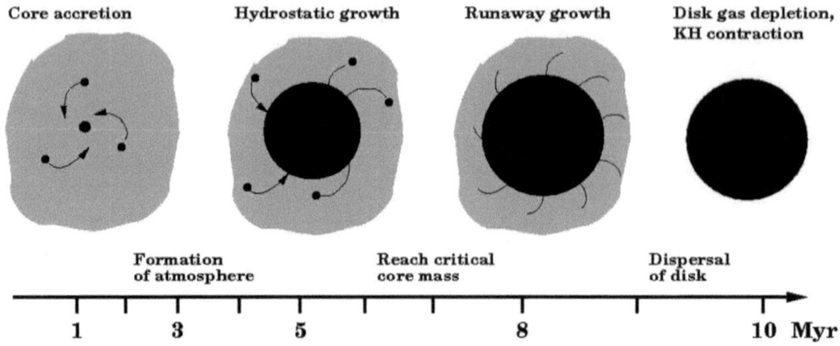

Fig. 13.11 Stages in the formation of giant planets in the core accretion model. The timescale is approximated from [1279] for the case of the formation of Jupiter

the *core accretion* model of giant planet formation, though it should be clear that giant exoplanets are so diverse that more than one mode of formation seems likely.

Figure 13.11 shows the principal phases in the core accretion model of giant gas planet formation. These include rapid core assembly though collisions very similar to terrestrial planet growth but more rapid and violent. After about 2 Myr this seed core begins to form an atmosphere and is growing hydrostatically while continuing to assemble the core through collisions. Once the core reaches a critical mass, runaway growth is triggered that sweeps up the surrounding gas. Eventually gas is depleted and accretion terminates and the giant protoplanet cools and contracts.

In the *core formation phase* a core protoplanet forms via rapid successive collisions until it is massive enough to retain an atmosphere. Even though this is much like the core model for terrestrial planets, timescales are radically different. In terrestrial planet formation the final $M_\oplus$is assembled during several 10^7 yr, the timescale in Fig. 13.11 has this happen in a few 10^6 yr. Terrestrial planet formation, ignoring possible migrations, proceeds at locations inside the snow line at 2.7 AU [1268] which are much more starved in terms of solid and icy bodies. This has consequences for the formation of more massive planets as the feeding radius also grows with separation from the central star (see Eqn. 13.20), as do surface densities. Condensations of ices increase the surface density of planetesimals and since the timescale for planet growth is inversely proportional to surface density (see [1269] and references therein) it also leads to shorter growth times.

Figure 13.12 (left) shows the dependence of the isolation mass on the separation from the central star. In this exercise [1227] estimate the minimum mass $M_{c,min}$ required by a protoplanetary core to accrete some gas from the disk and sustain an atmospheric envelope for a significant time. In this estimate the M_{iso} inside the snow line is much smaller than $M_{c,min}$, arguing that terrestrial planets can never capture a gas envelope before the disk disperses. Outside the snow line the opposite is making giant planet formation more likely. Note, that in the given estimate M_{iso} is grossly underestimated beyond the snow line. For a Jovian-type planet more precise calculations of M_{iso} lead to core masses between 5 and 10 $M_\oplus$ [1227, 1275] and

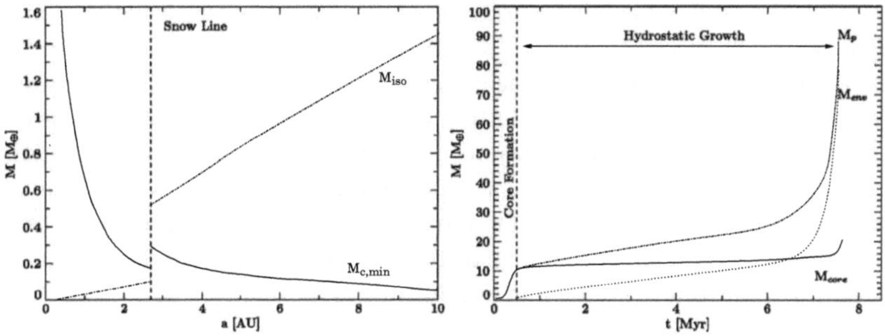

Fig. 13.12 *Left*: The dependence of M_{iso} (dot-dash line) in comparison with the minimum mass required to sustain a massive envelope (solid line) from [1227]. The central solar mass star is surrounded by a protoplanetary disk with z/a = 0.05. The disk follows the minimum mass Solar Nebula model (see Sect. 13.3.2) with a snow line at 2.7 AU. *Right*: The evolution of core mass (solid line), envelope mass (dotted line), and total mass (dot-dash line) (from [1280, 1281])

the critical core mass for runaway envelope growth might sometimes be reached even before all solid bodies inside the feeding zone have been swept up. One should also note that there are still large variations in the calculation of parameters such as M_{iso}, $M_{c,min}$, or the envelope mass M_{env}, the critical mass M_{crit} as well as growth timescales, even though trends now generally agree in most models (see [1227, 1275, 1279, 1282–1284]).

Core formation outside the snow line proceeds rapidly through runaway accretion within the feeding zone and over 50% of the final core mass ($\sim$10–12 $M_{\oplus}$) is reached after about 0.5 Myr for a Jupiter-like planet (see Fig. 13.12, right). The core now starts a period of *hydrostatic growth* in which gas is slowly accreted in an envelope of mass M_{env}, pushing the radius of the feeding zone outward. This allows further core mass growth, but at a much lower rate. At some point M_{env} reaches a critical mass M_{crit} beyond which it exceeds the core mass M_{core}, as shown in Fig. 13.12. This likely triggers runaway growth of the envelope allowing also for the accretion of hydrogen and helium. This period is now short as the gas supply quickly runs out when the disk disperses or growing tidal forces disrupt the flow (see Sect. 13.3.6).

Ikoma et al. [1283] investigates the dependences on core accretion rate and grain opacity within a set of assumptions. Here the protoplanet has a spherically symmetric structure with an envelope which is always in quasi-hydrostatic equilibrium. The core remains an inert body of constant density of 5.5 g cm^{-3}. The envelope has a uniform chemical composition with mass fractions of 0.74 for hydrogen, 0.24 for helium, and 0,02 for heavier elements. The internal energy arises from impacting planetesimals on the core surface with a mass accretion rate for $a > 2.7$ AU formulated after [1285] of

$$\dot{M}_{core} \approx 1.1 \times 10^{-7} \left(\frac{M_{core}}{[M_{\oplus}]} \right)^{2/3} \left(\frac{a}{[5 \text{ AU}]} \right)^{-31/12} f_{\Sigma} \; M_{\oplus}\text{yr}^{-1}, \qquad (13.26)$$

where f_Σ is the ratio of the initial surface density of solid materials to the minimum-mass Solar nebula model (see Sect. 13.3.2). This gives an average core growth rate of about $10^{-6}\ M_\oplus \mathrm{yr}^{-1}$. Note, that the dependency of this growth rate is very similar at $a < 2.7$ AU with a much lower starting core mass and average absolute growth.

The envelope continues smoothly to the nebular gas at the accretion radius at which the potential energy of the gas is equal to the thermal energy of the envelope. The radial temperature gradient in the envelope then efficiently transports the core surface luminosity via either radiative diffusion or convection. In the radiative case the resulting temperature gradient in equilibrium is

$$\frac{dT}{dr} = -\frac{3\kappa_R \rho}{16\sigma T^3}\frac{L}{4\pi r^2}. \tag{13.27}$$

Should the radiative temperature gradient in hydrostatic equilibrium be larger than the adiabatic temperature gradient, i.e.

$$\left(\frac{d\ln T}{d\ln P}\right)_{rad} < \left(\frac{d\ln T}{d\ln P}\right)_{ad} \tag{13.28}$$

some convective transport can be expected. For envelope opacities exceeding $10^{-2}\ \mathrm{cm}^2\ \mathrm{g}^{-1}$, the critical core mass, i.e. where the envelope mass equals the core mass, can be approximated as

$$\dot{M}_{crit} \approx 7\left(\frac{\dot{M}_{core}}{[10^{-7} M_\oplus\ \mathrm{yr}^{-1}]}\right)^q \left(\frac{\kappa_r}{[\mathrm{cm}^2\ \mathrm{g}^{-1}]}\right)^s M_\oplus, \tag{13.29}$$

where the indices q, s are between 0.2 and 0.3 [1283]. The characteristic growth time of the envelope mass τ_g depends on the core mass and the grain opacity and can be expressed as

$$\tau_g \approx 10^8 \left(\frac{M_{core}}{[M_\oplus]}\right)^{-2.5} \left(\frac{\kappa_r}{[\mathrm{cm}^2\ \mathrm{g}^{-1}]}\right) \mathrm{yr}. \tag{13.30}$$

Based on these model results it is debatable how much variance there is for producing different types of giant planets. Core masses exceeding $\sim 20\ M_\oplus$ are possible for larger surface densities which implicitly enhance $\dot{M}_{core}$. It is also a matter of debate how much play the two parameters, opacity and rapid initial core growth really have. This is a very simple treatment and more realistic simulations model spiral density waves that focus the gas in the planet's Hill sphere, while transferring angular momentum both inward and outward (see Sect. 13.3.7 and Fig. 13.13).

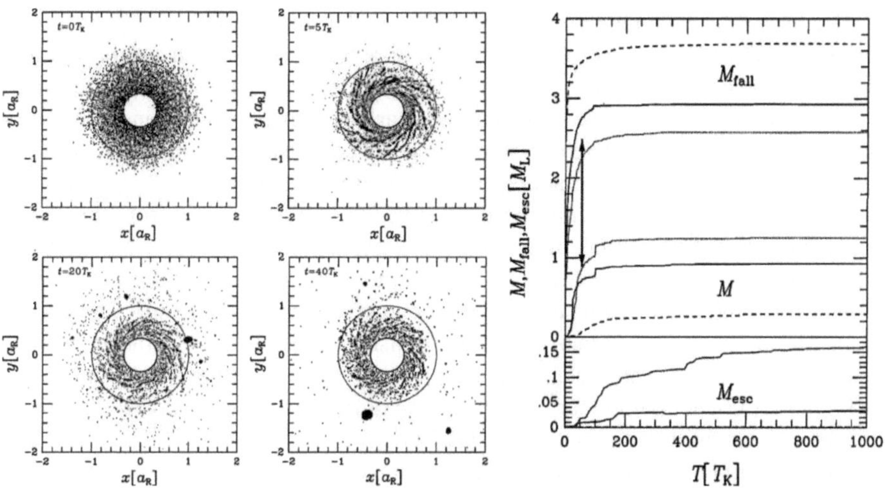

Fig. 13.13 *Left*: Snapshots of the circumterrestrial disk projected on the x–y plane at $t = 0, 5, 20, 40T_K$. The solid circle centered at the coordinate origin resembles proto-Earth and the dotted line the Roche lobe radius. The disk initially has a mass of 4 M_L, a $\Sigma_d \propto a^{-3}$, and a RMS eccentricity of 0.3 for lunartesimals. *Right*: The time evolution of the lunar mass M, the mass fallen to Earth M_{fall}, and the mass of the escapes from Earth M_{esc} for different sets of initial conditions (dashed/solid/dotted curve), The solid curve approximately resembles conditions in the snapshots to the left. The location of the double-ended arrow marks the location of the final snapshot shown (from Kokubo et al. [1289])

13.3.6 Resonances and Migration

Once planets have been formed and most of the gas in the disk has been depleted planetary systems evolve and undergo various structural changes. The observational account of exoplanets is much at odds with orbits where these planets are supposed to form. Hardly any full-scale planetary system model winds up with a planetary configuration as we observe it in the Solar System and it is certain that over time it took several forms of energy and angular momentum exchanges to make it happen. For example, the fact that trans-Neptunian objects are found in orbits which are in resonance with Neptune suggests that migration had to take place at some time. Having Neptune and Uranus migrate outward at some time in the history of the Solar System makes sense because it solves a timescale problem in the formation of the two planets. The outward migration drove the Kuiper Belt objects outward, bringing them into resonance with Neptune. In general, effects like migration through tidal interactions would also damp any orbital inclination, planet–planet scattering amplifies inclination, the *Kozai effect* due to distant planets can produce misalignment. Even today there is no consensus on how to apply these mechanisms on the length of time they affected the evolution of planetary systems. The following is designed to introduce some of these mechanismas.

The possible exchange of energy and angular momentum via gravitational torques between planet and protoplanetary disk can only occur if the protoplanet excites some non-axisymmetric motion that results in changes in a, eccentricity e, and inclination i. Migration in the very early stages of giant planet formation is likely, as here a sufficiently massive core co-exists and interacts with the protoplanetary disk. The interaction of the planet with the gas *exterior* to the orbit *increases* the angular momentum of the gas which will be repelled from the planet, which in turn loses angular momentum and moves inward. Interaction with gas *interior* to the orbit *decreases* the angular momentum of the gas which also will be repelled, but the planet gains angular momentum and moves outward. The change in angular momentum depends on surface density, planet mass, separation, angular velocity, and impact parameter. For a typical growing core mass at a Jupiter-like orbit and an impact parameter of $0.05a$ the exterior torque would drive the planet inward on a timescale of about 1 Myr [1227].

Gravitational perturbations can excite resonance in the planetary disk. The perturbation to the gravitational protential can be decomposed into frequencies which vary azimuthally. In general a resonance occurs when a characteristic frequency of the planet matches the frequency of the disk. There are several resonances to consider. The *co-rotation* resonance is co-orbital with the planet. *Lindblad* resonances occur when gas in the disk is at its natural frequency for radial oscillations with nominal locations given by

$$r_L = \left(1 \pm \frac{1}{m}\right)^{2/3} a, \tag{13.31}$$

where a is the semi-major axis, and m is an interger. For a planet in a circular orbit there is then one co-rotation resonance blended with an entire family of Lindblad resonances closely spaced with the planet's orbit. However, once the planet has an eccentric orbit, resonance modes of the pertubing gravitational potential become an important element in migration dynamics. With a mean angular velocity of the planet Ω_p, orbital resonance frequencies are now defined as

$$\Omega_p^{l,m} \equiv \Omega_p + \frac{l-m}{m}\Omega_p, \tag{13.32}$$

with l becoming a second integer, and the condition now is that any of the $\Omega_p^{l,m}$ matches one of the natural frequencies of the disk. Whether it is the capture of a planet or planetesimal, or the migration of a planet, the final location is determined by these orbital frequencies.

One usually distinguishes between two types of migration. *Type I* migration applies for low-mass planets which usually have weak interactions with the disk, and are not changing the disk structure. Any local exchange of angular momentum between planet and disk with respect to the redistribution of angular momentum is due to disk viscosity. Torques are governed by the sum of discrete resonances provided by the inner and outer Lindblad resonances and the co-rotation resonance. *Type II* migration applies for giant planets where the gravitational torque now dominates the angular momentum transport in the disk. The inner and outer Lindblad resonances close to the planet hardly matter as gas is repelled and the region around the planet is depleted.

Mean motion resonances then are dynamic relationships between objects orbiting the same central star such that the ratio of their orbital periods is expressed by small integers; e.g. for two objects in combinations of 2:1, 3:1, 4:1, 3:2, or 4:3 as observed in our Solar System or even higher numbers in exoplanet systems (see below). In the Solar System, three of the Galilean moons of Jupiter are in such mean motion resonances. The orbital periods of the three inner moons, i.e. *Io, Europa*, and *Ganymede*, appear in the ratio of 1:2:4. The fourth moon, *Callisto*, lies just outside a 2:1 resonance with *Ganymede*. *Neptune* and *Pluto* appear in a 3:2 resonance and some of the larger moons of *Saturn* appear in mean motion resonances.

Migration and resonances are a physical reality in protoplanetary and mature planetary disks. There is evidence for multiple exoplanet systems locked into resonances. Lee and Thommes [1286] (see references therein) summarize a number of systems where exoplanets are found in these resonances; 33 known systems with multiple planets exist, about 25% can be related to 2:1, 3:2, 3:1, and 5:1 resonances. And of course one of the successes of migration models is the explanation of the structure of the Kuiper Belt in our own Solar System, providing compelling evidence for the outward migration of Neptune. One model that evolved by these mechanisms is the *Nice Model* which, with respect to the Solar System, fosters the idea that a planetary system like the Solar System with initial quasi-circular, co-planar orbits, would have evolved to the current orbital configuration, provided that Jupiter and Saturn crossed resonance [1287].

13.3.7 The Making of the Moon

Detailed comparisons of the properties of Lunar and Earth rock samples have produced very strong evidence that the Earth and the Moon likely originated from the same precursor planet. Abundances of Lunar and Terrestrial materials are suffciently different to practically exclude a scenario where the Moon formed directly from the Earth. It is now highly unlikely that the Moon was formed at a different location and was captured by the Earth.

The Moon is now thought to have formed and evolved through a single (or a series of) catastrophic heating events, during which most of the highly volatile elements were lost, hydrogen likely entirely. In the giant impact hypothesis, a single collision event, an about Martian-sized planetoid collided with a proto-Earth about 40 Myr into the formation of the planet. In the process both bodies got destroyed and reassembled with the materials of both impact bodies leaving a ring of matter ejected by the collision around the new proto-Earth. In this process heavy elements, mostly iron, settled in the planet's core, while silicate-rich materials settled in the crust and in the debris ring which coagulated into the Moon. Simulations by [1288, 1289] showed that most of the disk material reassembles into a proto-Moon in less than a year's time. This is confirmed by N-body simulations which allow for the formation of a single moon just outside the Roche limit with an efficiency of incorporation of material of 10–55%. Kokubo et al. [1289] decribe the lunar disk evolution through contraction via collisional damping and while the velocity dispersion of the disk particles decreases, particle clumps grow inside the Roche limit. Particles are transferred to the outside of the Roche limit through the gravitational torques. When matter spills beyond the Roche limit, it collapses into a small moonlet. A lunar seed is formed through rapid accretion of small moonlets and grows by sweeping up particles transferred over the Roche limit. When the moon becomes large enough to gravitationally dominate the disk, it pushes the rest of the inner disk toward the proto-Earth. The formation timescale of the moon under various assumptions then is of the order of 1 month.

However, recent re-evaluations of lunar-volcanic glasses and highland basalts favor a much wetter Moon, which is not easily explained by the above mechanism. During these catastrophic heating events, highly volatile elements were were lost, and in the case of hydrogen it is believed to have been completely evaporated. Analyses of moon rocks show that the Moon might not be entirely depeleted in highly volatile substances, including water [1290]. From lunar melt inclusions ([1341] and references therein) it ia now found that high contents of water, fluorine, sulfur, and chlorine are very similar to primitive mid-ocean ridge basalts and it is concluded that some of the lunar interior contains as much water as Earth's upper mantle.

Appendix A
Gas Dynamics

The nature of stars is complex and involves almost every aspect of modern physics. In this respect the historical fact that it took mankind about half a century to understand stellar structure and evolution (see Sect. 2.2.3) seems quite a compliment to researchers. Though this statement reflects the advances in the first half of the 20th century it has to be admitted that much of stellar physics still needs to be understood, even now in the first years of the 21st century. For example, many definitions and principles important to the physics of mature stars (i.e., stars that are already engaged in their own nuclear energy production) are also relevant to the understanding of stellar formation. Though not designed as a substitute for a textbook about stellar physics, the following sections may introduce or remind the reader of some of the very basic but most useful physical concepts. It is also noted that these concepts are merely reviewed, not presented in a consistent pedagogic manner.

The physics of clouds and stars is ruled by the laws of thermodynamics and follows principles of ideal, adiabatic, and polytropic gases. Derivatives in gas laws are in many ways critical in order to express stability conditions for contracting and expanding gas clouds. It is crucial to properly define gaseous matter. In the strictest sense a *monatomic ideal gas* is an ensemble of the same type of particles confined to a specific volume. The only particle–particle interactions are fully elastic collisions. In this configuration it is the number of particles and the available number of degrees of freedom that are relevant. An ensemble of molecules of the same type can thus be treated as a monatomic gas with all its internal degrees of freedom due to modes of excitation. Ionized gases or *plasmas* have electrostatic interactions and are discussed later.

N.S. Schulz, *The Formation and Early Evolution of Stars*, Astronomy and Astrophysics Library, DOI 10.1007/978-3-642-23926-7, © Springer-Verlag Berlin Heidelberg 2012

A.1 Temperature Scales

One might think that a *temperature* is straightforward to define as it is an everyday experience. For example, temperature is felt outside the house, inside at the fireplace or by drinking a cup of hot chocolate. However, most of what is experienced is actually a temperature *difference* and commonly applied scales are relative. In order to obtain an absolute temperature one needs to invoke statistical physics. Temperature cannot be assigned as a property of isolated particles as it always depends on an entire ensemble of particles in a specific configuration described by its equation of state. In an ideal gas, for example, temperature T is defined in conjunction with an ensemble of non-interacting particles exerting pressure P in a well-defined volume V. Kinetic temperature is a statistical quantity and a measure for internal energy U. In a monatomic gas with N_{tot} particles the temperature relates to the internal energy as:

$$U = \tfrac{3}{2} N_{tot} k \, T \tag{A.1}$$

where $k = 1.381 \times 10^{-16}$ erg K^{-1}.

The Celsius scale defines its zero point at the freezing point of water and its scale by assigning the boiling point to 100. Lord Kelvin in the mid-1800s developed a temperature scale which sets the zero point to the point at which the pressure of all dilute gases extrapolates to zero from the triple point of water. This scale defines a *thermodynamic* temperature and relates to the Celsius scale as:

$$T = T_K = T_C + 273.15^o \tag{A.2}$$

It is important to realize that it is impossible to cool a gas down to the zero point (Nernst Theorem, 1926) of Kelvin's scale. In fact, given the presence of the 3 K background radiation that exists throughout the Universe, the lowest temperatures of the order of *nano-Kelvins* are only achieved in the laboratory. The coldest known places in the Universe are within our Galaxy, deeply embedded in molecular clouds – the very places where stars are born. Cores of Bok Globules can be as cold as a few K. On the other hand, temperatures in other regions within the Galaxy may even rise to 10^{14} K. Objects this hot are associated with very late evolutionary stages such as pulsars and γ-ray sources. Fig. A.1 illustrates examples of various temperature regimes as we know them today. The scale in the form of a thermometer highlights the range specifically related to early stellar evolution: from 1 K to 100 MK. The conversion relations between the scales are:

$$E = kT = 8.61712 \times 10^{-8} \frac{T}{[\mathrm{K}]} \ \mathrm{keV}$$

$$E = \frac{12.3985}{\lambda[\text{Å}]} \ \mathrm{keV} \tag{A.3}$$

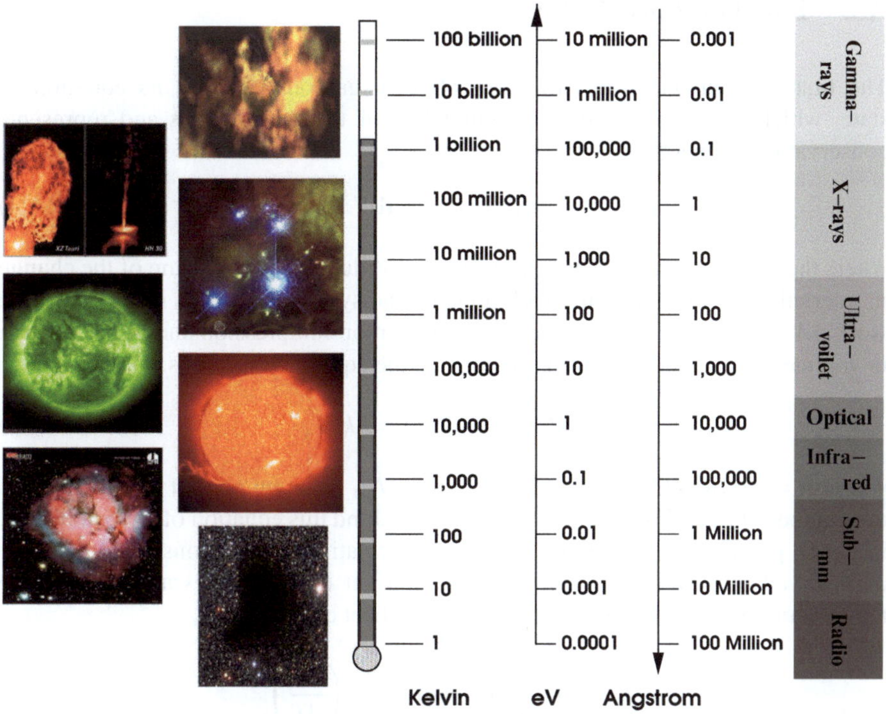

Fig. A.1 The temperature scale in the Universe spans over ten orders of magnitude ranging from the coldest cores of molecular clouds to hot vicinities of black holes. The scale highlights the range to be found in early stellar evolution, which roughly spans from 10 K (e.g., Barnard 68) to 100 MK (i.e. outburst and jet in XZ Tauri and HH30, respectively). Examples of temperatures in-between are ionized hydrogen clouds (e.g., NGC 5146, Cocoon Nebula), the surface temperatures of stars (e.g., our Sun in visible light), plasma temperatures in stellar coronal loops (e.g., our Sun in UV light), magnetized stars (i.e., hot massive stars at the core of the Orion Nebula). The hottest temperatures are usually found at later stages of stellar evolution in supernovas or the vicinity of degenerate matter (e.g., magnetars). Credits for insets: NASA/ESA/ISAS; R. Mallozzi, Burrows et al. [387], Bally et al. [1159], Schulz et al. [532]

For young stars the highest temperatures (of the order of 100 MK) observed occur during giant X-ray flares that usually last for hours or sometimes a few days, and jets. The coolest places (with 1 to 100 K) are molecular clouds. Ionized giant hydrogen clouds usually have temperatures around 1,000 K. The temperatures of stellar photospheres range between 3,000 and 50,000 K. In stellar coronae the plasma reaches 10 MK, almost as high as in stellar cores where nuclear fusion requires temperatures of about 15 MK. The temperature range involved in stellar formation and evolution thus spans many orders of magnitudes. In stellar physics the high temperature is only topped by temperatures of shocks in the early phases of a supernova, the death of a massive star, or when in the vicinity of gravitational powerhouses like neutron stars and black holes.

A.2 The Adiabatic Index

The first relation one wants to know about a gaseous cloud is its equation of state, which is solely based on the first law of thermodynamics and represents conservation of energy:

$$dQ = dU + dW \tag{A.4}$$

where the total amount of energy absorbed or produced dQ is the sum of the change in internal energy dU and the work done by the system dW. In an ideal gas where work $dW = PdV$ directly relates to expansion or compression and thus a change in volume dV against a uniform pressure P, the equation of state is:

$$PV = N_m RT = nkT \tag{A.5}$$

where $R = 8.3143435 \times 10^7$ erg mole^{-1} K^{-1}, N_m is the number of moles, and n is the number of particles per cm^3. The physics behind this equation of state, however, is better perceived by looking at various derivatives under constant conditions of involved quantities. For example, the amount of heat necessary to raise the temperature by one degree is expressed by the heat capacities:

$$C_v = \left(\frac{dQ}{dT}\right)_V \text{ and } C_p = \left(\frac{dQ}{dT}\right)_P \tag{A.6}$$

where d/dT denotes the differentiation with respect to temperature and the indices P and V indicate constant pressure or volume. The ratio of the two heat capacities:

$$\gamma = C_P/C_V \tag{A.7}$$

is called the *adiabatic index* and has a value of 5/3 or 7/5 depending on whether the gas is monatomic or diatomic. For polyatomic gases the ratio would be near 4/3 (i.e., if the gas contains significant amounts of elements other than H and He or a mix of atoms, molecules, and ions). The more internal degrees of freedom to store energy that exist, the more C_P is reduced, allowing the index to approach unity. A mix of neutral hydrogen with a fraction of ionized hydrogen is in this respect no longer strictly monatomic, because energy exchange between neutrals and ions is different. Similarly, mixes of H and He and their ions are to be treated as polyatomic if the ionization fractions are large. Deviations from the ideal gas assumption scale with n^2/V^2, which, however, in all phases of stellar formation is a very small number. Thus the ideal gas assumption is quite valid throughout stellar evolution.

A very important aspect with respect to idealized gas clouds is the case in which the radiated heat is small. For many gas clouds it is a good approximation to assume that no heat is exchanged with its surroundings. The changes in P, T, and V in the *adiabatic* case are then:

$$PV^{\gamma} = const; \quad TV^{\gamma-1} = const; \quad \text{and} \quad TP^{(1-\gamma)/\gamma)} = const \tag{A.8}$$

Together with equations A.4 and A.5 these relations lead to a set of three adiabatic exponents by requiring that $\mathrm{d}Q = 0$. The importance of these exponents had been realized by Eddington in 1918 and Chandrasekhar in 1939 [1291]. They are defined as:

$$\Gamma_1 = -\left(\frac{\mathrm{d}\ln P}{\mathrm{d}\ln V}\right)_{ad} = \left(\frac{\mathrm{d}\ln P}{\mathrm{d}\ln \rho}\right)_{ad} \tag{A.9}$$

$$\frac{\Gamma_2}{(\Gamma_2 - 1)} = \left(\frac{\mathrm{d}\ln P}{\mathrm{d}\ln T}\right)_{ad} \tag{A.10}$$

and

$$\Gamma_3 - 1 = -\left(\frac{\mathrm{d}\ln T}{\mathrm{d}\ln V}\right)_{ad} \tag{A.11}$$

In the classical (non-adiabatic) limit for a monatomic gas with no internal degrees of freedom, the three exponents are the same and equal to 5/3. For a typical non-interacting gas the specific internal energy U (internal energy per unit mass) is proportional to P/ρ (where ρ is the mass density). Thus pressure of such a system is related to density by:

$$P = K\rho^{\gamma} \tag{A.12}$$

where K is a proportionality factor determined by the gas considered and γ is generally Γ_1 . These exponents carry crucial information about the stability of the system against various types of perturbations. In the case of an isothermal cloud, it is primarily the first adiabatic exponent that defines the stability of a gas against external forces (such as gravity), and changes in Γ_2 and Γ_3 are negligible. The latter only have more significance once convection occurs and if processes are strongly non-isothermal.

A.3 Polytropes

Stellar interiors are frequently characterized by *polytropic* processes, where the adiabatic condition of $\mathrm{d}Q = 0$ is now substituted by a constant thermal capacity:

$$C = \frac{\mathrm{d}Q}{\mathrm{d}T} = const. > 0 \quad \text{and} \quad \gamma' = \frac{C_p - C}{C_v - C} \tag{A.13}$$

With the (A.12) the polytropic equation of state has then the index γ'. The polytropic index is then defined as $n = 1/(\gamma' - 1)$.

A.4 Thermodynamic Equilibrium

For an ideal gas of temperature T of n particles of a certain kind there are various excited states. Atoms in an excited state i with an excitation energy χ_i distribute relative to their ground states o following the *Boltzmann formula*:

$$\frac{n_i}{n_o} = \frac{g_i}{g_o} \mathrm{e}^{-\chi_i/kT} \tag{A.14}$$

where g_i and g_o are the statistical weights describing the degeneracy of states. However, at larger temperatures ground states are increasingly depleted. In the extreme case that all ground states are depleted n_o is substituted by the total number of states n, and g_o by the *partition function* $g = \sum_i g_i \, \mathrm{e}^{\chi_i/kT}$. The temperature T is then the equilibrium temperature. If the gas consists of a mix of many atoms and molecules in various states of excitation, all particle species have the same temperature in *thermodynamic equilibrium.* However, such a global equilibrium may not be applicable in molecular clouds and stars where temperatures may depend on spatial coordinates. Here the concept of thermal equilibrium is still valid for small volume elements. This is then called *local thermodynamic equilibrium (LTE).*

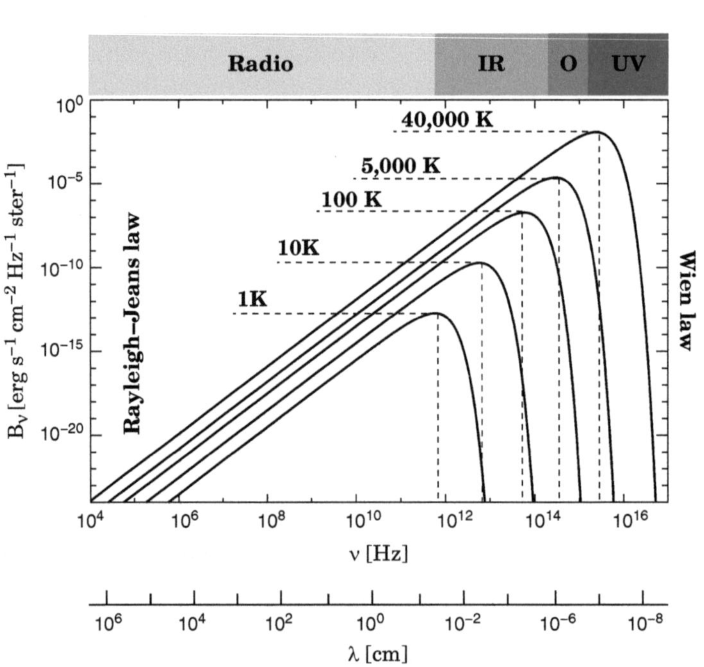

Fig. A.2 Spectra of blackbody radiation at various temperatures ranging from 1 to 40,000 K from the radio band to the UV band. Note that the lowest temperature spectra peak is at sub-mm wavelengths

In thermal equilibrium every process occurs at the same rate as its inverse process, meaning there is as much absorption of photons as there is emission. Under such conditions the intensity of the radiation field can be described as:

$$B_\nu(T) = \frac{2h\nu^3}{c^2}\frac{1}{e^{h\nu/kT} - 1} \tag{A.15}$$

It was G. Kirchhoff (1860) and M. Planck (1900) who realized that the intensity of this *blackbody* radiation is a universal function of T and ν. The energy of the photon is $h\nu$, where h is Planck's constant = 6.625 × 10^{27} erg s. The two limiting cases are *Wien's law* $\left(\frac{h\nu}{kT} \gg 1\right)$ and *Rayleigh–Jeans law* $\left(\frac{h\nu}{kT} \ll 1\right)$. The mean photon energy in local thermal equilibrium is equivalent to the temperature of the radiating body by:

$$\langle h\nu \rangle = 2.7012kT \tag{A.16}$$

The energy and wavelength scales in Figs. A.1 and A.2 are calculated through this equivalency. Integration of (A.15) yields the total radiation flux F of a blackbody radiator, which J. Stefan (1884 experimentally) and L. Boltzmann (1886 theoretically) determined as:

$$F = \int_0^\infty B_\nu(T)\,\mathrm{d}\nu = \sigma T^4 \tag{A.17}$$

where $\sigma = 5.67 \times 10^{-5}$ erg cm^{-2} s^{-1} K^{-4}.

A.5 Gravitational Potential and Mass Density

The biggest player in the process of stellar evolution is without doubt *gravity*. Just as simply as Newton's apple falls from the height of the tree towards the ground, gravity rules everything in our Universe that possesses mass. This applies to the apple, the orbits of the planets around the Sun, the stability of stars, and even to the light emitted by the farthest quasar at the edge of the Universe.

The source of the gravitational force is the gravitational potential, which is generally described by the *Poisson equation*:

$$\nabla^2\phi = 4\pi G\rho \tag{A.18}$$

where ϕ is the gravitational potential. This leads to a contribution to the external force $\mathbf{F_{ext}}$ on a cloud of gas of density ρ:

$$\mathbf{F_{ext}} = -\rho\nabla\phi \tag{A.19}$$

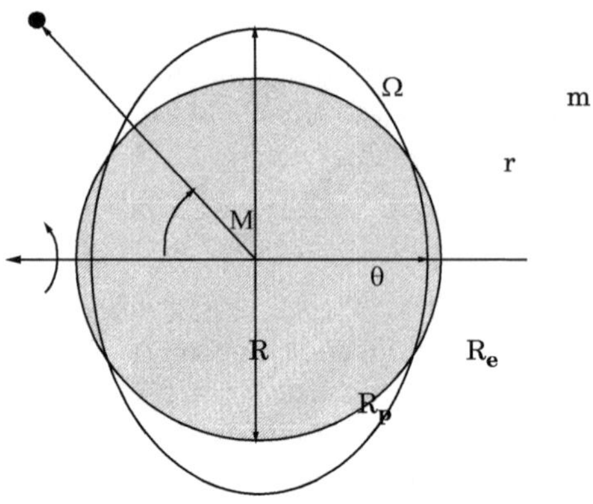

Fig. A.3 Geometry for an oblique rotator with obliqueness $(R_e - R_p)/R_e$

In the case of spherical symmetry, where mass is distributed such that the total gravitational potential only depends on the distance r towards the geometrical center of the mass distribution, ϕ can be expressed analytically as:

$$\phi(r) = -G\frac{M_r}{r} \tag{A.20}$$

where M_r is the enclosed mass within a sphere of radius r. In fact, spherical symmetry is the only case where an exact analytical evaluation of the gravitational potential is possible. Thus the situation changes dramatically once there are deviations from spherical symmetry. Imagine an isothermal gas cloud with no forces acting other than internal pressure and the gravitational force. Once it rotates it redistributes itself into a more oblique shape breaking the symmetry (see Fig. A.3). This means that the gravitational potential now maintains a cylindrical symmetry and an azimuthal angle θ dependence is added. In the case of slow rotation, the gravitational potential can be expanded as an infinite series of the form:

$$\phi(r,\theta) = -G\frac{M}{r}\left(1 - \left(\frac{R_e}{r}\right)^2 \mathcal{J}_2\mathcal{P}_2(\cos\,\theta) - \left(\frac{R_e}{r}\right)^4 \mathcal{J}_4\mathcal{P}_4(\cos\,\theta) - \ldots\right) \tag{A.21}$$

where $\mathcal{P}_n$ and $\mathcal{J}_n$ are *Legendre Polynomials* and *gravitational moments*, respectively, R_e the large radius of the oblique body; and θ the angle with respect to the axis of rotation (see Fig. A.3). While Legendre Polynomials are mathematical functions, the gravitational moments carry the physics of the corrections. In this respect $\mathcal{J}_2$ is related to the clouds moment of inertia, and $\mathcal{J}_4$ is sensitive to deviations from and changes in mass distribution.

In fact, mass distribution within the gas spheroid is critical. While the above potentials assume a roughly homogeneous density distribution, this is usually not the case in molecular clouds and stellar collapse situations where one assumes a power law dependence like:

$$\rho(r) = \rho_0 \left(\frac{r_0}{r} \right)^{\alpha} \tag{A.22}$$

where the density drops off as some power α of the radius from a homogeneous sphere inside. In the case of molecular cores it is generally assumed that $\alpha = 2$ throughout the cloud.

A.6 Conservation Laws

So far everything that has been considered applies to static systems. Realistically clouds and cores change with time. The concepts developed in the following, however, still rely on the assumption that most matter considered is gaseous and particle interactions are limited to direct *collisions*. Additional electrostatic interactions become important in plasmas (i.e., ionized matter) and will be discussed later.

The first basic law of gas dynamics to be considered is the *equation of continuity*, which describes the conservation of mass in a non-interacting flow of velocity $\mathbf{v}$. For a gas element at a fixed spatial location this is:

$$\frac{\partial \rho}{\partial t} - \nabla(\rho \mathbf{v}) = 0 \tag{A.23}$$

The second law is the equation of motion, which is based on *momentum conservation*. Gradients in the gas pressure imply forces acting on the gas element. A continuous flow of gas elements is generally described by *Euler's Equation*:

$$\rho \frac{\partial \mathbf{v}}{\partial t} + \rho \mathbf{v} \cdot \nabla \mathbf{v} = -\nabla P + \mathbf{F} \tag{A.24}$$

where the first term is mass acceleration, the second term is advection, the third term the pressure gradient, and $\mathbf{F}$ the sum of all external forces. Assuming that the second term is zero with no pressure gradient and that gravity is the only external force, this equation reduces to:

$$\frac{\mathrm{d}^2 r}{\mathrm{d}t^2} = -\frac{GM_r}{r^2} = -\frac{4}{3}\pi G \rho r \tag{A.25}$$

assuming the uniform sphere ρ here is an average density defined by $M_r = (4/3)\pi r^3 \rho$. This is the equation of motion of a harmonic oscillator and allows one to define a dynamical time $T/4$ (T is the oscillation period), where a gas element travels halfway across the gas sphere (i.e., synonymous to the situation of a collapsing sphere):

$$t_{dyn} = \sqrt{\frac{3\pi}{32G\rho}} \tag{A.26}$$

Note that this timescale is independent of r and resembles the scale defined in Sect. A.9. Other sources for external forces can be magnetic fields and rotation.

The third dynamic equation is the one for *conservation of energy*. A gas element carries kinetic energy as well as internal energy. The latter critically depends on the available number of degrees of freedom in the gas, in other words, equipartition assigns a mean energy of $\eta_i = 1/2kT$ for each degree of freedom i and, thus, the specific internal energy (i.e., internal energy per volume) of a monatomic gas (three degrees of freedom) is given by:

$$U = \tfrac{1}{2}\rho v^2 + \tfrac{3}{2}\rho kT \tag{A.27}$$

The energy equation for an isothermal sphere is then:

$$\frac{\partial U}{\partial t} + \nabla[(U + P)\mathbf{v}] = \mathbf{F}\,\mathbf{v} - \nabla F_{rad} \tag{A.28}$$

where the first term is the change in specific internal energy, the second term the total work performed during either expansion or contraction of the system, the third term the rate of energies provided by external forces, and the fourth term the loss of energy due to the irradiated flux F_{rad}. In general, there would be a fifth term describing the energy flux due to the heat conductivity in the gas. This flux, however, is near zero as long as there are roughly isothermal conditions and low ionization fractions. It has to be realized that the complexity of this equation is not only due to the many contributing terms but also to the fact that each of these terms is sensitive to the composition and state of the gas. The specific internal energy depends on the number of degrees of freedom of the gas, the irradiated flux on the opacity of the gas and the energy flux exhibited through external forces such as magnetic fields that depend on the ionization fraction of the gas. In particular, the irradiated radiation flux breaks the symmetry of the three dynamic equations in that energy is permanently lost from the system and, more importantly, a fourth equation is necessary to account for its amount.

A.7 Hydrostatic Equilibrium

The simple balance between *internal (thermal) pressure* and gravitational pressure forces is called *hydrostatic equilibrium*. In this state it is assumed that there are no macroscopic motions or, in other words, motion on extremely slow timescales. In this case and using $dM_r/dr = 4\pi r^2\rho$, (A.24) reduces to:

$$\frac{\mathrm{d}P}{\mathrm{d}M_r} = -\frac{GM_r}{4\pi r^4} \tag{A.29}$$

The mass M_r again is the enclosed mass inside a sphere of radius r. Within this gas sphere pressure is maximal inside and decreases outwards. Multiplying this equation by the volume of the sphere and integrating over the enclosed mass one gets a relation between the gravitational potential energy of the star:

$$\phi = -\int_0^{M_r} \frac{GM_r'}{r}\,\mathrm{d}M_r' \tag{A.30}$$

and its total energy (see (A.27))

$$U = \frac{3}{2}\int_0^{M} \frac{P}{\rho}\,\mathrm{d}M_r = -\frac{1}{2}\phi \tag{A.31}$$

also called the *virial theorem*. The internal energy of the system can be one half of the configuration's gravitational energy.

Typically, interstellar and molecular clouds are found to be mostly in hydrostatic equilibrium. Furthermore, many calculations assume or require a form of hydrostatic equilibrium for newly formed protostellar cores as well.

A.8 The Speed of Sound

An important measure of the dynamic properties of a gas flow is the speed at which sound waves can propagate through the gas. This speed can be evaluated considering small density and pressure perturbations subject to the hydrostatic equilibrium condition [549]. These perturbations may either occur under isothermal or adiabatic conditions (i.e., with an adiabatic exponent of either 1 or 5/3, respectively). Euler's equation (A.24) and the equation of continuity (A.23) yields a wave equation:

$$\frac{\partial^2\rho}{\partial t^2} = c_s^2\nabla^2\rho \tag{A.32}$$

where $c_s = (\gamma\,\mathrm{d}P/\mathrm{d}\rho)^{1/2}$ is the speed of sound (P and ρ are measured at equilibrium). For a monatomic gas (i.e., the particle density $n = \rho/\mu m_H$, where μ is the atomic weight and m_H is the hydrogen mass) the isothermal case results in:

$$c_s = \left(\gamma\frac{kT}{\mu m_H}\right)^{1/2} \tag{A.33}$$

For a mean thermal speed in an ideal gas one has:

$$c_s \simeq 0.19(T/[10\ \mathrm{K}])^{1/2}\ \mathrm{km\ s^{-1}} \tag{A.34}$$

Thus if a density (or pressure) wave travels faster than c_s, the gas flow is called *supersonic* and the gas does not have enough time to respond to local changes. In this case pressure gradients have little or no effect on the flow. If the wave travels more slowly than c_s, then the gas flow has time to adjust to local changes and remains in hydrostatic equilibrium.

Most astrophysical plasmas are dynamic and involve magnetic fields. Interactions are then described by invoking fluid dynamics. Often the ideal assumption of infinite conductivity (i.e., (B.14) is equal to zero), is made which makes it possible to determine the propagation of magnetic disturbances in the plasma (see Sect. B.4). In analogy to the speed of sound, H. Alfvén in 1942 defined the *Alfvén velocity* as:

$$v_A = \frac{B}{(4\pi\rho)^{1/2}} \tag{A.35}$$

Whenever magnetic forces dominate, perturbations travel with velocities faster than the speed of sound.

A.9 Timescales

Gas dynamics in stellar evolution is governed by several major timescales. Specifically relevant for formation and early stellar evolution are the following three times:

Free-fall time: During the early phases of collapse, matter falls inward under free-fall conditions as there is nothing to counter the increasing gravitational pull. This reduces A.25 to:

$$\frac{\mathrm{d}^2 r}{\mathrm{d}t^2} = -\frac{GM_r}{r^2} \tag{A.36}$$

Solving this equation [76] leads to a free-fall time of:

$$t_{ff} = \sqrt{\frac{3\pi}{32G\overline{\rho}}} \sim 2.1 \times 10^3 \sqrt{\frac{\mathrm{g\ cm^{-3}}}{\overline{\rho}}}\ \mathrm{s} \tag{A.37}$$

where $\overline{\rho}$ is the initial mean density of the collapsing cloud. This is the mean time in which the cloud collapses entirely. For an initial density of 10^{-19} g cm^{-3} this would take ~200,000 yr.

Thermal time: The free-fall phase halts around a matter density of 10^{-13} g cm^{-3} as internal pressure builds up. This changes the timescale. Once the first stable core is sustained by thermal pressure, a thermal or Kelvin–Helmholtz timescale can be defined for a core of radius R as

$$t_{KH} = \frac{|W|}{L_R} \sim 7 \times 10^{-5} \kappa_R \frac{M_R^2}{R^3 T^4} \text{ s} \tag{A.38}$$

i.e., by equating the energy of the gravitational contraction to the radiated energy. W is the gravitational energy $\left(\frac{GM^2}{R}\right)$, L_R the luminosity across the core surface, and κ_R the mean opacity (see Appendix C). The quasi-static protostars thermally adjust to gravitation on this timescale. Assuming an average opacity of stellar material with solar composition of $\sim$1.2 cm^2 g^{-1} one finds that a star like the Sun requires about 3×10^7 yr to contract towards the main sequence. Thus, the thermal time exceeds the free-fall time by orders of magnitude. *Accretion time*: Such a thermal adjustment only happens if t_{KH} is significantly smaller than t_{acc}, which is defined by the relation:

$$t_{acc} = \frac{M_{core}}{\dot{M}} \tag{A.39}$$

and which reflects the situation where accretion of matter does not dominate the evolution of the core. Like t_{KH}, t_{acc} significantly exceeds t_{ff}. If t_{KH} is larger than t_{acc}, then the core evolves adiabatically and the luminosity of the protostar is dominated by accretion shocks. More details are discussed in Chaps. 5 and 6.

A.10 Spherically Symmetric Accretion

The three main equations for gas dynamics, mass continuity, momentum, and energy conservation, are sufficient under the assumption that there are no energy losses due to radiation and that there is no heat conduction. Most of the time gas flows are assumed to be *steady*, which implies that:

$$P\rho^{\gamma} = const. \tag{A.40}$$

The case of a mass M accreting spherically from a large gas cloud is considered (rotation, magnetic fields and bulk motions of the gas are neglected). It is now necessary to define conditions, such as density and temperature of an ambient gas, far away from that mass as well as boundary conditions at the surface of the mass. The notion of spherical symmetry relieves the treatment of a dependence on azimuthal and circumferential angles [1292]. The velocity of infalling material is assumed negative ($v_r < 0$, it would be >0 in case of an outflowing wind) and has only a radial component. For a steady radial flow the three equations reduce to:

$$\frac{1}{r^2}\frac{\mathrm{d}}{\mathrm{d}r}(r^2 \rho v) = 0$$

$$v\frac{\mathrm{d}v}{\mathrm{d}r} + \frac{1}{\rho}\frac{\mathrm{d}P}{\mathrm{d}r} + \frac{GM}{r^2} = 0 \tag{A.41}$$
$$P = K\rho^{\gamma}$$

where gravity is the only external force and the energy equation is substituted by the equation of state for a polytrope. Integrating the momentum equation and using the definition of the sound speed yields the Bernoulli integral:

$$\frac{v^2}{2} + \frac{c_s^2}{\gamma - 1} - \frac{GM}{r} = const. \tag{A.42}$$

Note that this integration is not mathematically valid for $\gamma = 1$, the strict isothermal case. Here the integral has to be evaluated logarithmically. This however does not change the physical content of this integral. There is a critical radius within which the gas flow changes from subsonic to supersonic. This is called the *sonic radius* [549]:

$$r_s = \frac{GM}{2c_s^2(r_s)} \simeq 7.5 \times 10^{13}\left(\frac{T(r_s)}{[10^4\ \mathrm{K}]}\right)^{-1}\left(\frac{M}{[M_\odot]}\right)\ \mathrm{cm} \tag{A.43}$$

For a protostellar core accreting from a 1 $M_\odot$ molecular cloud of 10 K temperature this radius would be about 7.5×10^{10} cm. Below this radius the gas flow becomes increasingly supersonic and effectively free falling. In terms of a cloud size of 0.1 pc this means that throughout most of the cloud the gas flow will stay subsonic.

The above equations also allow one to derive an accretion rate from conditions at the outer boundary of a molecular cloud [549]. This derivation leads to:

$$\dot{M} \simeq 1.4 \times 10^{11}\left(\frac{M}{[M_\odot]}\right)^2\left(\frac{\rho(\infty)}{[10^{-24}\mathrm{g\ cm^{-3}}]}\right)\left(\frac{c_s(\infty)}{[10\ \mathrm{km\ s^{-1}}]}\right)^{-3}\ \mathrm{g\ s^{-1}} \tag{A.44}$$

For the case described above and typical values for $\rho(\infty)$ (10^{-20} g cm^{-3}) and $c_s(\infty)$ (0.35 km s^{-1}) this yields an accretion rate of the order of 10^{-7} $M_\odot$ yr^{-1}. Note that in the spherical Bondi case the mass accretion rate depends on M^2, whereas for disk accretion (see Chap. 8) it is independent of mass.

A.11 Rotation

Rotation has a profound effect on the stability of an ideal gas cloud. Since *angular momentum* remains conserved, any cloud will rotate faster as it collapses and centrifugal forces will eventually balance and even surpass gravity everywhere. Thus the cloud's certain fate is dispersion into the interstellar medium. To investigate the equation of motion of particles in a uniformly rotating cloud it is convenient to

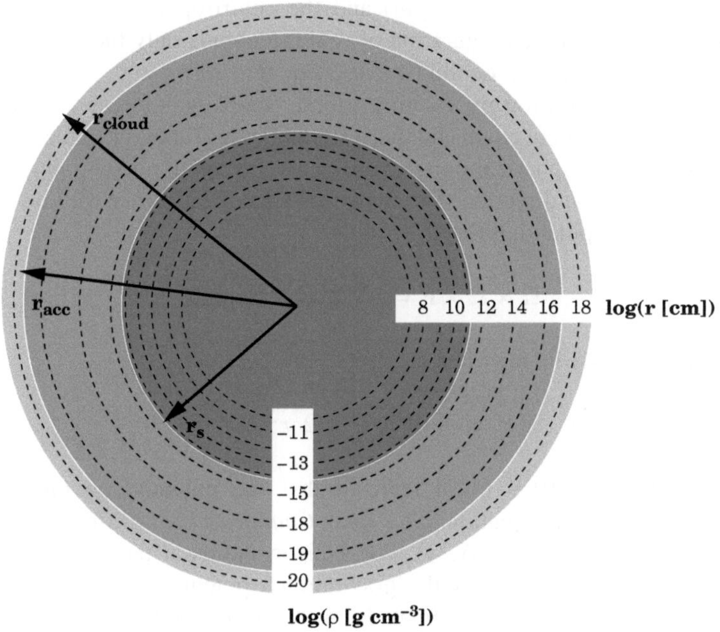

Fig. A.4 During spherical accretion a central star draws matter from an ambient cloud or atmosphere. Such a basic scenario may apply once an isothermal sphere of gas (like a molecular cloud core) collapses and a small core accretes from the outer envelope. The dark inner region marks the area of a free-falling envelope with supersonic infall velocities and a density $\rho \propto r^{-3/2}$. The lighter shaded outer envelopes are regions where infall happens at subsonic velocities in a more static envelope of density $\rho \propto r^{-2}$. The difference between the light shades indicates envelopes which are under the influence of gravity (*darker shade*) or not (*lightest shade*). Note the logarithmic radius and density scales in the diagram

operate in a frame moving with the rotating cloud implying that the initial velocity of the cloud element is zero. The operator for the rate of change in the inertial frame to the change as measured in the rotating frame is:

$$\frac{D\mathbf{u}}{Dt} = \left(\frac{\mathrm{d}\mathbf{u}}{\mathrm{d}t} + \Omega \times \mathbf{r}\right) \tag{A.45}$$

where $\mathbf{u} = \mathbf{u}_{\mathrm{rot}} + \Omega \times \mathbf{r}$ and Ω is the angular velocity of the rotating frame. Applying rotation to the equation of hydrostatic equilibrium one finds for the equation of motion in the rotating frame:

$$\frac{\mathrm{d}\mathbf{v}}{\mathrm{d}t} = -\frac{1}{\rho}\nabla P - \nabla\phi - 2\Omega \times \mathbf{v} - \Omega \times (\Omega \times \mathbf{r}) \tag{A.46}$$

where Ω is the angular velocity of the rotating frame. Most of the terms in the equation are familiar. There are two new terms due to rotation: the Coriolis acceleration term (second from right) and the centrifugal acceleration term. To describe the equilibrium configuration of rotating clouds only the centrifugal term is of interest since ideally in equilibrium **u** is equal to zero.

In a *slowly* and uniformly rotating cloud centrifugal forces will break the spherical symmetry of the cloud and the system may find another stable equilibrium configuration. Using spherical polar coordinates the rotation axis points along the unit vector in the polar direction. The centrifugal acceleration can then be expressed as the gradient of a potential; that is:

$$\Omega \times \Omega \times \mathbf{r} = -\nabla\left(\tfrac{1}{2}\Omega^2 r^2 \sin^2\,\theta\right) \tag{A.47}$$

and (A.46) can finally be rewritten as:

$$\nabla P = -\rho\nabla\left(\phi - \tfrac{1}{2}\Omega^2 r^2\,\sin^2\,\theta\right) \tag{A.48}$$

Chandrasekhar in 1969 [322] realized that the potential on the right-hand side satisfies Poisson's equation. The final *Chandrasekhar–Milne expansion* for a distorted star is shown in (A.21). For a given radius at the poles ($\theta = 0, \pi$) the effective potential is simply the gravitational potential; at the equator ($\theta = \pi/2$) the centrifugal pull is maximal. Remarkable in (A.48) is the fact that the gravitational potential is reduced by the effect of rotation with the square of the angular velocity. For typical molecular cloud average densities of 10^{-20} g cm^{-3} this means that rotational velocities cannot exceed 10^{-14} cm s^{-1} by much because then the centrifugal force would outweigh gravity.

The case of the slowly and uniformly rotating cloud considered above represents a solid body motion and may not be directly applicable for molecular clouds. Although the case is therefore a bit academic it still exhibits valid insights into the effects of cloud rotation. For the case of gravitational collapse of a molecular cloud, angular velocity has to increase as radius decreases. The collapse soon will come to a halt as centrifugal forces surpass gravity. One way out of this problem is to transport angular momentum out of the system; another is to break the cloud up into fragments, thus distributing some of the momentum. In the simple ideal gas cloud configuration considered this is not possible unless one reconsiders the assumptions that all particles in the cloud are neutral, collisions between particles are purely elastic and no external fields are involved. The final sections of Chap. 4 and much of Chap. 5 deal with this problem.

A.12 Ionized Matter

So far the description of gaseous matter has been based entirely on the assumption that there are no fractions of ion species and no interactions stemming from the fact that matter elements carry a net charge. In reality, there is hardly such an entity as a gas cloud that consists entirely of neutral particles. Within the Galaxy there is always the *interstellar radiation field* as well as *Cosmic Rays*. In contrast, the *intergalactic medium* outside the Galaxy is considered to be entirely ionized. Thus, gas and molecular clouds within the Galaxy always carry *non-zero* ionization fractions. As long as clouds remain electrically neutral over a large physical extent they can stably exist as a *plasma cloud*. Clouds have to be neutral as a whole, since electrostatic forces attract opposite charges and neutralize the cloud. Gaseous matter thus consists not only of neutral atoms and molecules but also of ions, radicals and free electrons. Similarly important with respect to the distribution of these ions are the corresponding electrons. The properties of a plasma are determined by the sum of the properties of its constituents. The density of a plasma is given by:

$$\rho = \sum_{k=i,e,n} \rho_k = \sum_{k=i,e,n} n_k m_k \tag{A.49}$$

where the subscripts i, e and n refer to various ions, all electrons, and various neutral particles. The kinetic energy of the plasma in thermal equilibrium is:

$$\frac{3}{2} kT = \frac{1}{2} \sum_{k=i,e,n} m_k \langle v_k^2 \rangle \tag{A.50}$$

where v_k is the mean square velocity of each constituent. Collisions between different particles ensure that the mean energies of all particles are the same. The velocity distribution $f(v)$ of each plasma constituent is Maxwellian:

$$f(v)\,\mathrm{d}v = 4\pi \left[\left(\frac{m}{2\pi kT}\right]\right)^{3/2} v^2\, \mathrm{e}^{-mv^2/kT}\,\mathrm{d}v \tag{A.51}$$

The peak of this distribution is then $v_{\text{peak}} = \sqrt{\frac{2kT}{m}}$.

A.13 Thermal Ionization

Section A.4 dealt with thermal excitation and (A.14) expressed the distribution of excited states relative to the ground state. If collisions transfer energies E larger than a specific *ionization energy* χ_{ion} the atom will become ionized. The electrons then have a kinetic energy $E_e = E - \chi_{ion}$. The *Saha equation* specifies the fraction of ionized atoms with respect to neutral atoms:

$$\frac{n_i}{n_n} = \frac{G_i}{G_n} \frac{2}{n_e} \frac{(2\pi m_e kT)^{3/2}}{h^3}\, \mathrm{e}^{-\chi_{ion}/kT} \tag{A.52}$$

where $\frac{G_i}{G_o}$ is the partition function ratio between ionized and neutral atoms. This equation was first developed by M. N. Saha in 1920. The important issue in (A.52) is that the ionization fraction is a function of temperature only, as all other ingredients are ionization properties of the gas. The necessary partition functions for all major elements of interest are tabulated [54] via the *temperature measure* $\Theta = \frac{5040}{T[K]}$. In the case that all atoms are at least partially ionized, (A.52) is still valid to determine the fraction between two ionization states. Implicitly, there is also a dependence on electron pressure $P_e = n_e kT$, and thus (A.52) can be expressed numerically as:

$$\frac{n_i}{n_n} = \frac{\Phi(T)}{P_e},$$
$$\text{where } \Phi(T) = 1.2020 \times 10^9 \frac{G_i}{G_n} T^{5/2} 10^{-\Theta\chi_{ion}} \quad \text{(A.53)}$$

Today's spectral analysis is routine and uses tabulated values for partition functions and ionization potentials [54].

To thermally ionize a gas cloud requires high collision rates, such as those in stellar atmospheres at temperatures ranging from 3,500 to 40,000 K. For example, at an electron pressure of 10 Pa in the atmospheres of main sequence stars, matter is completely ionized at temperatures above 10,000 K and the electron pressure is of the order of the gas pressure since $n_p \sim n_e$. Note that the relation between gas pressure and electron pressure depends on the metallicity of the gas cloud. At low temperatures most electrons come from elements with first ionization potentials of less than 10 eV. Applied to a cold gas cloud of less than 100 K it is clear that thermal ionization hardly contributes to the ionization fraction and electron pressure.

A.14 Ionization Balance

At the low temperatures of molecular clouds, the effect of thermal ionization is small compared with photoionization. Gaseous matter exposed to external and internal radiation fields is heated. The temperature in a static ionized gas cloud is determined by the balance between heating through photoionization events and cooling through successive recombination. Clouds lose energy through radiation which has to be accounted for in the energy balance. Absorption of the radiation field produces a population of free electrons which are rapidly thermalized. The mean energy of a photoelectron does not depend on the strength of the incident radiation field but on its shape. For example, if the radiation field does not provide photons of energies 13.6 eV (912 Å) and higher, then a hydrogen cloud is unlikely to be effectively ionized. The rate of creation of photoelectrons critically depends on the absolute strength of the field and the ability of the gas to recombine. To establish the energy balance of a gas cloud and its ambient radiation field one has to consider the heating rate through photoionization R_{photo} against cooling through recombination R_{re}, free–free (bremsstrahlung) radiation R_{ff} and collisionally excited line radiation R_{col}. Usually, an effective heating rate R_{eff} is defined as:

$$R_{eff} = R_{photo} - R_{re} = R_{ff} + R_{col} \tag{A.54}$$

where it should be noted that R_{photo} and R_{re} depend on the number densities of electrons and ions, and, as a valid approximation, elements heavier than He may be omitted in these rates.

Appendix B
Magnetic Fields and Plasmas

Some aspects concerning the modern treatment of stellar magnetic fields are revisited here in greater detail. Magnetic fields and their interactions with matter are very crucial elements in the study of the formation and early evolution of stars. This appendix highlights the interaction of magnetic fields in a much more fundamental way than presented in the previous chapters and much of what is presented has a wide range of applications, likely much beyond the scope of this book. The material is collected from a wide variety of publications, though only in rare cases will references be given. The appendix makes heavy use of material presented in the books by F. H. Shu: *Gas Dynamics, Vol. II* [1293], and E. Priest and T. Forbes: *Magnetic Reconnection* [1294]. In this respect it should be noted that the following is merely to support the reader to understand certain subtleties in this book. For a full recourse to magnetism and its interaction with matter the reader should consult the primary literature.

B.1 Magnetohydrodynamics

The study of the global properties of plasmas in a magnetic fields is called MHD. The most basic equations are *Maxwell's equations* (after J. Maxwell 1872):

$$\nabla \cdot \mathbf{E} = 4\pi q_e \tag{B.1}$$

$$\nabla \times \mathbf{E} = -\frac{1}{c}\frac{\partial \mathbf{B}}{\partial t} \tag{B.2}$$

$$\nabla \cdot \mathbf{B} = 0 \tag{B.3}$$

$$\nabla \times \mathbf{B} = \frac{4\pi}{c}\mathbf{j} + \frac{1}{c}\frac{\partial \mathbf{E}}{\partial t}, \tag{B.4}$$

Ihich, by ignoring the displacement current $\frac{1}{c}\frac{\partial \mathbf{E}}{\partial t}$, contain the laws of the *conservation of charge* (B.1), *Faraday's equation* (B.2), *Gauss's law* (B.3), and *Ampere's*

N.S. Schulz, *The Formation and Early Evolution of Stars*, Astronomy and Astrophysics Library, DOI 10.1007/978-3-642-23926-7, © Springer-Verlag Berlin Heidelberg 2012

law (B.4), respectively. In order to include magnetic and electric fields in fluid mechanics one has to re-formulate some of the conservation laws as stated in Sect. A.6. In this respect, mass conservation remains unchanged but all the other equations have to be modified.

The *momentum equation* of a MHD fluid element then reads:

$$\rho\frac{\partial \mathbf{v}}{\partial t} + \rho(\mathbf{v}\cdot\nabla)\mathbf{v} = -\nabla P + \frac{1}{c}(\mathbf{j}\times\mathbf{B}) + \nabla\mathbf{S} + \mathbf{F_g} \tag{B.5}$$

This equation shows a few important modifications with respect to the stress-free Euler equation (A.24). For the magnetic field it includes the magnetic force in the form of the vector product of the current density **J** and the magnetic field strength **B** and, besides identifying the gravitational force $\mathbf{F_g}$ as external force, considers the stress term $\nabla\mathbf{S}$. By invoking Ampere's law one can see that this magnetic force consists of a magnetic pressure force and a magnetic tension force:

$$\frac{1}{c}\mathbf{j}\times\mathbf{B} = -\frac{\mathbf{1}}{\mathbf{8\pi}}\nabla\mathbf{B^2} + \frac{\mathbf{1}}{\mathbf{4\pi}}(\mathbf{B}\cdot\nabla)\mathbf{B} \tag{B.6}$$

MHD fluids are usually not stress-free and instead of Equation A.24 one has to consider the general *Navier–Stokes equation* for a viscous fluid element. Magnetic fields under these conditions also permeate into the stress tensor, which, depending on the field strength and plasma conditions is an expression of considerable complexity. For a magnetic field strength of 50 G, a plasma density of 3×10^9 cm^{-3} and a proton temperature of 2.5×10^6 K, for example, the stress tensor **S** can be expressed in component form as [1295, 1296]:

$$S_{ij} = 3\eta_v\left(\frac{\delta_{ij}}{3} - \frac{B_iB_j}{B^2}\right)\left(\frac{\mathbf{B}\cdot\mathbf{B}\nabla\mathbf{v}}{B^2} - \frac{\nabla\mathbf{v}}{3}\right) \tag{B.7}$$

Here $\eta_v = 10^{-16}\,T^{5/2}$ g cm^{-1} s^{-1} is a viscosity coefficient and δ_{ij} is the Kronecker delta function.

Similarly complex is the Ansatz for the *energy equation* (A.28), which now reads like:

$$\frac{\partial U}{\partial t} + \nabla[(U+P)\mathbf{v}] = \nabla(\kappa_\mathbf{c}\nabla\mathbf{T}) + (\eta_\mathbf{e}\mathbf{j})\cdot\mathbf{j} + Q_\nu - Q_{rad} \tag{B.8}$$

Again the left-hand side is the change in specific energy plus the work performed by the system either by expansion or contraction. The right-hand side in (A.28) was kept rather general. In the MHD case of (B.8) these terms are now further identified. In this respect, $Q_{rad} = \nabla\mathbf{F_{rad}}$ is radiative energy loss, Q_ν is heating by viscous dissipation with tight relation to the stress tensor. The first term on the right-hand side is to describe thermal conductivity characterized by the thermal conductivity tensor $\kappa_\mathbf{c}$ and the temperature T at each location. The second term on the right is the magnetic energy.

Ultimately these equations represent a time-dependent set of equations which, together with the equations in Sect. A.6, define the MHD properties of a fluid (i.e., the equation of state, the density, and gas pressure) and provide a sufficient system to solve for the main variables: **v, B, j, E**, ρ, P, and T.

To solve for these quantities is only one aspect in dealing with MHD fluids. Similarly important is any knowledge about the actual electro-magnetic properties of the MHD fluid itself. This information is encapsulated in a few coefficients, which in uniform environments appear as scalars and in non-uniform environments as tensors. One is the which appears in the *induction equation* (see below), and which in its general form can be written as (by combining Faraday's, Ampere's and Ohm's laws):

$$\frac{\partial \mathbf{B}}{\partial t} = \nabla \times (\mathbf{v} \times \mathbf{B}) - \nabla \times \eta \nabla \times \mathbf{B} \tag{B.9}$$

Magnetic diffusivity (*electrical resistivity*) itself is defined in Table B.1. In case diffusivity is uniform, (B.9) reduces to:

$$\frac{\partial \mathbf{B}}{\partial t} = \nabla \times (\mathbf{v} \times \mathbf{B}) + \eta \nabla^2 \mathbf{B} \tag{B.10}$$

The first term on the right-hand side in Equation B.9 states that the magnetic flux within a closed circuit is constant. The second term on the right-hand side describes the diffusion of plasma particles across the magnetic field. If there is no diffusion, in which case this second term is zero, then the field is practically *frozen-in* to the plasma and moves with it. The diffusivity η can vary by quite large amounts depending temperature. In the Sun's corona η is about 0.5, whereas in the chromosphere is reaches up to 1,000.

The induction equation above is one of the most important relations in MHD as it allows us to derive the time evolution of **B** for any initial field configuration once **v** and η are known. Specifically, quantities like the *magnetic diffusivity* or the *electrical conductivity* are important proportionality constants that scale with and describe the field's response to the plasma. Heat production through Ohmic dissipation is described by these quantities and can be expressed as:

$$Q_o = \frac{|\mathbf{j}|^2}{\sigma} = \frac{\eta}{4\pi} |\nabla \times \mathbf{B}|^2. \tag{B.11}$$

In complex non-uniform configurations they have to be treated as tensor quantities. Table B.1 lists a few electrodynamic quantities ($\epsilon, \sigma, \mu, \eta$) and their value in terms of fundamental quantities.

For MHD flow properties one is primarily interested in solving for $\mathbf{v}, \mathbf{B}$, and P. For example, in most astrophysical applications plasmas are electrically neutral (see Sect. A.12) and from (B.6) one can deduce an expression for magnetic pressure depending only on knowledge of **B**:

Table B.1 Characterization of MHD plasmas

Symbol	Name	Form	Expression
ϵ	dielectric constant	scalar	$1 + (\pi\rho c^2/B^2)$
σ	electric conductivity	tensor	$n_e e^2/m_e \nu_c$ $\omega_{pe}^2/4\pi\nu_c$
μ	magnetic permeability	scalar	$(1 + (4\pi\rho\mathcal{E}_\perp/B^2))^{-1}$
η	magnetic diffusivity electrical resistivity	tensor	$c^2/4\pi\sigma$
Rm	magnetic Reynolds Number	scalar[a]	$v_o L_o/\eta$
P_m	magnetic Prandtl Number	scalar[a]	ν/η
L_u	Lundquist Number	scalar[a]	$v_A L_o/\eta$
$\mathcal{M}_A$	Alfvén Mach Number	scalar[a]	v_o/v_A
L_D	Debye length	scalar	$\left(\frac{kT}{4\pi e^2(n_e+Z_i^2 n_i)}\right)^{1/2}$
L_{De}	electron Debye length	scalar	$\left(\frac{kT}{4\pi e^2 n_e}\right)^{1/2}$
ω_{pe}	electron plasma frequency	scalar	$(4\pi n_e e^2/m_e)^{1/2}$
ω_L	Larmor frequency	scalar	$\frac{qB}{mc}$

a) dimensionless

$$P_{mag} = \frac{|\mathbf{B}|^2}{8\pi} \tag{B.12}$$

In the descriptions of non-magnetized hydrodynamics there exists an ensemble of characteristic and dimensionless numbers which qualitatively classify the physical regimes of a flow. Such are the *Reynolds number (Re)* (see (8.9)), the *Prandtl number* relating the coefficient of dynamic viscosity to the thermal conductivity, and the *Mach number* (see (3.12)). In magnetized plasmas they have their counterparts as listed in Table B.1. These numbers rely on the principle that MHD plasmas have characteristic speeds (v_o), viscosities (ν), conductivities (σ), and diffusivities (η). These numbers are based on a characteristic length-scale (L_o) and within this scale allow for approximations once certain regimes can be identified. For example an ideal MHD state is realized when $R_m \gg 1$ as here the characteristic speed of the flows dominates over magnetic diffusion. More detailed explanations can be found in [1294] and references therein.

B.2 Charged Particles in Magnetic Fields

There are a few basic properties of plasmas that are helpful to recall. One may consider a plasma composed of ions of charge $Z_i e$ mixed with electrons of charge $-e$ with overall charge neutrality (i.e., $Z_i n_i = n_e$). There exists a characteristic length-scale L_D by means of which, statistically through the attraction of electrons and the repulsion of ions, the Coulomb potential of any ion gets shielded such as:

$$\phi_C = \frac{Z_i e}{r} \exp(-r/L_D) \tag{B.13}$$

where L_D is the *Debye length* defined in Table B.1. It is sometimes useful to decouple the electron contribution and define a separate *electron Debye length*, which allows one to deduce a characteristic electron *plasma frequency* ω_{pe}. In other words, the electronic portion of the plasma can be described by the dynamics of a harmonic oscillator with a natural frequency of ω_{pe}.

The presence of magnetic fields in gas clouds with substantial ionization fractions has a profound effect on the dynamics of the plasma. The gas or plasma pressure is then expressed as the average momentum of gas particles $P = Nm\langle v^2_{ijk}\rangle$, where v_{ijk} and is now treated as a *tensor*. In the simple ideal gas case this pressure tensor is diagonal and since collisions dominate, all directions are equal and $\langle v^2_{ijk}\rangle = v^2$. In the case of a non-negligible magnetic pressure one needs to distinguish between pressure components across and along magnetic field lines. The magnetic field thus directs the charged particles within the three orthogonal spatial coordinates (x_i, x_j, x_k) and the pressure tensor possesses off-diagonal terms.

The motions of individual particles with charge q and mass m in a vector magnetic field $\mathbf{B}$ are described by the force:

$$\mathbf{F} = \frac{\mathrm{d}\mathbf{p}}{\mathrm{d}t} = q\mathbf{E} + \frac{q}{c}\mathbf{v} \times \mathbf{B} \tag{B.14}$$

where $\mathbf{p}$ is the momentum vector and $\mathbf{v}$ the velocity vector of the particle. Called the *Lorentz-force* it was first formulated by H. A. Lorentz in 1892. The particle then gyrates in a clockwise path when q is a negative charge, and counterclockwise when it is a positive charge. The gyro-radius in the non-relativistic case (also called *Larmor radius*) is given by:

$$r_L = \frac{mcv_\perp}{qB} = \frac{v_\perp}{\omega_L} \tag{B.15}$$

where $v_\perp$ is the velocity perpendicular ($\perp$) to the field and ω the gyro-frequency. The thermal energy per unit volume associated with the gyromotion about the field lines ($\rho\mathcal{E}_\perp$) is given by the integral over a velocity distribution function (see Sect. A.12) and a sum over all particles as:

$$\rho\mathcal{E}_\perp = \sum \int \frac{1}{2} m v_\perp^2 f \, \mathrm{d}^3 v \tag{B.16}$$

B.3 Bulk and Drift Motions

For the bulk of particles the average motion in the presence of uniform electric and magnetic fields is perpendicular to both fields, and for weak electric fields implies an average velocity due to an $\mathbf{E} \times \mathbf{B}$ drift of:

$$\mathbf{v_{bulk}} = \frac{c\mathbf{E} \times \mathbf{B}}{B^2} \tag{B.17}$$

In general, drift currents are the sources of many astrophysical magnetic fields. Note that the resulting bulk motion is always perpendicular to the force. The following extends the above consideration to the inclusion of external force fields, in this case gravitational force. The equation of motion then reads:

$$m\dot{\mathbf{v}} = q\mathbf{E} + \frac{q}{c}\mathbf{v} \times \mathbf{B} + m\mathbf{g} \tag{B.18}$$

Under the simplified situation where **g** and **E** are perpendicular to **B** – something that is realized in many astrophysical cases and which many times is referred to as crossed field cases – one can rewrite (B.18) such that:

$$m\dot{\mathbf{v}} = \frac{q}{c}\left(\mathbf{v} - v_{bulk} - \frac{mc}{q}\frac{\mathbf{g} \times \mathbf{B}}{B^2}\right) \tag{B.19}$$

The $\mathbf{g} \times \mathbf{B}$ drift is well known from the near-Earth environment as it causes charged particles in the radiation belts to circulate in an azimuthal direction as as well as to shuttle back and forth between the Earth's magnetic poles [1293]. Besides the normal gyro motion, the crossed field case then provides several drift motions including the electric (B.17) and gravitational drift (B.19), and the polarization drift in time-dependent fields.

Drifts and drags are also present in partially ionized media such as molecular clouds or near neutral winds. In this case a relative drift arises between the neutral and ionized particle species of the medium simply due to the fact that ions feel electromagnetic forces directly, whereas neutral particles have to engage in collisions with the ions. Neutral matter in general is not affected by the presence of magnetic fields. This relative drag is called *ambipolar diffusion* (see Sect. 4.3.4). The Lorentz force felt by the charged particles in a magnetic field within a unit volume reads:

$$F_L = \frac{1}{4\pi}(\nabla \times \mathbf{B}) \times \mathbf{B} \tag{B.20}$$

Since this force is not felt by neutrals, ions will drift with a different mean velocity. Resisting this drift will result in a frictional drag force created by mutual collisions. The drag force on neutral species by ion species can be expressed as:

$$\mathbf{F_d} = \gamma_d n_n n_i m_n m_i (\mathbf{v_i} - \mathbf{v_n}) \tag{B.21}$$

where γ_d is the drag coefficient:

$$\gamma_d = \frac{\langle u\sigma \rangle}{m_n + m_i} \tag{B.22}$$

and where n, m, and v (index n for neutral, i for ionized) are number density, mass, and mean velocity of neutral (n) and ionized (i) species. For simplicity it was assumed that the mass densities of these species can be expressed by $\rho = nm$. The term $n_i \langle u\sigma \rangle$ is the rate of collisions of ions with any neutral of comparable size. The bracketed expression is a mean of the elastic scattering cross section for neutral-ion collisions and the relative velocity of the ions in the rest frame of the neutrals. The ambipolar drift velocity can then be determined by:

$$w_D = \mathbf{v_i} - \mathbf{v_n} = \frac{1}{4\pi \gamma_d n_n n_i m_n m_i} (\nabla \times \mathbf{B}) \times \mathbf{B} \tag{B.23}$$

which in the case of magnetic uniformity is equivalent to the expression given in (B.17). The sign of the drift depends on a definition of the drag direction (i.e., whether it is the drag on the neutral by the ions or vice versa).

B.4 MHD Waves

Another important application of MHD is the propagation of waves through magnetized plasma. The subject of MHD waves is considerably more complex than in ordinary hydrodynamics and, therefore, most treatments are performed numerically. However, the propagation of shocks and disturbances in magnetized plasmas has many astrophysical applications specifically in stellar formation research and a summary of a few main properties is warranted.

A simple case is an ideal gas in static, uniform equilibrium with velocity $v_o = 0$, density $\rho = \rho_o$, pressure $P = P_o$, and a constant uniform magnetic field $|\mathbf{B_o}| = constant$. By introducing a small perturbation these conditions then read:

$$\rho = \rho_o(1+\epsilon), \quad \mathbf{v} = \mathbf{v_1}, \quad P = P_o(1+\psi), \quad \mathbf{B} = B_o(\mathbf{n} + \mathbf{n}') \tag{B.24}$$

The standard MHD equations under these conditions reduce to a set of equations describing the response of the perturbations to the magnetic field:

$$\begin{aligned} \frac{\partial \epsilon}{\partial t} + \nabla \mathbf{v_1} &= 0 \\ \frac{\partial v_1}{\partial t} + \frac{P_o}{\rho_o}\nabla\psi - \frac{B_o^2}{4\pi}(\nabla \times \mathbf{n}') \times \mathbf{n} &= 0 \\ \frac{\partial \mathbf{n}'}{\partial t} + \nabla \times (\mathbf{n} \times \mathbf{v}_1) &= 0 \\ \frac{\partial \psi}{\partial t} - \gamma \frac{\partial \epsilon}{\partial t} &= 0 \end{aligned} \tag{B.25}$$

For the last relation the equation of state for ideal gas (see Sect. A.2) was applied. It thus describes the propagation of an adiabatic perturbation where the coefficient γ is the adiabatic index. Combining these equations and applying the definitions to the speed of sound and the Alfvén speed given in Sect. A.8 leads to a second-order differential equation for the velocity of the perturbation with a solution of the form:

$$v_1 = \exp[i(\omega_1 t - \mathbf{k} \cdot \mathbf{x})], \tag{B.26}$$

where ω_1 is the perturbation propagation frequency and $\mathbf{k}$ the momentum vector plane of propagation. By splitting the wave equation into coordinate components one can identify various wave modes with respect to the plane of propagation and the magnetic field.

One of these modes is a transverse wave with wavefronts perpendicular to $\mathbf{k} = k\mathbf{e_z}$ and $\mathbf{n} = \mathbf{e_x} \cos\,\phi + \mathbf{e_y} \cos\,\phi$, i.e., with $\mathbf{v_1} = (v_x = 0, v_y = 0, v_z)$:

$$\omega^2 - k^2 v_A^2 \cos^2\,\phi = 0 \tag{B.27}$$

This wave mode is generally referred to as an *Alfvén wave*. In the case of $\phi = 0$ the wave speed is equal to the Alfvén velocity (see Sect. A.8). In the case of $\phi \neq 0$ the wavefronts may be considered to be inclined with respect to the unperturbed magnetic field. Thus, Alfvén waves always travel at Alfvén speed and $v_A \cos\,\phi$ merely represents the projection with respect to the magnetic field vector. Other modes under these conditions are identified as *slow* and *fast* MHD waves. It should also be noted that this simple picture, though approximately valid for many applications, is based on uniform and stress-free environments and, thus, in detail the treatment of MHD waves is highly complex.

B.5 Magnetic Reconnection

In many astrophysical situations it appears that one way or another the topology of the present magnetic field configuration is not conserved and subject to what is called *magnetic reconnection*. Such events have been mentioned in connection with molecular clouds for star-disk field configurations and as the prime underlying mechanism for stellar flaring activity (see Chap. 10).

Most treatments of magnetic reconnection are two dimensional and their descriptions involve many aspects, some of which will be described here. By definition, reconnection cannot take place under ideal MHD conditions, because it needs resistivity in the medium. Such resistivity is usually provided though collisions. However, nature proves a bit more resilient towards these arguments as reconnection events happen in the terrestrial magnetosphere within almost collisionless environments. In order to classify reconnection processes, it is thus necessary to determine the reconnection rate in various configurations even if they contain collisionless plasmas.

There are quite a number of reconnection processes and not all are of interest to stellar evolution research. Some of the basic ideas on the effect of magnetic reconnection were described by P. A. Sweet 1956 and E. Parker in 1957 [1297–1299]. Central to this are magnetic field configurations that feature parallel and antiparallel field lines separated by a $\mathbf{B} = 0$ surface, and an incompressible fluid where the total pressure remains constant. Ohmic dissipation near this null surface causes field annihilation. In 2-D treatments, this location is sometimes also called *null sheet*. Changes in the gravitational potential are regarded to be negligible within a characteristic volume $V_o = L_x L_y L_z$. For this simple configuration (see also the top diagram in Fig. B.1), the reconnection speeds of the plasma can be expressed in terms of the Alfvén speed. Essentially, the plasma pushes the field lines into the null sheet from both vertical directions, squeezing plasma out in the horizontal direction. The energy dissipation rate per unit volume (see (B.11)) is then given by:

$$Q_o = \frac{\eta}{4\pi}\left(\frac{B}{L_y}\right)^2. \tag{B.28}$$

This allows us to calculate the heat produced within V_o, which should be equal to the annihilation energy pressed into the region; that is:

$$\frac{\eta B^2}{4\pi L_y} = \frac{B^2}{8\pi} L_x L_z v_y \tag{B.29}$$

where v_y represents the reconnection speed. It is also the plasma speed in the y-direction with which the plasma flow is pushing the field lines into the null sheet. Using the expression for the magnetic Reynolds number in Table B.1 it is clear that the relation between the reconnection speed and the Alfvén velocity is a function of the magnetic Reynolds number of the horizontal outflow R_m. The reconnection rate $\mathcal{M}_e$ in this case yields:

$$\mathcal{M}_e = \frac{v_y}{v_A} = 2R_m^{-1/2} \tag{B.30}$$

On the other hand, the magnetic Reynolds number in astrophysical plasmas is usually a very large number. For a collisional plasma with a strong magnetic (but weak electric) field, magnetic diffusivity can be written as [12, 1300]:

$$\eta = 1.05 \times 10^{12} T^{-3/2} \ln \Lambda \text{ cm}^2 \text{ s}^{-1} \tag{B.31}$$

where $\ln \Lambda$ is the Coulomb logarithm [8]. For typical molecular cloud parameters as well as most laboratory applications $\ln \Lambda$ has a value of $\sim$10, and for collisional plasmas in stellar coronae and the Earth's magnetosphere it is higher than 20 but not more than 35. For most cases it is safe to assume that the characteristic velocity of the magnetized plasma is the Alfvén velocity and, thus, the magnetic Reynolds number can be approximated by the Lundquist number L_u. Thus, for

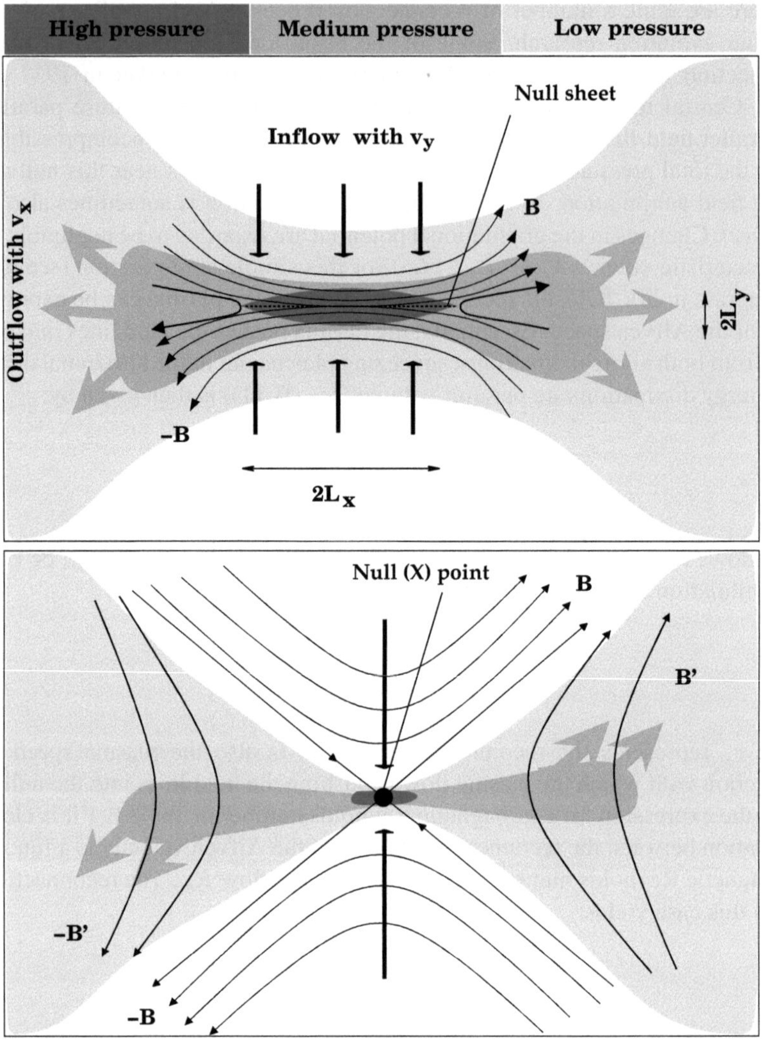

Fig. B.1 Schematic drawing of two magnetic reconnection scenarios under the influence of a collisional plasma. At the tops a vertical plasma inflow compresses a parallel and anti-parallel magnetic field line configuration into annihilation in a $B = 0$ sheet forcing plasma out horizontally. The different shades loosely indicate that plasma pressure is highest around the null sheet. At the bottom is a similar initial configuration, but now the field lines are compressed into a neutral X-point which in comparison with the scenario above results in a significant reduction of the diffusion area, which greatly enhances diffusion and thus the reconnection rate

active regions in the solar corona, where the temperature can get beyond 10^7 K, assuming loop lengths of 10^7 cm, an approximate Alfvén speed of $\sim 10^7$ cm s^{-1}, $\eta \sim 7 \times 10^2$ cm^2 s^{-1}, the L_u number is about 10^{11}. Above a more active region it can even reach 10^{14}.

From (B.30) it appears that reconnection takes place at a fraction of the dynamical speed of magnetized plasma. In turn this means that magnetic diffusion, with a timescale defined as:

$$\tau_{md} = \frac{L_o^2}{\eta} \tag{B.32}$$

requires very long timespans. Although in rare cases this may be what happens, it certainly cannot explain the sunspot and flaring activity on the Sun. Similarily, it is not feasible to describe magnetic young stellar activity. However, the Sweet–Parker scenario of field annihilation of a null sheet of limited size doesn't really maximize the possible reconnection rate. In 1963 H. E. Petschek thus proposed a slightly different scenario in which the null sheet is compressed into an X-point [677] (see bottom of Fig. B.1). Here magnetic diffusion is confined to an area around the neutral point, which as a consequence enhances diffusion rates and allows field annihilation to accelerate. There are two novel features about this reconnection geometry. First, it generates a maximum annihilation rate of the form:

$$\mathcal{M}_e^{max} = \frac{v_y}{v_A} \sim \frac{\pi}{8\ \ln R_m} \tag{B.33}$$

where the index e stands for *external* (i.e., for the inflow which is not near the X-point). Second, it allows the magnetic field to diffuse into the plasma where it propagates as an MHD wave outward [677]. This generates reconnection rates between 0.01 and 0.1, leading to much shorter timescales consistent with observed flaring activity (see Chaps. 8 and 10).

B.6 Dynamos

In the most general sense, the *magnetic dynamo theory* describes MHD flows within stellar bodies that feature differential rotation and convection. It is a common perception among astrophysicists that all stellar magnetic fields have their origins one way or another through self-excited magnetic dynamo activity. This may even be true for the interstellar or galactic field through some form of galactic dynamo. Though a complete dynamo theory is relatively complicated and not at all in every aspect understood, there are a few basic dynamo functions and relations one should recall when dealing with stellar and proto-stellar dynamos.

Complex motions of a plasma with a weak seed magnetic field can generate strong magnetic fields on larger scales. One would first attempt a simplified approach and try to solve the MHD problem axisymmetrically. Herein lies the first big problem: dynamos cannot maintain either a poloidal or a toroidal magnetic field against Ohmic dissipation. As a consequence, exact axisymmetric dynamos cannot be realized. This phenomenon is also known as *Cowling's theorem* after T. G. Cowling, formulated in 1965. In other words, the induction equation (B.10) does not allow for axisymmetric fluid motions that yield non-decaying and, similarily, axisymmetric configurations for the **B**-field. A quite comprehensible illustration of this effect can be found in [1293]. Observed phenomena in the Sun (i.e., sunspots, flares and prominences) indicate instead that magnetic reconnection is required [1294]. It should also be realized that the dynamo effect does not generate magnetic fields but amplifies existing ones. The following is an attempt to outline some of the groundwork for the realization of MHD dynamos.

Cowlings's theorem mandates that in order to maintain a seed magnetic field one needs to offset Ohmic dissipation. In addition, in order to get the dynamo going, a cycle has to build up which converts toroidal fields into poloidal ones and regenerates toroidal fields from converted poloidal ones. The mechanisms proposed include radial convection, magnetic instabilities such as the magnetostatic *Parker Instability* and various levels of turbulence. The current standard theory for Sun-like stars features various modifications of the so called $\alpha - \Omega$ *dynamo*, a concept that took shape in the mid-1950s [1301]. In simple terms, this may be described by a uniform magnetic field being deformed locally into Ω-shaped loops through cyclonic turbulence-generating eddies which have a cyclonic velocity α. The theoretical basis to describe the toroidal field is the induction equation. By neglecting other forces this equation can be simplistically formulated as [1301, 1302]:

$$\frac{\partial \mathbf{B}}{\partial t} = \eta \nabla^2 \mathbf{B} - \alpha \nabla \times \mathbf{B} \tag{B.34}$$

where α is defined as:

$$\alpha \mathbf{B} = \mathbf{u} \times \mathbf{b} \tag{B.35}$$

Where **u** and **b** are the velocity and induction field of the generated standing MHD wave solution. Within a characteristic length scale L_o the characteristic dynamo period is tied to the *eddy diffusivity* η_{ed} as:

$$\tau_{dy} = \frac{L_o^2}{4\eta_{ed}} \tag{B.36}$$

Eddy diffusivity characterizes the stability of the eddies creating the Ω-loops. For an azimuthal field flux written as $\Phi = L_o B$ the dynamical period can be expressed in terms of B and Φ. For the solar period of 22 yr this has the consequence that as long as η_{ed} is not suppressed by more than $1/B^2$, such a long period can be maintained in the presence of strong azimuthal fields [1303].

Modern versions of this dynamo recognize the fact that helioseismological data suggest that the Solar dynamo operates in only a small layer within the Sun's interior (see [1303–1305] and references therein). This layer is located at about 0.7 $R_{\odot}$ from the center between the boundary of the convective outer zone and the radiative core and has a thickness of less than 3 percent of the stellar radius. The vertical shear $\partial\Omega/\partial r$ is also confined to this layer, meaning there is no significant change of

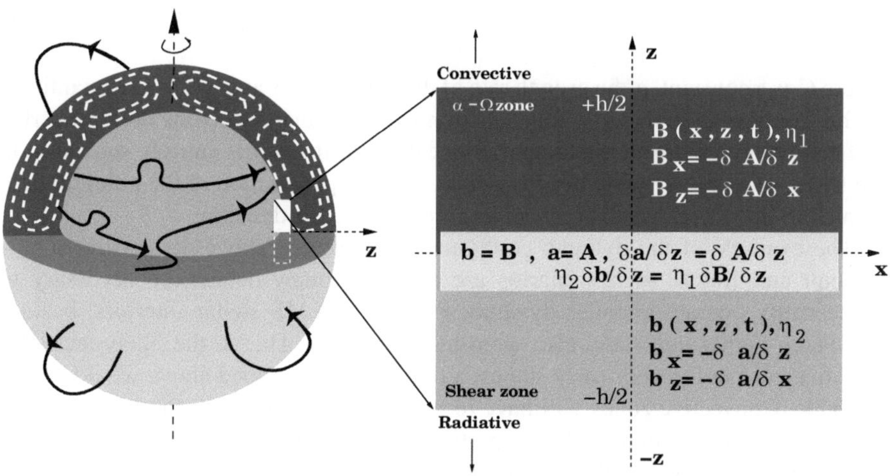

Fig. B.2 (*left*) Illustration of the $\alpha - \Omega$ dynamo effect in the Sun. The inner core represents the radiative zone, the outer shell the convective zone. In-between is the thin layer in which the actual dynamo operates. The thick lines with arrows are wound up and surfaced magnetic field lines. (*right*) A diagram that highlights the separation of the layer into the $\alpha-\Omega$ part below the convective zone and the shear part above the radiative zone together with the corresponding field specifications (see dynamo equation in the text). Highlighted at z = 0 are the necessary boundary conditions. The configuration described here is called the *interface dynamo*. The magnetic diffusivity is η_1 in the $\alpha - \Omega$ zone, η_2 in the shear zone

radial velocity between the top and the bottom of the convection zone. Figure B.2 illustrates schematically the set-up for such an *interface dynamo*. One configuration of such a dynamo was proposed by Parker in 1993 and is illustrated in Fig. B.2. The actual dynamo equations according to the specifications in Fig. B.2 read [1303]:

$$\left[\frac{\partial}{\partial t} - \eta_1\left(\frac{\partial^2}{\partial x^2} + \frac{\partial^2}{\partial z^2}\right)\right]\mathbf{B} = 0 \tag{B.37}$$

$$\left[\frac{\partial}{\partial t} - \eta_1\left(\frac{\partial^2}{\partial x^2} + \frac{\partial^2}{\partial z^2}\right)\right]\mathbf{A} = \alpha\mathbf{B} \tag{B.38}$$

$$\left[\frac{\partial}{\partial t} - \eta_2\left(\frac{\partial^2}{\partial x^2} + \frac{\partial^2}{\partial z^2}\right)\right]\mathbf{b} = G\,\frac{\partial a}{\partial x} \tag{B.39}$$

$$\left[\frac{\partial}{\partial t} - \eta_2\left(\frac{\partial^2}{\partial x^2} + \frac{\partial^2}{\partial z^2}\right)\right]\mathbf{a} = 0 \tag{B.40}$$

Where **A** and **a** are the azimuthal vector potentials describing the poloidal field for $z > 0$ and $z < 0$, respectively; G is the uniform shear dv_y/dz; α once again is the mean rotational velocity, $\eta_1\alpha$ is a measure for the mean helicity of the convective flow. Since these equations are linear, their solution for $z > 0$ is that of a plane wave:

$$\mathbf{B} = C \ \exp(\sigma t - Sz) \ \exp[i(\omega t + kx - Qz)] \tag{B.41}$$

where C is a constant, σ the growth rate of the wave, and S and Q are real quantities. In this configuration the $\alpha - \Omega$ production zone and the shear zone do not overlap and the dynamo seems less efficient. Since eddy diffusivity is entirely supressed in the shear zone, the dynamo depends on downward penetration of the poloidal field created in the $\alpha - \Omega$ zone, practically jump-starting the dynamo.

The example of the dynamo configuration presented above demonstrates how difficult and fragile such scenarios are and how many details are necessary to successfully operate a stellar dynamo. Proto- and PMS stellar interiors, besides still being highly uncertain, also seem more complex. Due to the likely lack of a radiative core in the very early stages, a dynamo as described above would not be operable. Convective zones will have to reach much deeper into the star and their structure will likely impact the properties of PMS stellar dynamos. In fact, there is a quantity in dynamo theory that actually links the transport properties of the convection zone with properties of the interface layer, the *dynamo number*. This quantity is a direct product of the dispersion relations which result from (B.40) and can be written as:

$$N_{dy} = \frac{\Gamma G}{\eta_1^2 k^3} = C \left(\frac{\tau_c}{P} \right)^2 \tag{B.42}$$

This number has to be sufficiently large to drive the surface wave along the interface and may be used as a measure of magnetic activity. The right part indicates that it is also proportional to the square of the characteristic convective turnover time and, thus, implicitly to the *Rossby number* defined in (10.3) once τ_c is set to the local convective overturn time at the base of the convection zone [1306]. C is once more an arbitrary proportionality constant. Scaled by the mixing length $(1/2)\alpha H_P$ (see also references in Sect. 6.1.3) the Rossby number is also the ratio of the characteristic convective velocity v to the rotational velocity of the star Ω:

$$R_o = \frac{2\pi v}{\alpha \Omega H_P} \tag{B.43}$$

where α is the mixing length ratio and H_P is the local pressure scale height. Although it seems that dynamo mechanisms scale with R_o, their relevance for magnetic field generation specifically for young stars is not clear and here further research is necessary.

Other discussed possibilities involve *turbulent* or *distributed* dynamos. These scenarios are based on the assumption that turbulent velocities in convection cells may generate small-scale magnetic fields. These small-scale turbulent fields may even co-exist with the dominant $\alpha - \Omega$ dynamo at the interface layer. In stars with deep convective zones, these turbulent fields could be the source for large-scale fields. Most of these concepts need further development [689, 1307, 1308].

B.7 Magnetic Disk Instabilities

The notion to treat protostellar disks as magnetized rotating fluids has changed the view of angular momentum transport. MHD waves and turbulence introduce perturbations in otherwise smooth flows. Central to the treatment of magnetized rotating fluids are either fluctuations of velocities or densities or both. The basic MHD equations to describe such fluctuations are again the same as introduced above, now applied towards an accretion flow. Conservation laws are used to illustrate how differential rotation in a disk frees magnetic energy to turbulent fluctuations in combination with angular momentum transport. Most of the material in this section was taken from a recent review on angular momentum transport in accretion disks by S. A. Balbus [565]. Thus, in addition to the standard MHD set of equations introduced above, it is useful to introduce cylindrical coordinates (see below) to satisfy the geometrical constraints of the disk and separate the azimuthal component of the momentum equation. This equation then can be written as *angular momentum conservation*:

$$\frac{\partial(\rho r v_\phi)}{\partial t} + \nabla\left(\rho r v_\phi \mathbf{v} - \frac{r B_\phi}{4\pi}\mathbf{B} + \left(P + \frac{B^2}{8\pi}\right)\mathbf{e}_\phi\right) = 0 \tag{B.44}$$

Note that the dissipative term $\rho\mathbf{v}\nabla\mathbf{v}$ has been ignored as it carries only a negligible amount of angular momentum. It also helps to simplify the energy equation and write it explicitly for contributing terms only and assume that matter is polytropic:

$$\frac{\partial U}{\partial t} + \nabla\mathcal{F}_e = -Q_{rad} \tag{B.45}$$

where the energy density U is:

$$U = \frac{1}{2}\rho v^2 + \frac{P}{\gamma - 1} + \rho\Phi + \frac{B^2}{8\pi} \tag{B.46}$$

and similarly for the energy flux $\mathcal{F}_e$ are:

$$\mathcal{F}_e = \mathbf{v}\left(\frac{1}{2}\rho v^2 + \frac{\gamma P}{\gamma - 1} + \rho\Phi\right) + \frac{\mathbf{B}}{4\pi} \times (\mathbf{v} \times \mathbf{B}) \tag{B.47}$$

The terms for kinetic, thermal, gravitational, and magnetic components are now clearly indicated from left to right, respectively. Again, the heating term Q_ν has been dropped as it does not contribute enough overall with respect to energy conservation.

In order to introduce turbulent fluctuations one defines an azimuthal velocity perturbation, such as:

$$\mathbf{u} = \mathbf{v} - r\Omega(r)e_\phi \tag{B.48}$$

where $\Omega(r)$ is approximated by the underlying rotational velocity profile. Rewriting the energy equation in terms of the perturbed velocity isolates the disk stress tensor responsible for wave and turbulent transport:

$$\frac{\partial U}{\partial t} + \nabla \mathcal{F}_e = -S_{r\phi} - Q_{rad} \tag{B.49}$$

consisting of Reynolds and Maxwell stresses:

$$S_{r\phi} = \rho u_r u_\phi - \frac{B_r B_\phi}{4\pi} \tag{B.50}$$

These stresses directly transport angular momentum and support turbulence by freeing energy from differential rotation. For protostellar disks $S_{r\phi}$ always has to be positive in order to keep the turbulence from dying out.

Reviews on how angular momentum is transported in (proto-) planetary disks in the absence of a magnetic field can be found in [290, 565]. Some of the first calculations on *magneto-rotational instability* (MRI) can be traced back to the early 1950s when Chandrasekhar calculated the factors involved in dissipative Couette flows, though the regime where the gradients of angular velocity and specific angular momentum oppose each other were not particularly pursued [1309]. *MRI* has its roots in the fact that the magnetic field in a differentially rotating fluid acts in a destabilizing way. Figure B.3 illustrates the underlying principle in analogy to nearby fluid elements coupled with each other by a spring-like force. Due to

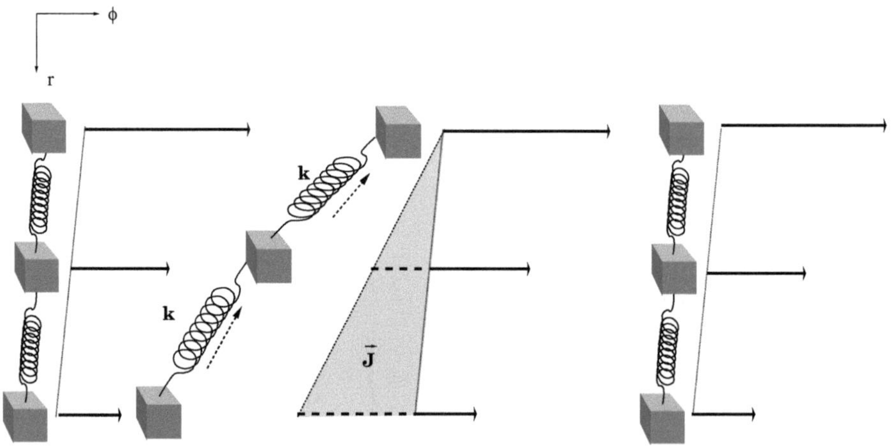

Fig. B.3 In analogy to the MRI effect one may want to consider fluid elements that are attached with springs of spring constant k. The fluid array (*left*) moves at a differential azimuthal velocity (thick black arrows). Since the top elements move faster than the element immediately following in direction of r, this element feels a drag that corresponds to the spring constant k. The next element experiences a similar drag and so on (*middle* array). Imagine the system springing back to its original differential profile (*right* array), the shaded area in the middle array demonstrates the accumulation of angular momentum that is transported in the radial direction differential rotation slower rotating elements experience a drag. At the end the angular momenta of these slower rotating elements increase at the expense of faster rotating ones. The spring tension grows with increasing element separation and angular momentum transport outward cascades away. In a protostellar systems this analogy roughly describes a very simple fluid system moving in an axisymmetric disk in the presence of a weak magnetic field. In fact, for a fluid element which is displaced in the orbital plane by some amount ξ with a spatial dependence of e^{ikz}, the induction equation in the case of frozen-in fields (see above) gives a displacement of the magnetic field of $\partial\mathbf{B} = ik\mathbf{B}\xi$ leading to a magnetic tension force, which can be written in the form:

$$\frac{ik\mathbf{B}}{4\pi\rho}\partial\mathbf{B} = -(kv_A)^2\xi \tag{B.51}$$

This equation has the form of the equation of motion for a spring-like force. The result is a linear displacement with a spring constant $(kv_A)^2$.

B.8 Expressions

In most astrophysical applications either *cylindrical* (r, ϕ, z) or *spherical* (r, θ, ϕ) polar coordinates are adopted according to the geometrical constraints of the MHD fluid. The following most common expressions involving the potential A and the field $\mathbf{B}$ are spelled out in component form for the two coordinate systems. The unit vectors for the cylindrical case are $\mathbf{e_r}$, $\mathbf{e}_\phi$ and $\mathbf{e_z}$, whereas for the spherical case they are $\mathbf{e_r}$, $\mathbf{e}_\theta$ and $\mathbf{e}_\phi$.

Cylindrical polar coordinates:

$$\begin{aligned}
\nabla A &= \frac{\partial A}{\partial r}\mathbf{e}_r + \frac{1}{r}\frac{\partial A}{\partial\phi}\mathbf{e}_\phi + \frac{\partial A}{\partial z}\mathbf{e}_z \\
\nabla\cdot\mathbf{B} &= \frac{1}{r}\frac{\partial}{\partial r}(rB_r) + \frac{1}{r}\frac{\partial B_\phi}{\partial\phi} + \frac{\partial B_z}{\partial z} \\
\nabla\times\mathbf{B} &= \left(\frac{1}{r}\frac{\partial B_z}{\partial\phi} - \frac{\partial B_\phi}{\partial z}\right)\mathbf{e}_r + \left(\frac{\partial B_r}{\partial z} - \frac{\partial B_z}{\partial r}\right)\mathbf{e}_\phi + \left(\frac{1}{r}\frac{\partial}{\partial r}(rB_\phi) - \frac{1}{r}\frac{\partial B_r}{\partial\phi}\right)\mathbf{e}_z \\
\nabla^2 A &= \frac{1}{r}\frac{\partial}{\partial r}\left(r\frac{\partial A}{\partial r}\right) + \frac{1}{r^2}\frac{\partial^2 A}{\partial\phi^2} + \frac{\partial^2 A}{\partial z^2} \\
(\mathbf{B}\cdot\nabla)\mathbf{B} &= \left(B_r\frac{\partial B_r}{\partial r} + \frac{B_\phi}{r}\frac{\partial B_r}{\partial\phi} - \frac{B_\phi^2}{r} + B_z\frac{\partial B_r}{\partial z}\right)\mathbf{e}_r \\
&\quad + \left(\frac{B_r}{r}\frac{\partial}{\partial r}(rB_\phi) + \frac{B_\phi}{r}\frac{\partial B_\phi}{\partial\phi} + B_z\frac{\partial B_\phi}{\partial z}\right)\mathbf{e}_\phi \\
&\quad + \left(B_z\frac{\partial B_z}{\partial z} + B_r\frac{B_z}{\partial r} + \frac{B_\phi}{r}\frac{\partial B_z}{\partial\phi}\right)\mathbf{e}_z
\end{aligned} \tag{B.52}$$

Spherical polar coordinates:

$$\nabla A = \frac{\partial A}{\partial r}\mathbf{e}_r + \frac{1}{r}\frac{\partial A}{\partial \theta}\mathbf{e}_\theta + \frac{1}{r\sin\theta}\frac{\partial A}{\partial \phi}\mathbf{e}_\phi$$

$$\nabla \cdot \mathbf{B} = \frac{1}{r^2}\frac{\partial}{\partial r}(r^2 B_r) + \frac{1}{r\sin\theta}\frac{\partial}{\partial \theta}(\sin\theta B_\theta) + \frac{1}{r\sin\theta}\frac{\partial B_\phi}{\partial \phi}$$

$$\nabla \times \mathbf{B} = \frac{1}{r\sin\theta}\left(\frac{\partial}{\partial \theta}(\sin\theta B_\phi) - \frac{\partial B_\theta}{\partial \phi}\right)\mathbf{e}_r + \left(\frac{1}{r\sin\theta}\frac{\partial B_r}{\partial \phi} - \frac{1}{r}\frac{\partial}{\partial r}(rB_\phi)\right)\mathbf{e}_\theta + \left(\frac{1}{r}\frac{\partial}{\partial r}(rB_\theta) - \frac{1}{r}\frac{\partial B_r}{\partial \theta}\right)\mathbf{e}_\phi$$

$$\nabla^2 A = \frac{1}{r2}\frac{\partial}{\partial r}\left(r^2\frac{\partial A}{\partial r}\right) + \frac{1}{r\sin\theta}\frac{\partial}{\partial \theta}\left(\sin\theta\frac{\partial A}{\partial \theta}\right) + \frac{1}{r^2\sin^2\theta}\frac{\partial^2 A}{\partial \phi^2}$$

$$\begin{aligned}(\mathbf{B}\cdot\nabla)\mathbf{B} = &\left(B_r\frac{\partial B_r}{\partial r} + \frac{B_\theta}{r}\frac{B_r}{\partial \theta} - \frac{B_\theta^2 + B_\phi^2}{r} + \frac{B_\phi}{r\sin\theta}\frac{\partial B_r}{\partial \phi}\right)\mathbf{e}_r \\ &+ \left(B_r\frac{B_\theta}{\partial r} + \frac{B_\theta}{r}\frac{B_\theta}{\partial \theta} + \frac{B_r B_\phi}{r} + \frac{B_\phi}{r\sin\theta}\left(\frac{\partial B_\theta}{\partial \phi} - \cos\theta B_\phi\right)\right)\mathbf{e}_\theta \\ &+ \left(B_r\frac{B_\phi}{\partial r} + \frac{B_r B_\phi}{r} + \frac{B_\phi}{r\sin\theta}\frac{\partial B_\phi}{\partial \phi} + \frac{B_\phi}{r\sin\theta}\frac{\partial}{\partial \theta}(B_\phi\sin\theta)\right)\mathbf{e}_\phi\end{aligned} \tag{B.53}$$

Appendix C
Radiative Interactions with Matter

This appendix describes the basic radiative processes that have to be considered when radiation interacts with gas clouds, atmospheres and dust. It contains a few useful relations which are specifically of interest for the calculation of opacities. The treatment of radiative transport requires all scattering, absorption, and emission processes to be accounted for.

Stellar formation is characterized by episodes of variable radiative properties and the interaction of radiation with various matter and plasma states. Radiative processes are key ingredients to making stars and as such are the only source of information to diagnose these processes. The following sections outline the basic physics [1310, 1311] relevant to model calculations and observational diagnostics. Specific emphasis is given to expressions of the r relevant radiative coefficients.

Intensity can simply be defined as:

$$I_\nu = \frac{\mathrm{d}E}{\mathrm{d}t\ \mathrm{d}\Omega\ \mathrm{d}A\ \mathrm{d}\nu}, \tag{C.1}$$

which is the radiated energy $\mathrm{d}E$ per unit time $\mathrm{d}t$, unit angle $\mathrm{d}\Omega$, unit area $\mathrm{d}A$, and frequency $\mathrm{d}\nu$. The general equation for radiative transport can be written as:

$$\frac{\mathrm{d}I_\nu}{\mathrm{d}s} = -\chi_\nu^{tot} I_\nu + j_\nu \tag{C.2}$$

j_ν is the emission coefficient describing local emissivity or spectral energy distribution. If κ_ν is the frequency-dependent absorption coefficient, σ_ν the frequency-dependent scattering coefficient, then the total extinction coefficient χ_ν^{tot} is defined as:

$$\chi_\nu^{tot} = \kappa_\nu + \sigma_\nu \tag{C.3}$$

Most interactions between stellar matter and radiation in stars happen at high temperatures and involve electrons rather than heavier nuclei. Although this is

N.S. Schulz, *The Formation and Early Evolution of Stars*, Astronomy and Astrophysics Library, DOI 10.1007/978-3-642-23926-7, © Springer-Verlag Berlin Heidelberg 2012

also true for radiative processes during the early evolution of stars, star formation additionally involves processes with atoms, molecules, and dust at low temperatures.

C.1 Radiative Equilibrium

A central issue in the theory of stellar atmospheres is finding solutions to the basic equation of radiative transfer. The model of a slab (=plane-parallel) of gas is the usual approach to the problem. The following treatment follows a description presented in [290, 1312]. Dusty protostellar envelopes cannot be considered as plane-parallel and the transfer equation has to take a spherical form. However, in most cases the assumption of azimuthal symmetry is valid, which eases the structure of the sphericali form of (C.2) considerably as the integral over the solid angle reduces to:

$$\oint \frac{\Omega}{4\pi} = \frac{1}{2} \int_{-1}^{+1} \mathrm{d}\mu \tag{C.4}$$

and the first three moments of the radiation distribution can be written per unit interval at the frequency ν as [1312, 1313]:

$$\begin{aligned} &\textit{Energy density} & J_\nu &= \frac{1}{2} \int_{-1}^{+1} I_\nu \, \mathrm{d}\mu \\ &\textit{Radiation flux} & F_\nu = 4\pi H_\nu &= 4\pi \frac{1}{2} \int_{-1}^{+1} I_\nu \mu \, \mathrm{d}\mu \\ &\textit{Radiation pressure} & P_\nu = \frac{4\pi}{c} K_\nu &= \frac{4\pi}{c} \frac{1}{2} \int_{-1}^{+1} I_\nu \mu^2 \, \mathrm{d}\mu \end{aligned} \tag{C.5}$$

For a gas sphere one then has three equations. The transfer equation:

$$\mu \frac{\partial I_\nu}{\partial r} + \frac{(1-\mu^2)}{r} \frac{\partial I_\nu}{\partial \mu} = -\kappa_\nu I_\nu + j_\nu \tag{C.6}$$

An equation for radiation flux moment H_ν:

$$\frac{\partial H_\nu}{\partial r} + \frac{2 H_\nu}{r} = -\kappa_\nu I_\nu + j_\nu \tag{C.7}$$

And an equation for the radiation pressure moment K_ν:

$$\frac{\partial K_\nu}{\partial r} + \frac{(3K_\nu - J_\nu)}{r} = -\kappa_\nu H_\nu \tag{C.8}$$

In the outer dusty envelopes around protostars, which at low frequencies are optically thin, and in the limit of $r \gg R$, where R is the radius of the stellar core, (C.6–C.8) reduce enormously as all radiative moments approach:

$$K_\nu \rightarrow H_\nu \rightarrow J_\nu \rightarrow \frac{I_{star} R^2}{4\pi r^2} \tag{C.9}$$

The luminosity emitted at a frequency within $d\nu$ is simply dominated by the direct radiation from the star with negligible contributions from all other directions. In other words, the luminosity then is expressed by (C.11), with $I_{star} = F_{obs}$ and $R = D$, since the envelope is optically thin.

For the optically thick case one expects the radiation field to be isotropic and in LTE. When in such a case the second term in (C.8) vanishes, the energy density approaches that for a blackbody. Integration over frequency using (C.18) for the opacity finally leads to the equilibrium luminosity:

$$L = -\frac{64\pi\sigma r^2 T^3}{3\kappa_R}\frac{\mathrm{d}T}{\mathrm{d}r} \tag{C.10}$$

C.2 Radiation Flux and Luminosity

There are several ways to describe amounts of radiation. Most common is the use of *energy flux* measured in erg cm^{-2} s^{-1}. Sometimes it is advantageous to use a *photon flux* which has units of photons cm^{-2} s^{-1}. The former is mainly used to account for macroscopic flux properties, such as broad-band spectra; the latter is useful to describe microscopic radiative properties such as line emission. The specific intensity or I_ν, as defined above, measures the amount of energy passing through an area element within fixed solid angle, time, and frequency intervals. Integration over all directions normal to the emitting surface gives the specific flux F_ν and subsequent integration over frequency would result in the total flux F. Note that in most cases it is sufficient to assume that the emitter is a sphere of uniform brightness (i.e., $I_\nu = B_\nu$). Thus when measured directly at the emitting sphere its *luminosity* amounts to $4\pi F$ assuming that it radiates across the entire sphere.

In the case of a distant star this, however, is not what is measured on Earth since F is reduced by the square of the distance resulting in an *observed* energy flux F_{obs}. The luminosity then can be expressed as:

$$\begin{aligned} L &= 4\pi F_{obs} D^2 \\ &= 1.196 \times 10^{32} \left(\frac{F_{obs}}{[10^{-12}\ \mathrm{erg\ cm^{-2}\ s^{-1}}]}\right)\left(\frac{D}{[1\ \mathrm{kpc}]}\right)^2 \mathrm{erg\ s^{-1}} \end{aligned} \tag{C.11}$$

Very often a term called *bolometric luminosity* (L_{bol}) is used. The bolometric system of magnitudes is defined as the total energy flux as measured above the Earth's atmosphere. For the Sun this is sufficient, but for distant stars one must also correct for interstellar extinction (see Sect. 3.4.3). Thus for stars L_{bol} is the total radiated energy flux in LTE at their surface. Often the Sun's bolometric luminosity ($L_\odot = 3.845 \times 10^{33}$ erg s^{-1}) is used as a reference value. One also needs to realize that there may be other sources for radiation, thermal and non-thermal. The luminosities from these sources of radiation, which are based on physically very different emitting regions and mechanisms, are either restricted to a characteristic bandpass or to a frequency range where all excluded frequencies are considered to be negligible.

C.3 Opacities

In a most general sense, *opacity* $\chi(\mathbf{r}, t, \nu)$ is a measure for the removal of energy from a radiation field by matter. As such it depends on the position $\mathbf{r}$ of the active material, time, and frequency. Position and time are dependent on geometrical flow models and not relevant for discussing the basic properties of the interaction of radiation and matter; thus, the following sections only concentrate on frequency dependence. The reader should be aware that in realistic astrophysical calculations all dependencies need to be included, since it is always necessary to evaluate the opacity at a specific frequency and then sum, average or integrate it over all contributing frequencies. Note that in the following the notation to index the frequency is used rather than express its functional form.

There are two basic components to opacity: one is described by the absorption coefficient κ_ν and the other by the scattering coefficient σ_ν. It is also necessary to distinguish between various phases of matter and to assume an opacity for gaseous matter χ_ν^g and an opacity for solid dust particles χ_ν^d:

$$\chi_\nu^{tot} = \chi_\nu^g + \chi_\nu^d \tag{C.12}$$

where

$$\chi_\nu^g = \kappa_\nu^g + \sigma_\nu^g \tag{C.13}$$

and

$$\chi_\nu^d = \kappa_\nu^d + \sigma_\nu^d \tag{C.14}$$

Dust opacities are treated in Sect. C.9 and, unless specifically in conflict, the index g for the gas portion of the opacity treatment is dropped in the following for convenience.

For monochromatic photons there are several types of interactions that contribute to κ_ν and j_ν (see Fig. C.1):

- *electron scattering,*
- *free–free (ff) absorption,*
- *bound–free (bf) absorption,*
- *bound–bound (bb) absorption.*

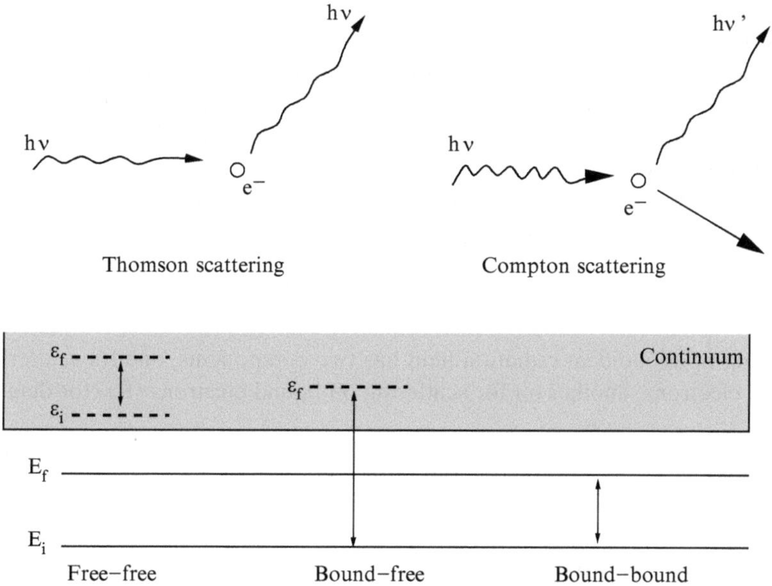

Fig. C.1 Various radiative processes interacting with matter (see text)

For the continuum absorption coefficient the first two items are most important. Electron scattering and free–free absorption are continuum processes, and free-bound absorption produces free–free continua and lines, whereas bound–bound absorption produces lines only. However, the latter may become important for continuum coefficients when many lines blend and crowd and thus act very much like continua. The following highlights a few important aspects of these processes. For in-depth studies the books by G. B. Rybicki and A. P. Lightman [1310] and D. F. Gray [54] are specifically recommended.

Electron scattering involves scattering of photons by free electrons. In the IR domain incident photon energies are small (i.e., $h\nu \ll m_e c^2$) and it is exclusively Thomson (Rayleigh) scattering that has to be considered.

Free–free absorption describes the absorption (emission) of photons by free electrons *near ions*. Such continuum states ϵ are illustrated in the gray-shaded area in Fig. C.1. Free–free transitions become important at high temperatures ($>10^6$ K) and/or long wavelengths (300 μm). Assuming LTE, which also implies a Maxwellian velocity distribution (see (A.51)) of free electrons, the absorption coefficient for such a transition can be expressed in a semi-classical form and depends on frequency and temperature (see below).

Bound–free absorption results from ionization and radiative recombination events in atoms or ions. In this case an electron is removed from its bound state E_i into a continuum state ϵ_f (see Fig. C.1). Similarly, an electron can lose its gained energy and transfer back to a bound state, either directly into E_i or by cascading through intermediate bound states. This can either happen through an incident radiation field (*photoionization*) or collisions with free electrons and ions (*collisional ionization*).

Bound–bound absorption describes the excitation and radiative recombination of matter. Here there are several fundamental processes at work, *spontaneous emission* from excited levels, *stimulated emission* by an external radiation field, *absorption* and *collisional excitation*.

The gas absorption coefficient incorporates free–free (ff), bound–free (bf) and bound–bound (bb) processes and is then written as:

$$\kappa_\nu = \kappa_\nu^{ff} + \kappa_\nu^{bf} + \kappa_\nu^{bb} \tag{C.15}$$

Scattering of an incident radiation field has two components, one for scattering on free (T) electrons, another on for scattering on bound electrons (R) (for details see below):

$$\sigma_\nu = \sigma_T + \sigma_{R,\nu} \tag{C.16}$$

C.4 Mean Opacities

The frequency dependent absorption coefficient κ_ν can be thought as the fractional decrease in flux density at a frequency ν per unit path length through an absorbing gas. Once integrated or averaged over an assumed flux density one gets the total opacity of the medium. Generally one adopts the so-called *diffusion approximation*, which assumes LTE and a mean opacity to be constant everywhere. Under LTE conditions the emitted and absorbed intensities are the same; that is:

$$j_\nu = \kappa_\nu B_\nu(T) \tag{C.17}$$

where $B_\nu(T)$ is the blackbody (Planck) spectrum.

The LTE condition allows one to treat locally emitted intensity as blackbody spectra. Averaging over all frequencies yields:

$$\frac{1}{\kappa_R} = \int_0^\infty \frac{1}{\kappa_\nu}\frac{\partial B_\nu}{\partial T}\,\mathrm{d}\nu \left[\int_0^\infty \frac{\partial B_\nu}{\partial T}\,\mathrm{d}\nu\right]^{-1} \tag{C.18}$$

The result is called the *Rosseland* mean opacity. At near the surface of a star the frequency-independent absorption coefficient can be expressed in terms of the *Planck* mean opacity as:

$$\kappa_P = \frac{\pi}{\sigma T^4}\int_0^\infty \kappa_\nu B_\nu(T)\,\mathrm{d}\nu \tag{C.19}$$

Were κ_ν in (C.19) to be independent of frequency (which is never realized, but it is a quite helpful approximation in various cases), the temperature structure of the absorbing medium is one of a standard gray atmosphere with an effective temperature of [1344]:

$$T_{eff}^4 = \tfrac{4}{3}(\tau + \tfrac{2}{3})^{-1}T^4 \tag{C.20}$$

In other words, the medium is taken to be radiating as a blackbody with effective temperature T_{eff} instead of T as long as $\tau \sim 0.7$.

This opacity is then a global quantitative measure of the ability of the gas or plasma to absorb radiation. For example, in a slab of gas of thickness dr the absorbed flux rate per unit area dF can be described as:

$$\mathrm{d}F = -F\mathrm{d}\tau = -\kappa_R \rho F\ \mathrm{d}r, \tag{C.21}$$

where τ is the optical depth. By integrating (C.21) the radiation flux density can be expressed as:

$$F = F_o\ \exp[-\kappa_R \rho r] \tag{C.22}$$

where F_o is the initial flux of the radiation source. The absorption length $\lambda_R = (\kappa_R \rho)^{-1}$ is the mean free path of the photon. If τ is larger than unity the gas is called *optically thick* for radiation. This definition may apply to a single or to all scattering and absorption categories (see above). In the case of Thomson scattering when $\tau = \tau_T > 1$, the gas is optically thick for electron scattering.

Large opacity values indicate that radiation will likely be absorbed by the gas, while low values allow for large radiative energy fluxes leaving the gas. Opacities at temperatures smaller than ~1,700 K are usually dominated by dust, between 1,700 and 5,000 K by molecules and neutral atoms, and above 5,000 K by ions and electrons.

C.5 Scattering Opacities

Always important to consider is the case where electromagnetic waves are scattered by free (stationary) electrons in the classical limit (i.e., $h\nu \ll m_e c^2$), which is usually referred to as *Thomson scattering* and which was first published by W. Thomson in 1906. The cross section for Thomson scattering is:

$$\sigma_T = \frac{8\pi}{3}\left[\frac{e^2}{mc^2}\right]^2 \approx 6.65 \times 10^{-25}\ \mathrm{cm}^2 \tag{C.23}$$

Note that σ_T is frequency independent. An isotropic and homogeneous layer of hydrogen gas at a density $10^{-13}\ \mathrm{g\,cm^{-3}}$ becomes optically thick for Thomson scattering (i.e., $\tau_T > 1$) at a thickness of:

$$d = \frac{\tau_T}{\sigma_T n} = 2.4 \times 10^{13}\ \mathrm{cm} \approx 2\ \mathrm{AU} \tag{C.24}$$

where n is the electron number density. At very high photon energies (i.e., $h\nu \lesssim m_e c^2$) the cross section becomes dependent on energy and is then described by *Compton scattering*.

Electromagnetic waves are also scattered by bound electrons, which is referred to as *Rayleigh scattering* after Lord Rayleigh's result from 1899. Rayleigh scattering differs from Thomson scattering when the characteristic transition frequency ν_{line} in atoms or molecules is much larger than ν by

$$\sigma_{R,\nu} = \sigma_T \frac{\nu^4}{(\nu^2 - \nu_{line}^2)^2} \quad \text{for } \nu < \nu_{line} \tag{C.25}$$

C.6 Continuum Opacities

Free–free transitions are generally associated with energy losses or gains by electrons in the field of ions. In common terms the emission from such events is called *bremsstrahlung*, and in the case of thermalized electrons it is also called *thermal* bremsstrahlung. The energy of the outgoing photon is always the difference of electron energy before and after the braking event and, thus, in this case the emitted energy spectrum is a smooth function with no characteristic edges. For H-like ions the free–free cross section can be expressed in a semi-classical way with a frequency and temperature dependence of:

$$\kappa_\nu^{ff} \propto \frac{G^{ff}(\nu, T)}{\nu^3 T^{1/2}} (1 - \exp(-h\nu/kT)) \tag{C.26}$$

where G^{ff} is the free–free Gaunt factor. For many applications it is useful to apply the Rosseland mean of the coefficient shown in (C.18). The free–free absorption coefficient then can be conveniently written as [1310]:

$$\kappa_R^{ff} = 1.7 \times 10^{-25} T^{-7/2} Z^2 n_e n_i G^{ff} \tag{C.27}$$

also known as Kramers' law of free–free absorption. Note that here the Gaunt factor G^{ff} is frequency averaged as well and the result is about unity. Some studies recommend 1.2 as a good approximation. The total loss rate of an *emitting* plasma is [849]:

$$-\frac{dE}{dt} = 1.435 \times 10^{-27} T^{1/2} Z^2 n_e n_i G^{ff} \tag{C.28}$$

Unlike for free–free processes, bound–free absorption cross sections are not smooth functions of ν but contain discrete features which, at the limit of all bound states E_i, transpose as edges (see Fig. C.2 and also Fig. 3.19). In practice, the calculation of exact photoionization cross sections is extremely difficult and most

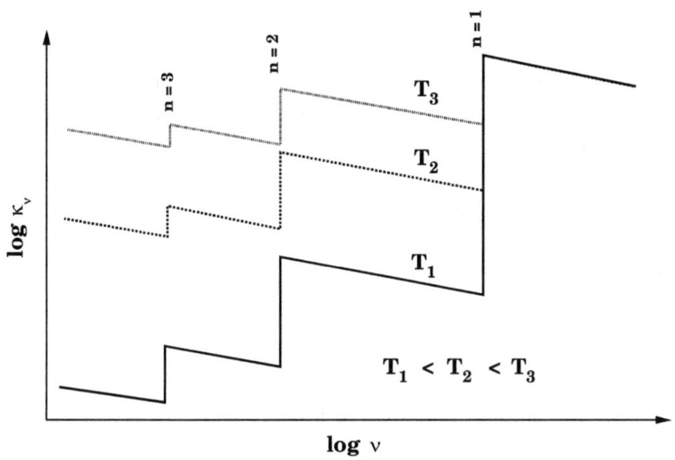

Fig. C.2 Schematic illustration of the temperature and frequency dependence of the bound–free absorption coefficient for H-like atoms. Adapted from [1311]

reliable only for H-like ions at best. In this case, the semi-classical expression for the bound–free absorption coefficient looks like [1311]:

$$\kappa_\nu^{bf} \propto \frac{Z^4}{\nu^3 G_a(T)} \exp(-Z^2\chi_H/kT) \sum_{n \geq n_c} \frac{1}{n^3} \exp(Z^2\chi_H/n^2kT) G_n \qquad \text{(C.29)}$$

where Z is the ion charge in numbers of electrons, $G_a(T)G_n$ are partition functions (a stands for all-absorbing atoms), n the principal quantum number and n_c the cutoff quantum number for which $n_c^2 \gg \frac{Z^2\chi_H}{h\nu}$, and χ_H the hydrogenic ionization potential. The term 'semi-classical' means that the validity the cross sections for bound–bound transitions (see below) is extended into the regime of virtual quantum numbers beyond the atomic series limit. The actual functional dependence of κ_ν^{bf} on ν and T is shown schematically in Fig. C.2. Were it not for the sum at the end, the frequency dependence of this expression is not so much different from (C.26) and in-between bound levels κ_ν^{bf} falls off as $1/\nu^3$. In this respect a Rosseland mean such as Kramers' law for the free–free case is possible, but makes sense only within limited frequency intervals.

C.7 Line Opacities

Compared with the derivations of expressions for continuum processes the bound–bound case seems comparatively simple. However, one has now to consider the determination of these coefficients for the large number of transitions involved. There are three fundamental processes to consider: spontaneous emission from an excited level E_f to a energetically lower atomic level E_i at a rate A_{fi}, stimulated emission by an external radiation field with intensity I_ν and a rate $B_{fi}I_\nu$, and absorption of a photon at a rate $B_{if}I_\nu$. The rate coefficients A_{fi}, B_{fi}, and B_{if} are the Einstein coefficients and in thermodynamic equilibrium are related by Kirchhoff's law through:

$$n_i B_{if} I = n_f A_{fi} + n_f B_{fi} I \tag{C.30}$$

where I is the mean radiation intensity:

$$I = \frac{1}{4\pi} \int_0^\infty \int I_\nu \, \mathrm{d}\Omega \, \mathrm{d}\nu \tag{C.31}$$

and n_i and n_f are the number densities of levels i and f with:

$$\frac{n_i}{n_f} = \frac{g_i}{g_f} \exp(h\nu_{if}/kT) \tag{C.32}$$

where g_i and g_f are the statistical weight of levels i and f, and $h\nu_{if}$ the transition (line) energy. The line absorption coefficient for this transition is then:

$$\kappa_{i,f,\nu}^{bb} = \frac{h\nu}{4\pi}(B_{if}n_i - B_{fi}n_f)\phi_{i,f}(\nu) \tag{C.33}$$

and the line emission coefficient, where $\phi(\nu)$ is the absorption line profile:

$$j_\nu^{bb} = \frac{h\nu}{4\pi} A_{fi} n_f \phi(\nu) \tag{C.34}$$

then can be evaluated using the Einstein relations:

$$A_{fi} = \frac{2h\nu^3}{c^2} B_{fi}; \quad g_i B_{if} = g_f B_{fi} \tag{C.35}$$

B_{if} can also be expressed in terms of the quantum mechanical oscillator strength f_{if} of a line:

$$B_{if} = \frac{4\pi}{h\nu} \frac{\pi e^2}{m_e c} f_{if} \tag{C.36}$$

At higher temperatures (i.e., higher than 1,000 K) excitations also happen through collisions with electrons, ions, or even molecules. In this case (C.30) has another term: $C_{fi} n_f$.

The line absorption coefficient must then incorporate all line contributions towards the frequency ν and thus:

$$\kappa_\nu^{bb} = \sum_{i<f} \kappa_{i,f,\nu}^{bb} \tag{C.37}$$

Note that in (C.33) it has already been implicitly assumed that $\phi_{i,f}(\nu) = \phi_{f,i}(\nu)$ again reflecting the precise LTE conditions. In that case one can assume that the occupation numbers n_i and n_f follow the Boltzmann distribution and with $g_i B_{i,f} = g_f B_{f,i}$ the final bound–bound absorption coefficient becomes:

$$\kappa_\nu^{bb} = \sum_{i<f} \frac{h\nu}{4\pi} g_i B_{i,f} \phi_{i,f}(\nu) n_a \frac{\exp(\frac{E_i}{kT})}{G_a(T)} \left[1 - \exp\left(-\frac{h\nu}{kT}\right)\right] \tag{C.38}$$

where n_a and $G_a(T)$ are again the density and partition function of the absorbing species, and the term in square brackets is the correction for stimulated emission. The line function $\phi(\nu)$ is usually assumed to be Gaussian, though in many cases a Lorentzian shape needs to be applied. In case of a delta function as a line profile, the absorption coefficient would primarily depend on the oscillator strength of the line and plasma density (see also Sect. D.3) with a temperature dependence at very high temperatures. Physical environments affect the line shapes. Additional broadenings to the natural, such as thermal, turbulent, and collisional broadening, have to be taken into account as well (see Appendix F).

C.8 Molecular Excitations

The subject of excitation of molecules plays a central role in almost every aspect of stellar formation. Molecules contribute another quality to a gas by the addition of internal degrees of freedom. They not only play an important role in molecular cloud cores but also in cool (<5,000 K) stellar atmospheres. Like atoms and ions they are subject to line emission and absorption based on their electronic properties. The total energy of a molecule can be stated as:

$$E_{mol} = E_e + E_r + E_\nu \tag{C.39}$$

where E_e is the energy due to its electrons, E_r its rotational energy, and E_ν its vibrational energy. While transitions between electronic states usually occur at optical and ultraviolet wavelengths, rotational and vibrational lines appear at much longer wavelengths (i.e., vibrational lines are visible in the IR band, rotational lines in the radio band). The degrees of freedom provided by the molecules, on the one hand, not only contribute to the change in internal energy (see Chap. 5), but also play a major role in the observational diagnostics of collapsing clouds. The determination of molecular opacities is described in Sect. C.7. However, the problem now is that there are many more transitions for molecules to consider than there are in the atomic case. The following few paragraphs are designed to remind the reader of what is involved in these transitions. Much can be found in standard textbooks (the review in [1311] is highly recommended).

For a rigid di-atomic molecule the energies of rotational states can be described by:

$$E_r = \frac{h^2}{8\pi^2 \mathcal{I}} J(J+1) = hcBJ(J+1) \tag{C.40}$$

where $J = 0, 1, 2, 3, \ldots$ is the rotational quantum number and B the rotational constant. This constant is governed by the moment of inertia $\mathcal{I}$ with respect to the molecular rotation axis. The *selection rules* for electrical dipole radiation in general imply $\Delta J = 0, \pm 1$ (except for $0 \leftrightarrow 0$) but in detail become very complex through various molecular angular and orbital momentum couplings producing a large number of transitions. The inverse relationship with the moment of inertia pushes the frequency range for rotational transitions into the mega- and gigahertz domain. The fact that the rotational energies of molecules strongly depend on their moments of inertia also helps to distinguish between molecules that contain various isotopes of an element, thus handing the observer a precious diagnostical tool. However, molecules are not really rigid but have a finite force constant giving rise to vibrational excitations. The energies of these excitations then follow from the equation for an harmonic oscillator. Together rotational and vibrational energy levels add a vast amount of absorbing power to interactions with molecules. Special cases are symmetric molecules such as H_2, as here rotational–vibrational transitions are forbidden for electric dipole radiation. The reason for this is that in this molecule the center of mass always coincides with the center of electric charge and thus no dipole moment can develop. Consequently transition probabilities are all based on quadrupole moments and this molecule therefore has hardly any significant radiative signatures in the IR or radio bandpass.

Vibrational excitations are a natural result of the fact that bonds between atoms in molecules are not rigid but have a finite force constant k. The energy is thus that of a harmonic oscillator of frequency ν_0 with vibrational quantum numbers v:

$$E_v = h\nu_0 \left(v + \tfrac{1}{2}\right) \tag{C.41}$$

where quantum numbers $v = 0, 1, 2, \ldots$ and $\Delta v = \pm 1$. However, selection rules for v are even more complex than for the rotational case and not as straightforward to express as they depend on the shape of the potential energy function, such as the *Morse potential* for a specific molecule. The combined case (i.e., the Morse potential plus the rotational potential) leads to exceedingly diverse rotational–vibrational energy levels [1310].

Because of the very different orders of magnitude of the spacings between the electronic, vibrational, and rotational energy levels (see Fig. C.3), molecular transitions occur in very different regions of the spectrum. But, in addition there exists a large number of rotational–vibrational transitions, which for molecules such as O_3, CO_2 and CH_4 amount to line listings of the order of 10^5 lines [6]. Many previous hydrodynamic models that related to stellar atmospheres for a long time had a constant gas opacity value of $\chi^g = 2 \times 10^{-4}\ \mathrm{cm^2\ g^{-1}}$, which was based on the

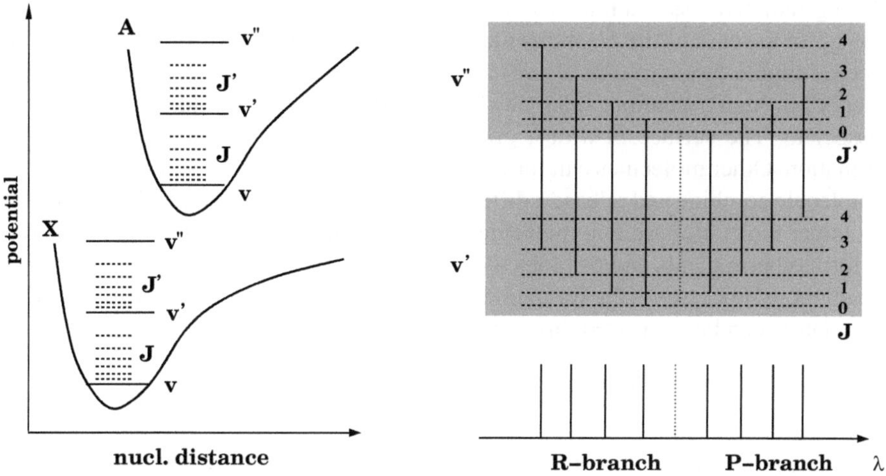

Fig. C.3 (*left*) Schematic illustration of locations and splittings of vibrational (v), rotational (J) and electronic (A, X) levels. The latter are shaped similar to Morse potentials. (*right*) Vibrational–rotational bands with transition of the R-branch (lines to the left of the $J = 0 \leftrightarrow J' = 0$ transition, which is actually forbidden) and the P-branch (lines to the right). Adapted from [1311]

Rosseland mean published in the early 1980s [1314]. Today this value is considered merely as a lower limit. More modern calculations of molecular mean opacities can be found in [1315, 1316] and references therein. General information on opacity databases may also be obtained though *OP* and *OPAL* as listed in Appendix E.

In thermal equilibrium the number of molecules in a specific state is given by (A.14). For each molecule all degrees of freedom, i.e., translational, rotational, vibrational, and electronic, have to be accounted for. The partition function for each molecule is then a complex product:

$$G = G_{trans} \times G_{rot} \times G_{vib} \times G_{el} \tag{C.42}$$

The total partition function in a molecular cloud is of great importance in the early stages of stellar collapse (see Chap. 5).

Formation and destruction of molecules in interstellar clouds and throughout stellar collapse are significant but extremely complex processes. The main reactions that lead to the formation, destruction, or rearrangement of molecular bonds involve [1317, 1318]:

- *direct radiative association*,
- *photodissociation* processes,
- *gas exchange reactions*,
- *catalytic formation* on the surfaces of dust grains,
- *dissociative recombination*,
- *ion–molecule reactions*.

During radiative association two atoms combine under irradiation of the binding energy. In the case of the H_2 molecule this does not work since this molecule has no electric dipole moment (see above) and thus is not able to get rid of the binding energy through radiation. The newly formed molecule would instantaneously dissociate. The surface of a dust grain, however, acts as a catalyst for molecule formation. Other molecules can form by gas exchange reactions, where an atom and a molecule combine and release a different atom. Similar and specifically interesting for larger molecules are ion–molecule reactions, where radicals and atoms interact to form other radicals and/or molecules through collisions. The quantity of interest for all these processes is the formation rate constant k, which in the case of radiative association can be expressed through:

$$\frac{\mathrm{d}n_{AB}}{\mathrm{d}t} = k n_A n_B \quad \text{or} \quad \frac{\mathrm{d}n_{AB}}{\mathrm{d}t} = k n_{AB} n_{gr} \tag{C.43}$$

where n_{AB}, n_A, n_B and n_{gr} are the number densities of the molecule AB, the atoms A and B, and the dust grains. Values for a k range from as low as $10^{-20}\,\mathrm{cm}^3\ \mathrm{s}^{-1}$ for OH formation at $T \geq 50\,\mathrm{K}$ [1319] and as high as $10^9\,\mathrm{cm}^3\ \mathrm{s}^{-1}$ for $CH_3NH_3^+$ formation [1320].

The ultimate destruction of molecules is predominantly achieved by *photodissociation*, *photoionization* and at sufficiently high temperatures effectively by *collisions*. These processes necessarily have to provide energies exceeding a molecule's dissociation energy χ_{dis}, where energy is defined as the amount needed to dissociate the molecule from the lowest rotational–vibrational level to produce neutral atoms in their ground states. In other words, if the molecule is excited beyond a limiting vibrational state it will break apart. The photodissociation rates k_{pd} can be calculated once the intensity distribution of the radiation field I_λ and the $\sigma_{AB}(\lambda)$ for the specific molecule AB are known:

$$k_{pd} = \int_{\lambda_h}^{\lambda_d} I_\lambda \sigma_{AB}(\lambda)\mathrm{d}\lambda \tag{C.44}$$

The absorption cross sections can be calculated as the sum over all possible rotational–vibrational excitations and is determined from molecular absorption spectra. The lower limit of the intensity distribution is marked by $\lambda_h = 912$ Å, the wavelength at which hydrogen is photoionized. The upper limit (λ_d) corresponds to the equivalent wavelength of dissociation energy. Dissociation energies typically range within a few eV and have been measured for many existing molecules. Photodissociation rates usually have values between 10^{-13} and 10^{-9} with a trend toward higher molecular complexity also meaning higher destruction rates.

Photoionization may not be as destructive as photodissociation, and the necessary ionization potentials χ_{ion} (a few 10 eVs) are usually higher than dissociation energies. However, turning the molecule into an ion makes it quite susceptible to ion–molecule reactions that can lead to the destruction of the molecule or to a different molecule. Clearly, as temperatures rise above ~2,000 K, collisions should have destroyed most of the molecular population.

C.9 Dust Opacities

The calculation of dust opacities involves assumptions about the shape and distribution of dust particles. Most generally one can express the dust opacity for a single species as:

$$\chi^d(\mathbf{r}, \nu, t) = \pi \int_{a_{min}}^{a_{max}} a^2 Q(a, \nu) f(\mathbf{r}, a, t)\, \mathrm{d}a \tag{C.45}$$

where a is the grain radius; a_{min} ($\sim 10^{-3}$ μm) is the lower limit to the grain size to have macroscopic properties; a_{max} is the convenient upper limit which most of the time is set to ∞; $Q(a, \nu)$ is the grain-size-dependent extinction efficiency and incorporates the structural properties of dust grains. These efficiencies are usually calculated using the *Mie theory* [1321, 1322]. Last, but not least, $f(\mathbf{r}, a, t)$ denotes the local size distribution function of the grains. A few basic assumptions can simplify (C.45). First, one may assume steady-state conditions and eliminate time dependence. This is in order in many situations and for the sake of understanding some of the basic physics involved in opacity calculations. For stellar collapse calculation, time is essential and has to be included. Second, one may assume that dust species are homogeneously distributed, meaning all dust species are well mixed. Third, there is only one single temperature for all dust grains. Fourth, the dust shapes are spherical and the shape is described by its radius a. For collapse calculations, the assumption about the temperature seems feasible, but that about the shape is more stringent as ellipsoid shapes are more likely. In this case, opacity can be evaluated for all involved dust species with separate contributions from absorption and scattering [1323]:

$$\begin{pmatrix} \kappa_\nu^d \\ \sigma_\nu^d \end{pmatrix} = \sum_i \alpha_i \int_a \begin{pmatrix} Q_i^{abs}(a, \nu) \\ Q_i^{sca}(a, \nu) \end{pmatrix} \pi a^2 n_i(a)\, \mathrm{d}a \tag{C.46}$$

where index i refers to the i-th dust species, α_i is a weighting factor based on the abundance of the i-th species. $n_i a$ is the normalized dust size distribution function for the ith species. In the small particle limit of Mie theory (i.e., $\lambda \gg 2\pi\, a$), the size dependence of the extinction coefficient reduces to a and by ignoring scattering for now one can write (see also [1324]):

$$\kappa_\nu^d = \pi \sum_i \alpha_i Q_{i,\nu} \int_a a^3 n_i(a)\, \mathrm{d}a \tag{C.47}$$

Normalized dust size distribution functions can have various forms. Many times the following two forms are used:

$$n_i(a) \propto \begin{cases} a^{-\gamma_i} & a_{min} \le a \le a_{max} \\ a^{-\gamma_i} e^{-a/a_o} & a_{min} \le a \end{cases} \tag{C.48}$$

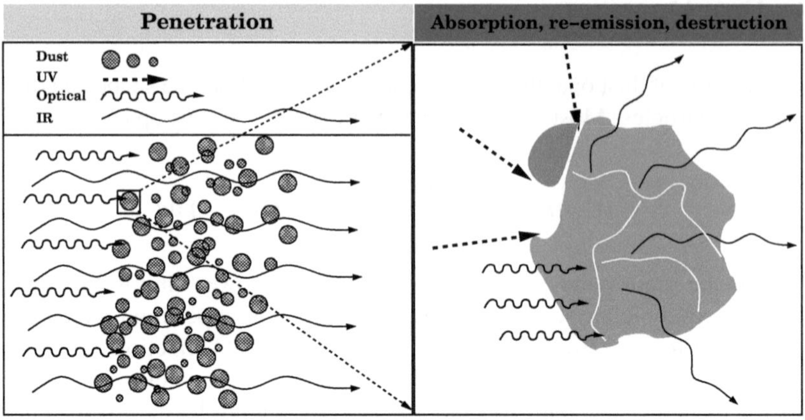

Fig. C.4 Illustration of the main processes involving dust and radiation. The left diagram shows the penetrative properties of long-wavelength radiation. The size of dust particles may usually be assumed to be 1 μm and smaller. Visible and UV radiation cannot penetrate a typical cloud of dust. The right diagram shows dust absorbing optical and UV radiation and re-emitting IR light. The pounding of short-wavelength light causes the particle to crack and disintegrate

where the top functions follow the MRN approach [128], while the bottom choose the KMH approach [1325]. Recent calculations can be found, for example, in [1323, 1326]. Calculated dust opacities for coagulated dust grains of size between 1 μm and 1.3 mm for protostellar cores are tabulated in [420].

One of the most important ingredients of interstellar matter is dust. Like molecules the presence of dust adds additional degrees of freedom to increase the internal energy of inter- and circumstellar matter. As important as these properties are for star formation and early evolution, even today they are very little understood. A key point is that absorption and emissivity properties need to be known over a very large wavelength range. Dust also appears in very different environments adding to uncertainties in the modeling of dust properties. Figure C.4 illustrates the major interactive processes of dust particles with radiation, which involve the *penetration* of long-wavelength radiation, *absorption* of short-wavelengths, *thermal emission* at long wavelengths and *destruction* of dust by short wavelength radiation.

More intriguing is the absorption and re-emission of radiation. Here the absorption coefficient dramatically depends on the composition of the dust; the thorough review of interstellar dust properties by F. Boulanger and colleagues [548] is recommended. Absorption may not only happen at short wavelengths. Some models, for example, involve dust cores with ice mantles leading to vibrational IR absorption bands from trapped molecules. Another factor in models involves the so-called *fluffiness* of dust clusters allowing for enhanced absorptivity at a few hundred μm wavelengths [1327]. Actual opacities were modeled by V. Ossenkopf and T. Henning for dust between 1 μm and 1.3 mm [420].

Appendix D
Spectroscopy

The analysis of electromagnetic light from stars, dust and nebulae is the only source of information available to study star-forming regions. Observations today are performed throughout the entire electromagnetic spectrum and, although this constitutes 15 orders of magnitude in wavelength range, spectral signatures and analysis methods throughout the entire spectrum are rather similar. This appendix offer some additional information on spectroscopy issues and describes some novel diagnostic tools, that are rarely found in other spectroscopy resources. Many items specifically apply to high-energy spectra. These are included because high-quality X-ray spectra have become only recently available.

D.1 Line Profiles

One of the most powerful tools to diagnose the dynamics of line-emitting or line-absorbing media is the analysis of basic line properties. The shape of a line is often approximated by a Gaussian distribution:

$$\Phi_\nu = \frac{1}{\sqrt{\pi}\,\Delta\nu_D}\,\mathrm{e}^{-\left(\frac{\nu-\nu_O}{\Delta\nu_D}\right)^2} \tag{D.1}$$

where ν_o is the line frequency in the rest laboratory frame, ν is the frequency in the observers' frame and $\Delta\nu_D$ is the *Doppler width* of the line defined as:

$$\Delta\nu D = \sqrt{\nu_o/c^2(v_{th}^2 + v_{turb}^2)} \tag{D.2}$$

where the thermal velocity contribution $v_{th}^2 = 2kT/m_i$ and turbulent velocity contribution is v_{turb}. The former accounts for the movement of the gas atoms, ions, or molecules of mass m_i due to their kinetic temperature T; the latter accounts for the micro-turbulent motions of gas particles.

N.S. Schulz, *The Formation and Early Evolution of Stars*, Astronomy and Astrophysics Library, DOI 10.1007/978-3-642-23926-7, © Springer-Verlag Berlin Heidelberg 2012

Generally, many processes are involved such as natural broadening, Stark broadening, van der Waals broadening, and Zeeman broadening, all of which take place simultaneously and contribute to the line profile. The result is a convolution of several distribution functions. *Natural* broadening, for example, represents the inverse of the radiative lifetime of an excited state. Its shape is that of a *Lorentzian* (or *Cauchy*) distribution, while its convolution with various Doppler distributions is that of a *Voigt* profile. The latter convolution of the dispersion and Gaussian profiles has a fairly complex construct of the form:

$$\Phi_{V,\nu} = \frac{1}{\pi^{1/2}\Delta\nu_D}\frac{a}{\pi}\int_{-\infty}^{+\infty}\frac{\exp(-u'^2)}{(u-u')^2+a}\,du' \tag{D.3}$$

where $u = \Delta\nu/\Delta\nu_D$, $a = (\Gamma/4\pi)/\Delta\nu_D$, and Γ is the line damping constant. The line shape is then represented by a Gaussian core and Lorentzian wings (see also below).

D.2 Zeeman Broadening

Emission and absorption lines are sensitive to a variety of physical environments such as enhanced pressures, rotation and velocity fields, to name a few. Their impacts on line shapes hand the observer powerful diagnostical tools. The line diagnostics in these cases have been part of standard textbooks for manyyears [54, 1311, 1312, 1344] (the reader is generally referred to these). From studies of atomic structure it is also well known that the presence of electrical and magnetic fields affects the atomic level structure in decisive ways, which in the case of electric fields is the *Stark effect* and in the case of magnetic fields it is the *Zeeman effect*. Once again, despite these being rather standard effects, recent successful measurements of *Zeeman broadening* of lines from proto- and PMS stars have brought renewed attention towards its importance.

The Zeeman analysis of line emission from T Tauri stars requires well-modeled stellar atmospheres (for a review see [54]). There were extensive model calculations in the 1990s that set standards for LTE model atmospheres [1328–1330]. The shift in wavelength due to the magnetic field for a particular Zeeman component is [663]:

$$\Delta\lambda = \pm\frac{e}{4\pi m_e c^2}\lambda^2 g B = \pm 4.67\times10^{-7}\lambda^2 g B \ \ \text{mÅ kG}^{-1} \tag{D.4}$$

where λ is the line wavelength in Å. The relation shows that Zeeman shift scales with λ^2, while Doppler broadenings scale only with λ. The longer the observed wavelength, the more sensitive the line shape to Zeeman broadening as it dominates over Doppler-related broadenings, such as those due to rotation, pressure, and turbulence (see above). The use of IR lines to measure magnetic fields in MS stars

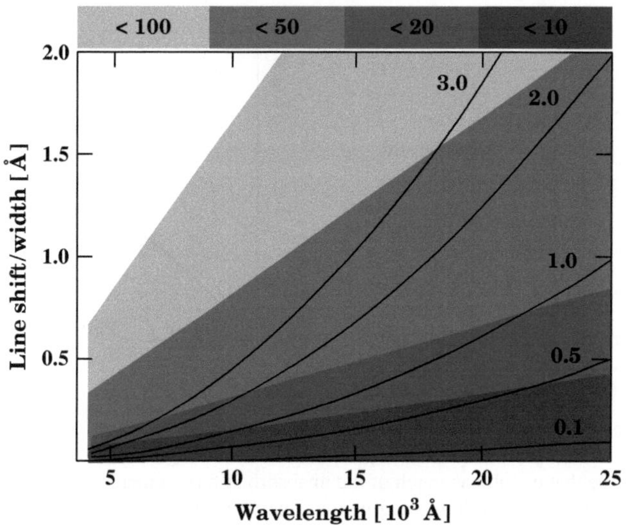

Fig. D.1 Weighing Zeeman broadening against Doppler broadening. The shades cover areas of Doppler widths in km s^{-1} that have values less than 100 (*lightest shade*) to less than 10 (*darkest shade*). The lines represent Zeeman shifts for magnetic field lines of 0.1 to 3.0 kG as labeled. These shifts were calculated using log g = 3.67 for BP Tau [663]

has been around for quite some time [1331, 1332]. Figure D.1 demonstrates this effect for typical T Tauri atmospheric parameters, while at optical wavelengths possible Zeeman shifts for fields up to and likely above 3 kG are dominated by or compete with Doppler shifts. Towards IR wavelengths this picture changes dramatically and fields above 1 kG clearly dominate Doppler shifts of 10 km s^{-1}, which are typical for T Tauri stars.

D.3 Equivalent Widths and Curve of Growth

The strength of an absorption line with respect to its continuum is measured in terms of the equivalent width W_λ, which is defined as:

$$W_\lambda = \int \frac{I_0 - I(\lambda)}{I_0} d\lambda \tag{D.5}$$

where I_0 is the continuum intensity at the center of the line and I(λ) the intensity of the observed spectrum (see Fig. D.2).

The curve of growth (COG) for a given spectral line describes the behavior of W_λ as the number of absorbing atoms in the line-of-sight increases. The right part of Fig. D.2 illustrates this behavior. The components describing the three regions

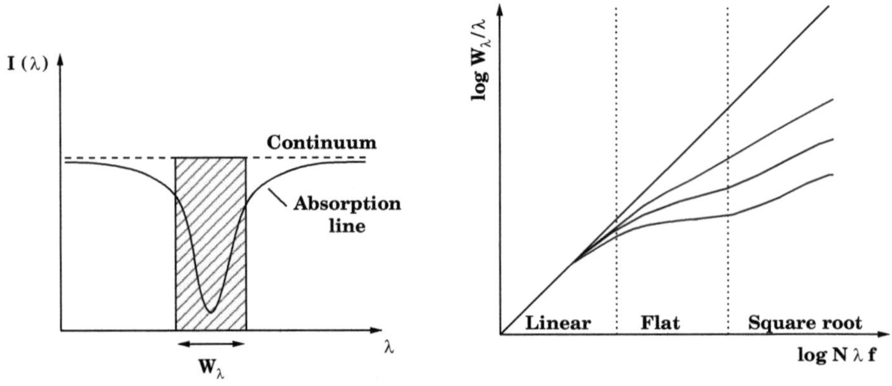

Fig. D.2 Definition of line equivalent width and a schematic concept of the curve of growth properties. (*left*) The equivalent width of a line is the wavelength range under the continuum of that particular line that engulfs as much of the unabsorbed flux as there is in the line itself (shaded area). (*right*) A schematic plot indicating various analytic solutions in different parts of the curve of growth

of the cog are described in the following. For an optically thin, homogeneous layer (i.e., unsaturated lines) W_λ is related to the column density N_j of ion j as:

$$\frac{W_\lambda}{\lambda} = \frac{\pi e^2}{m_e c^2} N_j \lambda f_{jk} \tag{D.6}$$

where f_{jk} is the oscillator strength for transition k to j, e is the elementary charge and m_e the electron mass. Thus for small N_j the number of absorbed photons increases in proportion to the number of atoms. We call this the *linear* part of the curve of growth. The absorption line is not yet saturated and here one wouldn't see any change in W_λ from velocity broadening.

Once the line saturates it gets dominated by the Doppler parameter b describing a broadening of the line through relative velocities. The simplest case is a Gaussian velocity distribution, $b = \sqrt{2} v_{rms}$. The width can then be approximated by:

$$\frac{W_\lambda}{\lambda} = \frac{2b\lambda}{c} \sqrt{\ln\left(\frac{\sqrt{\pi} e^2}{m_e c^2} \frac{N_j \lambda f_{jk}}{b}\right)} \tag{D.7}$$

This is called the *flat* part of the COG, where the dependence of the equivalent width on the column density is not very strong.

In saturated lines when the column density is high enough, the Lorentzian wings of the line begin to dominate:

$$\frac{W_\lambda}{\lambda} = \frac{\lambda^{3/2}}{c} \sqrt{\frac{e^2}{m_e c^2} N_j \lambda f_{jk} \Gamma} \tag{D.8}$$

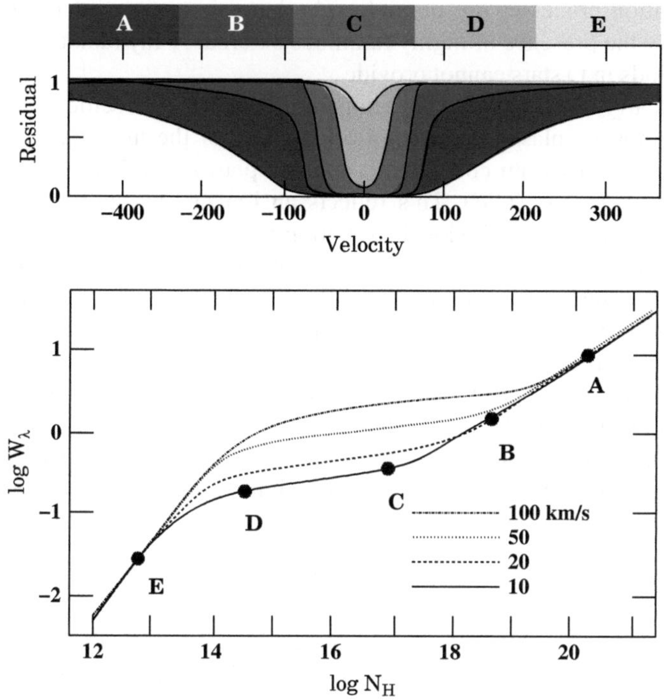

Fig. D.3 Different regimes of the curve of growth showing the relationship between the Doppler parameter, the column density and the shape of an absorption line. Adapted from Charlton [1333]

where Γ is the damping constant of the Lorentz profile (e.g., [1312]). This is called the *square root* portion of the COG, where the equivalent width is again strongly dependent on the column density.

Although the flat part of the COG has little dependence on N_j, it is a powerful tool because of its high sensitivity to the Doppler parameter. Figure D.3 again illustrates the different regimes for H I Lyman α transition [1333]. The upper panel shows absorption profiles as they appear for various Doppler parameters. The lower panel shows a calculated COG for neutral hydrogen column densities between $N(\mathrm{H\,I}) \sim 10^{12} - 10^{20}$ cm^{-2} versus W_λ for various Doppler parameters identifying the locations of the widths from the top panel.

D.4 Spectra from Collisionally Ionized Plasmas

Spectral emissions from the far-UV to X-rays originate from hot plasmas of temperatures between 10^5 K and 10^8 K. The origins of these emissions are generally collisionally ionized plasmas, though at temperatures below a few 10^5 K significant photoionization processes are possible. The latter is not considered here as this would imply the presence of highly luminous external X-ray radiation fields which even the winds in O stars cannot provide.

Spectra carry information about temperature, density, velocities, geometry, and whether or not the plasma is in equilibrium. One of the most basic assumptions about the emissions from collisionally ionized plasmas is that it is in *collisional ionization equilibrium (CIE)*. This reflects an ionization equillibrium where the dominant processes are collisional excitation from the ground state and includes corresponding radiative and di-electronic recombination. The referral to the ground state only in combination with an optically thin plasma is also called the *coronal approximation* in analogy to the conditions believed to exist in the Solar corona. Optically thin in this respect means particle densities of 10^{10} cm^{-3} or less. The only exceptions (see Sect. D.5) to transitions from the ground state are transitions involving a few metastable line levels.

A typical X-ray spectrum from a hot collisionally ionized plasma is shown in Fig. D.4. The emission measure of an electron ion plasma of volume V is defined as:

$$\mathcal{EM} = \int n_e n_H \, \mathrm{d}V \tag{D.9}$$

where the product of electron and ion density is due to the fact that single collisions are considered. In reality, however, the line emissivity is also a function of temperature with both peak temperature and finite temperature distribution depending on the ion species. In other words, O VIII ions reach their maximum emissivity at $\log T = 6.3$ K, whereas Si XIV ions reach their maximum at $\log T = 7.2$ K and Fe XXV ions at $\log T = 7.8$ K. The emitted flux in each line of a collisionally ionized plasma is then expressed in the form [1334]:

$$F_{line} = \frac{A_{line,Z}}{4\pi d^2} \int G_{line}(T_e, n_e) n_e n_H \, \mathrm{d}V \tag{D.10}$$

where F_{line} is the line flux, $A_{line,Z}$ is the elemental abundance, d the distance of the plasma, V is the emission volume, and n_e and n_H are the electron and hydrogen densities, respectively. The core of this function is the line emissivity $G_{line}(T_e, n_e)$ in units of photons cm^3 s^{-1}. The function G depends on the plasma's electron temperature and density and contains all fundamental atomic data. It also reflects the ionization balance. The function G also inherits the temperature distribution of ion emissivity and, since this function is usually sharply peaked over a small temperature range, one replaces $G_{line}(T_e, n_e) = \overline{G}_{line}\mathcal{EM}(T_{max})$. Furthermore, one ignores the dependence on density n_e. Then $\overline{G}_{line}$ is the mean emissivity and T_{max} is the temperature of maximum emissivity. The line flux then can be simply expressed by:

$$F_{line} = \frac{A_{line,Z}}{4\pi d^2} \overline{G}_{line} \mathcal{EM}(T_{max}) \tag{D.11}$$

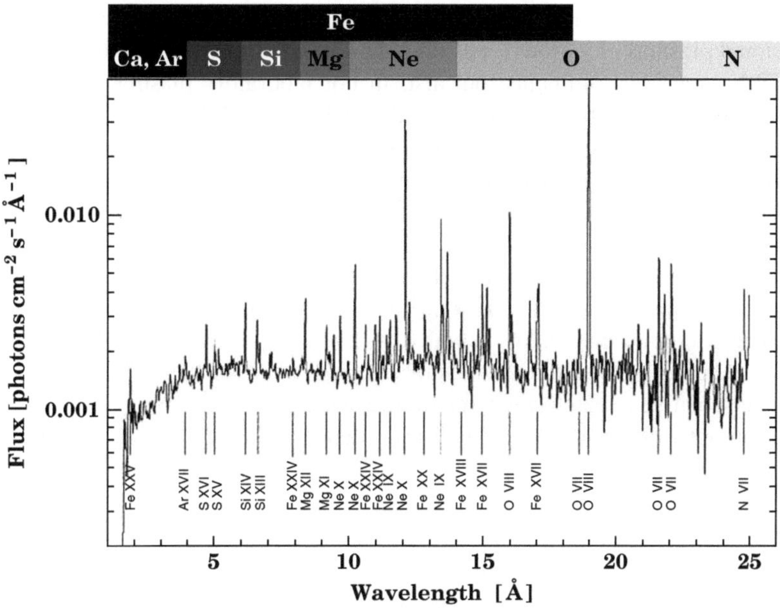

Fig. D.4 High resolution X-ray spectrum of the binary star II Peg. The binary harbors a star with an active corona. The X-ray luminosity exceeds that of the Sun by many orders of magnitude (similar to what is observed in young PMS stars). Marked are the domains of H-like and He-like lines from various ion species ranging from O ($Z = 8$) to Fe ($Z = 26$). Iron ions from Fe XVII to Fe XXV are present. The peak temperature of this spectrum is several 10 MK hot. Credit: from Huenemoerder et al. [683]

The emission measure distributions shown in Fig. 10.15 were computed in a slightly different manner as it is often more useful to apply a differential form of the emission measure, where $DV = (\mathrm{d}V/\mathrm{d}\log T)\mathrm{d}\log T$ and the line flux can then be written as

$$F_{line} = \frac{A_{line,Z}}{4\pi d^2} \int \overline{G}_{line} \left[\left(n_e n_H \frac{\mathrm{d}V}{\mathrm{d}\log T} \right) \right] \mathrm{d}\log T \qquad \text{(D.12)}$$

where the term in brackets in now called the *differential emission measure (DEM)*. The DEM is a powerful construct as it describes the plasma distribution with respect to temperature [683, 1335, 1336] under the assumption of uniform abundances.

D.5 X-ray Line Diagnostics

High-resolution X-ray spectra provide a variety of lines from mostly H-like and He-like ions and thus offer novel plasma diagnostical tools to study young stellar objects. Most of these tools have been developed since the late 1960s through observations of the solar corona [971]. Ratios of these lines are widely used for collisional plasmas of various types ranging not only from flares in the solar corona, the ISM, supernova remnants, but also in hot laboratory plasmas, such as those confined in tokamak fusion plasmas [981, 983, 1337, 1338]. Central to X-ray line diagnostics are the He-like line triplets of C V, N VI, O VII, Ne IX, Mg XI and Si XIII corresponding to their ion production temperature range between 2 MK and 10 MK. He-like line triplets consist of transitions involving the $1s^2\,{}^1S_0$ ground state, the metastable *forbidden* $1s2s\,{}^3S_1$ level and the $1s2p\,{}^3P_{1,2}$ *intercombination* level. These transitions form three lines, the resonance (r) line ($1s^2\,{}^1S_0$– $1s2p\,{}^1P_1$), the intercombination (i) line ($1s^2\,{}^1S_0$–$1s2p\,{}^3P_{(2,1)}$) and forbidden (f) line ($1s^2\,{}^1S_0$–$1s2s\,{}^3S_1$). The intercombination line actually has two components, a detail which in most cases can be ignored. In the absence of external radiation fields, the ratios of these lines are sensitive to electronic temperature and electron density.

For an optically thin plasma the *G-ratio* depends on electron temperature; and is the flux ratio:

$$G = \frac{(f_{i1} + f_{i2}) + f_f}{f_r} \tag{D.13}$$

where f_{i1}, f_{i2}, f_f, f_r are the corresponding line fluxes of the intercombination, forbidden, and resonance line respectively. As long as CIE holds the change in G-ratio follows from the fact that the excitation of the ${}^3P_{(2,1)}$ and ${}^3S_1$ levels decrease faster than collisional excitation to the ${}^1P_1$ level with decreasing electron temperature [968]. The functional dependence of G on temperature is complex and may be best described as a subtle increase of G with temperature at low temperatures followed by a rapid decrease above a critical value. This critical value increases with Z, for C V this temperature is around 3×10^5 K, and for Si XIII it is above 1×10^6 K (see [984]).

The density sensitivity of the He-like triplet involves the metastable forbidden line. Since the $1s2s\,{}^3S_1$ level does not possess a significant dipole moment it is more likely to be depopulated through collisional excitation into the intercombination levels before decaying into the ground state, once the electron density becomes high enough. This effect is reflected in the R-ratio which is simply:

$$R = \frac{f_f}{f_i} \tag{D.14}$$

The R-ratio has a density sensitivity that can be parameterized as [1339]:

$$R = \frac{R_0}{1 + n_e/n_{crit}} \tag{D.15}$$

Here the ratio in the low density limit is R_0. As long as the electron density n_e falls well short of the critical density n_{crit}, the measured ratio of these lines will always be near R_0. Theoretical values have been calculated [983, 1340] for each of the corresponding elements. Figure D.5 shows the dependence of these values with wavelength and atomic number Z. The density paradigm breaks down if strong

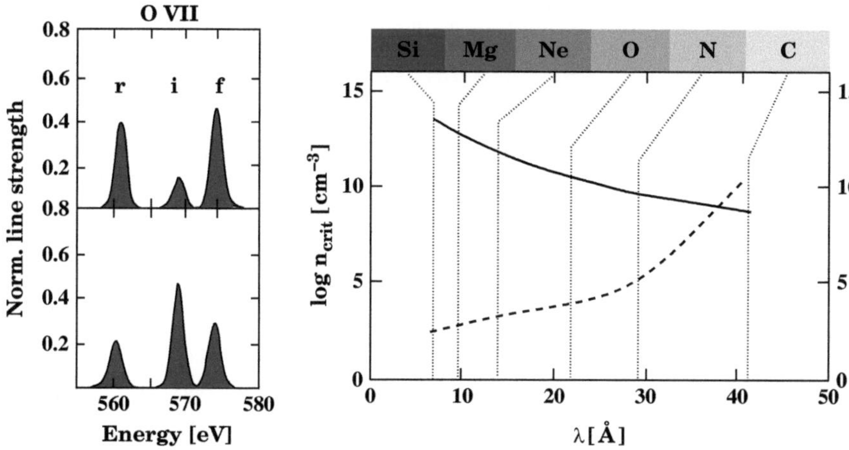

Fig. D.5 (*right*) Dependence of the low-density limit R_0 (*right scale*) and the critical density n_{crit} (left scale) of the R-ratio with wavelength and Z. (*left*) The relative strengths of the components of the O VII He-like triplet. The top spectrum shows the appearance of the low density limit at $n_e = 10^{10}$ cm^{-3}, and the bottom spectrum shows the triplet at $n_e = 10^{12}$ cm^{-3} when no external radiation fields are present. Adapted from Proquet and Dubau [984] and Ness et al. [978]

external radiation fields are present that can also depopulate the metastable level by excitation to $^3P_{2,1}$. Then the density and radiation effect compete and the ratio has additional terms for the radiation field:

$$R = \frac{R_0}{1 + n_e/n_{crit} + \phi/\phi_{crit}} \tag{D.16}$$

where ϕ is the field flux at the necessary pumping wavelength and ϕ_{crit} the critical field for each triplet [978, 984, 1339].

Appendix E
Abbreviations

This appendix contains short explanations of the acronyms and abbreviations used throughout the book. the descriptions given are minimal and only designed for look-up purposes. In the book acronyms are spelled out only once when they appear for the first time. There are no references or other referrals attached to these explanations with the exception of links to websites in some cases (where a fuller explanation is given). In all other cases, the main text of this book or a standard textbook will provide further information if necessary.

A.C.: Time count in years *After* the birth of *Christ*.

AGB: *Asymptotic Giant Branch* stars resemble an advanced stage in the MS evolution of stars in the HRD. these are stars with degenerate C–O cores and with H- and He-burning shells.

APEC: the *Astrophysical Plasma Emission Code* and *Database (APED)* was developed to model thermal spectra of stellar coronae and other hot astrophysical plasmas in the Far-UV and X-ray wavelength range (see also *CHIANTI, HULLAC, MEKAL, SPEX*). For more information see http://cxc.harvard.edu/atomdb/sources_apec.html.

AU: the *Astronomical Unit* is a length measure in terms of the nominal Earth–Sun distance.

B.C.: Time count in years *Before* the birth of *Christ*.

bb: *bound–bound* processes in radiative tranfer.

bf: *bround–free* processes in radiative tranfer.

N.S. Schulz, *The Formation and Early Evolution of Stars*, Astronomy and Astrophysics Library, DOI 10.1007/978-3-642-23926-7, © Springer-Verlag Berlin Heidelberg 2012

CAI: *Ca- and Al-rich Inclusions* appear in meteorites, specifically in chondrites, as enriched short-lived isotope daughter products.

CCCP: the *Chandra Carina Complex Project* is a high-resolution X-ray survey lead by L. Tounsley with 38 *Chandra* pointings towards the Carina star forming region.

CCD: *Charge-Coupled Device* is a light-sensitive, areal solid state detector designed for IR, optical, UV, and now also for X-ray imaging.

CHIANTI: *CHIANTI* is an atomic database for the spectroscopic diagnostics of astrophysical plasmas (see also *APEC, HULLAC, MEKAL, SPEX*). For more information see http://www.solar.nrl.navy.mil/chianti.html.

CIE: *Collisional Ionization Equilibrium* describes processes involving collisional excitation from ground states and the corresponding radiative and di-electronic recombination.

CMD: the *Color-Magnitude Diagram* plots the color index versus the visual magnitude of a stellar object.

CME: A *Coronal Mass Ejection* marks a highly accelerated ejection of particles into interplanetary space from the Sun.

CMF: the *Cluster Mass Function* consists of the initial distribution of stellar masses after current cluster formation within a GMC (see below). The *CMF* is thought to be close or identical to the *IMF* (see below).

CNO: the *Carbon–Nitrogen–Oxygen* cycle is the standard nuclear fusion chain process in the Sun's core.

COG: the *Curve of Growth* analysis is described in Sect. D.3.

COUP: *Chandra Orion Ultradeep Project* is a consortium of scientists, led by E. Feigelson, analyzing the vast *Chandra* X-ray database of the ONC.

CR: *Cosmic Rays* is high-energy particle radiation with the bulk in the range between 0.1 and 10 GeV consisting primarily of protons, α particles, and a small fraction of heavier particles.

CTTS: *Classical T Tauri Stars* are very young low-mass PMS stars. In the IR their SEDs are classified as class II, in the optical band they show strong H_α lines.

DEM: the *Differential Emission Measure* distribution of a star shows the emissivity of a specific abundance distribution of ions in a radiating astrophysical plasma with respect to temperature.

EGG: *Evaporated Gaseous Globule* is a dense core in molecular clouds which appears when a photoevaporating radiation field exposes them to the observer.

EGO: *Extended Green Objects* are identified based on their extended 4.5 μm emission in *Spitzer* images. These sources have been labeled from the common coding of the 4.5 μm band as green in 3-color *IRAC* (see below) images.

ELP: the *Emission Line Project* is a loose consortium of scientists to support X-ray and EUV line databases for astrophysical analysis.

EM: the *Emission Measure* (see also *DEM*) describes the emissivity of a hot astrophysical plasma within a specific volume.

ESC: *Embedded Stellar Clusters* are very young (<3 Myr) stellar clusters still embedded in their natal molecular cloud.

EUV: *Extreme Ultraviolet* wavelengths range between X-rays and the Far-UV, $\sim$700 Å $\lesssim \lambda \lesssim\sim$70 Å.

ff: *free–free* processes in radiative transfer.

FUV: *Far-Ultraviolet* wavelengths in the range 1,000 Å $\lesssim \lambda \lesssim$ 150 Å.

GC: *Globular Clusters* are compact stellar clusters harboring old stellar populations off the Galactic plane.

GL: the *Ghosh and Lamb* model is an accretion disk model devised by P. Gosh and F. K. Lamb based on the principle of star–disk locking.

GMC: *Giant Molecular Clouds* are dense interstellar clouds with masses larger than $10^5\ M_\odot$.

HAEBE: *Herbig Ae/Be* stars are very young intermediate (2–8 $M_\odot$) mass PMS stars.

HAe: *Herbig Ae* stars are a sub-group of *HAEBE* stars.

HBe: *Herbig Be* stars are a sub-group of *HAEBE* stars.

HDF: the *Hubble Deep Field* is a Director's Discretionary program on HST in Cycle 5 to image a typical field at high galactic latitude in four wavelength passbands as deeply as reasonably possible. For more information see http://www.stsci.edu/ftp/science/hdf/hdf.html.

HH: *Herbig–Haro* objects are outflow signatures where either a wind or jet interacts with inter- and circumstellar material.

HR: Stands for *Hertzsprung–Russell* after E. Hertzsprung and H. N. Russell (see below).

HRD: *Hertzsprung–Russell Diagram* is a tool to classify stars and describe their evolution by plotting the effective stellar surface temperature T_{eff} versus bolometric luminosity L_{bol}.

HUDF: the *Hubble Ultra-Deep Field* is a million-second-long exposure that reveals the first galaxies consisting of two separate images taken by *Hubble's Advanced Camera for Surveys (ACS)* and the *Near Infrared Camera and Multi-object Spectrometer (NICMOS)*. For more information see http://hubblesite.org/newscenter/archive/releases/2004/07/.

HULLAC: the *Hebrew University Lawrence Livermore Atomic Code* is a suite of programs to generate collisional radiative models for astrophysical plasmas (see also *APEC, CHIANTI, MEKAL, SPEX*).

IFDC: *IR Dark Clouds* appear highly absorbed with $A_V > 25$ and in silhouette against the Galactic mid-infrared background.

IGM the *Intergalactic Medium* consists of ionized gas around and in the vicinity of the Milky Way and in between galaxies in general.

IMF: the *Initial Mass Function* consists of the initial distribution of stellar masses after the first cluster formation within a GMC (see above). The expression also applies to initial distributions in galaxies as well as is being used in cosmology in the context of primordial star formation.

IR: *Infra-Red* is the wavelength range in the electromagnetic spectrum spanning from 0.8 μm to 350 μm.

ISM: the *Interstellar Medium* consists of all gas phases, interstellar dust, force fields, cosmic rays, and radiation fields between the stars.

ISRF: the *Interstellar Radiation Field* is the total emission from stars interacting with the ISM, its spectrum ranges from X-ray, FUV, UV, optical to the Far-IR and sub-mm bands.

KH: *Kelvin–Helmholtz*. Named after Lord Kelvin (W. Thomson) and H. von Helmholtz, who in the 19th century among others formulated the laws of thermodynamics. *KH* is mainly used in the expression for the thermal contraction timescale of stars.

KLF: the *K-band Luminosity Function* is computed from K-band (2.2 μm) observations of stellar clusters.

LG: the *Local Group* of galaxies in the neighborhood of the *Milky Way* within 100 Mpc.

LISM: *Local Interstellar Medium* is the same as ISM (see above) but distinctly for the neighborhood of the Sun within roughly 1,500 pc.

LMC: *Large Magellanic Cloud* is a nearby ($\sim$55 kpc) neighbor galaxy visible in the southern hemisphere.

LTE: *Local Thermodynamic Equilibrium* is the same as TE (see below) but for the case where the equilibrium condition is not valid everywhere except for a specified boundary condition.

MC: *Molecular Cloud*s are dense interstellar clouds with masses between 10^3 and 10^5 $M_\odot$.

MEKAL: *Mewe–Kaastra–Liedahl* code calculates emission spectra from hot diffuse plasmas (see also *APEC, CHIANTI, HULLAC, SPEX*).

MHD: *MagnetoHydroDynamics* is the theory describing the (hydro-)dynamics of gas flows under the influence of magnetic fields.

MIPSGAL: is a 278-degree Galactic plane survey conducted through the *Multiband Infrared Photometer for Spitzer (MIPS)* instrument on *Spitzer*. For more information see http://mipsgal.ipac.caltech.edu/.

MRI: *Magneto-Rotational Instability* (also called *Balbus–Hawley Instability*) transports angular momentum through an accretion disk via magnetic torques.

MS: the *Main Sequence* of stars in the HRD represents the evolutionary stage where nuclear fusion cycles fully supply internal energy.

MW: our own Galaxy, the *Milky Way*.

NGC: the *New General Catalog* was first published by J. L. E. Dreyer in 1888 based on J. Herschel's General Catalog and contained about 6,000 sources. For more information see http://www.ngcic.org/.

NTTS: the expression for *Naked T Tauri Stars* designates very young PMS stars without circumstellar disks.

OMC: the *Orion Molecular Cloud* denominates the molecular clumps in the Orion star-forming region.

ONC: the *Orion Nebula Cluster* is embedded in the Orion Nebula M42.

OP: the *Opacity Project* refers to an international collaboration formed in 1984 to calculate the extensive atomic data required to estimate stellar envelope opacities. For more information see http://vizier.u-strasbg.fr/OP.html.

OPAL: the *Opacity Project at Livermore* code was developed at the Lawrence Livermore National Laboratory to compute the opacities of low- to mid-Z elements. For more information see http://www-phys.llnl.gov/Research/OPAL/.

PAH: *Polycyclic Aromatic Hydrocarbon*s are cyclic rings of carbon and hydrogen and are found in their simplest forms in the ISM.

PDR: *Photodissociation Region*s are created in the ISM through the ionizing radiation field of O-stars.

PMF: the *Point Mass Formation* denotes the event when the first seed of a star forms after stellar collapse.

PMS: *Pre-Main Sequence* stars have evolved beyond the birthline and contract towards the MS.

PTTS: *Post T Tauri Stars* have evolved beyond the T Tauri stage but have not yet reached the MS.

Proplyd: *Protoplanetary Disk*s, specifically as found irradiated in the Orion Nebula.

PUCH: *Precursor Ultra Compact H II* regions.

RASS: the *ROSAT All-Sky Survey* was the first all-sky X-ray survey ever performed. For more information see http://www.xray.mpe.mpg.de/rosat.

SED: the *Spectral Energy Distribution*s at IR to radio wavelengths.

SFR: the *Star Formation Rate* in galaxies and on cosmological scales.

SIMBAD: astronomical database run by *CDS* (see Appendix F).

SMC: *Small Magellanic Cloud* is a small nearby (∼55 kpc) neighboring galaxy visible in the southern hemisphere.

SNR: *Supernova Remnants* are remnant clouds from supernova explosions. A second meaning depending on context is the *Supernova Rate* of occurrence.

SPEX: *SPEctral X-ray and UV modeling, analysis and fitting* software package (see also *APEC, CHIANTI, HULLAC, MEKAL*). For more information see http://www.sron.nl/divisions/hea/spex/.

TE: *Thermodynamic Equilibrium* is a thermodynamic condition in which a system absorbs as much radiation as it emits. A system in thermodynamic equilibrium radiates a blackbody spectrum.

TWA *TW Hydra Association*

UBV: optical photometry filters: U = ultraviolet ($\lambda_{peak} \sim 3,650$ Å), B = blue ($\lambda_{peak} \sim 4,400$ Å), V = visual ($\lambda_{peak} \sim 5,480$ Å).

UCH: *Ultra Compact H II* regions.

UIB: *Unidentified Infrared Band*s.

UV: *Ultraviolet* wavelengths, 4,000 Å $\lesssim \lambda \lesssim$ 150 Å.

WTTS: *Weak-lined T Tauri Stars* are very young low-mass PMS stars. In the IR their SEDs are classified as class III, in the optical band they show weak H_α lines.

XEST: the *XMM-Newton Extended Survey* of the *Taurus* Molecular Cloud led by M. Güdel.

YSO: *Young Stellar Object* denominates a stellar system during stellar formation and throughout all early evolutionary phases.

ZAMS: the *Zero-Age Main Sequence* is reached when nuclear fusion in the core of the star dominates the energy production in the star.

Appendix F
Institutes, Observatories, and Instruments

Star-formation research has made vast advances in a few decades and, besides rapid development in computing power as well as the development of numerical methods, the field has benefited from unprecedented progress in instrument development and technologies as well as new optics, telecopes, and space observatories to put them to work. The following is a glossary of the instruments and observatories that contributed towards star-formation research and which in one way or another are mentioned in this book. The entries for each item are minimal as full information can be readily acquired through website listings. Where such web-links are absent, pointers to main sites within the list are indicated by *see above* and *see below* instructions.

ACIS: the *Advanced Camera for Imaging and Spectroscopy*, built by the Massachusetts Institute of Technology is one of two cameras in the focal plane of *Chandra* (see *CXO* below).

ACS: the Advanced Camera for Surveys was a wide-field, high-resolution optical and UV camera onboard *HST* (see below). It was also equipped with a coronagraph to block out the glare of quasars and stars.

AIA: the Atmospheric Imaging Assembly on *SDO* images the solar atmosphere in multiple EUV wavelengths (see *SDO* below).

AIP: the *Astronomisches Institut Potsdam* is located near Berlin/Germany. For more information see http://www.aip.de/.

ALMA: the *Atacama Large Millimeter Array* is operated by *NRAO* (see below).

ASCA: *ASCA* (1993–2000) was the fourth cosmic X-ray astronomy satellite flown by the Japanese space agency and for which the United States provided part of the scientific payload. The observatory possessed medium resolution X-ray imaging

N.S. Schulz, *The Formation and Early Evolution of Stars*, Astronomy and Astrophysics Library, DOI 10.1007/978-3-642-23926-7, © Springer-Verlag Berlin Heidelberg 2012

capacity and several X-ray imaging detectors with medium spectral resolution (see Table 2.1). For detailed information see http://heasarc.gsfc.nasa.gov/docs/asca.

AURA: the *Association of Universities for Research in Astronomy* is a consortium of universities, and other educational and non-profit institutions, which operate astronomical observatories. For detailed information see http://www.aura-astronomy.org.

BIMA: the *Berkeley Illinois Maryland Association* is a consortium of radio- and sub-mm research groups located at the University of California, Berkeley, at the University of Illinois, Urbana, and at the University of Maryland. The consortium operated a mm-wave radio interferometer at Hat Creek in California, which was shut down in June 2004. It was superseded by the *Combined Array for Research in Millimeter-wave Astronomy (CARMA)* in conjunction with the California Institute of Technology. For detailed information see http://bima.astro.umd.edu.

CARMA: see *BIMA*.

CDS: the *Centre de Données astronomiques de Strasbourg*. For more information see http://cdsweb.u-strasbg.fr/.

CGRO: the *Compton Gamma Ray Observatory* was the second of *NASA's* four Great Observatories after *HST* (see below) was launched in 1991. The mission featured various experiments covering an energy range from 30 keV to 30 GeV. One of these experiments was *EGRET* (see below). For detailed information see http://cossc.gsfc.nasa.gov.

CSO: the *Caltech Submillimeter Observatory* is operated by the California Institute of Technology since 1988. It consists of a 10.4 m Leighton radio dish and is located near the summit of Mauna Kea, Hawaii. For detailed information see http://www.submm.caltech.edu/cso/.

CXC: *Chandra X-ray Center* is the designated *Chandra* (see *CXO* below) science and operations center. For more information see http://cxc.harvard.edu/.

CXO: the *Chandra X-ray Observatory* (most of the time simply referred to as *Chandra*) was the third of *NASA's* four Great Observatories after *HST* (see below) and *CGRO* (see above) and was launched in 1999 (see Table 2.1). The observatory carries the first high-resolution imaging telecope (0.5 arcsec spatial resolution between 0.1 and 10 keV). The main focal plane camera is *ACIS* (see above), which also operates with two high-resolution X-ray gratings spectrometers, *LETGS* and *HETGS* (see below). The *CXO* Science Operations Center is located at the Harvard–Smithsonian Center for Astrophysics in Massachusetts. For detailed information see http://chandra.harvard.edu.

DENIS: *Deep Near Infrared Survey* is a European joint project that aims to obtain a complete survey of the southern hemisphere at wavelengths of 0.82 μm, 1.25 μm, and 2.15 μm. For detailed information see http://www-denis.iap.fr/.

EGRET: the *Energetic Gamma Ray Experiment Telescope* onboard *CGRO* (see above) covering an energy range of 20 MeV to 30 GeV.

ESA: the *European Space Agency* is the European counterpart of the American *NASA*. It currently coordinates space programs involving 15 member states. For more information see http://www.esa.int/.

EUVE: the *Extreme Ultraviolet Explorer* (1992–2001) was funded by *NASA* and built and operated by the University of California, Berkeley. In its first 6 months of operation *EUVE* carried out an all-sky survey in the wavelength range between 70 and 760 Å (see Table 2.1). For detailed information see http://ssl.berkeley.edu/euve.

FUSE: the *Far Ultraviolet Satellite Explorer*, funded by *NASA*, was developed and is being operated by the Johns Hopkins University (see Table 2.1). *FUSE* was funded by *NASA* in collaboration with the space agencies of Canada and France. It operates from 900 to 1,200 Å. For detailed information see http://fuse.pha.jhu.edu.

GAIA: is an *ESA* (see above) mission related to astrometry following the footsteps of *Hipparchos*. It is to be launched in early 2010. For detailed information see http://sci.esa.int/Gaia.

GBT: the *Green Bank Telescope* is operated by *NRAO* (see below).

GLIMPSE: the *Galactic Legacy Infrared Mid-Plane Survey Extraordinaire* is a *Spitzer* legacy program surveying the Galactic plane. For detailed information see http://www.astro.wisc.edu/sirtf/.

HAWK-I: the new *High-Acuity Wide-field K-band Imager* for ESO's *VLT*, which is equipped with a mosaic of four 2 k × 2 k arrays and operates from 0.9–2.4 μm over 7.5 arcmin × 7.5 arcmin with 0.1″ pixels. (see *VLT* below).

Herschel: formerly the *Far-InfraRed and Sub-millimetre Telescope (FIRST)*, *Herschel* is a development of *ESA* and launched 2009 (see Table 2.1). For detailed information see http://sci.esa.int under category *missions*

HETGS: The *High-Energy Transmission Grating Spectrometer*, built by the Massachusetts Institute of Technology, is one of two high-resolution X-ray spectrometers onboard *Chandra* (see *CXO* above) operating at wavelengths between 1.5 and 45 Å. For detailed information see http://space.mit.edu/ASC and http://asc.harvard.edu/cal.

HESS: the *High Energy Stereoscopic System* is a system of imaging atmospheric Cherenkov telescopes that investigates cosmic gamma rays in the 100 GeV to 100 TeV energy range, located in Namibia, near the Gamsberg mountain. For detailed information see http://www.mpi-hd.mpg.de/hfm/HESS/.

HHT: the 10 m *Heinrich Hertz Telecope* on *JCMT* (see below).

Hi-GAL: is the *Herschel* open time key project that maps out the inner Galaxy within $-60° < l < +60°$ and $-1° < b < +1°$ in 5 bands between 60 and 600 μm. For detailed information see https://hi-gal.ifsi-roma.inaf.it/higal/.

HST: the *Hubble Space Telescope* was the first of *NASA's* four great observatories in Earth's orbit launched in 1999 (see Table 2.1). During *HST*'s several refurbishing missions, several cameras have so far been operated at the telescope's focus: *WFPC2*, *STIS*, *NICMOS* (see below) and and *ACS* (see above). For detailed information see http://hubblesite.org.

IRAC: the *Infrared Array Camera* is one of *SST*'s (see below) infrared cameras.

IRAM: the *Institut de Radio Astronomie Millimétrique* is a French–German–Spanish collaboration. *IRAM* operates two observatories at mm-wavelengths: a 30-m single dish telescope on Pico Veleta in Spain and a 6-antenna telescope on the Plateau de Bure in France. For detailed information see http://www.iram.es.

IRAS: the *Infrared Astronomical Satellite* was a joint IR space mission involving the United States, the Netherlands, and the United Kingdom. It conducted the first IR all-sky survey between 8 and 120 μm. For detailed information see http://lambda.gsfc.nasa.gov/product/iras.

ISO: the *Infrared Satellite Observatory* was an *ESA* (see above) IR space mission in cooperation with Japan and the United States (see Table 2.1). For detailed information see http://www.iso.vilspa.esa.es.

IUE: the *International Ultraviolet Explorer* was a collaboration between *ESA* (see above) and *NASA* (see below and Table 2.1).

For detailed information see http://sci.esa.int under category *missions*

JCMT: the *James Clerk Maxwell Telescope*, with a diameter of 15 m, is one of the largest telescopes for sub-mm observations. It is operated by astronomical research groups from Hawaii, the United Kingdom, Canada, and the Netherlands. For detailed information see http://www.jach.hawaii.edu/JACpublic/JCMT/.

JAXA: the *Japanese Aerospace Exploration Agency* is Japan's merged independent i administrative institution to perform all activities in aerospace as one organization, from basic research and development to utilization. For detailed information see http://www.jaxa.jp/

JWST: the *James Webb Space Telecope* will be an orbiting optical/IR observatory to replace the *HST* (see above) by about 2011 (see Table 2.1). For detailed information see http://ngst.gsfc.nasa.gov

LAT: The *Fermi Observatory* carries two instruments: the *Large Area Telescope (LAT)* and the *GLAST Burst Monitor (GBM)*. The *LAT* is *Fermi's* primary instrument. For detailed information see http://www.nasa.gov/mission_pages/GLAST/

LETGS: the *Low-Energy Transmission Grating Spectrometer* is one of two high-resolution X-ray spectrometers onboard *Chandra* (see *CXO* above) operating at wavelengths between 2 and 180 Å.
For detailed information see http://asc.harvard.edu/cal

2MASS: the *Two Micron All-Sky Survey* is a more recent and ground-based near-IR digital imaging survey of the entire sky conducted at 1.25 μm, 1.65 μm and 2.17 μm wavelengths. The survey is conducted by the University of Massachusetts, Amherst. For detailed information see http://pegasus.phast.umass.edu

MPE: the *Max Planck Institut für Extraterrestrische Physik* is located in Garching near Munich/Germany. For more information see http://www.mpe.mpg.de/

MSX: the *Midcourse Space Experiment* was an IR and sub mm space probe that operated from 1996–1997 in four wavelength bands (8.28, 12.13, 14.65, 21.3 μm). For more information see http://irsa.ipac.caltech.edu/Missions/msx.html

NASA: the *National Aeronautic and Space Administration* is the US counterpart of the European *ESA* (see above) and administers America's space program. For more information see http://www.nasa.gov/home/

NICMOS: the *Near Infrared Camera and Multi-Object Spectrometer* is a highly sensitive near-IR camera onboard the *HST* (see above).

NRAO: the *National Radio Astronomical Observatory* is a long-term working facility for radio astronomy. It operates many facilities around the world including the *ALMA*, the *GBT*, the *VLA* and the *VLBA*. For detailed information see http://www.nrao.edu

OVRO: the *Owons Valley Radio Observatory* mm-array telescope is operated by the California Institute of Technology. Like *BIMA* it will be superseded by *CARMA* (see *BIMA* above). For detailed information see http://www.ovro.caltech.edu/mm/main.html

PSPC: the *Position Sensitive Proportional Counter* was an energy-sensitive X-ray detector onboard *ROSAT* (see below).

RGS: the *Reflection Grating Spectrometer* is the high-resolution X-ray spectrometer onboard the European X-ray observatory *XMM* (see below).

ROSAT: the *Röntgensatellit (ROSAT)* was a German–United Kingdom–United States collaboration, built and operated under the leadership of the *MPE* (see above). During its first year of operation it conducted the first-ever X-ray all-sky survey (*RASS*, see Appendix E) in the energy band between 0.1 and 2.4 keV. For detailed information see http://wave.xray.mpe.mpg.de/rosat

SCUBA: the *Submillimeter Common Users Bolometer Array* is operated at the focus of the *JCMT* (see above).

SDO: the *Solar Dynamics Observatory* launched on February 11, 2010, and studies the interior of the Sun, the Sun's magnetic field, the hot plasma of the solar corona, and the irradiance that creates the ionospheres of the planets at EUV wavelengths. For detailed information see http://sdo.gsfc.nasa.gov/

SHARC: the *Submillimetre High Angular Resolution Camera* is operated by the *CSO* (see above).

SMM: the *Solar Maximum Mission* spacecraft monitored solar activity between 1980 and 1989. For detailed information see http://umbra.nascom.nasa.gov/smm/

SMTO: the Heinrich Hertz *Submillimeter Telescope Observatory* is located on Emerald Peak of Mt. Graham in Arizona. The facility is operated as a joint venture of the Steward Observatory of the University of Arizona and the *Max-Planck-Institut für Radioastronomie(MPfR)* of Germany. For detailed information see http://maisel.as.arizona.edu:8080

SOFIA: the *Stratospheric Observatory for Infrared Astronomy* project is an airborne IR project in preparation by the United States's *NASA* (see above) and Germany. For detailed information see http://sofia.arc.nasa.gov

SST: the *Spitzer Space Telescope (SST)* (formerly *SIRTF*) was the last of *NASA's* (see above) four great observatories to be launched after *HST*, *CGRO* and *CXO* (see above). It produces high-resolution images of the universe at IR wavelengths. For detailed information see http://sirtf.caltech.edu

STIS: the *Space Telescope Imaging Spectrograph* was an optical and UV spectrometer on board emphHST (see above).

STScI: the *Space Telescope Science Institute* is the science center for the *HST* (see above) and all related issues. For detailed information see http://www.stsci.edu

SWAS: the *Submillimeter Wave Astronomy Satellite* is one of *NASA's* (see above) small explorer projects. Its science operations center is located at the Harvard–Smithsonian Center for Astrophysics in Massachusetts.

VLA: the *Very Large Array* (see Fig. 4.3) is operated by *NRAO* (see above).

VLBA: the *Very Long Baseline Array* is operated by *NRAO* (see above).

VLT: the *Very Large Telescope* is *ESO*'s flagship facility consisting of four Unit Telescopes with main mirrors of 8.2 m diameter and four movable 1.8 m diameter Auxiliary Telescopes. For detailed information see http://www.eso.org/public/teles-instr/vlt.html

WFPC2: the *Wide Field Planetary Camera 2* was the first replacement of the original *HST* (see above) optical focal plane camera *WFPC*.

WHAM: the *Wisconsin H-Alpha Mapper* is a custom-built observatory designed for studying the warm ionized medium of the ISM in detail. Its primary mission is to produce the first spectral, all-sky survey of this emission from the Milky Way. For detailed information see http://www.astro.wisc.edu/wham/

WMAP: the *Wilkinson Microwave Anisotropy Probe* is a NASA Explorer mission that launched June 2001 to make fundamental measurements of cosmology. For detailed information see http://map.gsfc.nasa.gov/

XMM: the *X-ray Multi-mirror Mission (XMM)* is the latest X-ray observatory launched by *ESA* (see above). It carries three telescopes and various spectrometers operating between 0.2 and 15 keV. For detailed information see http://xmm.vilspa.esa.es/

Appendix G
Variables, Constants, and Units

The following tables lists the variables and constants used in the course of the book. The first table provides useful unit conversions. The second table lists the values of most of the constants that appear in the text. The top half is devoted to universal constants and the bottom half shows such values as the Sun's mass and radius, which are often used as benchmarks. The first table lists variables accompanied by their name(s) and units. The first listing is generally in SI units, followed by the CGS units that are predominantly used in literature. Some of the symbols have double listings as a consequence of the many contexts that have been combined in this book. In such instances these contexts provide enough information to distinguish between them.

The values in Tables G.1 and G.2 are rounded to three digits and, thus, do not represent the most precisely known values. They are represented as used in the text. The sources of these values and conversions are:

Table G.1 Useful unit conversions

Quantity	Unit	Value
Energy	1.0 J	6.242×10^{18} eV
	1.0 J	1.000×10^{7} erg
	1.0 eV	1.602×10^{-12} erg
Force	1.0 N	1.000×10^{5} dyne
Frequency	Hz (equivalent)	$2.418 \times 10^{14} \times E[eV]$
Length	1.0 lyr	3.262 pc
	1.0 AU	4.848×10^{-6} pc
	1.0 Å	1.000×10^{-10} m
Magnetic field	1.0 T	1.000×10^{4} G
Mass	1.0 kg	6.022×10^{26} amu
Power	1.0 W	1.000×10^{7} erg s^{-1}
Temperature	1.0 K (equivalent)	8.617×10^{-5} eV
Time	1.0 yr	3.156×10^{7} s
Wavelength	Å (equivalent)	$12.3985/E[eV]$

N.S. Schulz, *The Formation and Early Evolution of Stars*, Astronomy and Astrophysics Library, DOI 10.1007/978-3-642-23926-7, © Springer-Verlag Berlin Heidelberg 2012

- *Allen's Astrophysical Quantities*, 4th ed., AIP Press, 2000;
- *Astronomy Methods* by *H. Bradt*, 1st ed., Cambridge University Press, 2004;
- *Handbook of Space Astronomy & Astrophysics* by *M. Zombeck*, 2nd ed., Cambridge University Press, 1990.

Table G.2 List of constants

Constant	Description	SI Units	CGS Units
c	Vacuum light speed	$2.998 \times 10^{8}\,\mathrm{m\,s^{-1}}$	$2.998 \times 10^{10}\,\mathrm{m\,s^{-1}}$
G	Gravitation constant	$6.673 \times 10^{-11}\,\mathrm{m^3\,kg^{-1}\,s^{-2}}$	$6.673 \times 10^{-8}\,\mathrm{dyn\,cm^2 g^{-3}}$
h	Planck constant	$6.626 \times 10^{-34}\,\mathrm{J\,s}$	$6.626 \times 10^{-27}\,\mathrm{erg\,s}$
e	Electron charge	$1.602 \times 10^{-19}\,\mathrm{C}$	$4.803 \times 10^{-10}\,\mathrm{esu}$
m_e	Electron mass	$9.109 \times 10^{-31}\,\mathrm{kg}$	$9.109 \times 10^{-28}\,\mathrm{g}$
m_p	Proton mass	$1.673 \times 10^{-27}\,\mathrm{kg}$	$1.673 \times 10^{-24}\,\mathrm{g}$
N_A	Avogadro constant	$6.022 \times 10^{23}\,\mathrm{mol^{-1}}$	Same as SI
m_u	Atomic mass unit	$1.661 \times 10^{-27}\,\mathrm{kg}$	$1.661 \times 10^{-24}\,\mathrm{g}$
k_b	Boltzmann constant	$1.381 \times 10^{-23}\,\mathrm{J\,K^{-1}}$	$1.381 \times 10^{-16}\,\mathrm{erg\,K^{-1}}$
R	Molar gas constant	$8.315\,\mathrm{J\,mol^{-1}\,K^{-1}}$	$8.314 \times 10^{7}\,\mathrm{erg\,mol^{-1}\,K^{-1}}$
σ	Stefan–Boltzmann constant	$5.671 \times 10^{-8}\,\mathrm{W\,m^{-2}\,K^{-4}}$	$5.671 \times 10^{-5}\,\mathrm{erg\,cm^{-2}\,K^{-4}\,s^{-1}}$
AU	Astronomical unit	$1.496 \times 10^{11}\,\mathrm{m}$	$1.496 \times 10^{13}\,\mathrm{cm}$
lyr	Light year	$9.461 \times 10^{15}\,\mathrm{m}$	$9.461 \times 10^{17}\,\mathrm{cm}$
pc	Parsec	$3.086 \times 10^{16}\,\mathrm{m}$	$3.086 \times 10^{18}\,\mathrm{cm}$
$M_\odot$	Solar mass	$1.989 \times 10^{30}\,\mathrm{kg}$	$1.989 \times 10^{33}\,\mathrm{g}$
$R_\odot$	Solar radius	$6.955 \times 10^{8}\,\mathrm{m}$	$6.955 \times 10^{10}\,\mathrm{cm}$
$L_\odot$	Solar luminosity	$3.845 \times 10^{26}\,\mathrm{W}$	$3.845 \times 10^{33}\,\mathrm{erg\,s^{-1}}$
$L_{\odot,x}$	Solar X-ray luminosity	$3.98 \times 10^{20}\,\mathrm{W}$	$3.98 \times 10^{27}\,\mathrm{erg\,s^{-1}}$
$T_\odot$	Solar surface temperature	$5,779\,\mathrm{K}$	same as SI
M_{moon}	Moon mass	$7.348 \times 10^{22}\,\mathrm{kg}$	$7.348 \times 10^{25}\,\mathrm{g}$
R_{moon}	Moon radius	$1.737 \times 10^{6}\,\mathrm{m}$	$1.737 \times 10^{8}\,\mathrm{cm}$
$M_\oplus$	Earth mass	$5.972 \times 10^{24}\,\mathrm{kg}$	$5.972 \times 10^{27}\,\mathrm{g}$
$R_\oplus$	Earth radius	$6.367 \times 10^{6}\,\mathrm{m}$	$6.367 \times 10^{8}\,\mathrm{cm}$
M_{Jup}	Jupiter mass	$1.899 \times 10^{27}\,\mathrm{kg}$	$1.899 \times 10^{30}\,\mathrm{g}$
R_{Jup}	Jupiter radius	$6.991 \times 10^{7}\,\mathrm{m}$	$6.991 \times 10^{9}\,\mathrm{cm}$

Table G.3 List of variables

Variable	Description	Units
a, b	Radiative or geometric constants	–
$\mathbf{A}, \mathbf{a}$	Magnetic vector potentials	T m, G cm
A_λ, A_V	Extinction	Magnitudes
A_Z	Abundance fraction	–
α, α_{IR}	Spectral index, IR SED index	–
B_ν, B_λ	Surface brightness	W m^{-2} Hz^{-1} $ster^{-1}$ erg s^{-1} cm^{-2} Hz^{-1} $ster^{-1}$
$\mathbf{B}, \overline{B}, B_{ism}, B_o$	Various magnetic field strengths	Tesla, G, kG, μG
B_{surf}	Stellar surface field	Tesla, G, kG, μG
C_p, C_v	Heat capacities	Joule K^{-1}, erg K^{-1}
c_s	Sound speed	ms^{-1}, km s^{-1}
χ_i, χ_H	Ionization energies	eV, keV
χ, χ^d, χ^g	Total opacity, dust opacity, gas opacity	m^2 kg^{-1}, cm^2 g^{-1}
D	Distance	kpc, pc, lyr
$\mathbf{E}$	Electric field strength	Vm^{-1}, statvolt cm^{-1}
$E, E_e, E_r, E_v, \rho\mathcal{E}_\perp$	Various energies	Joule, erg, eV, keV
$E(\lambda - V), E(B - V)$	Color excesses	Magnitudes
ϵ	Dielectric constant	farad m^1
ϵ_{SFE}	Star-formation rate	–
η, η_1, η_2	Magnetic diffusivity	m^2 s^{-1}, cm^2 s^{-1}
η_o	Ohmic resistivity	m^2 s^{-1}, cm^2 s^{-1}
η_v	Viscosity coefficient	kg m^{-1} s^{-1}, g cm^{-1} s^{-1}
$\mathbf{F}$	External force	N, dyne
$F_{inert}, F_{viscous}$	Inertial force, viscous force	N, dyne
F, F_{obs}, F_{rad}	Various radiation fluxes	W m^{-2} erg $s^{-1}cm^{-2}$
F_ν, F_λ	Flux density	W m^{-2} erg s^{-1} cm^{-2} erg s^{-1} cm^{-2} keV s^{-1} cm^{-2}
f_D	Deuterium fraction	–
$f(L)$	Luminosity function	–
f_{jk}	Atomic oscillator strength	–
$G, G_o, G_n, G_i, G_{trans}, G_{rot}, G_{vib}, G_{el}$	Various partition functions	–
G^{ff}	Free–free Gaunt factor	–
g_o, g_i	Statistical weights	–
γ	Adiabatic index	–
γ_d	Drag coefficient	–
Γ	Line damping constant	–
$\Gamma_1, \Gamma_2, \Gamma_3$	1st, 2nd, and 3rd adiabatic exponent	–
H, h_d, h_a	Height, various disk heights	m, cm, AU
I_ν, I_λ	Intensity	See brightness
$I_o, I(\lambda), I_{observed}, I_{source}$	Various intensities	Instrument counts
J	Angular momentum	kg m^2 s^{-1}, g cm^2 s^{-1}

(continued)

Table G.3 List of variables (continued)

Variable	Description	Units
$\mathbf{j}, \mathbf{j_e}$	Currents	A, statamp
j_ν, j_λ	Emission coefficient	m^2 kg^{-1}, cm^2 g^{-1}
k_{pd}	Photodissociation rate	–
$\kappa, \kappa_\nu, \kappa_\nu^d, \kappa_\nu^g, \kappa_P, \kappa_R$	Various absorption coefficients, mean absorption coefficients	m^2 kg^{-1} cm^2 g^{-1}
κ_c	Thermal conductivity	J s^{-1} m^{-1} K^{-1} kcal $s^{-1}cm^{-1}K^{-1}$
L_{grav}, L	(Gravitational) luminosity -	Joule s^{-1}, erg s^{-1}
L_{xs}, L_x	Soft X-ray luminosity, X-ray luminosity	Joule s^{-1}, erg s^{-1}
$L_{star}, L_{acc}, L_{shock}$	Stellar, accretion, and accretion shock luminosity	Joule s^{-1}, erg s^{-1}
L_r	Luminosity through sphere of r	Joule s^{-1}, erg s^{-1}
L_ν, L_λ	IR range luminosities	Joule s^{-1}, erg s^{-1}
L_{D_o}, L_D	Luminosities from D burning	Joule s^{-1}, erg s^{-1}
L_u	Lundquist number	–
L_D, L_{De}	Debye lengths	m, cm
L_x, L_y, L_z	Characteristic lengths	m, cm
l_{loop}	Magnetic loop length	cm
λ	Wavelength	m, cm, mm, μm, Å
$\lambda_R, \lambda_J, \lambda_d$	Absorption length, Jeans length, driving length	See wavelength
λ_r	Reconnection scaling factor	–
m	Mass	kg, g
M, M_{star}, M_{stars}	Stellar masses, stellar cluster mass	kg, g, $M_\odot$
$M_{gas}, M_g, M_u, M_\Phi, M_J$	Various ISM cloud masses	kg, g, $M_\odot$
$\dot{M}, \dot{M}_{acc}$	Mass accretion rate	kg s^{-1}, g s^{-1}, $M_\odot$ yr^{-1}
$\dot{M}_{wind}, \dot{M}_{ph}$	Mass loss rates	kg s^{-1}, g s^{-1}, $M_\odot$ yr^{-1}
$\dot{M}_{xwind}, \dot{M}_{stream}$	X-wind loss rates	kg s^{-1}, g s^{-1}, $M_\odot$ yr^{-1}
$\mathcal{M}, \mathcal{M}_{cr}$	Magnetic energy densities	Joule m^{-3}, erg cm^{-3}
$\mathcal{M}_A$	Alfvén Mach number	–
$\mathcal{M}_e, \mathcal{M}_e^{max}$	Magnetic reconnection rates	–
μ	Atomic mass	amu
μ	Magnetic permeability	N A^{-2}
μ_Z	Molecular weight	amu
n	Polytropic index	–
n, n_e	Electron number density	m^{-3}, cm^{-3}
n_i, n_H	Ion number density, hydrogen number density	– –
n_A, n_B, n_{AB}, n_{gr}	Various number densities	–
n_i	Number of i states	–
n_o	Number of ground states	–
$N_H, N_{H,Z}$	Column densities	cm^{-2}
N_{tot}	Total number of particles in an ensemble	–
$\mathcal{N}_{UV}$	UV ionizing rate	Photons s^{-1}

(continued)

Table G.3 List of variables (continued)

Variable	Description	Units
$\mathcal{N}_{XR}$	X-ray ionizing rate	Photons s^{-1}
ν	Frequency	Hz
Ω, Ω_x	Rotational speeds	s^{-1}
Ω_{surf}	Stellar surface speed	s^{-1}
Ω	Solid angle	ster
$\mathbf{\Omega}$	Rotational velocity	s^{-1}
ω	Gyrofrequency	s^{-1}
ω_{pe}, ω_L	Electron plasma frequency, Larmor frequency	s^{-1}
ω_D	Ambipolar diffusion speed	$m\ s^{-1}$, $km\ s^{-1}$
$P, P_e, \overline{P}, P_o, P_{ext}$	Various pressures	Pa($N\ m^{-2}$), bar, torr
P_m	mag. Prandl number	–
P	Stellar period	s, d
$\mathbf{p}$	Momentum	$kg\ m\ s^{-2}$, $g\ cm\ s^{-2}$
ϕ	Gravitational potential	Joule, erg, eV, keV
ϕ_C	Coulomb potential	V, statvolt
Φ	Cloud magnetic flux	Tesla m^2, G cm^2
Q, Q_o, Q_ν, Q_{rad}	Heats, viscous heat dissipation, radiative heat loss	Joule, erg, eV, keV
q, q_e	Charge	Coulomb
r, r_o, r_s, r_L	Various radii	m, cm
$r_c, r_b, r_m, r_{trunc}, r_w$	Various accretion disk radii	m, cm, AU, pc
r_{inf}, r_{out}	Circumstellar envelope radii	m, cm, AU, pc
R, R_{star}	Stellar radii	m, cm, AU, pc
R_A, R_J	Alfvén radius, Jeans radius	m, cm, AU, pc
R_e, R_p	Obliqueness radii	m, cm
$R_{\mathrm{H\,II}}$	Strömgren radius	m, cm, AU, pc
$R_{eff}, R_{ff}, R_{photo}, R_{re}, R_{col}$	Various heating rates	Joule s^{-1}, erg s^{-1}
R_V	Relative extinction	–
R_o	Rossby number	–
Re	Reynolds number	–
Rm	Magnetic Reynolds number	–
$\rho, \rho_o, \overline{\rho}$	Mass densities	$kg\ m^{-3}$, $g\ cm^{-3}$
s	Thickness	m, cm, μm
$\mathbf{S}, S_{ij}$	Stress tensor, tensor components	$N\ m^{-2}$, $dyn\ cm^{-2}$
S, S_λ	Spectral flux density	$W\ m^{-2}\ Hz^{-1}$ $erg\ s^{-1}\ cm^{-2}\ Hz^{-1}$ $erg\ s^{-1}\ cm^{-2}\ keV^{-1}$ $keV\ s^{-1}\ cm^{-2}\ keV^{-1}$
Σ	Mean surface density	$kg\ m^{-2}$, $g\ cm^{-2}$
σ	Electric conductivity	farad, cm
σ	Velocity dispersion	$m\ s^{-1}$, $km\ s^{-1}$
σ	Momentum transfer cross-section	m^2, cm^2
σ_T, σ_R	ff scattering-cross sections	m^2, cm^2
$\sigma_{AB}, \sigma_{ISM}, \sigma_{gas}, \sigma_{mol}, \sigma_{grains}$	Absorption cross-sections	m^2, cm^2

(continued)

Table G.3 List of variables (continued)

Variable	Description	Units
T, T_e	Temperature, electron temperature	K
T_{eff}	Stellar surface temperature	K
T_{surf}, T_{max}	Disk surface temperature, maximum disk temperature	K
T_{rec}	Reconnection temperature	K
Θ	Temperature measure	–
t	Time	s
$t_b, t_d, t_D, t_{dyn}, t_{ff}, t_{KH}, t_{acc}$ t_{th}, t_v, t_{br}	Various timescales	s
τ_{rh}, τ_{rc}	Reconnection times	s
τ_c	Convection turning time	s
τ_{dy}	Dynamo period	s
τ	Optical depth	–
τ_T	Thomson depth	–
U	Internal energy	Joule, erg, eV, keV
V, V_{cl}	Volume, cloud volume	m^3, cm^3
$v_{wind}, v_{in}, v_{\perp}, v_A, v_K, v_r, v_\phi$	Outflow and accretion flow velocities	$m\ s^{-1}$, $km\ s^{-1}$
$\mathbf{v}, \mathbf{u}, v, v_{rms}$	Various velocities	$m\ s^{-1}$, $km\ s^{-1}$
W_λ	Equivalent width	Å, eV, keV
Z	Atomic number	–

References

1. K. Schwarzschild, B. Meyermann, A. Kohlschütter, O. Birck, Astronomische Mitteilungen der Universitaets-Sternwarte zu Goettingen **14**, 1 (1910)
2. H. Rosenberg, Astronomische Nachrichten **193**, 357 (1913)
3. J.A. Ambartsumian: Stellar Evolution and Astrophysics In: *Acad. Sci. Armenian, Erevan, USSR*, (1947)
4. D. Batchelor: Solar data at NSSDC. In: *National Space Science Data Center NEWS* (Volume 12, Number 3, 1996)
5. Exhibition *www.dudleyobservatory.org/ Exhibits/bessel.htm* (The Dudley Observatory, Schenectady, NY 12308, 2003)
6. P.H. Hauschildt: in *The Limiting Effects of Dust in Brown Dwarf Model Atmospheres* (http://www.uni-hamburg.de/~stcd101/PAPERS/, 2001)
7. D. Hollenbach: Photoevaporating Disks Around Young Stars. In: *Cores, Disks, Jets and Outflows in Low and High Mass Star Froming Environments* (WWW presentations: http://www.ism.ucalgary.ca/meetings/banff/index.html)
8. E.H. Holt and R.E. Haskell: *Plasma Dynamics* (Macmillan, New York, 1965)
9. W.F. Huebner, A. Merts, N.H. Magee and M.F. Argo: Los Alamos Sci. Lab. Rep. **LA-6760-M**
10. J.H. Kastner: Imaging Science in Astronomy. In: *Encyclopedia of Imaging Science and Technology*, ed J.P. Hornak (Wiley, New York, 2 Volume Set, 2002) p. 682–96
11. R.S. Klessen: private communication
12. G. Schmidt: *Physics of High Temperature Plasmas* (Academic Press, London, 1966)
13. E.P. Velikov: Sov. Phys.-JETP **36**, 995 (see also JETP Letters **35**, 1398) (1959)
14. Web-based catalog only: http://obswww.unige.ch/webda
15. A. Wittmann: Feuer des Lebens–Vom Wunsch, die Sonne verstehen zu können, In: *Scheibe, Kugel, Schwarzes Loch*, ed Uwe Schultz (Verlag C.H.Beck, München 1990) p. 231–241
16. P. Guthnick, Astronomische Nachrichten **196**, 357 (1913). DOI 10.1002/asna.19131962602
17. J. Stebbins, Science **41**, 809 (1915). DOI 10.1126/science.41.1066.809
18. H.L. Johnson, W.W. Morgan, Ap. J. **114**, 522 (1951). DOI 10.1086/145496
19. A. Unsoeld, B. Baschek, *The new cosmos : an introduction to astronomy and astrophysics* (2001)
20. G. Kirchhoff, R. Bunsen, Annalen der Physik **186**, 161 (1860). DOI 10.1002/andp.18601860602
21. A.J. Cannon, E. Pickering, *The Henry Draper (HD) Catalogue. Vol.91: 0h, 1h, 2h and 3h* (1918)
22. W.W. Morgan, P.C. Keenan, E. Kellman, *An atlas of stellar spectra, with an outline of spectral classification* (1943)
23. I. Kant, *Allgemeine Naturgeschichte und Theorie des Himmels* (1755)

N.S. Schulz, *The Formation and Early Evolution of Stars*, Astronomy and Astrophysics Library, DOI 10.1007/978-3-642-23926-7, © Springer-Verlag Berlin Heidelberg 2012

24. A.S. Eddington, M.N.R.A.S. **77**, 596 (1917)
25. A.S. Eddington, *The Internal Constitution of the Stars* (1926)
26. A.S. Eddington, *Stars and atoms* (1927)
27. J.H. Jeans, *Astronomy and cosmogony* (1928)
28. H.A. Bethe, Physical Review **55**, 434 (1939). DOI 10.1103/PhysRev.55.434
29. S. Chandrasekhar, *An introduction to the study of stellar structure* (1939)
30. J. North, *The Norton history of astronomy and cosmology.* (1995)
31. V. Trimble, Sky Telegram **99**(1), 010000 (2000)
32. G. Herbig, in *Physics of Star Formation in Galaxies*, ed. by F. Palla, H. Zinnecker, A. Maeder, & G. Meynet (2002), pp. 1–7
33. A. Blaauw, B.A.I.N. **11**, 414 (1952)
34. A. Blaauw, A. J. **59**, 317 (1954). DOI 10.1086/107069
35. G.H. Herbig, Ap. J. **214**, 747 (1977). DOI 10.1086/155304
36. B.J. Bok, E.F. Reilly, Ap. J. **105**, 255 (1947). DOI 10.1086/144901
37. B.J. Bok, Harvard Observatory Monographs **7**, 53 (1948)
38. A.H. Joy, P.A.S.P. **54**, 15 (1942). DOI 10.1086/125386
39. A.H. Joy, Ap. J. **102**, 168 (1945). DOI 10.1086/144749
40. G.H. Herbig, Ap. J. **119**, 483 (1954). DOI 10.1086/145854
41. V.A. Ambartsumian, Reviews of Modern Physics **30**, 944 (1958). DOI 10.1103/RevModPhys.30.944
42. P.P. Parenago, VizieR Online Data Catalog **2171**, 0 (1997)
43. L.G. Henyey, R. Lelevier, R.D. Levée, P.A.S.P. **67**, 154 (1955). DOI 10.1086/126791
44. C.B. Haselgrove, F. Hoyle, M.N.R.A.S. **116**, 527 (1956)
45. C.J. Lada, in *Star Forming Regions, IAU Symposium*, vol. 115, ed. by M. Peimbert & J. Jugaku (1987), *IAU Symposium*, vol. 115, pp. 1–17
46. G.H. Herbig, K.R. Bell, *Third Catalog of Emission-Line Stars of the Orion Population : 3 : 1988* (1988)
47. P. Andre, in *Radio Emission from the Stars and the Sun, Astronomical Society of the Pacific Conference Series*, vol. 93, ed. by A. R. Taylor & J. M. Paredes (1996), *Astronomical Society of the Pacific Conference Series*, vol. 93, pp. 273–+
48. G.H. Herbig, Advances in Astronomy and Astrophysics **1**, 47 (1962)
49. G.H. Herbig, N. Kameswara Rao, Ap. J. **174**, 401 (1972). DOI 10.1086/151500
50. B.F. Jones, G.H. Herbig, A. J. **84**, 1872 (1979). DOI 10.1086/112619
51. M.T. Adams, K.M. Strom, S.E. Strom, Ap. J. Suppl. **53**, 893 (1983). DOI 10.1086/190913
52. M. Kun, L. Pasztor, A.P.S.S. **174**, 13 (1990). DOI 10.1007/BF00645653
53. L.A. Marschall, G.B. Karshner, N.F. Comins, A. J. **99**, 1536 (1990). DOI 10.1086/115437
54. D.F. Gray, *The observation and analysis of stellar photospheres.* (1992)
55. E.E. Mendoza V., Ap. J. **143**, 1010 (1966). DOI 10.1086/148584
56. S.E. Strom, K.M. Strom, J. Yost, L. Carrasco, G. Grasdalen, Ap. J. **173**, 353 (1972). DOI 10.1086/151425
57. M. Cohen, R.D. Schwartz, M.N.R.A.S. **174**, 137 (1976)
58. A.E. Rydgren, S.E. Strom, K.M. Strom, Ap. J. Suppl. **30**, 307 (1976). DOI 10.1086/190364
59. S.V.W. Beckwith, A.I. Sargent, R.S. Chini, R. Guesten, A. J. **99**, 924 (1990). DOI 10.1086/115385
60. P. Andre, D. Ward-Thompson, M. Barsony, Ap. J. **406**, 122 (1993). DOI 10.1086/172425
61. F. Berrilli, G. Corciulo, G. Ingrosso, D. Lorenzetti, B. Nisini, F. Strafella, Ap. J. **398**, 254 (1992). DOI 10.1086/171853
62. P. Andre, T. Montmerle, Ap. J. **420**, 837 (1994). DOI 10.1086/173608
63. C. Ceccarelli, E. Caux, L. Loinard, A. Castets, A.G.G.M. Tielens, S. Molinari, R. Liseau, P. Saraceno, H. Smith, G. White, Astron. Astrophys. **342**, L21 (1999)
64. C.H. Young, Y.L. Shirley, N.J. Evans, II, J.M.C. Rawlings, Ap. J. Suppl. **145**, 111 (2003). DOI 10.1086/345341
65. W.H.M. Ku, G.A. Chanan, Ap. J. Lettersf **234**, L59 (1979). DOI 10.1086/183109
66. G.F. Gahm, Ap. J. Lettersf **242**, L163 (1980). DOI 10.1086/183424

67. E.D. Feigelson, W.M. Decampli, Ap. J. Lettersf **243**, L89 (1981). DOI 10.1086/183449
68. F.D. Seward, W.R. Forman, R. Giacconi, R.E. Griffiths, F.R. Harnden, Jr., C. Jones, J.P. Pye, Ap. J. Lettersf **234**, L55 (1979). DOI 10.1086/183108
69. F.R. Harnden, Jr., G. Branduardi, P. Gorenstein, J. Grindlay, R. Rosner, K. Topka, M. Elvis, J.P. Pye, G.S. Vaiana, Ap. J. Lettersf **234**, L51 (1979). DOI 10.1086/183107
70. L.B. Lucy, R.L. White, Ap. J. **241**, 300 (1980). DOI 10.1086/158342
71. E.D. Feigelson, T. Montmerle, A.R.A.A. **37**, 363 (1999). DOI 10.1146/annurev.astro.37.1.363
72. J. Truemper, Advances in Space Research **2**, 241 (1982). DOI 10.1016/0273-1177(82)90070-9
73. M.F. Sterzik, J.M. Alcala, R. Neuhaeuser, J.H.M.M. Schmitt, Astron. Astrophys. **297**, 418 (1995)
74. G. Münch, Ap. J. **125**, 42 (1957). DOI 10.1086/146281
75. E.M. Burbidge, G.R. Burbidge, W.A. Fowler, F. Hoyle, Reviews of Modern Physics **29**, 547 (1957). DOI 10.1103/RevModPhys.29.547
76. L. Spitzer, *Physical processes in the interstellar medium* (1978)
77. M. Jura, S.G. Kleinmann, Ap. J. Suppl. **73**, 769 (1990). DOI 10.1086/191488
78. M. Jura, S.G. Kleinmann, Ap. J. **341**, 359 (1989). DOI 10.1086/167499
79. P.C. Frisch, Space Sci. Rev. **72**, 499 (1995). DOI 10.1007/BF00749006
80. P.C. Frisch, D.G. York, Ap. J. Lettersf **271**, L59 (1983). DOI 10.1086/184095
81. F. Paresce, in *NASA Conference Publication*, *NASA Conference Publication*, vol. 2345, ed. by Y. Kondo, F. C. Bruhweiler, & B. D. Savage (1984), *NASA Conference Publication*, vol. 2345, pp. 169–184
82. J.S. Mathis, A.R.A.A. **28**, 37 (1990). DOI 10.1146/annurev.aa.28.090190.000345
83. J.C. Weingartner, B.T. Draine, Ap. J. **553**, 581 (2001). DOI 10.1086/320963
84. J.S. Mathis, P.G. Mezger, N. Panagia, Astron. Astrophys. **128**, 212 (1983)
85. H.J. Habing, B.A.I.N. **19**, 421 (1968)
86. R.C. Henry, Ap. J. **570**, 697 (2002). DOI 10.1086/339623
87. J.D. Slavin, P.C. Frisch, Astron. Astrophys. **491**, 53 (2008). DOI 10.1051/0004-6361:20078101
88. F.J. Low, R.F. Kurtz, W.M. Poteet, T. Nishimura, Ap. J. Lettersf **214**, L115 (1977). DOI 10.1086/182454
89. D. Rouan, P.J. Lena, J.L. Puget, K.S. de Boer, J.J. Wijnbergen, Ap. J. Lettersf **213**, L35 (1977). DOI 10.1086/182405
90. P.G. Mezger, J.S. Mathis, N. Panagia, Astron. Astrophys. **105**, 372 (1982)
91. A.G.G.M. Tielens, *The Physics and Chemistry of the Interstellar Medium* (2005)
92. F. Comerón, N. Schneider, D. Russeil, Astron. Astrophys. **433**, 955 (2005). DOI 10.1051/0004-6361:20041586
93. B.T. Draine, C.F. McKee, A.R.A.A. **31**, 373 (1993). DOI 10.1146/annurev.aa.31.090193.002105
94. C.D. Anderson, S.H. Neddermeyer, Physical Review **50**, 263 (1936). DOI 10.1103/PhysRev.50.263
95. Y. Uchiyama, F.A. Aharonian, T. Tanaka, T. Takahashi, Y. Maeda, Nature **449**, 576 (2007). DOI 10.1038/nature06210
96. A.K. Leroy, A. Bolatto, K. Gordon, K. Sandstrom, P. Gratier, E. Rosolowsky, C.W. Engelbracht, N. Mizuno, E. Corbelli, Y. Fukui, A. Kawamura, ArXiv e-prints (2011)
97. B.T. Draine, D.A. Dale, G. Bendo, K.D. Gordon, J.D.T. Smith, L. Armus, C.W. Engelbracht, G. Helou, R.C. Kennicutt, Jr., A. Li, H. Roussel, F. Walter, D. Calzetti, J. Moustakas, E.J. Murphy, G.H. Rieke, C. Bot, D.J. Hollenbach, K. Sheth, H.I. Teplitz, Ap. J. **663**, 866 (2007). DOI 10.1086/518306
98. H.I. Ewen, E.M. Purcell, Nature **168**, 356 (1951). DOI 10.1038/168356a0
99. W.N. Christiansen, J.V. Hindman, Australian Journal of Scientific Research A Physical Sciences **5**, 437 (1952)
100. W.B. Burton, A.R.A.A. **14**, 275 (1976). DOI 10.1146/annurev.aa.14.090176.001423

101. J.M. Dickey, F.J. Lockman, A.R.A.A. **28**, 215 (1990). DOI 10.1146/annurev.aa.28.090190.001243
102. P. Reich, W. Reich, Astronomy Data Image Library pp. 1–+ (1995)
103. M.A. Gordon, W.B. Burton, Ap. J. **208**, 346 (1976). DOI 10.1086/154613
104. T.M. Dame, D. Hartmann, P. Thaddeus, Ap. J. **547**, 792 (2001). DOI 10.1086/318388
105. R.J. Allen, in *Gas and Galaxy Evolution, Astronomical Society of the Pacific Conference Series*, vol. 240, ed. by J. E. Hibbard, M. Rupen, & J. H. van Gorkom (2001), *Astronomical Society of the Pacific Conference Series*, vol. 240, pp. 331–+
106. L.M. Haffner, R.J. Reynolds, S.L. Tufte, G.J. Madsen, K.P. Jaehnig, J.W. Percival, Ap. J. Suppl. **149**, 405 (2003). DOI 10.1086/378850
107. T.M. Dame, H. Ungerechts, R.S. Cohen, E.J. de Geus, I.A. Grenier, J. May, D.C. Murphy, L.A. Nyman, P. Thaddeus, Ap. J. **322**, 706 (1987). DOI 10.1086/165766
108. W.L. Kraushaar, G.W. Clark, G.P. Garmire, R. Borken, P. Higbie, V. Leong, T. Thorsos, Ap. J. **177**, 341 (1972). DOI 10.1086/151713
109. C.E. Fichtel, R.C. Hartman, D.A. Kniffen, D.J. Thompson, H. Ogelman, M.E. Ozel, T. Tumer, G.F. Bignami, Ap. J. **198**, 163 (1975). DOI 10.1086/153590
110. F.W. Stecker, in *NASA Conference Publication, NASA Conference Publication*, vol. 2, ed. by C. E. Fichtel & F. W. Stecker (1977), *NASA Conference Publication*, vol. 2, pp. 315–346
111. F.J. Kerr, in *Star Formation, IAU Symposium*, vol. 75, ed. by T. de Jong & A. Maeder (1977), *IAU Symposium*, vol. 75, pp. 3–19
112. A.M. Juett, N.S. Schulz, D. Chakrabarty, Ap. J. **612**, 308 (2004). DOI 10.1086/422511
113. L. Spitzer, Jr., E.B. Jenkins, A.R.A.A. **13**, 133 (1975). DOI 10.1146/annurev.aa.13.090175.001025
114. B.D. Savage, E.B. Jenkins, Ap. J. **172**, 491 (1972). DOI 10.1086/151369
115. P.C. Frisch, *The interstellar medium of our Galaxy* (Kluwer Academic Publishers, 2002), pp. 647–+
116. J.B. Rogerson, D.G. York, Ap. J. Lettersf **186**, L95+ (1973). DOI 10.1086/181366
117. H.W. Moos, W.C. Cash, L.L. Cowie, A.F. Davidsen, A.K. Dupree, P.D. Feldman, S.D. Friedman, J.C. Green, R.F. Green, C. Gry, J.B. Hutchings, E.B. Jenkins, J.L. Linsky, R.F. Malina, A.G. Michalitsianos, B.D. Savage, J.M. Shull, O.H.W. Siegmund, T.P. Snow, G. Sonneborn, A. Vidal-Madjar, A.J. Willis, B.E. Woodgate, D.G. York, T.B. Ake, B.G. Andersson, J.P. Andrews, R.H. Barkhouser, L. Bianchi, W.P. Blair, K.R. Brownsberger, A.N. Cha, P. Chayer, S.J. Conard, A.W. Fullerton, G.A. Gaines, R. Grange, M.A. Gummin, G. Hebrard, G.A. Kriss, J.W. Kruk, D. Mark, D.K. McCarthy, C.L. Morbey, R. Murowinski, E.M. Murphy, W.R. Oegerle, R.G. Ohl, C. Oliveira, S.N. Osterman, D.J. Sahnow, M. Saisse, K.R. Sembach, H.A. Weaver, B.Y. Welsh, E. Wilkinson, W. Zheng, Ap. J. Lettersf **538**, L1 (2000). DOI 10.1086/312795
118. J.M. Shull, in *Massive Stars: Their Lives in the Interstellar Medium, Astronomical Society of the Pacific Conference Series*, vol. 35, ed. by J. P. Cassinelli & E. B. Churchwell (1993), *Astronomical Society of the Pacific Conference Series*, vol. 35, pp. 327–+
119. J. Wilms, A. Allen, R. McCray, Ap. J. **542**, 914 (2000). DOI 10.1086/317016
120. R. Morrison, D. McCammon, Ap. J. **270**, 119 (1983). DOI 10.1086/161102
121. J.B. Kortright, S.K. Kim, Phys. Rev. B **62**, 12216 (2000). DOI 10.1103/PhysRevB.62.12216
122. J.C. Lee, J. Xiang, B. Ravel, J. Kortright, K. Flanagan, Ap. J. **702**, 970 (2009). DOI 10.1088/0004-637X/702/2/970
123. N.S. Schulz, W. Cui, C.R. Canizares, H.L. Marshall, J.C. Lee, J.M. Miller, W.H.G. Lewin, Ap. J. **565**, 1141 (2002). DOI 10.1086/324482
124. M.V. Zombeck, *Handbook of space astronomy and astrophysics* (1990)
125. A. McWilliam, G.W. Preston, C. Sneden, L. Searle, A. J. **109**, 2757 (1995). DOI 10.1086/117486
126. E.B. Jenkins, B.D. Savage, Ap. J. **187**, 243 (1974). DOI 10.1086/152620
127. P. Predehl, J.H.M.M. Schmitt, Astron. Astrophys. **293**, 889 (1995)
128. J.S. Mathis, W. Rumpl, K.H. Nordsieck, Ap. J. **217**, 425 (1977). DOI 10.1086/155591
129. E.L. Fitzpatrick, D. Massa, Ap. J. **307**, 286 (1986). DOI 10.1086/164415

130. J.A. Cardelli, G.C. Clayton, J.S. Mathis, Ap. J. **345**, 245 (1989). DOI 10.1086/167900
131. B.D. Savage, J.S. Mathis, A.R.A.A. **17**, 73 (1979). DOI 10.1146/annurev.aa.17.090179.000445
132. A.N. Witt, R.C. Bohlin, T.P. Stecher, Ap. J. Lettersf **305**, L23 (1986). DOI 10.1086/184677
133. J.S. Mathis, Ap. J. **308**, 281 (1986). DOI 10.1086/164499
134. B.T. Draine, H.M. Lee, Ap. J. **285**, 89 (1984). DOI 10.1086/162480
135. J.W. Overbeck, Ap. J. **141**, 864 (1965). DOI 10.1086/148180
136. C.W. Mauche, P. Gorenstein, Ap. J. **302**, 371 (1986). DOI 10.1086/163996
137. B. Draine, in *Interstellar Dust, IAU Symposium*, vol. 135, ed. by L. J. Allamandola & A. G. G. M. Tielens (1989), *IAU Symposium*, vol. 135, pp. 313–+
138. L.G. Henyey, J.L. Greenstein, Ap. J. **93**, 70 (1941). DOI 10.1086/144246
139. K.D. Gordon, in *Astrophysics of Dust, Astronomical Society of the Pacific Conference Series*, vol. 309, ed. by A. N. Witt, G. C. Clayton, & B. T. Draine (2004), *Astronomical Society of the Pacific Conference Series*, vol. 309, pp. 77–+
140. A. Li, J.M. Greenberg, Astron. Astrophys. **323**, 566 (1997)
141. G.C. Clayton, M.J. Wolff, U.J. Sofia, K.D. Gordon, K.A. Misselt, Ap. J. **588**, 871 (2003). DOI 10.1086/374316
142. V. Zubko, E. Dwek, R.G. Arendt, Ap. J. Suppl. **152**, 211 (2004). DOI 10.1086/382351
143. M. Compiègne, N. Flagey, A. Noriega-Crespo, P.G. Martin, J.P. Bernard, R. Paladini, S. Molinari, Ap. J. Lettersf **724**, L44 (2010). DOI 10.1088/2041-8205/724/1/L44
144. S. Molinari, B. Swinyard, J. Bally, M. Barlow, J.P. Bernard, P. Martin, T. Moore, A. Noriega-Crespo, R. Plume, L. Testi, A. Zavagno, A. Abergel, B. Ali, P. André, J.P. Baluteau, M. Benedettini, O. Berné, N.P. Billot, J. Blommaert, S. Bontemps, F. Boulanger, J. Brand, C. Brunt, M. Burton, L. Campeggio, S. Carey, P. Caselli, R. Cesaroni, J. Cernicharo, S. Chakrabarti, A. Chrysostomou, C. Codella, M. Cohen, M. Compiegne, C.J. Davis, P. de Bernardis, G. de Gasperis, J. Di Francesco, A.M. di Giorgio, D. Elia, F. Faustini, J.F. Fischera, Y. Fukui, G.A. Fuller, K. Ganga, P. Garcia-Lario, M. Giard, G. Giardino, J.. Glenn, P. Goldsmith, M. Griffin, M. Hoare, M. Huang, B. Jiang, C. Joblin, G. Joncas, M. Juvela, J. Kirk, G. Lagache, J.Z. Li, T.L. Lim, S.D. Lord, P.W. Lucas, B. Maiolo, M. Marengo, D. Marshall, S. Masi, F. Massi, M. Matsuura, C. Meny, V. Minier, M.A. Miville-Deschênes, L. Montier, F. Motte, T.G. Müller, P. Natoli, J. Neves, L. Olmi, R. Paladini, D. Paradis, M. Pestalozzi, S. Pezzuto, F. Piacentini, M. Pomarès, C.C. Popescu, W.T. Reach, J. Richer, I. Ristorcelli, A. Roy, P. Royer, D. Russeil, P. Saraceno, M. Sauvage, P. Schilke, N. Schneider-Bontemps, F. Schuller, B. Schultz, D.S. Shepherd, B. Sibthorpe, H.A. Smith, M.D. Smith, L. Spinoglio, D. Stamatellos, F. Strafella, G. Stringfellow, E. Sturm, R. Taylor, M.A. Thompson, R.J. Tuffs, G. Umana, L. Valenziano, R. Vavrek, S. Viti, C. Waelkens, D. Ward-Thompson, G. White, F. Wyrowski, H.W. Yorke, Q. Zhang, P.A.S.P. **122**, 314 (2010). DOI 10.1086/651314
145. E.P. Hubble, Ap. J. **64**, 321 (1926). DOI 10.1086/143018
146. A. Sandage, *The Hubble atlas of galaxies* (1961)
147. A. Sandage, J. Bedke, *The Carnegie atlas of galaxies* (1994)
148. G. de Vaucouleurs, A. de Vaucouleurs, H.G. Corwin, Jr., R.J. Buta, G. Paturel, P. Fouque, *Third Reference Catalogue of Bright Galaxies* (1991)
149. M.S. Roberts, M.P. Haynes, A.R.A.A. **32**, 115 (1994). DOI 10.1146/annurev.aa.32.090194.000555
150. L.J. Sage, Astron. Astrophys. **272**, 123 (1993)
151. R.J. Allen, J.H. Knapen, R. Bohlin, T.P. Stecher, Ap. J. **487**, 171 (1997). DOI 10.1086/304590
152. R.J. Allen, P.D. Atherton, R.P.J. Tilanus, in *Birth and Evolution of Massive Stars and Stellar Groups, Astrophysics and Space Science Library*, vol. 120, ed. by W. Boland & H. van Woerden (1985), *Astrophysics and Space Science Library*, vol. 120, pp. 243–251
153. R.P.J. Tilanus, R.J. Allen, in *NASA Conference Publication, NASA Conference Publication*, vol. 2466, ed. by C. J. Lonsdale Persson (1987), *NASA Conference Publication*, vol. 2466, pp. 309–313

154. C.A. Norman, S. Ikeuchi, Ap. J. **345**, 372 (1989). DOI 10.1086/167912
155. D.J. Hollenbach, A.G.G.M. Tielens, A.R.A.A. **35**, 179 (1997). DOI 10.1146/annurev.astro.35.1.179
156. D.J. Hollenbach, A.G.G.M. Tielens, Reviews of Modern Physics **71**, 173 (1999). DOI 10.1103/RevModPhys.71.173
157. N.Z. Scoville, M. Polletta, S. Ewald, S.R. Stolovy, R. Thompson, M. Rieke, A. J. **122**, 3017 (2001). DOI 10.1086/323445
158. D. Calzetti, P.A.S.P. **113**, 1449 (2001). DOI 10.1086/324269
159. P. Kroupa, Science **295**, 82 (2002). DOI 10.1126/science.1067524
160. W. Gliese, Veroeffentlichungen des Astronomischen Rechen-Instituts Heidelberg **22**, 1 (1969)
161. G.E. Miller, J.M. Scalo, Ap. J. Suppl. **41**, 513 (1979). DOI 10.1086/190629
162. N. Reid, G. Gilmore, M.N.R.A.S. **201**, 73 (1982)
163. G. Gilmore, N. Reid, M.N.R.A.S. **202**, 1025 (1983)
164. J.M. Scalo, F.C.Ph. **11**, 1 (1986)
165. N. Bastian, K.R. Covey, M.R. Meyer, A.R.A.A. **48**, 339 (2010). DOI 10.1146/annurev-astro-082708-101642
166. P. Kroupa, C.A. Tout, G. Gilmore, M.N.R.A.S. **244**, 76 (1990)
167. P. Kroupa, C.A. Tout, G. Gilmore, M.N.R.A.S. **262**, 545 (1993)
168. D.F. Figer, Nature **434**, 192 (2005). DOI 10.1038/nature03293
169. P. Kroupa, M.N.R.A.S. **322**, 231 (2001). DOI 10.1046/j.1365-8711.2001.04022.x
170. G. Chabrier, P.A.S.P. **115**, 763 (2003). DOI 10.1086/376392
171. G. de Marchi, F. Paresce, S. Portegies Zwart, in *The Initial Mass Function 50 Years Later, Astrophysics and Space Science Library*, vol. 327, ed. by E. Corbelli, F. Palla, & H. Zinnecker (2005), *Astrophysics and Space Science Library*, vol. 327, pp. 77–+
172. C. Weidner, I.A. Bonnell, H. Zinnecker, Ap. J. **724**, 1503 (2010). DOI 10.1088/0004-637X/724/2/1503
173. R.C. Kennicutt, Jr., A.R.A.A. **36**, 189 (1998). DOI 10.1146/annurev.astro.36.1.189
174. S. Horiuchi, J.F. Beacom, C.S. Kochanek, J.L. Prieto, K.Z. Stanek, T.A. Thompson, ArXiv e-prints (2011)
175. H. Janka, K. Langanke, A. Marek, G. Martínez-Pinedo, B. Müller, Physical Review **442**, 38 (2007). DOI 10.1016/j.physrep.2007.02.002
176. A.J.T. Poelarends, F. Herwig, N. Langer, A. Heger, Ap. J. **675**, 614 (2008). DOI 10.1086/520872
177. R.C. Kennicutt, Jr., P. Tamblyn, C.E. Congdon, Ap. J. **435**, 22 (1994). DOI 10.1086/174790
178. J.S. Gallagher, D.A. Hunter, H. Bushouse, A. J. **97**, 700 (1989). DOI 10.1086/115015
179. R.C. Kennicutt, Jr., Ap. J. **388**, 310 (1992). DOI 10.1086/171154
180. P. Madau, H.C. Ferguson, M.E. Dickinson, M. Giavalisco, C.C. Steidel, A. Fruchter, M.N.R.A.S. **283**, 1388 (1996)
181. M. Rowan-Robinson, R.G. Mann, S.J. Oliver, A. Efstathiou, N. Eaton, P. Goldschmidt, B. Mobasher, S.B.G. Serjeant, T.J. Sumner, L. Danese, D. Elbaz, A. Franceschini, E. Egami, M. Kontizas, A. Lawrence, R. McMahon, H.U. Norgaard-Nielsen, I. Perez-Fournon, J.I. Gonzalez-Serrano, M.N.R.A.S. **289**, 490 (1997)
182. S.C. Chapman, D. Scott, C.C. Steidel, C. Borys, M. Halpern, S.L. Morris, K.L. Adelberger, M. Dickinson, M. Giavalisco, M. Pettini, M.N.R.A.S. **319**, 318 (2000). DOI 10.1046/j.1365-8711.2000.03866.x
183. D. Rosa-González, E. Terlevich, R. Terlevich, M.N.R.A.S. **332**, 283 (2002). DOI 10.1046/j.1365-8711.2002.05285.x
184. R.C. Kennicutt, Jr., J.C. Lee, S.J. Funes, José G., S. Sakai, S. Akiyama, Ap. J. Suppl. **178**, 247 (2008). DOI 10.1086/590058
185. R.C. Kennicutt, C.N. Hao, D. Calzetti, J. Moustakas, D.A. Dale, G. Bendo, C.W. Engelbracht, B.D. Johnson, J.C. Lee, Ap. J. **703**, 1672 (2009). DOI 10.1088/0004-637X/703/2/1672

186. D. Calzetti, R.C. Kennicutt, C.W. Engelbracht, C. Leitherer, B.T. Draine, L. Kewley, J. Moustakas, M. Sosey, D.A. Dale, K.D. Gordon, G.X. Helou, D.J. Hollenbach, L. Armus, G. Bendo, C. Bot, B. Buckalew, T. Jarrett, A. Li, M. Meyer, E.J. Murphy, M. Prescott, M.W. Regan, G.H. Rieke, H. Roussel, K. Sheth, J.D.T. Smith, M.D. Thornley, F. Walter, Ap. J. **666**, 870 (2007). DOI 10.1086/520082
187. N. Murray, M. Rahman, Ap. J. **709**, 424 (2010). DOI 10.1088/0004-637X/709/1/424
188. T.P. Robitaille, B.A. Whitney, Ap. J. Lettersf **710**, L11 (2010). DOI 10.1088/2041-8205/710/1/L11
189. L.L. Watkins, N.W. Evans, J.H. An, M.N.R.A.S. **406**, 264 (2010). DOI 10.1111/j.1365-2966.2010.16708.x
190. C.K.s. Young, M.J. Currie, R.J. Dickens, A.L. Luo, T.J. Zhang, C.J.A.A. **8**, 369 (2008). DOI 10.1088/1009-9271/8/4/01
191. C.J. Pritchet, S. van den Bergh, A. J. **118**, 883 (1999). DOI 10.1086/300977
192. E.K. Grebel, Origin and Evolution of the Elements pp. 234–+ (2004)
193. A.E. Dolphin, D.R. Weisz, E.D. Skillman, J.A. Holtzman, ArXiv Astrophysics e-prints (2005)
194. E.K. Grebel, in *Star Formation from the Small to the Large Scale*, *ESA Special Publication*, vol. 445, ed. by F. Favata, A. Kaas, & A. Wilson (2000), *ESA Special Publication*, vol. 445, pp. 87–+
195. J. Gallego, J. Zamorano, A. Aragon-Salamanca, M. Rego, Ap. J. Lettersf **455**, L1+ (1995). DOI 10.1086/309804
196. M.S. Povich, N. Smith, S.R. Majewski, K.V. Getman, L.K. Townsley, B.L. Babler, P.S. Broos, R. Indebetouw, M.R. Meade, T.P. Robitaille, K.G. Stassun, B.A. Whitney, Y. Yonekura, Y. Fukui, Ap. J. Suppl. **194**, 14 (2011). DOI 10.1088/0067-0049/194/1/14
197. C.J. Lada, E.A. Lada, A.R.A.A. **41**, 57 (2003). DOI 10.1146/annurev.astro.41.011802.094844
198. R.J. Bouwens, G.D. Illingworth, P.A. Oesch, M. Stiavelli, P. van Dokkum, M. Trenti, D. Magee, I. Labbé, M. Franx, C.M. Carollo, V. Gonzalez, Ap. J. Lettersf **709**, L133 (2010). DOI 10.1088/2041-8205/709/2/L133
199. S.J. Lilly, L. Tresse, F. Hammer, D. Crampton, O. Le Fevre, Ap. J. **455**, 108 (1995). DOI 10.1086/176560
200. S.J. Lilly, O. Le Fevre, F. Hammer, D. Crampton, Ap. J. Lettersf **460**, L1+ (1996). DOI 10.1086/309975
201. P. Schechter, Ap. J. **203**, 297 (1976). DOI 10.1086/154079
202. B.G. Elmegreen, in *Protostars and Planets III*, ed. by E. H. Levy & J. I. Lunine (1993), pp. 97–124
203. B.A. Wilking, C.J. Lada, Ap. J. **274**, 698 (1983). DOI 10.1086/161482
204. J.M. Shull, S. Beckwith, A.R.A.A. **20**, 163 (1982). DOI 10.1146/annurev.aa.20.090182.001115
205. R. Genzel, in *NATO ASIC Proc. 342: The Physics of Star Formation and Early Stellar Evolution*, ed. by C. J. Lada & N. D. Kylafis (1991), pp. 155–+
206. D.B. Sanders, N.Z. Scoville, J.S. Young, B.T. Soifer, F.P. Schloerb, W.L. Rice, G.E. Danielson, Ap. J. Lettersf **305**, L45 (1986). DOI 10.1086/184682
207. R.S. Cohen, T.M. Dame, P. Thaddeus, Ap. J. Suppl. **60**, 695 (1986). DOI 10.1086/191101
208. P.M. Solomon, A.R. Rivolo, J. Barrett, A. Yahil, Ap. J. **319**, 730 (1987). DOI 10.1086/165493
209. N.Z. Scoville, M.S. Yun, D.B. Sanders, D.P. Clemens, W.H. Waller, Ap. J. Suppl. **63**, 821 (1987). DOI 10.1086/191185
210. J.P. Williams, C.F. McKee, Ap. J. **476**, 166 (1997). DOI 10.1086/303588
211. B.G. Elmegreen, Ap. J. **577**, 206 (2002). DOI 10.1086/342177
212. D. Bazell, F.X. Desert, Ap. J. **333**, 353 (1988). DOI 10.1086/166751
213. L. Blitz, F.H. Shu, Ap. J. **238**, 148 (1980). DOI 10.1086/157968
214. R.B. Larson, M.N.R.A.S. **194**, 809 (1981)

215. T.M. Dame, B.G. Elmegreen, R.S. Cohen, P. Thaddeus, Ap. J. **305**, 892 (1986). DOI 10.1086/164304
216. D.A. Weintraub, A.A. Goodman, R.L. Akeson, Protostars and Planets IV pp. 247–+ (2000)
217. L. Davis, Jr., J.L. Greenstein, Ap. J. **114**, 206 (1951). DOI 10.1086/145464
218. B.T. Draine, in *Polarimetry of the Interstellar Medium, Astronomical Society of the Pacific Conference Series*, vol. 97, ed. by W. G. Roberge & D. C. B. Whittet (1996), *Astronomical Society of the Pacific Conference Series*, vol. 97, pp. 16–+
219. A. Lazarian, A.A. Goodman, P.C. Myers, Ap. J. **490**, 273 (1997). DOI 10.1086/304874
220. A.A. Goodman, T.J. Jones, E.A. Lada, P.C. Myers, Ap. J. **448**, 748 (1995). DOI 10.1086/176003
221. M.J. Wolff, G.C. Clayton, P.G. Martin, R.E. Schulte-Ladbeck, Ap. J. **423**, 412 (1994). DOI 10.1086/173817
222. P.A. Gerakines, D.C.B. Whittet, Plan. & Space Sci. **43**, 1325 (1995). DOI 10.1016/0032-0633(95)00027-3
223. H.G. Arce, A.A. Goodman, P. Bastien, N. Manset, M. Sumner, Ap. J. Lettersf **499**, L93+ (1998). DOI 10.1086/311337
224. B.A. Wilking, M.J. Lebofsky, J.C. Kemp, G.H. Rieke, A. J. **84**, 199 (1979). DOI 10.1086/112408
225. E.C. Reifenstein, T.L. Wilson, B.F. Burke, P.G. Mezger, W.J. Altenhoff, Astron. Astrophys. **4**, 357 (1970)
226. D.A. Schleuning, Ap. J. **493**, 811 (1998). DOI 10.1086/305139
227. R. Rao, R.M. Crutcher, R.L. Plambeck, M.C.H. Wright, Ap. J. Lettersf **502**, L75+ (1998). DOI 10.1086/311485
228. A. Schulz, R. Lenzen, T. Schmidt, K. Proetel, Astron. Astrophys. **95**, 94 (1981)
229. A.A. Goodman, in *The Physics and Chemistry of Interstellar Molecular Clouds, Lecture Notes in Physics, Berlin Springer Verlag*, vol. 459, ed. by G. Winnewisser & G. C. Pelz (1995), *Lecture Notes in Physics, Berlin Springer Verlag*, vol. 459, pp. 82–85
230. J.L. Dotson, Ap. J. **470**, 566 (1996). DOI 10.1086/177888
231. J.P. Vallee, P. Bastien, Astron. Astrophys. **313**, 255 (1996)
232. J.L. Dotson, J. Davidson, C.D. Dowell, D.A. Schleuning, R.H. Hildebrand, Ap. J. Suppl. **128**, 335 (2000). DOI 10.1086/313384
233. C.L. Brogan, T.H. Troland, Ap. J. **560**, 821 (2001). DOI 10.1086/322444
234. M. Houde, P. Bastien, J.L. Dotson, C.D. Dowell, R.H. Hildebrand, R. Peng, T.G. Phillips, J.E. Vaillancourt, H. Yoshida, Ap. J. **569**, 803 (2002). DOI 10.1086/339356
235. P.C. Myers, R.A. Linke, P.J. Benson, Ap. J. **264**, 517 (1983). DOI 10.1086/160619
236. P.C. Myers, G.A. Fuller, A.A. Goodman, P.J. Benson, Ap. J. **376**, 561 (1991). DOI 10.1086/170305
237. J.W.S. Vilas-Boas, P.C. Myers, G.A. Fuller, Ap. J. **532**, 1038 (2000). DOI 10.1086/308586
238. J.P. Williams, L. Blitz, C.F. McKee, Protostars and Planets IV pp. 97–+ (2000)
239. F. Bertoldi, C.F. McKee, Ap. J. **395**, 140 (1992). DOI 10.1086/171638
240. L. Blitz, P. Thaddeus, Ap. J. **241**, 676 (1980). DOI 10.1086/158379
241. R.J. Maddalena, P. Thaddeus, Ap. J. **294**, 231 (1985). DOI 10.1086/163291
242. J.P. Williams, E.J. de Geus, L. Blitz, Ap. J. **428**, 693 (1994). DOI 10.1086/174279
243. J.P. Williams, L. Blitz, A.A. Stark, Ap. J. **451**, 252 (1995). DOI 10.1086/176216
244. N.F.H. Tothill, G.J. White, H.E. Matthews, W.H. McCutcheon, M.J. McCaughrean, M.A. Kenworthy, Ap. J. **580**, 285 (2002). DOI 10.1086/343068
245. J. Stutzki, R. Guesten, Ap. J. **356**, 513 (1990). DOI 10.1086/168859
246. L. Blitz, in *Protostars and Planets III*, ed. by E. H. Levy & J. I. Lunine (1993), pp. 125–161
247. R. Simon, J.M. Jackson, D.P. Clemens, T.M. Bania, M.H. Heyer, Ap. J. **551**, 747 (2001). DOI 10.1086/320230
248. F. Motte, P. Andre, R. Neri, Astron. Astrophys. **336**, 150 (1998)
249. M.A. Reid, C.D. Wilson, Ap. J. **650**, 970 (2006). DOI 10.1086/507019
250. J. Alves, M. Lombardi, C.J. Lada, Astron. Astrophys. **462**, L17 (2007). DOI 10.1051/0004-6361:20066389

251. H. Beuther, H. Linz, T. Henning, A. Bik, F. Wyrowski, F. Schuller, P. Schilke, S. Thorwirth, K.. Kim, ArXiv e-prints (2011)
252. M.P. Egan, R.F. Shipman, S.D. Price, S.J. Carey, F.O. Clark, M. Cohen, Ap. J. Lettersf **494**, L199+ (1998). DOI 10.1086/311198
253. P. Hennebelle, M. Pérault, D. Teyssier, S. Ganesh, Astron. Astrophys. **365**, 598 (2001). DOI 10.1051/0004-6361:20000052
254. S.J. Carey, P.A. Feldman, R.O. Redman, M.P. Egan, J.M. MacLeod, S.D. Price, Ap. J. Lettersf **543**, L157 (2000). DOI 10.1086/317270
255. J.M. Rathborne, J.M. Jackson, E.T. Chambers, I. Stojimirovic, R. Simon, R. Shipman, W. Frieswijk, Ap. J. **715**, 310 (2010). DOI 10.1088/0004-637X/715/1/310
256. L. Blitz, L. Magnani, in *Birth and Evolution of Massive Stars and Stellar Groups, Astrophysics and Space Science Library*, vol. 120, ed. by W. Boland & H. van Woerden (1985), *Astrophysics and Space Science Library*, vol. 120, pp. 23–28
257. L. Magnani, L. Blitz, L. Mundy, Ap. J. **295**, 402 (1985). DOI 10.1086/163385
258. L. Magnani, J.P. Caillault, A. Buchalter, C.A. Beichman, Ap. J. Suppl. **96**, 159 (1995). DOI 10.1086/192116
259. E.L. Martin, M. Kun, Astron. Astrophys. Suppl. **116**, 467 (1996)
260. R. Neuhäuser, in *Reviews in Modern Astronomy, Reviews in Modern Astronomy*, vol. 12, ed. by R. E. Schielicke (1999), *Reviews in Modern Astronomy*, vol. 12, pp. 27–+
261. M.F. Sterzik, J.H.M.M. Schmitt, in *Cool Stars, Stellar Systems, and the Sun, Astronomical Society of the Pacific Conference Series*, vol. 154, ed. by R. A. Donahue & J. A. Bookbinder (1998), *Astronomical Society of the Pacific Conference Series*, vol. 154, pp. 1339–+
262. A. Heithausen, Astron. Astrophys. **349**, L53 (1999)
263. P.C. Myers, D. Mardones, M. Tafalla, J.P. Williams, D.J. Wilner, Ap. J. Lettersf **465**, L133+ (1996). DOI 10.1086/310146
264. D. Mardones, P.C. Myers, M. Tafalla, D.J. Wilner, R. Bachiller, G. Garay, Ap. J. **489**, 719 (1997). DOI 10.1086/304812
265. A. Heithausen, F. Bertoldi, F. Bensch, Astron. Astrophys. **383**, 591 (2002). DOI 10.1051/0004-6361:20011806
266. J.J. Hester, P.A. Scowen, R. Sankrit, T.R. Lauer, E.A. Ajhar, W.A. Baum, A. Code, D.G. Currie, G.E. Danielson, S.P. Ewald, S.M. Faber, C.J. Grillmair, E.J. Groth, J.A. Holtzman, D.A. Hunter, J. Kristian, R.M. Light, C.R. Lynds, D.G. Monet, E.J. O'Neil, Jr., E.J. Shaya, K.P. Seidelmann, J.A. Westphal, A. J. **111**, 2349 (1996). DOI 10.1086/117968
267. J.O. Kane, D.D. Ryutov, B.A. Remington, S.G. Glendinning, M. Pound, D. Arnett, in *American Astronomical Society Meeting Abstracts #198, Bulletin of the American Astronomical Society*, vol. 33 (2001), *Bulletin of the American Astronomical Society*, vol. 33, pp. 883–+
268. B. Lefloch, J. Cernicharo, L.F. Rodríguez, M.A. Miville-Deschênes, D. Cesarsky, A. Heras, Ap. J. **581**, 335 (2002). DOI 10.1086/344049
269. W.T. Reach, J. Rho, E. Young, J. Muzerolle, S. Fajardo-Acosta, L. Hartmann, A. Sicilia-Aguilar, L. Allen, S. Carey, J.C. Cuillandre, T.H. Jarrett, P. Lowrance, A. Marston, A. Noriega-Crespo, R.L. Hurt, Ap. J. Suppl. **154**, 385 (2004). DOI 10.1086/422193
270. B. Reipurth, Astron. Astrophys. **117**, 183 (1983)
271. B. Reipurth, B. Pettersson, Astron. Astrophys. **267**, 439 (1993)
272. N.S. Schulz, T.W. Berghoefer, H. Zinnecker, Astron. Astrophys. **325**, 1001 (1997)
273. C.H. Young, J.K. Jørgensen, Y.L. Shirley, J. Kauffmann, T. Huard, S.P. Lai, C.W. Lee, A. Crapsi, T.L. Bourke, C.P. Dullemond, T.Y. Brooke, A. Porras, W. Spiesman, L.E. Allen, G.A. Blake, N.J. Evans, II, P.M. Harvey, D.W. Koerner, L.G. Mundy, P.C. Myers, D.L. Padgett, A.I. Sargent, K.R. Stapelfeldt, E.F. van Dishoeck, F. Bertoldi, N. Chapman, L. Cieza, C.H. DeVries, N.A. Ridge, Z. Wahhaj, Ap. J. Suppl. **154**, 396 (2004). DOI 10.1086/422818
274. B.G. Elmegreen, Ap. J. Lettersf **342**, L67 (1989). DOI 10.1086/185486
275. E. Herbst, W. Klemperer, Ap. J. **185**, 505 (1973). DOI 10.1086/152436
276. H. Gerola, A.E. Glassgold, Ap. J. Suppl. **37**, 1 (1978). DOI 10.1086/190515
277. E. Herbst, C.M. Leung, Ap. J. Suppl. **69**, 271 (1989). DOI 10.1086/191314

278. S.W. Stahler, Ap. J. **281**, 209 (1984). DOI 10.1086/162090
279. J.P. Chieze, G. Pineau des Forets, in *Physical Processes in Fragmentation and Star Formation*, *Astrophysics and Space Science Library*, vol. 162, ed. by R. Capuzzo-Dolcetta, C. Chiosi, & A. di Fazio (1990), *Astrophysics and Space Science Library*, vol. 162, pp. 17–27
280. S.S. Prasad, in *Fragmentation of Molecular Clouds and Star Formation*, *IAU Symposium*, vol. 147, ed. by E. Falgarone, F. Boulanger, & G. Duvert (1991), *IAU Symposium*, vol. 147, pp. 93–+
281. Y.N. Efremov, B.G. Elmegreen, M.N.R.A.S. **299**, 588 (1998). DOI 10.1046/j.1365-8711.1998.01819.x
282. C.F. McKee, E.G. Zweibel, Ap. J. **399**, 551 (1992). DOI 10.1086/171946
283. C.F. McKee, E.G. Zweibel, A.A. Goodman, C. Heiles, in *Protostars and Planets III*, ed. by E. H. Levy & J. I. Lunine (1993), pp. 327–+
284. F.H. Shu, Ap. J. **214**, 488 (1977). DOI 10.1086/155274
285. F.H. Shu, F.C. Adams, S. Lizano, A.R.A.A. **25**, 23 (1987). DOI 10.1146/annurev.aa.25.090187.000323
286. P.C. Myers, A.A. Goodman, R. Gusten, C. Heiles, Ap. J. **442**, 177 (1995). DOI 10.1086/175433
287. R.M. Crutcher, Ap. J. **520**, 706 (1999). DOI 10.1086/307483
288. T.C. Mouschovias, L. Spitzer, Jr., Ap. J. **210**, 326 (1976). DOI 10.1086/154835
289. F.H. Shu, Z.Y. Li, Ap. J. **475**, 251 (1997). DOI 10.1086/303521
290. L. Hartmann, *Accretion Processes in Star Formation* (1998)
291. K. Tomisaka, S. Ikeuchi, T. Nakamura, Ap. J. **326**, 208 (1988). DOI 10.1086/166082
292. C.F. von Weizsäcker, Ap. J. **114**, 165 (1951). DOI 10.1086/145462
293. B.G. Elmegreen, S. Kim, L. Staveley-Smith, Ap. J. **548**, 749 (2001). DOI 10.1086/319021
294. R.S. Klessen, ArXiv Astrophysics e-prints (2003)
295. J.C. Tan, Ap. J. **536**, 173 (2000). DOI 10.1086/308905
296. Y. Lithwick, P. Goldreich, Ap. J. **562**, 279 (2001). DOI 10.1086/323470
297. P. Goldreich, S. Sridhar, Ap. J. **438**, 763 (1995). DOI 10.1086/175121
298. J.M. Stone, E.C. Ostriker, C.F. Gammie, Ap. J. Lettersf **508**, L99 (1998). DOI 10.1086/311718
299. M.M. Mac Low, Ap. J. **524**, 169 (1999). DOI 10.1086/307784
300. J. Leorat, T. Passot, A. Pouquet, M.N.R.A.S. **243**, 293 (1990)
301. R.S. Klessen, Ap. J. **556**, 837 (2001). DOI 10.1086/321626
302. M.M. Mac Low, R.S. Klessen, Reviews of Modern Physics **76**, 125 (2004). DOI 10.1103/RevModPhys.76.125
303. R.S. Klessen, F. Heitsch, M.M. Mac Low, Ap. J. **535**, 887 (2000). DOI 10.1086/308891
304. S.W. Stahler, Ap. J. **268**, 165 (1983). DOI 10.1086/160942
305. M. Kiguchi, S. Narita, S.M. Miyama, C. Hayashi, Ap. J. **317**, 830 (1987). DOI 10.1086/165332
306. K. Tomisaka, S. Ikeuchi, T. Nakamura, Ap. J. **341**, 220 (1989). DOI 10.1086/167487
307. L. Mestel, R.B. Paris, Astron. Astrophys. **136**, 98 (1984)
308. P.F. Goldsmith, R. Arquilla, in *Protostars and Planets II*, ed. by D. C. Black & M. S. Matthews (1985), pp. 137–149
309. C. Heiles, A.A. Goodman, C.F. McKee, E.G. Zweibel, in *Protostars and Planets III*, ed. by E. H. Levy & J. I. Lunine (1993), pp. 279–326
310. L.K. Townsley, E.D. Feigelson, T. Montmerle, P.S. Broos, Y.H. Chu, G.P. Garmire, Ap. J. **593**, 874 (2003). DOI 10.1086/376692
311. A. Wootten, R. Snell, A.E. Glassgold, Ap. J. **234**, 876 (1979). DOI 10.1086/157569
312. A. Dalgarno, S. Lepp, Ap. J. Lettersf **287**, L47 (1984). DOI 10.1086/184395
313. B.G. Elmegreen, Ap. J. **232**, 729 (1979). DOI 10.1086/157333
314. F. Palla, in *Massive Star Birth: A Crossroads of Astrophysics*, *IAU Symposium*, vol. 227, ed. by R. Cesaroni, M. Felli, E. Churchwell, & M. Walmsley (2005), *IAU Symposium*, vol. 227, pp. 196–205. DOI 10.1017/S1743921305004540

315. J.H. Oort, L. Spitzer, Jr., Ap. J. **121**, 6 (1955). DOI 10.1086/145958
316. A. Whitworth, M.N.R.A.S. **186**, 59 (1979)
317. H.W. Yorke, G. Tenorio-Tagle, P. Bodenheimer, M. Rozyczka, Astron. Astrophys. **216**, 207 (1989)
318. F. Bertoldi, C.F. McKee, Ap. J. **354**, 529 (1990). DOI 10.1086/168713
319. N.J. Evans, II, J.M.C. Rawlings, Y.L. Shirley, L.G. Mundy, Ap. J. **557**, 193 (2001). DOI 10.1086/321639
320. R. Ebert, Zeitschrift für Astrophysics **37**, 217 (1955)
321. W.B. Bonnor, M.N.R.A.S. **116**, 351 (1956)
322. S. Chandrasekhar, *Ellipsoidal figures of equilibrium* (1969)
323. R. Henriksen, P. Andre, S. Bontemps, Astron. Astrophys. **323**, 549 (1997)
324. S. Basu, T.C. Mouschovias, Ap. J. **452**, 386 (1995). DOI 10.1086/176310
325. M. Tafalla, P.C. Myers, P. Caselli, C.M. Walmsley, C. Comito, Ap. J. **569**, 815 (2002). DOI 10.1086/339321
326. J.F. Alves, C.J. Lada, E.A. Lada, Nature **409**, 159 (2001)
327. D.W.A. Harvey, D.J. Wilner, C.J. Lada, P.C. Myers, J.F. Alves, H. Chen, Ap. J. **563**, 903 (2001). DOI 10.1086/324076
328. G. Racca, M. Gómez, S.J. Kenyon, A. J. **124**, 2178 (2002). DOI 10.1086/342738
329. T.C. Mouschovias, E.V. Paleologou, Ap. J. **237**, 877 (1980). DOI 10.1086/157936
330. J. Gillis, L. Mestel, R.B. Paris, M.N.R.A.S. **187**, 311 (1979)
331. L.A. Hillenbrand, A. J. **113**, 1733 (1997). DOI 10.1086/118389
332. J.E. Gaustad, The Opacity of Diffuse Cosmic Matter and the Early Stages of Star Formation . Ph.D. thesis, PRINCETON UNIVERSITY. (1962)
333. R.B. Larson, M.N.R.A.S. **145**, 271 (1969)
334. R.B. Larson, M.N.R.A.S. **145**, 297 (1969)
335. M.V. Penston, M.N.R.A.S. **144**, 425 (1969)
336. P.N. Foster, R.A. Chevalier, Ap. J. **416**, 303 (1993). DOI 10.1086/173236
337. G. Wuchterl, W.M. Tscharnuter, Astron. Astrophys. **398**, 1081 (2003). DOI 10.1051/0004-6361:20021707
338. P.C. Myers, F.C. Adams, H. Chen, E. Schaff, Ap. J. **492**, 703 (1998). DOI 10.1086/305048
339. W.M. Tscharnuter, in *NATO ASIC Proc. 342: The Physics of Star Formation and Early Stellar Evolution*, ed. by C. J. Lada & N. D. Kylafis (1991), pp. 411–+
340. M. Balluch, Astron. Astrophys. **243**, 168 (1991)
341. C. Hayashi, A.R.A.A. **4**, 171 (1966). DOI 10.1146/annurev.aa.04.090166.001131
342. G. Wuchterl, Astron. Astrophys. **238**, 83 (1990)
343. A.N. Cox, J.E. Tabor, Ap. J. Suppl. **31**, 271 (1976). DOI 10.1086/190383
344. D.R. Alexander, G.C. Augason, H.R. Johnson, Ap. J. **345**, 1014 (1989). DOI 10.1086/167971
345. D.R. Alexander, J.W. Ferguson, Ap. J. **437**, 879 (1994). DOI 10.1086/175039
346. F. Palla, in *Physics of Star Formation in Galaxies*, ed. by F. Palla, H. Zinnecker, A. Maeder, & G. Meynet (2002), pp. 9–133
347. E. Dorfi, Astron. Astrophys. **114**, 151 (1982)
348. R.A. Fiedler, T.C. Mouschovias, Ap. J. **391**, 199 (1992). DOI 10.1086/171336
349. R.A. Fiedler, T.C. Mouschovias, Ap. J. **415**, 680 (1993). DOI 10.1086/173193
350. S. Basu, T.C. Mouschovias, Ap. J. **432**, 720 (1994). DOI 10.1086/174611
351. S. Basu, T.C. Mouschovias, Ap. J. **453**, 271 (1995). DOI 10.1086/176387
352. H.W. Yorke, P. Bodenheimer, Ap. J. **525**, 330 (1999). DOI 10.1086/307867
353. R. Krasnopolsky, A. Königl, Ap. J. **580**, 987 (2002). DOI 10.1086/343890
354. K. Tomisaka, Ap. J. **575**, 306 (2002). DOI 10.1086/341133
355. S. Terebey, F.H. Shu, P. Cassen, Ap. J. **286**, 529 (1984). DOI 10.1086/162628
356. F.C. Adams, C.J. Lada, F.H. Shu, Ap. J. **312**, 788 (1987). DOI 10.1086/164924
357. E.H. Scott, D.C. Black, Ap. J. **239**, 166 (1980). DOI 10.1086/158098
358. D. Galli, F.H. Shu, Ap. J. **417**, 220 (1993). DOI 10.1086/173305
359. D. Galli, F.H. Shu, Ap. J. **417**, 243 (1993). DOI 10.1086/173306

360. F. Nakamura, T. Matsumoto, T. Hanawa, K. Tomisaka, Ap. J. **510**, 274 (1999). DOI 10.1086/306553
361. G.E. Ciolek, T.C. Mouschovias, Ap. J. **454**, 194 (1995). DOI 10.1086/176477
362. S.J. Desch, T.C. Mouschovias, Ap. J. **550**, 314 (2001). DOI 10.1086/319703
363. C.J. Clarke, J.E. Pringle, M.N.R.A.S. **288**, 674 (1997)
364. K. Saigo, T. Hanawa, Ap. J. **493**, 342 (1998). DOI 10.1086/305093
365. F.H. Shu, S. Terebey, in *Cool Stars, Stellar Systems, and the Sun*, *Lecture Notes in Physics, Berlin Springer Verlag*, vol. 193, ed. by S. L. Baliunas & L. Hartmann (1984), *Lecture Notes in Physics, Berlin Springer Verlag*, vol. 193, pp. 78–+. DOI 10.1007/3-540-12907-3182
366. K. Tomisaka, Ap. J. Lettersf **502**, L163+ (1998). DOI 10.1086/311504
367. F.C. Adams, S.P. Ruden, F.H. Shu, Ap. J. **347**, 959 (1989). DOI 10.1086/168187
368. J.C. Papaloizou, G.J. Savonije, M.N.R.A.S. **248**, 353 (1991)
369. S.A. Balbus, J.F. Hawley, Ap. J. **376**, 214 (1991). DOI 10.1086/170270
370. G. Laughlin, P. Bodenheimer, Ap. J. **436**, 335 (1994). DOI 10.1086/174909
371. H.W. Yorke, P. Bodenheimer, G. Laughlin, Ap. J. **411**, 274 (1993). DOI 10.1086/172827
372. H.W. Yorke, P. Bodenheimer, G. Laughlin, Ap. J. **443**, 199 (1995). DOI 10.1086/175514
373. H. Masunaga, S.i. Inutsuka, Ap. J. **531**, 350 (2000). DOI 10.1086/308439
374. C. Hunter, Ap. J. **218**, 834 (1977). DOI 10.1086/155739
375. A. Whitworth, D. Summers, M.N.R.A.S. **214**, 1 (1985)
376. S. Ogino, K. Tomisaka, F. Nakamura, P.A.S.J. **51**, 637 (1999)
377. R.B. Larson, in *Galactic Star Formation Across the Stellar Mass Spectrum*, *Astronomical Society of the Pacific Conference Series*, vol. 287, ed. by J. M. De Buizer & N. S. van der Bliek (2003), *Astronomical Society of the Pacific Conference Series*, vol. 287, pp. 65–80
378. S. Basu, Ap. J. **509**, 229 (1998). DOI 10.1086/306494
379. K. Tomisaka, P.A.S.J. **48**, 701 (1996)
380. T. Matsumoto, T. Hanawa, F. Nakamura, Ap. J. **478**, 569 (1997). DOI 10.1086/303822
381. F. Nakamura, Ap. J. **543**, 291 (2000). DOI 10.1086/317113
382. J. Muzerolle, L. Hillenbrand, N. Calvet, C. Briceño, L. Hartmann, Ap. J. **592**, 266 (2003). DOI 10.1086/375704
383. T. Montmerle, N. Grosso, Y. Tsuboi, K. Koyama, Ap. J. **532**, 1097 (2000). DOI 10.1086/308611
384. C.A. Beichman, P.C. Myers, J.P. Emerson, S. Harris, R. Mathieu, P.J. Benson, R.E. Jennings, Ap. J. **307**, 337 (1986). DOI 10.1086/164421
385. C.A. Beichman, F. Boulanger, M. Moshir, Ap. J. **386**, 248 (1992). DOI 10.1086/171011
386. S.J. Kenyon, M. Gomez, R.O. Marzke, L. Hartmann, A. J. **108**, 251 (1994). DOI 10.1086/117064
387. C.J. Burrows, K.R. Stapelfeldt, A.M. Watson, J.E. Krist, G.E. Ballester, J.T. Clarke, D. Crisp, J.S. Gallagher, III, R.E. Griffiths, J.J. Hester, J.G. Hoessel, J.A. Holtzman, J.R. Mould, P.A. Scowen, J.T. Trauger, J.A. Westphal, Ap. J. **473**, 437 (1996). DOI 10.1086/178156
388. S.W. Stahler, F.H. Shu, R.E. Taam, Ap. J. **241**, 637 (1980). DOI 10.1086/158377
389. S.W. Stahler, F.H. Shu, R.E. Taam, Ap. J. **242**, 226 (1980). DOI 10.1086/158459
390. F. Palla, S.W. Stahler, Ap. J. Lettersf **360**, L47 (1990). DOI 10.1086/185809
391. E. Böhm-Vitense, Zeitschrift für Astrophysics **46**, 108 (1958)
392. S. Canuto, M.C. Zerner, G.H.F. Diercksen, Ap. J. **377**, 150 (1991). DOI 10.1086/170343
393. V.M. Canuto, I. Goldman, I. Mazzitelli, Ap. J. **473**, 550 (1996). DOI 10.1086/178166
394. P. Ventura, A. Zeppieri, I. Mazzitelli, F. D'Antona, Astron. Astrophys. **331**, 1011 (1998)
395. P. Bodenheimer, Ap. J. **142**, 451 (1965). DOI 10.1086/148310
396. S.W. Stahler, Ap. J. **332**, 804 (1988). DOI 10.1086/166694
397. F. Palla, S.W. Stahler, Ap. J. **392**, 667 (1992). DOI 10.1086/171468
398. F. Palla, S.W. Stahler, Ap. J. **418**, 414 (1993). DOI 10.1086/173402
399. M. Hayashi, P.A.S.J. **17**, 177 (1965)
400. J.E. Gaustad, Ap. J. **138**, 1050 (1963). DOI 10.1086/147705
401. R.J. Gould, Ap. J. **140**, 638 (1964). DOI 10.1086/147954
402. I. Iben, Jr., Ap. J. **141**, 993 (1965). DOI 10.1086/148193

403. P.A. Bernasconi, A. Maeder, Astron. Astrophys. **307**, 829 (1996)
404. G. Wuchterl, W.M. Tscharnuter, Astron. Astrophys. **398**, 1081 (2003). DOI 10.1051/0004-6361:20021707
405. M. Forestini, Astron. Astrophys. **285**, 473 (1994)
406. H.J.G.L.M. Lamers, J.P. Cassinelli, *Introduction to Stellar Winds* (1999)
407. N.S. Schulz, C. Canizares, D. Huenemoerder, K. Tibbets, Ap. J. **595**, 365 (2003). DOI 10.1086/377214
408. S.W. Stahler, Ap. J. **274**, 822 (1983). DOI 10.1086/161495
409. G. Wuchterl, R.S. Klessen, Ap. J. Lettersf **560**, L185 (2001). DOI 10.1086/324307
410. F. Palla, S.W. Stahler, Ap. J. **525**, 772 (1999). DOI 10.1086/307928
411. F. D'Antona, I. Mazzitelli, Ap. J. Suppl. **90**, 467 (1994). DOI 10.1086/191867
412. I. Baraffe, G. Chabrier, F. Allard, P.H. Hauschildt, Astron. Astrophys. **327**, 1054 (1997)
413. F. D'Antona, I. Mazzitelli, in *Brown Dwarfs and Extrasolar Planets, Astronomical Society of the Pacific Conference Series*, vol. 134, ed. by R. Rebolo, E. L. Martin, & M. R. Zapatero Osorio (1998), *Astronomical Society of the Pacific Conference Series*, vol. 134, pp. 442–+
414. S.J. Kenyon, L. Hartmann, Ap. J. Suppl. **101**, 117 (1995). DOI 10.1086/192235
415. C.J. Lada, B.A. Wilking, Ap. J. **287**, 610 (1984). DOI 10.1086/162719
416. F.C. Adams, F.H. Shu, Ap. J. **296**, 655 (1985). DOI 10.1086/163483
417. C.J. Lada, A.R.A.A. **23**, 267 (1985). DOI 10.1146/annurev.aa.23.090185.001411
418. B.A. Wilking, C.J. Lada, E.T. Young, Ap. J. **340**, 823 (1989). DOI 10.1086/167439
419. S. Beckwith, A.I. Sargent, N.Z. Scoville, C.R. Masson, B. Zuckerman, T.G. Phillips, Ap. J. **309**, 755 (1986). DOI 10.1086/164645
420. V. Ossenkopf, T. Henning, Astron. Astrophys. **291**, 943 (1994)
421. S. Bontemps, P. Andre, S. Terebey, S. Cabrit, Astron. Astrophys. **311**, 858 (1996)
422. P. Saraceno, P. Andre, C. Ceccarelli, M. Griffin, S. Molinari, Astron. Astrophys. **309**, 827 (1996)
423. S.T. Megeath, L.E. Allen, R.A. Gutermuth, J.L. Pipher, P.C. Myers, N. Calvet, L. Hartmann, J. Muzerolle, G.G. Fazio, Ap. J. Suppl. **154**, 367 (2004). DOI 10.1086/422823
424. L.E. Allen, N. Calvet, P. D'Alessio, B. Merin, L. Hartmann, S.T. Megeath, R.A. Gutermuth, J. Muzerolle, J.L. Pipher, P.C. Myers, G.G. Fazio, Ap. J. Suppl. **154**, 363 (2004). DOI 10.1086/422715
425. G. Haro, B. Iriarte, E. Chavira, Boletin de los Observatorios Tonantzintla y Tacubaya **1**, 3 (1953)
426. G.H. Herbig, Ap. J. **128**, 259 (1958). DOI 10.1086/146540
427. G.H. Herbig, L.V. Kuhi, Ap. J. **137**, 398 (1963). DOI 10.1086/147519
428. M. Cohen, L.V. Kuhi, Ap. J. Suppl. **41**, 743 (1979). DOI 10.1086/190641
429. T. Montmerle, L. Koch-Miramond, E. Falgarone, J.E. Grindlay, Ap. J. **269**, 182 (1983). DOI 10.1086/161029
430. M. Kun, A.P.S.S. **125**, 13 (1986). DOI 10.1007/BF00643968
431. F.M. Walter, Ap. J. **306**, 573 (1986). DOI 10.1086/164367
432. P. Andre, T. Montmerle, E.D. Feigelson, A. J. **93**, 1182 (1987). DOI 10.1086/114400
433. E.D. Feigelson, J.M. Jackson, R.D. Mathieu, P.C. Myers, F.M. Walter, A. J. **94**, 1251 (1987). DOI 10.1086/114561
434. F.M. Walter, A. Brown, R.D. Mathieu, P.C. Myers, F.J. Vrba, A. J. **96**, 297 (1988). DOI 10.1086/114809
435. P. Hartigan, S.J. Kenyon, L. Hartmann, S.E. Strom, S. Edwards, A.D. Welty, J. Stauffer, Ap. J. **382**, 617 (1991). DOI 10.1086/170749
436. E.D. Feigelson, S. Casanova, T. Montmerle, J. Guibert, Ap. J. **416**, 623 (1993). DOI 10.1086/173264
437. D.P. Huenemoerder, W.A. Lawson, E.D. Feigelson, M.N.R.A.S. **271**, 967 (1994)
438. F.M. Walter, F.J. Vrba, R.D. Mathieu, A. Brown, P.C. Myers, A. J. **107**, 692 (1994). DOI 10.1086/116889
439. E. Gullbring, L. Hartmann, C. Briceno, N. Calvet, Ap. J. **492**, 323 (1998). DOI 10.1086/305032

440. K.M. Strom, S.E. Strom, S. Edwards, S. Cabrit, M.F. Skrutskie, A. J. **97**, 1451 (1989). DOI 10.1086/115085
441. M. Cohen, J.P. Emerson, C.A. Beichman, Ap. J. **339**, 455 (1989). DOI 10.1086/167310
442. D. Lynden-Bell, J.E. Pringle, M.N.R.A.S. **168**, 603 (1974)
443. L. Hartmann, N. Calvet, E. Gullbring, P. D'Alessio, Ap. J. **495**, 385 (1998). DOI 10.1086/305277
444. L.A. Hillenbrand, ArXiv Astrophysics e-prints (2002)
445. L. Hartmann, R. Hewett, N. Calvet, Ap. J. **426**, 669 (1994). DOI 10.1086/174104
446. J. Muzerolle, N. Calvet, C. Briceño, L. Hartmann, L. Hillenbrand, Ap. J. Lettersf **535**, L47 (2000). DOI 10.1086/312691
447. J.H. Kastner, D.P. Huenemoerder, N.S. Schulz, C.R. Canizares, D.A. Weintraub, Ap. J. **567**, 434 (2002). DOI 10.1086/338419
448. S.W. Stahler, F.M. Walter, in *Protostars and Planets III*, ed. by E. H. Levy & J. I. Lunine (1993), pp. 405–428
449. G. Basri, C. Bertout, in *Protostars and Planets III*, ed. by E. H. Levy & J. I. Lunine (1993), pp. 543–566
450. R. Neuhaeuser, M.F. Sterzik, G. Torres, E.L. Martin, Astron. Astrophys. **299**, L13+ (1995)
451. R. Neuhaeuser, M.F. Sterzik, J.H.M.M. Schmitt, R. Wichmann, J. Krautter, Astron. Astrophys. **297**, 391 (1995)
452. R. Wichmann, J. Krautter, J.H.M.M. Schmitt, R. Neuhaeuser, J.M. Alcala, H. Zinnecker, R.M. Wagner, R. Mundt, M.F. Sterzik, Astron. Astrophys. **312**, 439 (1996)
453. G.H. Herbig, Ap. J. **113**, 697 (1951). DOI 10.1086/145440
454. G.H. Herbig, J.R.A.S.C. **46**, 222 (1952)
455. G. Haro, Ap. J. **115**, 572 (1952). DOI 10.1086/145576
456. G. Haro, Ap. J. **117**, 73 (1953). DOI 10.1086/145669
457. R.D. Schwartz, Ap. J. **195**, 631 (1975). DOI 10.1086/153364
458. B. Reipurth, Astron. Astrophys. **220**, 249 (1989)
459. B. Reipurth, J. Cernicharo, in *Revista Mexicana de Astronomia y Astrofisica Conference Series, Revista Mexicana de Astronomia y Astrofisica Conference Series*, vol. 1, ed. by S. Lizano & J. M. Torrelles (1995), *Revista Mexicana de Astronomia y Astrofisica Conference Series*, vol. 1, pp. 43–+
460. B. Reipurth, J. Bally, A.R.A.A. **39**, 403 (2001). DOI 10.1146/annurev.astro.39.1.403
461. G.H. Herbig, Vistas in Astronomy **8**, 109 (1966). DOI 10.1016/0083-6656(66)90025-0
462. G.H. Herbig, Ap. J. **217**, 693 (1977). DOI 10.1086/155615
463. B. Reipurth, L. Hartmann, S.J. Kenyon, A. Smette, P. Bouchet, A. J. **124**, 2194 (2002). DOI 10.1086/342740
464. L. Hartmann, N. Calvet, A. J. **109**, 1846 (1995). DOI 10.1086/117411
465. G.H. Herbig, P.P. Petrov, R. Duemmler, Ap. J. **595**, 384 (2003). DOI 10.1086/377194
466. R.B. Larson, M.N.R.A.S. **190**, 321 (1980)
467. L. Hartmann, K. Hinkle, N. Calvet, Ap. J. **609**, 906 (2004). DOI 10.1086/421317
468. G.H. Herbig, Ap. J. Suppl. **4**, 337 (1960). DOI 10.1086/190050
469. L.A. Hillenbrand, S.E. Strom, F.J. Vrba, J. Keene, Ap. J. **397**, 613 (1992). DOI 10.1086/171819
470. P.S. The, D. de Winter, M.R. Perez, Astron. Astrophys. Suppl. **104**, 315 (1994)
471. T. Preibisch, H. Zinnecker, in *Roentgenstrahlung from the Universe*, ed. by H. U. Zimmermann, J. Trümper, & H. Yorke (1996), pp. 17–20
472. J.K. Davies, A. Evans, M.F. Bode, D.C.B. Whittet, M.N.R.A.S. **247**, 517 (1990)
473. L.B.F.M. Waters, C. Waelkens, A.R.A.A. **36**, 233 (1998). DOI 10.1146/annurev.astro.36.1.233
474. L.B.F.M. Waters, J.M. Marlborough, in *Pulsation; Rotation; and Mass Loss in Early-Type Stars, IAU Symposium*, vol. 162, ed. by L. A. Balona, H. F. Henrichs, & J. M. Le Contel (1994), *IAU Symposium*, vol. 162, pp. 399–+
475. M.E. van den Ancker, D. de Winter, H.R.E. Tjin A Djie, Astron. Astrophys. **330**, 145 (1998)

476. F. Berrilli, G. Corciulo, G. Ingrosso, D. Lorenzetti, B. Nisini, F. Strafella, Ap. J. **398**, 254 (1992). DOI 10.1086/171853
477. H. Zinnecker, H.W. Yorke, A.R.A.A. **45**, 481 (2007). DOI 10.1146/annurev.astro.44.051905.092549
478. H. Zinnecker, A.P.S.S. **99**, 41 (1984). DOI 10.1007/BF00650229
479. H.A. Abt, A.R.A.A. **21**, 343 (1983). DOI 10.1146/annurev.aa.21.090183.002015
480. H. Zinnecker, in *Physics of Star Formation in Galaxies*, ed. by F. Palla, H. Zinnecker, A. Maeder, & G. Meynet (2002), pp. 135–225
481. P. Bodenheimer, A. Burkert, R.I. Klein, A.P. Boss, Protostars and Planets IV pp. 675–+ (2000)
482. B. Reipurth, H. Zinnecker, Astron. Astrophys. **278**, 81 (1993)
483. M. Simon, A.M. Ghez, C. Leinert, L. Cassar, W.P. Chen, R.R. Howell, R.F. Jameson, K. Matthews, G. Neugebauer, A. Richichi, Ap. J. **443**, 625 (1995). DOI 10.1086/175554
484. D.S. Evans, Q.J.R.A.S. **9**, 388 (1968)
485. R.D. Mathieu, A.R.A.A. **32**, 465 (1994). DOI 10.1146/annurev.aa.32.090194.002341
486. H. Zinnecker, in *Modes of Star Formation and the Origin of Field Populations, Astronomical Society of the Pacific Conference Series*, vol. 285, ed. by E. K. Grebel & W. Brandner (2002), *Astronomical Society of the Pacific Conference Series*, vol. 285, pp. 131–+
487. W.D. Heintz, J.R.A.S.C. **63**, 275 (1969)
488. A.H. Batten, *Binary and multiple systems of stars* (1973)
489. H.A. Abt, A.R.A.A. **21**, 343 (1983). DOI 10.1146/annurev.aa.21.090183.002015
490. B. Reipurth, H. Zinnecker, Astron. Astrophys. **278**, 81 (1993)
491. C. Leinert, H. Zinnecker, N. Weitzel, J. Christou, S.T. Ridgway, R. Jameson, M. Haas, R. Lenzen, Astron. Astrophys. **278**, 129 (1993)
492. A.M. Ghez, G. Neugebauer, K. Matthews, A. J. **106**, 2005 (1993). DOI 10.1086/116782
493. S.P. Goodwin, P. Kroupa, A. Goodman, A. Burkert, Protostars and Planets V pp. 133–147 (2007)
494. A. Duquennoy, M. Mayor, Astron. Astrophys. **248**, 485 (1991)
495. A.M. Ghez, D.W. McCarthy, J.L. Patience, T.L. Beck, Ap. J. **481**, 378 (1997). DOI 10.1086/304031
496. R. Kohler, C. Leinert, Astron. Astrophys. **331**, 977 (1998)
497. M.G. Petr, V. Coudé du Foresto, S.V.W. Beckwith, A. Richichi, M.J. McCaughrean, Ap. J. **500**, 825 (1998). DOI 10.1086/305751
498. W. Brandner, R. Koehler, Ap. J. Lettersf **499**, L79+ (1998). DOI 10.1086/311338
499. R. Köhler, M.G. Petr-Gotzens, M.J. McCaughrean, J. Bouvier, G. Duchêne, A. Quirrenbach, H. Zinnecker, Astron. Astrophys. **458**, 461 (2006). DOI 10.1051/0004-6361:20054561
500. S. Bontemps, P. André, A.A. Kaas, L. Nordh, G. Olofsson, M. Huldtgren, A. Abergel, J. Blommaert, F. Boulanger, M. Burgdorf, C.J. Cesarsky, D. Cesarsky, E. Copet, J. Davies, E. Falgarone, G. Lagache, T. Montmerle, M. Pérault, P. Persi, T. Prusti, J.L. Puget, F. Sibille, Astron. Astrophys. **372**, 173 (2001). DOI 10.1051/0004-6361:20010474
501. W. Brandner, H. Zinnecker, Astron. Astrophys. **321**, 220 (1997)
502. P. Hartigan, K.M. Strom, S.E. Strom, Ap. J. **427**, 961 (1994). DOI 10.1086/174203
503. O.P. Lay, J.E. Carlstrom, R.E. Hills, Ap. J. Lettersf **452**, L73+ (1995). DOI 10.1086/309713
504. G.A. Fuller, E.F. Ladd, K.W. Hodapp, Ap. J. Lettersf **463**, L97+ (1996). DOI 10.1086/310066
505. L.W. Looney, L.G. Mundy, W.J. Welch, Ap. J. Lettersf **484**, L157+ (1997). DOI 10.1086/310795
506. S. Terebey, D. van Buren, D.L. Padgett, T. Hancock, M. Brundage, Ap. J. Lettersf **507**, L71 (1998). DOI 10.1086/311671
507. G. Torres, R. Neuhäuser, E.W. Guenther, A. J. **123**, 1701 (2002). DOI 10.1086/339178
508. D.L. Looper, J.J. Bochanski, A.J. Burgasser, S. Mohanty, E.E. Mamajek, J.K. Faherty, A.A. West, M.A. Pitts, A. J. **140**, 1486 (2010). DOI 10.1088/0004-6256/140/5/1486
509. R. Teixeira, C. Ducourant, G. Chauvin, A. Krone-Martins, I. Song, B. Zuckerman, Astron. Astrophys. **489**, 825 (2010). DOI 10.1051/0004-6361:200810133

510. J.H. Kastner, B. Zuckerman, P. Hily-Blant, T. Forveille, Astron. Astrophys. **492**, 469 (2008). DOI 10.1051/0004-6361:200810815
511. D.R. Rodriguez, J.H. Kastner, D. Wilner, C. Qi, Ap. J. **720**, 1684 (2010). DOI 10.1088/0004-637X/720/2/1684
512. D. Fabrycky, S. Tremaine, Ap. J. **669**, 1298 (2007). DOI 10.1086/521702
513. B. Zuckerman, A.R.A.A. **39**, 549 (2001). DOI 10.1146/annurev.astro.39.1.549
514. B. Zuckerman, E.E. Becklin, Ap. J. **414**, 793 (1993). DOI 10.1086/173123
515. J. Muzerolle, N. Calvet, C. Briceño, L. Hartmann, L. Hillenbrand, Ap. J. Lettersf **535**, L47 (2000). DOI 10.1086/312691
516. R. Jayawardhana, L. Hartmann, G. Fazio, R.S. Fisher, C.M. Telesco, R.K. Piña, Ap. J. Lettersf **520**, L41 (1999). DOI 10.1086/312148
517. S.M. Andrews, I. Czekala, D.J. Wilner, C. Espaillat, C.P. Dullemond, A.M. Hughes, Ap. J. **710**, 462 (2010). DOI 10.1088/0004-637X/710/1/462
518. D.P. Huenemoerder, J.H. Kastner, P. Testa, N.S. Schulz, D.A. Weintraub, Ap. J. **671**, 592 (2007). DOI 10.1086/522921
519. R.D. Mathieu, K. Stassun, G. Basri, E.L.N. Jensen, C.M. Johns-Krull, J.A. Valenti, L.W. Hartmann, A. J. **113**, 1841 (1997). DOI 10.1086/118395
520. G. Basri, C.M. Johns-Krull, R.D. Mathieu, A. J. **114**, 781 (1997). DOI 10.1086/118510
521. P. Artymowicz, S.H. Lubow, Ap. J. Lettersf **467**, L77+ (1996). DOI 10.1086/310200
522. D.M. Salter, Á. Kóspál, K.V. Getman, M.R. Hogerheijde, T.A. van Kempen, J.M. Carpenter, G.A. Blake, D. Wilner, Astron. Astrophys. **521**, A32+ (2010). DOI 10.1051/0004-6361/201015197
523. K.V. Getman, E.D. Feigelson, P.S. Broos, G. Micela, G.P. Garmire, Ap. J. **688**, 418 (2008). DOI 10.1086/592033
524. K.V. Getman, P.S. Broos, D.M. Salter, G.P. Garmire, M.R. Hogerheijde, Ap. J. **730**, 6 (2011). DOI 10.1088/0004-637X/730/1/6
525. N.S. Schulz, P. Testa, D.P. Huenemoerder, K. Ishibashi, C.R. Canizares, Ap. J. **653**, 636 (2006). DOI 10.1086/508625
526. A.I. Gómez de Castro, Ap. J. Lettersf **698**, L108 (2009). DOI 10.1088/0004-637X/698/2/L108
527. S.H.P. Alencar, C.H.F. Melo, C.P. Dullemond, J. Andersen, C. Batalha, L.P.R. Vaz, R.D. Mathieu, Astron. Astrophys. **409**, 1037 (2003). DOI 10.1051/0004-6361:20031229
528. L. Siess, E. Dufour, M. Forestini, Astron. Astrophys. **358**, 593 (2000)
529. G. Schaller, D. Schaerer, G. Meynet, A. Maeder, Astron. Astrophys. Suppl. **96**, 269 (1992)
530. D.P. Huenemoerder, N.S. Schulz, P. Testa, A. Kesich, C.R. Canizares, Ap. J. **707**, 942 (2009). DOI 10.1088/0004-637X/707/2/942
531. G.H. Herbig, R.F. Griffin, A. J. **132**, 1763 (2006). DOI 10.1086/507769
532. N.S. Schulz, C. Canizares, D. Huenemoerder, J.H. Kastner, S.C. Taylor, E.J. Bergstrom, Ap. J. **549**, 441 (2001). DOI 10.1086/319078
533. C. Schröder, J.H.M.M. Schmitt, Astron. Astrophys. **475**, 677 (2007). DOI 10.1051/0004-6361:20077429
534. A.P. Boss, Ap. J. **410**, 157 (1993). DOI 10.1086/172734
535. R.B. Larson, M.N.R.A.S. **214**, 379 (1985)
536. R.B. Larson, M.N.R.A.S. **184**, 69 (1978)
537. J.E. Pringle, M.N.R.A.S. **239**, 361 (1989)
538. I. Bonnell, H. Martel, P. Bastien, J.P. Arcoragi, W. Benz, Ap. J. **377**, 553 (1991). DOI 10.1086/170384
539. C.J. Clarke, J.E. Pringle, M.N.R.A.S. **249**, 588 (1991)
540. A.P. Boss, Ap. J. **346**, 336 (1989). DOI 10.1086/168014
541. I.A. Bonnell, M.R. Bate, M.N.R.A.S. **271**, 999 (1994)
542. I. Bonnell, in *The Third Pacific Rim Conference on Recent Development on Binary Star Research, Astronomical Society of the Pacific Conference Series*, vol. 130, ed. by K.-C. Leung (1997), *Astronomical Society of the Pacific Conference Series*, vol. 130, pp. 1–+

543. F. Verbunt, E.P.J. van den Heuvel, in *X-ray Binaries*, ed. by W. H. G. Lewin, J. van Paradijs, & E. P. J. van den Heuvel (1995), pp. 457–494
544. B. Paczyński, A.R.A.A. **9**, 183 (1971). DOI 10.1146/annurev.aa.09.090171.001151
545. P.P. Eggleton, Ap. J. **268**, 368 (1983). DOI 10.1086/160960
546. Z. Han, C.A. Tout, P.P. Eggleton, M.N.R.A.S. **319**, 215 (2000). DOI 10.1046/j.1365-8711.2000.03839.x
547. M.R. Krumholz, T.A. Thompson, Ap. J. **661**, 1034 (2007). DOI 10.1086/515566
548. F. Boulanger, P. Cox, A.P. Jones, in *Infrared Space Astronomy, Today and Tomorrow*, ed. by F. Casoli, J. Lequeux, & F. David (2000), pp. 251–+
549. J. Frank, A. King, D. Raine, *Accretion power in astrophysics.* (1992)
550. C.D. Koresko, Ap. J. Lettersf **507**, L145 (1998). DOI 10.1086/311694
551. K.R. Stapelfeldt, C.J. Burrows, J.E. Krist, A.M. Watson, G.E. Ballester, J.T. Clarke, D. Crisp, R.W. Evans, J.S. Gallagher, III, R.E. Griffiths, J.J. Hester, J.G. Hoessel, J.A. Holtzman, J.R. Mould, P.A. Scowen, J.T. Trauger, J.A. Westphal, Ap. J. **508**, 736 (1998). DOI 10.1086/306422
552. J.E. Krist, K.R. Stapelfeldt, F. Ménard, D.L. Padgett, C.J. Burrows, Ap. J. **538**, 793 (2000). DOI 10.1086/309170
553. J. Silber, T. Gledhill, G. Duchêne, F. Ménard, Ap. J. Lettersf **536**, L89 (2000). DOI 10.1086/312731
554. D.J. Wilner, P.T.P. Ho, J.H. Kastner, L.F. Rodríguez, Ap. J. Lettersf **534**, L101 (2000). DOI 10.1086/312642
555. M. Osterloh, S.V.W. Beckwith, Ap. J. **439**, 288 (1995). DOI 10.1086/175172
556. L.W. Looney, Unveiling the Envelope and Disk: a Sub-Arcsecond Survey of Young Stellar Systems. Ph.D. thesis, UNIVERSITY OF MARYLAND COLLEGE PARK (1998)
557. L.W. Looney, L.G. Mundy, W.J. Welch, Ap. J. **529**, 477 (2000). DOI 10.1086/308239
558. S.V.W. Beckwith, A.I. Sargent, Ap. J. **402**, 280 (1993). DOI 10.1086/172131
559. D. Hollenbach, D. Johnstone, F. Shu, in *Massive Stars: Their Lives in the Interstellar Medium, Astronomical Society of the Pacific Conference Series*, vol. 35, ed. by J. P. Cassinelli & E. B. Churchwell (1993), *Astronomical Society of the Pacific Conference Series*, vol. 35, pp. 26–+
560. N.I. Shakura, R.A. Sunyaev, Astron. Astrophys. **24**, 337 (1973)
561. L.G. Mundy, L.W. Looney, W.J. Welch, Protostars and Planets IV pp. 355–+ (2000)
562. N. Ohashi, M. Hayashi, P.T.P. Ho, M. Momose, M. Tamura, N. Hirano, A.I. Sargent, Ap. J. **488**, 317 (1997). DOI 10.1086/304685
563. A.I. Sargent, S.V.W. Beckwith, Ap. J. Lettersf **382**, L31 (1991). DOI 10.1086/186207
564. A. Natta, in *Infrared Space Astronomy, Today and Tomorrow*, ed. by F. Casoli, J. Lequeux, & F. David (2000), pp. 193–+
565. S.A. Balbus, A.R.A.A. **41**, 555 (2003). DOI 10.1146/annurev.astro.41.081401.155207
566. D.N.C. Lin, J. Papaloizou, M.N.R.A.S. **191**, 37 (1980)
567. P.J. Armitage, C.S. Reynolds, J. Chiang, Ap. J. **548**, 868 (2001). DOI 10.1086/318990
568. S.A. Balbus, J.F. Hawley, Reviews of Modern Physics **70**, 1 (1998). DOI 10.1103/RevModPhys.70.1
569. S. Chandrasekhar, Proceedings of the National Academy of Science **46**, 253 (1960). DOI 10.1073/pnas.46.2.253
570. J.F. Hawley, C.F. Gammie, S.A. Balbus, Ap. J. **440**, 742 (1995). DOI 10.1086/175311
571. A. Brandenburg, A. Nordlund, R.F. Stein, U. Torkelsson, Ap. J. **446**, 741 (1995). DOI 10.1086/175831
572. P.J. Armitage, Ap. J. Lettersf **501**, L189+ (1998). DOI 10.1086/311463
573. J.F. Hawley, Ap. J. **528**, 462 (2000). DOI 10.1086/308180
574. S. Fromang, C. Terquem, S.A. Balbus, M.N.R.A.S. **329**, 18 (2002). DOI 10.1046/j.1365-8711.2002.04940.x
575. O.M. Blaes, S.A. Balbus, Ap. J. **421**, 163 (1994). DOI 10.1086/173634
576. S.A. Balbus, J.F. Hawley, Space Sci. Rev. **92**, 39 (2000). DOI 10.1023/A:1005293132737

577. T. Sano, S.M. Miyama, T. Umebayashi, T. Nakano, Ap. J. **543**, 486 (2000). DOI 10.1086/317075
578. C.F. Gammie, Ap. J. **457**, 355 (1996). DOI 10.1086/176735
579. A.E. Glassgold, J. Najita, J. Igea, Ap. J. **480**, 344 (1997). DOI 10.1086/303952
580. A.E. Glassgold, E.D. Feigelson, T. Montmerle, Protostars and Planets IV pp. 429–+ (2000)
581. S. Hirose, N.J. Turner, Ap. J. Lettersf **732**, L30+ (2011). DOI 10.1088/2041-8205/732/2/L30
582. C.P. Dullemond, D. Hollenbach, I. Kamp, P. D'Alessio,
583. S.J. Kenyon, L. Hartmann, Ap. J. **323**, 714 (1987). DOI 10.1086/165866
584. T. Fleming, J.M. Stone, Ap. J. **585**, 908 (2003). DOI 10.1086/345848
585. D. Semenov, D. Wiebe, T. Henning, Astron. Astrophys. **417**, 93 (2004). DOI 10.1051/0004-6361:20034128
586. D.J. Hollenbach, H.W. Yorke, D. Johnstone, Protostars and Planets IV pp. 401–+ (2000)
587. F.H. Shu, S. Lizano, S.P. Ruden, J. Najita, Ap. J. Lettersf **328**, L19 (1988). DOI 10.1086/185152
588. F.C. Adams, D. Hollenbach, G. Laughlin, U. Gorti, Ap. J. **611**, 360 (2004). DOI 10.1086/421989
589. K.E. Haisch, Jr., E.A. Lada, C.J. Lada, A. J. **121**, 2065 (2001). DOI 10.1086/319951
590. A.G.W. Cameron, Icarus **18**, 407 (1973). DOI 10.1016/0019-1035(73)90153-X
591. B.G. Elmegreen, Astron. Astrophys. **80**, 77 (1979)
592. G.P. Horedt, Astron. Astrophys. **110**, 209 (1982)
593. C.J. Clarke, J.E. Pringle, M.N.R.A.S. **261**, 190 (1993)
594. S.M. Hall, C.J. Clarke, J.E. Pringle, M.N.R.A.S. **278**, 303 (1996)
595. D. Hollenbach, D. Johnstone, S. Lizano, F. Shu, Ap. J. **428**, 654 (1994). DOI 10.1086/174276
596. H. Störzer, D. Hollenbach, Ap. J. **515**, 669 (1999). DOI 10.1086/307055
597. C.J. Clarke, A. Gendrin, M. Sotomayor, M.N.R.A.S. **328**, 485 (2001). DOI 10.1046/j.1365-8711.2001.04891.x
598. A.S. Font, I.G. McCarthy, D. Johnstone, D.R. Ballantyne, Ap. J. **607**, 890 (2004). DOI 10.1086/383518
599. I.D. Howarth, R.K. Prinja, Ap. J. Suppl. **69**, 527 (1989). DOI 10.1086/191321
600. S. Richling, H.W. Yorke, Astron. Astrophys. **327**, 317 (1997)
601. R. Banerjee, R.E. Pudritz, Ap. J. **641**, 949 (2006). DOI 10.1086/500496
602. R. Banerjee, R.E. Pudritz, Ap. J. **660**, 479 (2007)
603. R. Mundt, in *Protostars and Planets II*, ed. by D. C. Black & M. S. Matthews (1985), pp. 414–433
604. J.S. Richer, D.S. Shepherd, S. Cabrit, R. Bachiller, E. Churchwell, Protostars and Planets IV pp. 867–+ (2000)
605. Y. Wu, M. Huang, J. He, Astron. Astrophys. Suppl. **115**, 283 (1996)
606. M. Camenzind, in *Reviews in Modern Astronomy*, *Reviews in Modern Astronomy*, vol. 3, ed. by G. Klare (1990), *Reviews in Modern Astronomy*, vol. 3, pp. 234–265
607. C. Fendt, M. Čemeljić, Astron. Astrophys. **395**, 1045 (2002). DOI 10.1051/0004-6361:20021442
608. P. Hartigan, J. Bally, B. Reipurth, J.A. Morse, Protostars and Planets IV pp. 841–+ (2000)
609. J. Eislöffel, in *Reviews in Modern Astronomy*, *Reviews in Modern Astronomy*, vol. 13, ed. by R. E. Schielicke (2000), *Reviews in Modern Astronomy*, vol. 13, pp. 81–+
610. A. Konigl, R.E. Pudritz, Protostars and Planets IV pp. 759–+ (2000)
611. E. Schatzman, Annales d'Astrophysique **25**, 18 (1962)
612. E.J. Weber, L. Davis, Jr., Ap. J. **148**, 217 (1967). DOI 10.1086/149138
613. L. Mestel, M.N.R.A.S. **138**, 359 (1968)
614. R.D. Blandford, D.G. Payne, M.N.R.A.S. **199**, 883 (1982)
615. R. Mundt, T.P. Ray, in *The Nature and Evolutionary Status of Herbig Ae/Be Stars*, *Astronomical Society of the Pacific Conference Series*, vol. 62, ed. by P. S. The, M. R. Perez, & E. P. J. van den Heuvel (1994), *Astronomical Society of the Pacific Conference Series*, vol. 62, pp. 237–+

616. M. Corcoran, T.P. Ray, Astron. Astrophys. **321**, 189 (1997)
617. G.V. Ustyugova, A.V. Koldoba, M.M. Romanova, V.M. Chechetkin, R.V.E. Lovelace, Ap. J. Lettersf **439**, L39 (1995). DOI 10.1086/187739
618. R. Ouyed, R.E. Pudritz, Ap. J. **482**, 712 (1997). DOI 10.1086/304170
619. M.M. Romanova, G.V. Ustyugova, A.V. Koldoba, V.M. Chechetkin, R.V.E. Lovelace, Ap. J. **482**, 708 (1997). DOI 10.1086/304199
620. D.L. Meier, S. Edgington, P. Godon, D.G. Payne, K.R. Lind, Nature **388**, 350 (1997). DOI 10.1038/41034
621. M. Heinemann, S. Olbert, Journal of Geophysical Research **83**, 2457 (1978). DOI 10.1029/JA083iA06p02457
622. R. Ouyed, R.E. Pudritz, Ap. J. **484**, 794 (1997). DOI 10.1086/304355
623. R. Ouyed, R.E. Pudritz, M.N.R.A.S. **309**, 233 (1999). DOI 10.1046/j.1365-8711.1999.02828.x
624. R.E. Pudritz, C.A. Norman, Ap. J. **301**, 571 (1986). DOI 10.1086/163924
625. G. Pelletier, R.E. Pudritz, Ap. J. **394**, 117 (1992). DOI 10.1086/171565
626. P. Bodenheimer, J.P. Ostriker, Ap. J. **161**, 1101 (1970). DOI 10.1086/150610
627. D.L. Moss, M.N.R.A.S. **161**, 225 (1973)
628. S.N. Vogel, L.V. Kuhi, Ap. J. **245**, 960 (1981). DOI 10.1086/158872
629. L. Hartmann, K.B. MacGregor, Ap. J. **259**, 180 (1982). DOI 10.1086/160158
630. A.S. Endal, L.W. Twigg, Ap. J. **260**, 342 (1982). DOI 10.1086/160259
631. L. Hartmann, R. Hewett, S. Stahler, R.D. Mathieu, Ap. J. **309**, 275 (1986). DOI 10.1086/164599
632. F. van Leeuwen, P. Alphenaar, J.J.M. Meys, Astron. Astrophys. Suppl. **67**, 483 (1987)
633. J.R. Stauffer, L. Hartmann, D.R. Soderblom, N. Burnham, Ap. J. **280**, 202 (1984). DOI 10.1086/161986
634. P. Bodenheimer, Ap. J. **224**, 488 (1978). DOI 10.1086/156396
635. M.M. Casali, C. Eiroa, Astron. Astrophys. **306**, 427 (1996)
636. L. Hartmann, J.R. Stauffer, A. J. **97**, 873 (1989). DOI 10.1086/115033
637. S. Matt, R.E. Pudritz, Ap. J. Lettersf **607**, L43 (2004). DOI 10.1086/421351
638. R.P. Kraft, *Stellar Rotation* (1970), pp. 385–+
639. J. Bouvier, C. Bertout, W. Benz, M. Mayor, Astron. Astrophys. **165**, 110 (1986)
640. D.R. Soderblom, C.A. Pilachowski, S.B. Fedele, B.F. Jones, A. J. **105**, 2299 (1993). DOI 10.1086/116608
641. D. Queloz, S. Allain, J.C. Mermilliod, J. Bouvier, M. Mayor, Astron. Astrophys. **335**, 183 (1998)
642. J. Bouvier, M. Forestini, S. Allain, Astron. Astrophys. **326**, 1023 (1997)
643. E. Covino, L. Terranegra, A. Magazzu, J.M. Alcala, S. Allain, J. Bouvier, J. Krautter, R. Wichmann, in *Cool Stars, Stellar Systems, and the Sun, Astronomical Society of the Pacific Conference Series*, vol. 109, ed. by R. Pallavicini & A. K. Dupree (1996), *Astronomical Society of the Pacific Conference Series*, vol. 109, pp. 421–+
644. J. Bouvier, M. Forestini, in *Circumstellar Dust Disks and Planet Formation*, ed. by R. Ferlet & A. Vidal-Madjar (1994), pp. 347–+
645. J. Bouvier, R. Wichmann, K. Grankin, S. Allain, E. Covino, M. Fernandez, E.L. Martin, L. Terranegra, S. Catalano, E. Marilli, Astron. Astrophys. **318**, 495 (1997)
646. D.P. Rowse, I.W. Roxburgh, Solar Physics **74**, 165 (1981). DOI 10.1007/BF00151286
647. L. Mestel, in *Cool Stars, Stellar Systems, and the Sun*, *Lecture Notes in Physics, Berlin Springer Verlag*, vol. 193, ed. by S. L. Baliunas & L. Hartmann (1984), *Lecture Notes in Physics, Berlin Springer Verlag*, vol. 193, pp. 49–+. DOI 10.1007/3-540-12907-3179
648. P. Ghosh, F.K. Lamb, Ap. J. **232**, 259 (1979). DOI 10.1086/157285
649. A. Koenigl, Ap. J. Lettersf **370**, L39 (1991). DOI 10.1086/185972
650. F. Shu, J. Najita, E. Ostriker, F. Wilkin, S. Ruden, S. Lizano, Ap. J. **429**, 781 (1994). DOI 10.1086/174363
651. E.F. Guinan, S.G. Engle, in *IAU Symposium*, *IAU Symposium*, vol. 258, ed. by E. E. Mamajek, D. R. Soderblom, & R. F. G. Wyse (2009), *IAU Symposium*, vol. 258, pp. 395–408. DOI 10.1017/S1743921309032050

652. C. Bertout, G. Basri, J. Bouvier, Ap. J. **330**, 350 (1988). DOI 10.1086/166476
653. C. Bertout, S. Harder, F. Malbet, C. Mennessier, O. Regev, A. J. **112**, 2159 (1996). DOI 10.1086/118171
654. W. Herbst, C.A.L. Bailer-Jones, R. Mundt, Ap. J. Lettersf **554**, L197 (2001). DOI 10.1086/321706
655. S.H.P. Alencar, C.M. Johns-Krull, G. Basri, A. J. **122**, 3335 (2001). DOI 10.1086/323914
656. C.M. Johns-Krull, A.D. Gafford, Ap. J. **573**, 685 (2002). DOI 10.1086/340898
657. B. Stelzer, R. Neuhäuser, Astron. Astrophys. **377**, 538 (2001). DOI 10.1051/0004-6361:20011093
658. T. Montmerle, in *Stellar Coronae in the Chandra and XMM-NEWTON Era, Astronomical Society of the Pacific Conference Series*, vol. 277, ed. by F. Favata & J. J. Drake (2002), *Astronomical Society of the Pacific Conference Series*, vol. 277, pp. 173–+
659. A. Collier Cameron, C.G. Campbell, Astron. Astrophys. **274**, 309 (1993)
660. D.N. Brown, J.D. Landstreet, Ap. J. **246**, 899 (1981). DOI 10.1086/158982
661. R.M. Johnstone, M.V. Penston, M.N.R.A.S. **227**, 797 (1987)
662. I. Appenzeller, R. Oestreicher, I. Jankovics, Astron. Astrophys. **141**, 108 (1984)
663. C.M. Johns-Krull, J.A. Valenti, C. Koresko, Ap. J. **516**, 900 (1999). DOI 10.1086/307128
664. E.C. Ostriker, F.H. Shu, Ap. J. **447**, 813 (1995). DOI 10.1086/175920
665. S. Hirose, Y. Uchida, K. Shibata, R. Matsumoto, P.A.S.J. **49**, 193 (1997)
666. J. Ferreira, G. Pelletier, S. Appl, M.N.R.A.S. **312**, 387 (2000). DOI 10.1046/j.1365-8711.2000.03215.x
667. J.M. Ferreira, M.N.R.A.S. **316**, 647 (2000). DOI 10.1046/j.1365-8711.2000.03540.x
668. P.P. Petrov, A.P.S.S. **169**, 61 (1990). DOI 10.1007/BF00640687
669. F.H. Shu, H. Shang, A.E. Glassgold, T. Lee, Science **277**, 1475 (1997). DOI 10.1126/science.277.5331.1475
670. P. Hartigan, S.J. Kenyon, L. Hartmann, S.E. Strom, S. Edwards, A.D. Welty, J. Stauffer, Ap. J. **382**, 617 (1991). DOI 10.1086/170749
671. G.H.J. van den Oord, R. Mewe, Astron. Astrophys. **213**, 245 (1989)
672. K. Shibata, T. Yokoyama, Ap. J. Lettersf **526**, L49 (1999). DOI 10.1086/312354
673. Y. Tsuboi, K. Koyama, H. Murakami, M. Hayashi, S. Skinner, S. Ueno, Ap. J. **503**, 894 (1998). DOI 10.1086/306024
674. Y. Tsuboi, K. Koyama, in *Japanese-German Workshop on High Energy Astrophysics*, ed. by W. Becker & M. Itoh (1999), pp. 3–+
675. T. Yokoyama, K. Shibata, Ap. J. Lettersf **494**, L113+ (1998). DOI 10.1086/311174
676. T. Yokoyama, K. Shibata, Ap. J. **549**, 1160 (2001). DOI 10.1086/319440
677. H.E. Petschek, NASA Special Publication **50**, 425 (1964)
678. H. Isobe, T. Yokoyama, M. Shimojo, T. Morimoto, H. Kozu, S. Eto, N. Narukage, K. Shibata, Ap. J. **566**, 528 (2002). DOI 10.1086/324777
679. R. Mewe, J.S. Kaastra, C.J. Schrijver, G.H.J. van den Oord, F.J.M. Alkemade, Astron. Astrophys. **296**, 477 (1995)
680. R. Mewe, in *X-Ray Spectroscopy in Astrophysics, Lecture Notes in Physics, Berlin Springer Verlag*, vol. 520, ed. by J. van Paradijs & J. A. M. Bleeker (1999), *Lecture Notes in Physics, Berlin Springer Verlag*, vol. 520, pp. 109–+
681. U. Feldman, J.M. Laming, G.A. Doschek, Ap. J. Lettersf **451**, L79+ (1995). DOI 10.1086/309695
682. K. Shibata, T. Yokoyama, Ap. J. **577**, 422 (2002). DOI 10.1086/342141
683. D.P. Huenemoerder, C.R. Canizares, N.S. Schulz, Ap. J. **559**, 1135 (2001). DOI 10.1086/322419
684. D.P. Huenemoerder, C.R. Canizares, J.J. Drake, J. Sanz-Forcada, Ap. J. **595**, 1131 (2003). DOI 10.1086/377490
685. L. Mestel, *Stellar magnetism* (1999)
686. D. Moss, Astron. Astrophys. **403**, 693 (2003). DOI 10.1051/0004-6361:20030431
687. D. Moss, Astron. Astrophys. **305**, 140 (1996)
688. L.L. Kitchatinov, Astronomy Reports **45**, 816 (2001). DOI 10.1134/1.1408982

689. C.J. Schrijver, C. Zwaan, *Solar and Stellar Magnetic Activity* (2000)
690. E.D. Feigelson, J.A. Gaffney, III, G. Garmire, L.A. Hillenbrand, L. Townsley, Ap. J. **584**, 911 (2003). DOI 10.1086/345811
691. E. Churchwell, A.R.A.A. **40**, 27 (2002). DOI 10.1146/annurev.astro.40.060401.093845
692. H.W. Yorke, C. Sonnhalter, Ap. J. **569**, 846 (2002). DOI 10.1086/339264
693. M.R. Krumholz, R.I. Klein, C.F. McKee, Ap. J. **656**, 959 (2007). DOI 10.1086/510664
694. P. Norberg, A. Maeder, Astron. Astrophys. **359**, 1025 (2000)
695. R. Behrend, A. Maeder, Astron. Astrophys. **373**, 190 (2001). DOI 10.1051/0004-6361:20010585
696. C.F. McKee, J.C. Tan, Ap. J. **585**, 850 (2003). DOI 10.1086/346149
697. I.A. Bonnell, M.R. Bate, C.J. Clarke, J.E. Pringle, M.N.R.A.S. **285**, 201 (1997)
698. I.A. Bonnell, M.R. Bate, C.J. Clarke, J.E. Pringle, M.N.R.A.S. **323**, 785 (2001). DOI 10.1046/j.1365-8711.2001.04270.x
699. I.A. Bonnell, M.R. Bate, H. Zinnecker, M.N.R.A.S. **298**, 93 (1998). DOI 10.1046/j.1365-8711.1998.01590.x
700. S.W. Stahler, Ap. J. **496**, 869 (1998). DOI 10.1086/305404
701. J.E. Dale, M.B. Davies, M.N.R.A.S. **366**, 1424 (2006). DOI 10.1111/j.1365-2966.2005.09937.x
702. P.A. Crowther, in *EAS Publications Series*, *EAS Publications Series*, vol. 13, ed. by M. Heydari-Malayeri, P. Stee, & J.-P. Zahn (2004), *EAS Publications Series*, vol. 13, pp. 1–20. DOI 10.1051/eas:2004048
703. D.R. Gies, in *A Massive Star Odyssey: From Main Sequence to Supernova*, *IAU Symposium*, vol. 212, ed. by K. van der Hucht, A. Herrero, & C. Esteban (2003), *IAU Symposium*, vol. 212, pp. 91–+
704. R.W. Hilditch, T.J. Harries, S.A. Bell, Astron. Astrophys. **314**, 165 (1996)
705. P. Massey, Ap. J. Suppl. **141**, 81 (2002). DOI 10.1086/338286
706. D.C. Abbott, in *Mass Loss and Evolution of O-Type Stars*, *IAU Symposium*, vol. 83, ed. by P. S. Conti & C. W. H. De Loore (1979), *IAU Symposium*, vol. 83, pp. 237–239
707. J.I. Castor, D.C. Abbott, R.I. Klein, Ap. J. **195**, 157 (1975). DOI 10.1086/153315
708. R.P. Kudritzki, J. Puls, A.R.A.A. **38**, 613 (2000). DOI 10.1146/annurev.astro.38.1.613
709. R.P. Kudritzki, A. Pualdrach, J. Puls, in *New Insights in Astrophysics. Eight Years of UV Astronomy with IUE*, *ESA Special Publication*, vol. 263, ed. by E. J. Rolfe (1986), *ESA Special Publication*, vol. 263, pp. 247–258
710. R.P. Kudritzki, A. Pauldrach, J. Puls, D.C. Abbott, Astron. Astrophys. **219**, 205 (1989)
711. J. Puls, R.P. Kudritzki, A. Herrero, A.W.A. Pauldrach, S.M. Haser, D.J. Lennon, R. Gabler, S.A. Voels, J.M. Vilchez, S. Wachter, A. Feldmeier, Astron. Astrophys. **305**, 171 (1996)
712. A.G.A. Brown, A. Blaauw, R. Hoogerwerf, J.H.J. de Bruijne, P.T. de Zeeuw, in *NATO ASIC Proc. 540: The Origin of Stars and Planetary Systems*, ed. by C. J. Lada & N. D. Kylafis (1999), pp. 411–+
713. D. Gouliermis, M. Kontizas, E. Kontizas, R. Korakitis, Astron. Astrophys. **405**, 111 (2003). DOI 10.1051/0004-6361:20030483
714. N.R. Walborn, Ap. J. Lettersf **167**, L31+ (1971). DOI 10.1086/180754
715. J. Maíz-Apellániz, E. Pérez, J.M. Mas-Hesse, A. J. **128**, 1196 (2004). DOI 10.1086/422925
716. Q. Zhang, S.M. Fall, Ap. J. Lettersf **527**, L81 (1999). DOI 10.1086/312412
717. F.J. Selman, J. Melnick, Astron. Astrophys. **443**, 851 (2005). DOI 10.1051/0004-6361:20042173
718. A. Stolte, W. Brandner, B. Brandl, H. Zinnecker, A. J. **132**, 253 (2006). DOI 10.1086/504589
719. J. Pflamm-Altenburg, P. Kroupa, M.N.R.A.S. **373**, 295 (2006). DOI 10.1111/j.1365-2966.2006.11028.x
720. A. Bik, H.J.G.L.M. Lamers, N. Bastian, N. Panagia, M. Romaniello, Astron. Astrophys. **397**, 473 (2003). DOI 10.1051/0004-6361:20021384
721. R. de Grijs, N. Bastian, H.J.G.L.M. Lamers, Ap. J. Lettersf **583**, L17 (2003). DOI 10.1086/367928

722. W.J. de Wit, L. Testi, F. Palla, L. Vanzi, H. Zinnecker, Astron. Astrophys. **425**, 937 (2004). DOI 10.1051/0004-6361:20040454
723. W.J. de Wit, L. Testi, F. Palla, H. Zinnecker, Astron. Astrophys. **437**, 247 (2005). DOI 10.1051/0004-6361:20042489
724. M.S. Roberts, P.A.S.P. **69**, 59 (1957). DOI 10.1086/127013
725. C. Weidner, P. Kroupa, M.N.R.A.S. **365**, 1333 (2006). DOI 10.1111/j.1365-2966.2005.09824.x
726. R.J. Parker, S.P. Goodwin, M.N.R.A.S. **380**, 1271 (2007). DOI 10.1111/j.1365-2966.2007.12179.x
727. P. Kroupa, in *IAU Symposium, IAU Symposium*, vol. 237, ed. by B. G. Elmegreen & J. Palous (2007), *IAU Symposium*, vol. 237, pp. 230–237. DOI 10.1017/S1743921307001524
728. J. Pflamm-Altenburg, P. Kroupa, M.N.R.A.S. **404**, 1564 (2010). DOI 10.1111/j.1365-2966.2010.16376.x
729. V.V. Gvaramadze, D.J. Bomans, Astron. Astrophys. **490**, 1071 (2008). DOI 10.1051/0004-6361:200810411
730. N.J. Irvine, Ap. J. Lettersf **337**, L33 (1989). DOI 10.1086/185372
731. R. Selier, M. Heydari-Malayeri, D.A. Gouliermis, Astron. Astrophys. **529**, A40+ (2011). DOI 10.1051/0004-6361/201016100
732. J. Maíz Apellániz, Ap. J. **677**, 1278 (2008). DOI 10.1086/533525
733. M.S. Oey, C.J. Clarke, Ap. J. Lettersf **620**, L43 (2005). DOI 10.1086/428396
734. P.A. Crowther, O. Schnurr, R. Hirschi, N. Yusof, R.J. Parker, S.P. Goodwin, H.A. Kassim, M.N.R.A.S. **408**, 731 (2010). DOI 10.1111/j.1365-2966.2010.17167.x
735. K.M. Menten, T. Pillai, F. Wyrowski, in *Massive Star Birth: A Crossroads of Astrophysics, IAU Symposium*, vol. 227, ed. by R. Cesaroni, M. Felli, E. Churchwell, & M. Walmsley (2005), *IAU Symposium*, vol. 227, pp. 23–34. DOI 10.1017/S1743921305004321
736. A. Bik, L. Kaper, M.M. Hanson, M. Smits, Astron. Astrophys. **440**, 121 (2005). DOI 10.1051/0004-6361:20042597
737. R.D. Blum, A. Damineli, P.S. Conti, A. J. **121**, 3149 (2001). DOI 10.1086/321088
738. D.O.S. Wood, E. Churchwell, Ap. J. Suppl. **69**, 831 (1989). DOI 10.1086/191329
739. K. Qiu, Q. Zhang, S.T. Megeath, R.A. Gutermuth, H. Beuther, D.S. Shepherd, T.K. Sridharan, L. Testi, C.G. De Pree, Ap. J. **685**, 1005 (2008). DOI 10.1086/591044
740. T. Pillai, F. Wyrowski, K.M. Menten, E. Krügel, Astron. Astrophys. **447**, 929 (2006). DOI 10.1051/0004-6361:20042145
741. N. Peretto, G.A. Fuller, Astron. Astrophys. **505**, 405 (2009). DOI 10.1051/0004-6361/200912127
742. E.T. Chambers, J.M. Jackson, J.M. Rathborne, R. Simon, Ap. J. Suppl. **181**, 360 (2009). DOI 10.1088/0067-0049/181/2/360
743. G. Kim, C.W. Lee, J. Kim, Y. Lee, J. Ballesteros-Paredes, P.C. Myers, S. Kurtz, Journal of Korean Astronomical Society **43**, 9 (2010)
744. N. Peretto, G.A. Fuller, Ap. J. **723**, 555 (2010). DOI 10.1088/0004-637X/723/1/555
745. N. Peretto, G.A. Fuller, R. Plume, L.D. Anderson, J. Bally, C. Battersby, M.T. Beltran, J.P. Bernard, L. Calzoletti, A.M. Digiorgio, F. Faustini, J.M. Kirk, C. Lenfestey, D. Marshall, P. Martin, S. Molinari, L. Montier, F. Motte, I. Ristorcelli, J.A. Rod6n, H.A. Smith, A. Traficante, M. Veneziani, D. Ward-Thompson, L. Wilcock, Astron. Astrophys. **518**, L98+ (2010). DOI 10.1051/0004-6361/201014652
746. S. Molinari, B. Swinyard, J. Bally, M. Barlow, J.P. Bernard, P. Martin, T. Moore, A. Noriega-Crespo, R. Plume, L. Testi, A. Zavagno, A. Abergel, B. Ali, L. Anderson, P. André, J.P. Baluteau, C. Battersby, M.T. Beltrán, M. Benedettini, N. Billot, J. Blommaert, S. Bontemps, F. Boulanger, J. Brand, C. Brunt, M. Burton, L. Calzoletti, S. Carey, P. Caselli, R. Cesaroni, J. Cernicharo, S. Chakrabarti, A. Chrysostomou, M. Cohen, M. Compiegne, P. de Bernardis, G. de Gasperis, A.M. di Giorgio, D. Elia, F. Faustini, N. Flagey, Y. Fukui, G.A. Fuller, K. Ganga, P. Garcia-Lario, J. Glenn, P.F. Goldsmith, M. Griffin, M. Hoare, M. Huang, D. Ikhenaode, C. Joblin, G. Joncas, M. Juvela, J.M. Kirk, G. Lagache, J.Z. Li, T.L. Lim, S.D. Lord, M. Marengo, D.J. Marshall, S. Masi, F. Massi, M. Matsuura, V. Minier,

M.A. Miville-Deschênes, L.A. Montier, L. Morgan, F. Motte, J.C. Mottram, T.G. Müller, P. Natoli, J. Neves, L. Olmi, R. Paladini, D. Paradis, H. Parsons, N. Peretto, M. Pestalozzi, S. Pezzuto, F. Piacentini, L. Piazzo, D. Polychroni, M. Pomarès, C.C. Popescu, W.T. Reach, I. Ristorcelli, J.F. Robitaille, T. Robitaille, J.A. Rodón, A. Roy, P. Royer, D. Russeil, P. Saraceno, M. Sauvage, P. Schilke, E. Schisano, N. Schneider, F. Schuller, B. Schulz, B. Sibthorpe, H.A. Smith, M.D. Smith, L. Spinoglio, D. Stamatellos, F. Strafella, G.S. Stringfellow, E. Sturm, R. Taylor, M.A. Thompson, A. Traficante, R.J. Tuffs, G. Umana, L. Valenziano, R. Vavrek, M. Veneziani, S. Viti, C. Waelkens, D. Ward-Thompson, G. White, L.A. Wilcock, F. Wyrowski, H.W. Yorke, Q. Zhang, Astron. Astrophys. **518**, L100+ (2010). DOI 10.1051/0004-6361/201014659
747. J.C. Tan, in *Massive Star Birth: A Crossroads of Astrophysics*, *IAU Symposium*, vol. 227, ed. by R. Cesaroni, M. Felli, E. Churchwell, & M. Walmsley (2005), *IAU Symposium*, vol. 227, pp. 318–327. DOI 10.1017/S1743921305004692
748. D. Johnstone, J.D. Fiege, R.O. Redman, P.A. Feldman, S.J. Carey, Ap. J. Lettersf **588**, L37 (2003). DOI 10.1086/375524
749. S.P. Ellingsen, Ap. J. **638**, 241 (2006). DOI 10.1086/498673
750. V.L. Fish, M.J. Reid, Ap. J. Suppl. **164**, 99 (2006). DOI 10.1086/502650
751. W.H.T. Vlemmings, G. Surcis, K.J.E. Torstensson, H.J. van Langevelde, M.N.R.A.S. **404**, 134 (2010). DOI 10.1111/j.1365-2966.2010.16297.x
752. A.P. Sarma, E. Momjian, Ap. J. Lettersf **730**, L5+ (2011). DOI 10.1088/2041-8205/730/1/L5
753. J.D. Pandian, E. Momjian, Y. Xu, K.M. Menten, P.F. Goldsmith, Ap. J. **730**, 55 (2011). DOI 10.1088/0004-637X/730/1/55
754. K.M. Menten, M.J. Reid, P. Pratap, J.M. Moran, T.L. Wilson, Ap. J. Lettersf **401**, L39 (1992). DOI 10.1086/186665
755. E.C. Sutton, A.M. Sobolev, S.P. Ellingsen, D.M. Cragg, D.M. Mehringer, A.B. Ostrovskii, P.D. Godfrey, Ap. J. **554**, 173 (2001). DOI 10.1086/321349
756. D.M. Cragg, A.M. Sobolev, P.D. Godfrey, M.N.R.A.S. **360**, 533 (2005). DOI 10.1111/j.1365-2966.2005.09077.x
757. G. Garay, L.F. Rodriguez, J.M. Moran, E. Churchwell, Ap. J. **418**, 368 (1993). DOI 10.1086/173396
758. S. Goedhart, M.J. Gaylard, D.J. van der Walt, M.N.R.A.S. **339**, L33 (2003). DOI 10.1046/j.1365-8711.2003.06426.x
759. D.J. van der Walt, S. Goedhart, M.J. Gaylard, M.N.R.A.S. **398**, 961 (2009). DOI 10.1111/j.1365-2966.2009.15147.x
760. S. Goedhart, V. Minier, M.J. Gaylard, D.J. van der Walt, M.N.R.A.S. **356**, 839 (2005). DOI 10.1111/j.1365-2966.2004.08519.x
761. E.D. Araya, P. Hofner, W.M. Goss, S. Kurtz, A.M.S. Richards, H. Linz, L. Olmi, M. Sewiło, Ap. J. Lettersf **717**, L133 (2010). DOI 10.1088/2041-8205/717/2/L133
762. D.J. van der Walt, Bulletin de la Societe Royale des Sciences de Liege **80**, 260 (2011)
763. A.H. Barrett, P.T.P. Ho, P.C. Myers, Ap. J. Lettersf **211**, L39 (1977). DOI 10.1086/182338
764. E.E. Becklin, G. Neugebauer, Ap. J. **147**, 799 (1967). DOI 10.1086/149055
765. D.E. Kleinmann, F.J. Low, Ap. J. Lettersf **149**, L1+ (1967). DOI 10.1086/180039
766. D.A. Allen, M.G. Burton, Nature **363**, 54 (1993). DOI 10.1038/363054a0
767. H. Beuther, T.K. Sridharan, M. Saito, Ap. J. Lettersf **634**, L185 (2005). DOI 10.1086/498867
768. J. Bally, N.J. Cunningham, N. Moeckel, M.G. Burton, N. Smith, A. Frank, A. Nordlund, Ap. J. **727**, 113 (2011). DOI 10.1088/0004-637X/727/2/113
769. E.F. van Dishoeck, G.A. Blake, A.R.A.A. **36**, 317 (1998). DOI 10.1146/annurev.astro.36.1.317
770. S. Kurtz, R. Cesaroni, E. Churchwell, P. Hofner, C.M. Walmsley, Protostars and Planets IV pp. 299–326 (2000)
771. C.J. Cyganowski, B.A. Whitney, E. Holden, E. Braden, C.L. Brogan, E. Churchwell, R. Indebetouw, D.F. Watson, B.L. Babler, R. Benjamin, M. Gomez, M.R. Meade, M.S. Povich, T.P. Robitaille, C. Watson, A. J. **136**, 2391 (2008). DOI 10.1088/0004-6256/136/6/2391

772. C.J. Cyganowski, C.L. Brogan, T.R. Hunter, E. Churchwell, Ap. J. **702**, 1615 (2009). DOI 10.1088/0004-637X/702/2/1615
773. J.M. De Buizer, W.D. Vacca, A. J. **140**, 196 (2010). DOI 10.1088/0004-6256/140/1/196
774. N. Smith, J. Bally, R.Y. Shuping, M. Morris, M. Kassis, A. J. **130**, 1763 (2005). DOI 10.1086/432912
775. C.J. Davis, M.S.N. Kumar, G. Sandell, D. Froebrich, M.D. Smith, M.J. Currie, M.N.R.A.S. **374**, 29 (2007). DOI 10.1111/j.1365-2966.2006.11163.x
776. J.E. Ybarra, E.A. Lada, Z. Balog, S.W. Fleming, R.L. Phelps, Ap. J. **714**, 469 (2010). DOI 10.1088/0004-637X/714/1/469
777. J.M. Rathborne, R. Simon, J.M. Jackson, Ap. J. **662**, 1082 (2007). DOI 10.1086/513178
778. G. Pineau des Forets, D.R. Flower, J.P. Chieze, in *Herbig-Haro Flows and the Birth of Stars, IAU Symposium*, vol. 182, ed. by B. Reipurth & C. Bertout (1997), *IAU Symposium*, vol. 182, pp. 199–212
779. G. Pineau des Forêts, D. Flower, in *IAU Symposium, IAU Symposium*, vol. 178, ed. by E. F. van Dishoeck (1997), *IAU Symposium*, vol. 178, pp. 113–127
780. Q. Zhang, T.R. Hunter, H. Beuther, T.K. Sridharan, S.Y. Liu, Y.N. Su, H.R. Chen, Y. Chen, Ap. J. **658**, 1152 (2007). DOI 10.1086/511381
781. D.S. Shepherd, E. Churchwell, Ap. J. **472**, 225 (1996). DOI 10.1086/178057
782. N.A. Ridge, T.J.T. Moore, Astron. Astrophys. **378**, 495 (2001). DOI 10.1051/0004-6361:20011180
783. C.O. Johnson, C.G. Depree, W.M. Goss, Ap. J. **500**, 302 (1998). DOI 10.1086/305717
784. R.A. Gaume, W.M. Goss, H.R. Dickel, T.L. Wilson, K.J. Johnston, Ap. J. **438**, 776 (1995). DOI 10.1086/175122
785. T. Murphy, M. Cohen, R.D. Ekers, A.J. Green, R.M. Wark, V. Moss, M.N.R.A.S. **405**, 1560 (2010). DOI 10.1111/j.1365-2966.2010.16589.x
786. A. Bik, W.F. Thi, Astron. Astrophys. **427**, L13 (2004). DOI 10.1051/0004-6361:200400088
787. A.J. Walsh, M.G. Burton, A.R. Hyland, G. Robinson, M.N.R.A.S. **301**, 640 (1998). DOI 10.1046/j.1365-8711.1998.02014.x
788. G. Garay, S. Lizano, P.A.S.P. **111**, 1049 (1999). DOI 10.1086/316416
789. K.T. Kim, B.C. Koo, Ap. J. **549**, 979 (2001). DOI 10.1086/319447
790. E. Churchwell, C.M. Walmsley, R. Cesaroni, Astron. Astrophys. Suppl. **83**, 119 (1990)
791. H. Beuther, D. Shepherd, in *Cores to Clusters: Star Formation with Next Generation Telescopes*, ed. by M. S. N. Kumar, M. Tafalla, & P. Caselli (2005), pp. 105–119
792. H. Beuther, Q. Zhang, T.K. Sridharan, C.F. Lee, L.A. Zapata, Astron. Astrophys. **454**, 221 (2006). DOI 10.1051/0004-6361:20064887
793. C.L. Dobbs, I.A. Bonnell, P.C. Clark, M.N.R.A.S. **360**, 2 (2005). DOI 10.1111/j.1365-2966.2005.08941.x
794. R.B. Larson, M.N.R.A.S. **359**, 211 (2005). DOI 10.1111/j.1365-2966.2005.08881.x
795. A.K. Jappsen, R.S. Klessen, R.B. Larson, Y. Li, M.M. Mac Low, Astron. Astrophys. **435**, 611 (2005). DOI 10.1051/0004-6361:20042178
796. M.R. Krumholz, Ap. J. Lettersf **641**, L45 (2006). DOI 10.1086/503771
797. T. Peters, R. Banerjee, R.S. Klessen, M.M. Mac Low, Ap. J. **729**, 72 (2011). DOI 10.1088/0004-637X/729/1/72
798. P. Hennebelle, B. Commerçon, M. Joos, R.S. Klessen, M. Krumholz, J.C. Tan, R. Teyssier, Astron. Astrophys. **528**, A72+ (2011). DOI 10.1051/0004-6361/201016052
799. M.R. Krumholz, C.F. McKee, Nature **451**, 1082 (2008). DOI 10.1038/nature06620
800. G. Suttner, H.W. Yorke, Ap. J. **551**, 461 (2001). DOI 10.1086/320061
801. N.J. Turner, E. Quataert, H.W. Yorke, Ap. J. **662**, 1052 (2007). DOI 10.1086/513179
802. R. Kuiper, H. Klahr, H. Beuther, T. Henning, Ap. J. **722**, 1556 (2010). DOI 10.1088/0004-637X/722/2/1556
803. M.R. Krumholz, R.I. Klein, C.F. McKee, S.S.R. Offner, A.J. Cunningham, Science **323**, 754 (2009). DOI 10.1126/science.1165857
804. E. Jacquet, M.R. Krumholz, Ap. J. **730**, 116 (2011). DOI 10.1088/0004-637X/730/2/116

805. M.R. Krumholz, C.F. McKee, R.I. Klein, Ap. J. Lettersf **618**, L33 (2005). DOI 10.1086/427555
806. A.J. Cunningham, R.I. Klein, M.R. Krumholz, C.F. McKee, ArXiv e-prints (2011)
807. M.R. Krumholz, I.A. Bonnell, ArXiv e-prints (2007)
808. M.R. Bate, I.A. Bonnell, V. Bromm, M.N.R.A.S. **339**, 577 (2003). DOI 10.1046/j.1365-8711.2003.06210.x
809. I.A. Bonnell, M.R. Bate, S.G. Vine, M.N.R.A.S. **343**, 413 (2003). DOI 10.1046/j.1365-8711.2003.06687.x
810. I.A. Bonnell, M.R. Bate, M.N.R.A.S. **370**, 488 (2006). DOI 10.1111/j.1365-2966.2006.10495.x
811. I.A. Bonnell, R.J. Smith, P.C. Clark, M.R. Bate, M.N.R.A.S. **410**, 2339 (2011). DOI 10.1111/j.1365-2966.2010.17603.x
812. R. Indebetouw, T.P. Robitaille, B.A. Whitney, E. Churchwell, B. Babler, M. Meade, C. Watson, M. Wolfire, Ap. J. **666**, 321 (2007). DOI 10.1086/520316
813. W.H. Hsu, L. Hartmann, F. Heitsch, G.C. Gómez, Ap. J. **721**, 1531 (2010). DOI 10.1088/0004-637X/721/2/1531
814. E. Churchwell, in *NATO ASIC Proc. 540: The Origin of Stars and Planetary Systems*, ed. by C. J. Lada & N. D. Kylafis (1999), pp. 515–+
815. T. Henning, K. Schreyer, R. Launhardt, A. Burkert, Astron. Astrophys. **353**, 211 (2000)
816. B. Vaidya, C. Fendt, H. Beuther, Ap. J. **702**, 567 (2009). DOI 10.1088/0004-637X/702/1/567
817. R. Banerjee, R.E. Pudritz, D.W. Anderson, M.N.R.A.S. **373**, 1091 (2006). DOI 10.1111/j.1365-2966.2006.11089.x
818. J. Bally, H. Zinnecker, A. J. **129**, 2281 (2005). DOI 10.1086/429098
819. D. Shepherd, in *Massive Star Birth: A Crossroads of Astrophysics, IAU Symposium*, vol. 227, ed. by R. Cesaroni, M. Felli, E. Churchwell, & M. Walmsley (2005), *IAU Symposium*, vol. 227, pp. 237–246. DOI 10.1017/S174392130500459X
820. T. Preibisch, Y. Balega, K.H. Hofmann, G. Weigelt, H. Zinnecker, New A. **4**, 531 (1999). DOI 10.1016/S1384-1076(99)00042-1
821. D. Schertl, Y.Y. Balega, T. Preibisch, G. Weigelt, Astron. Astrophys.
822. S. Kraus, G. Weigelt, Y.Y. Balega, J.A. Docobo, K.H. Hofmann, T. Preibisch, D. Schertl, V.S. Tamazian, T. Driebe, K. Ohnaka, R. Petrov, M. Schöller, M. Smith, Astron. Astrophys. **497**, 195 (2009). DOI 10.1051/0004-6361/200810368
823. C. Allen, A. Poveda, A. Hernández-Alcántara, in *Revista Mexicana de Astronomia y Astrofisica Conference Series, Revista Mexicana de Astronomia y Astrofisica Conference Series*, vol. 21, ed. by C. Allen & C. Scarfe (2004), *Revista Mexicana de Astronomia y Astrofisica Conference Series*, vol. 21, pp. 195–199
824. S.T. Megeath, T.L. Wilson, M.R. Corbin, Ap. J. Lettersf **622**, L141 (2005). DOI 10.1086/429720
825. M.R. Krumholz, ArXiv Astrophysics e-prints (2006)
826. I.A. Bonnell, M.R. Bate, M.N.R.A.S. **362**, 915 (2005). DOI 10.1111/j.1365-2966.2005.09360.x
827. A.C. Fabian, J.E. Pringle, M.J. Rees, M.N.R.A.S. **172**, 15P (1975)
828. W.C. Saslaw, D. Zipoy, Nature **216**, 976 (1967). DOI 10.1038/216976a0
829. P.J.E. Peebles, R.H. Dicke, Ap. J. **154**, 891 (1968). DOI 10.1086/149811
830. H.M.P. Couchman, M.J. Rees, M.N.R.A.S. **221**, 53 (1986)
831. F. Palla, E.E. Salpeter, S.W. Stahler, Ap. J. **271**, 632 (1983). DOI 10.1086/161231
832. M.M. Mac Low, J.M. Shull, Ap. J. **302**, 585 (1986). DOI 10.1086/164017
833. M. Tegmark, J. Silk, M.J. Rees, A. Blanchard, T. Abel, F. Palla, Ap. J. **474**, 1 (1997). DOI 10.1086/303434
834. T. Abel, P. Anninos, M.L. Norman, Y. Zhang, Ap. J. **508**, 518 (1998). DOI 10.1086/306410
835. T. Abel, G.L. Bryan, M.L. Norman, Ap. J. **540**, 39 (2000). DOI 10.1086/309295
836. C. Chiappini, U. Frischknecht, G. Meynet, R. Hirschi, B. Barbuy, M. Pignatari, T. Decressin, A. Maeder, Nature **472**, 454 (2011). DOI 10.1038/nature10000

837. J. Tumlinson, Nature **472**, 426 (2011). DOI 10.1038/472426a
838. T. Greif, V. Springel, S. White, S. Glover, P. Clark, R. Smith, R. Klessen, V. Bromm, ArXiv e-prints (2011)
839. P.C. Clark, S.C.O. Glover, R.S. Klessen, V. Bromm, Ap. J. **727**, 110 (2011). DOI 10.1088/0004-637X/727/2/110
840. M.A. Alpar, in *Timing Neutron Stars*, ed. by H. Ögelman & E. P. J. van den Heuvel (1989), pp. 431–+
841. G.F. Gahm, K. Fredga, R. Liseau, D. Dravins, Astron. Astrophys. **73**, L4 (1979)
842. I. Appenzeller, Astron. Astrophys. **71**, 305 (1979)
843. C.L. Imhoff, M.S. Giampapa, Ap. J. Lettersf **239**, L115 (1980). DOI 10.1086/183305
844. I. Appenzeller, C. Chavarria, J. Krautter, R. Mundt, B. Wolf, in *2nd European IUE Conference, ESA Special Publication*, vol. 157, ed. by B. Battrick & J. Mort (1980), *ESA Special Publication*, vol. 157, pp. 209–211
845. R. Neuhauser, M.F. Sterzik, J.H.M.M. Schmitt, in *Cool Stars, Stellar Systems, and the Sun, Astronomical Society of the Pacific Conference Series*, vol. 64, ed. by J.-P. Caillault (1994), *Astronomical Society of the Pacific Conference Series*, vol. 64, pp. 113–+
846. L. Carkner, E. Mamajek, E. Feigelson, R. Neuhauser, R. Wichmann, J. Krautter, Ap. J. **490**, 735 (1997). DOI 10.1086/304885
847. J.H. Kastner, D.P. Huenemoerder, N.S. Schulz, C.R. Canizares, J. Li, D.A. Weintraub, Ap. J. Lettersf **605**, L49 (2004). DOI 10.1086/420769
848. M. Gagné, M.E. Oksala, D.H. Cohen, S.K. Tonnesen, A. ud-Doula, S.P. Owocki, R.H.D. Townsend, J.J. MacFarlane, Ap. J. **628**, 986 (2005). DOI 10.1086/430873
849. M.S. Longair, in *X-Ray Spectroscopy in Astrophysics, Lecture Notes in Physics, Berlin Springer Verlag*, vol. 520, ed. by J. van Paradijs & J. A. M. Bleeker (1999), *Lecture Notes in Physics, Berlin Springer Verlag*, vol. 520, pp. 1–+
850. A. Porras, M. Christopher, L. Allen, J. Di Francesco, S.T. Megeath, P.C. Myers, A. J. **126**, 1916 (2003). DOI 10.1086/377623
851. E. Bica, C.M. Dutra, B. Barbuy, Astron. Astrophys. **397**, 177 (2003). DOI 10.1051/0004-6361:20021479
852. E.D. Feigelson, L.K. Townsley, Ap. J. **673**, 354 (2008). DOI 10.1086/524031
853. M. Gagne, in *Chandra Proposal* (2006), pp. 2112–+
854. K.V. Getman, E.D. Feigelson, L. Townsley, J. Bally, C.J. Lada, B. Reipurth, Ap. J. **575**, 354 (2002). DOI 10.1086/341219
855. T. Preibisch, Astron. Astrophys. **401**, 543 (2003). DOI 10.1051/0004-6361:20030128
856. E. Winston, S.T. Megeath, S.J. Wolk, B. Spitzbart, R. Gutermuth, L.E. Allen, J. Hernandez, K. Covey, J. Muzerolle, J.L. Hora, P.C. Myers, G.G. Fazio, A. J. **140**, 266 (2010). DOI 10.1088/0004-6256/140/1/266
857. T. Preibisch, H. Zinnecker, A. J. **123**, 1613 (2002). DOI 10.1086/338851
858. S.J. Wolk, E. Winston, T.L. Bourke, R. Gutermuth, S.T. Megeath, B.D. Spitzbart, R. Osten, Ap. J. **715**, 671 (2010). DOI 10.1088/0004-637X/715/1/671
859. M. Tsujimoto, K. Koyama, Y. Tsuboi, M. Goto, N. Kobayashi, Ap. J. **566**, 974 (2002). DOI 10.1086/338110
860. G. Garmire, E.D. Feigelson, P. Broos, L.A. Hillenbrand, S.H. Pravdo, L. Townsley, Y. Tsuboi, A. J. **120**, 1426 (2000). DOI 10.1086/301523
861. E.D. Feigelson, K. Getman, L. Townsley, G. Garmire, T. Preibisch, N. Grosso, T. Montmerle, A. Muench, M. McCaughrean, Ap. J. Suppl. **160**, 379 (2005). DOI 10.1086/432512
862. J.A. Caballero, J.F. Albacete-Colombo, J. López-Santiago, Astron. Astrophys. **521**, A45+ (2010). DOI 10.1051/0004-6361/201014861
863. S. Skinner, M. Gagné, E. Belzer, Ap. J. **598**, 375 (2003). DOI 10.1086/378085
864. H. Nakajima, K. Imanishi, S.I. Takagi, K. Koyama, M. Tsujimoto, P.A.S.J. **55**, 635 (2003)
865. E. Flaccomio, G. Micela, S. Sciortino, F. Damiani, F. Favata, F.R. Harnden, Jr., J. Schachter, Astron. Astrophys. **355**, 651 (2000)
866. J. Wang, L.K. Townsley, E.D. Feigelson, P.S. Broos, K.V. Getman, C.G. Román-Zúñiga, E. Lada, Ap. J. **675**, 464 (2008). DOI 10.1086/526406

867. S. Wolk, B. Spitzbart, T. Bourke, J. Alves, in *American Astronomical Society Meeting Abstracts*, *Bulletin of the American Astronomical Society*, vol. 35 (2003), *Bulletin of the American Astronomical Society*, vol. 35, pp. 1219–+
868. L.K. Townsley, P.S. Broos, M.F. Corcoran, E.D. Feigelson, M. Gagné, T. Montmerle, M.S. Oey, N. Smith, G.P. Garmire, K.V. Getman, M.S. Povich, N. Remage Evans, Y. Nazé, E.R. Parkin, T. Preibisch, J. Wang, S.J. Wolk, Y.H. Chu, D.H. Cohen, R.A. Gruendl, K. Hamaguchi, R.R. King, M.M. Mac Low, M.J. McCaughrean, A.F.J. Moffat, L.M. Oskinova, J.M. Pittard, K.G. Stassun, A. ud-Doula, N.R. Walborn, W.L. Waldron, E. Churchwell, J.S. Nichols, S.P. Owocki, N.S. Schulz, Ap. J. Suppl. **194**, 1 (2011). DOI 10.1088/0067-0049/194/1/1
869. P.S. Broos, L.K. Townsley, E.D. Feigelson, K.V. Getman, G.P. Garmire, T. Preibisch, N. Smith, B.L. Babler, S. Hodgkin, R. Indebetouw, M. Irwin, R.R. King, J. Lewis, S.R. Majewski, M.J. McCaughrean, M.R. Meade, H. Zinnecker, Ap. J. Suppl. **194**, 2 (2011). DOI 10.1088/0067-0049/194/1/2
870. K. Imanishi, H. Nakajima, M. Tsujimoto, K. Koyama, Y. Tsuboi, P.A.S.J. **55**, 653 (2003)
871. M. Gagné, S.L. Skinner, K.J. Daniel, Ap. J. **613**, 393 (2004). DOI 10.1086/422828
872. J. Rho, S.V. Ramírez, M.F. Corcoran, K. Hamaguchi, B. Lefloch, Ap. J. **607**, 904 (2004). DOI 10.1086/383081
873. M. Gagne, in *Chandra Proposal* (2002), pp. 1141–+
874. J.E. McCleary, S.J. Wolk, A. J. **141**, 201 (2011). DOI 10.1088/0004-6256/141/6/201
875. J. Harnden, S.S. Murray, F. Damiani, G. Micela, S. Sciortino, APS Meeting Abstracts pp. 17,063–+ (2002)
876. J.L. Linsky, M. Gagne, A. Mytyk, in *IAU Symposium*, *IAU Symposium*, vol. 221 (2003), *IAU Symposium*, vol. 221, pp. 207P–+
877. E.D. Feigelson, A.L. Martin, C.J. McNeill, P.S. Broos, G.P. Garmire, A. J. **138**, 227 (2009). DOI 10.1088/0004-6256/138/1/227
878. S.J. Wolk, B.D. Spitzbart, T.L. Bourke, R.A. Gutermuth, M. Vigil, F. Comerón, A. J. **135**, 693 (2008). DOI 10.1088/0004-6256/135/2/693
879. E.D. Feigelson, P. Broos, J.A. Gaffney, III, G. Garmire, L.A. Hillenbrand, S.H. Pravdo, L. Townsley, Y. Tsuboi, Ap. J. **574**, 258 (2002). DOI 10.1086/340936
880. P. Guillout, M.F. Sterzik, J.H.M.M. Schmitt, C. Motch, D. Egret, W. Voges, R. Neuhaeuser, Astron. Astrophys. **334**, 540 (1998)
881. M. Gagne, J.P. Caillault, J.R. Stauffer, Ap. J. **445**, 280 (1995). DOI 10.1086/175694
882. M.F. Sterzik, J.H.M.M. Schmitt, A. J. **114**, 1673 (1997). DOI 10.1086/118597
883. M.J. Freyberg, J.H.M.M. Schmitt, Astron. Astrophys. **296**, L21+ (1995)
884. K. Imanishi, M. Tsujimoto, K. Koyama, Ap. J. **563**, 361 (2001). DOI 10.1086/323697
885. H. Beuther, J. Kerp, T. Preibisch, T. Stanke, P. Schilke, Astron. Astrophys. **395**, 169 (2002). DOI 10.1051/0004-6361:20021261
886. J. Rho, W.T. Reach, B. Lefloch, G.G. Fazio, Ap. J. **643**, 965 (2006). DOI 10.1086/503245
887. A. Sicilia-Aguilar, T. Henning, A. Juhász, J. Bouwman, G. Garmire, A. Garmire, Ap. J. **687**, 1145 (2008). DOI 10.1086/591932
888. A.J. Delgado, O. González-Martín, E.J. Alfaro, J. Yun, Ap. J. **646**, 269 (2006). DOI 10.1086/504828
889. H. Ozawa, N. Grosso, T. Montmerle, Astron. Astrophys. **429**, 963 (2005). DOI 10.1051/0004-6361:20040480
890. M. Kissler-Patig, J.F. Pirard, M. Casali, A. Moorwood, N. Ageorges, C. Alves de Oliveira, P. Baksai, L.R. Bedin, E. Bendek, P. Biereichel, B. Delabre, R. Dorn, R. Esteves, G. Finger, D. Gojak, G. Huster, Y. Jung, M. Kiekebush, B. Klein, F. Koch, J.L. Lizon, L. Mehrgan, M. Petr-Gotzens, J. Pritchard, F. Selman, J. Stegmeier, Astron. Astrophys. **491**, 941 (2008). DOI 10.1051/0004-6361:200809910
891. M.S. Povich, N. Smith, S.R. Majewski, K.V. Getman, L.K. Townsley, B.L. Babler, P.S. Broos, R. Indebetouw, M.R. Meade, T.P. Robitaille, K.G. Stassun, B.A. Whitney, Y. Yonekura, Y. Fukui, Ap. J. Suppl. **194**, 14 (2011). DOI 10.1088/0067-0049/194/1/14

892. M. Felli, G.B. Taylor, M. Catarzi, E. Churchwell, S. Kurtz, Astron. Astrophys. Suppl. **101**, 127 (1993)
893. G. Garay, J.M. Moran, M.J. Reid, Ap. J. **314**, 535 (1987). DOI 10.1086/165084
894. P.C. Stine, E.D. Feigelson, P. Andre, T. Montmerle, A. J. **96**, 1394 (1988). DOI 10.1086/114889
895. S.M. White, R. Pallavicini, M.R. Kundu, Astron. Astrophys. **259**, 149 (1992)
896. E. Chiang, R.B. Phillips, C.J. Lonsdale, A. J. **111**, 355 (1996). DOI 10.1086/117788
897. S.L. Skinner, A. Brown, R.T. Stewart, Ap. J. Suppl. **87**, 217 (1993). DOI 10.1086/191803
898. M. Cohen, J.H. Bieging, A. J. **92**, 1396 (1986). DOI 10.1086/114273
899. F. Yusef-Zadeh, C. Law, M. Wardle, Q.D. Wang, A. Fruscione, C.C. Lang, A. Cotera, Ap. J. **570**, 665 (2002). DOI 10.1086/340058
900. A.F.J. Moffat, M.F. Corcoran, I.R. Stevens, G. Skalkowski, S.V. Marchenko, A. Mücke, A. Ptak, B.S. Koribalski, L. Brenneman, R. Mushotzky, J.M. Pittard, A.M.T. Pollock, W. Brandner, Ap. J. **573**, 191 (2002). DOI 10.1086/340491
901. M. Güdel, K.R. Briggs, T. Montmerle, M. Audard, L. Rebull, S.L. Skinner, Science **319**, 309 (2008). DOI 10.1126/science.1149926
902. B. Stelzer, E. Flaccomio, T. Montmerle, G. Micela, S. Sciortino, F. Favata, T. Preibisch, E.D. Feigelson, Ap. J. Suppl. **160**, 557 (2005). DOI 10.1086/432375
903. F. Favata, E. Flaccomio, F. Reale, G. Micela, S. Sciortino, H. Shang, K.G. Stassun, E.D. Feigelson, Ap. J. Suppl. **160**, 469 (2005). DOI 10.1086/432542
904. T. Preibisch, H. Zinnecker, G.H. Herbig, Astron. Astrophys. **310**, 456 (1996)
905. T. Preibisch, R. Neuhaeuser, T. Stanke, Astron. Astrophys. **338**, 923 (1998)
906. E.D. Feigelson, J.A. Gaffney, III, G. Garmire, L.A. Hillenbrand, L. Townsley, Ap. J. **584**, 911 (2003). DOI 10.1086/345811
907. E. Flaccomio, G. Micela, S. Sciortino, Astron. Astrophys. **402**, 277 (2003). DOI 10.1051/0004-6361:20030203
908. N.S. Schulz, J.H. Kastner, in *The Hot Universe*, *IAU Symposium*, vol. 188, ed. by K. Koyama, S. Kitamoto, & M. Itoh (1998), *IAU Symposium*, vol. 188, pp. 220–+
909. J.H. Kastner, B. Zuckerman, D.A. Weintraub, T. Forveille, Science **277**, 67 (1997). DOI 10.1126/science.277.5322.67
910. E. Flaccomio, F. Damiani, G. Micela, S. Sciortino, F.R. Harnden, Jr., S.S. Murray, S.J. Wolk, Ap. J. **582**, 382 (2003). DOI 10.1086/344535
911. C.J. Lada, A.A. Muench, K.E. Haisch, Jr., E.A. Lada, J.F. Alves, E.V. Tollestrup, S.P. Willner, A. J. **120**, 3162 (2000). DOI 10.1086/316848
912. T. Preibisch, H. Zinnecker, J.H.M.M. Schmitt, Astron. Astrophys. **279**, L33 (1993)
913. J. Heise, A.C. Brinkman, J. Schrijver, R. Mewe, E.H.B.M. Gronenschild, A.J.F. den Boggende, J. Grindlay, Ap. J. Lettersf **202**, L73 (1975). DOI 10.1086/181984
914. J.H.M.M. Schmitt, P. Kahabka, J. Stauffer, A.J.M. Piters, Astron. Astrophys. **277**, 114 (1993)
915. K.M. Strom, S.E. Strom, Ap. J. **424**, 237 (1994). DOI 10.1086/173886
916. Y. Tsuboi, K. Hamaguchi, K. Koyama, S. Yamauchi, Astronomische Nachrichten **320**, 175 (1999)
917. E. Flaccomio, F. Damiani, G. Micela, S. Sciortino, F.R. Harnden, Jr., S.S. Murray, S.J. Wolk, in *Stellar Coronae in the Chandra and XMM-NEWTON Era*, *Astronomical Society of the Pacific Conference Series*, vol. 277, ed. by F. Favata & J. J. Drake (2002), *Astronomical Society of the Pacific Conference Series*, vol. 277, pp. 155–+
918. R. Pallavicini, L. Golub, R. Rosner, G.S. Vaiana, T. Ayres, J.L. Linsky, Ap. J. **248**, 279 (1981). DOI 10.1086/159152
919. M. Guedel, E.F. Guinan, S.L. Skinner, Ap. J. **483**, 947 (1997). DOI 10.1086/304264
920. E.J. Gaidos, P.A.S.P. **110**, 1259 (1998). DOI 10.1086/316251
921. T.W. Berghoefer, J.H.M.M. Schmitt, J.P. Cassinelli, Astron. Astrophys. Suppl. **118**, 481 (1996)
922. S. Randich, J.H.M.M. Schmitt, C.F. Prosser, J.R. Stauffer, Astron. Astrophys. **305**, 785 (1996)

923. B. Montesinos, J.H. Thomas, P. Ventura, I. Mazzitelli, M.N.R.A.S. **326**, 877 (2001). DOI 10.1046/j.1365-8711.2001.04476.x
924. T. Preibisch, H. Zinnecker, A. J. **122**, 866 (2001). DOI 10.1086/321177
925. R. Neuhäuser, C. Briceño, F. Comerón, T. Hearty, E.L. Martín, J.H.M.M. Schmitt, B. Stelzer, R. Supper, W. Voges, H. Zinnecker, Astron. Astrophys. **343**, 883 (1999)
926. R.E. Rutledge, G. Basri, E.L. Martín, L. Bildsten, Ap. J. Lettersf **538**, L141 (2000). DOI 10.1086/312817
927. F. Comerón, G.H. Rieke, R. Neuhäuser, Astron. Astrophys. **343**, 477 (1999)
928. N. Adams, S. Wolk, F.M. Walter, R. Jeffries, T. Naylor, in *American Astronomical Society Meeting Abstracts*, *Bulletin of the American Astronomical Society*, vol. 34 (2002), *Bulletin of the American Astronomical Society*, vol. 34, pp. 1176–+
929. K.L. Luhman, G.H. Rieke, E.T. Young, A.S. Cotera, H. Chen, M.J. Rieke, G. Schneider, R.I. Thompson, Ap. J. **540**, 1016 (2000). DOI 10.1086/309365
930. P.W. Lucas, P.F. Roche, M.N.R.A.S. **314**, 858 (2000). DOI 10.1046/j.1365-8711.2000.03515.x
931. A. Burrows, D. Saumon, T. Guillot, W.B. Hubbard, J.I. Lunine, Nature **375**, 299 (1995). DOI 10.1038/375299a0
932. A. Burrows, M. Marley, W.B. Hubbard, J.I. Lunine, T. Guillot, D. Saumon, R. Freedman, D. Sudarsky, C. Sharp, Ap. J. **491**, 856 (1997). DOI 10.1086/305002
933. I. Baraffe, G. Chabrier, F. Allard, P.H. Hauschildt, Astron. Astrophys. **337**, 403 (1998)
934. J.J. Drake, R.A. Stern, G. Stringfellow, M. Mathioudakis, J.M. Laming, D.L. Lambert, Ap. J. **469**, 828 (1996). DOI 10.1086/177830
935. J. Muzerolle, C. Briceño, N. Calvet, L. Hartmann, L. Hillenbrand, E. Gullbring, Ap. J. Lettersf **545**, L141 (2000). DOI 10.1086/317877
936. B.A. Wilking, T.P. Greene, M.R. Meyer, A. J. **117**, 469 (1999). DOI 10.1086/300674
937. Y. Tsuboi, Y. Maeda, E.D. Feigelson, G.P. Garmire, G. Chartas, K. Mori, S.H. Pravdo, Ap. J. Lettersf **587**, L51 (2003). DOI 10.1086/375017
938. N. Grosso, T. Montmerle, E.D. Feigelson, P. André, S. Casanova, J. Gregorio-Hetem, Nature **387**, 56 (1997). DOI 10.1038/387056a0
939. Y. Tsuboi, K. Imanishi, K. Koyama, N. Grosso, T. Montmerle, Ap. J. **532**, 1089 (2000). DOI 10.1086/308591
940. K.L. Luhman, G.H. Rieke, Ap. J. **525**, 440 (1999). DOI 10.1086/307891
941. N. Grosso, T. Montmerle, S. Bontemps, P. André, E.D. Feigelson, Astron. Astrophys. **359**, 113 (2000)
942. K. Imanishi, K. Koyama, Y. Tsuboi, Ap. J. **557**, 747 (2001). DOI 10.1086/321691
943. M. Kohno, K. Koyama, K. Hamaguchi, Ap. J. **567**, 423 (2002). DOI 10.1086/338381
944. Y. Tsuboi, K. Koyama, K. Hamaguchi, K. Tatematsu, Y. Sekimoto, J. Bally, B. Reipurth, Ap. J. **554**, 734 (2001). DOI 10.1086/321392
945. K. Koyama, K. Hamaguchi, S. Ueno, N. Kobayashi, E.D. Feigelson, P.A.S.J. **48**, L87 (1996)
946. S.H. Pravdo, E.D. Feigelson, G. Garmire, Y. Maeda, Y. Tsuboi, J. Bally, Nature **413**, 708 (2001)
947. L. Prisinzano, G. Micela, E. Flaccomio, J.R. Stauffer, T. Megeath, L. Rebull, M. Robberto, K. Smith, E.D. Feigelson, N. Grosso, S. Wolk, Ap. J. **677**, 401 (2008). DOI 10.1086/528842
948. W. Herbst, R. Racine, A. J. **81**, 840 (1976). DOI 10.1086/111963
949. T. Preibisch, Y.Y. Balega, D. Schertl, G. Weigelt, Astron. Astrophys. **392**, 945 (2002). DOI 10.1051/0004-6361:20021191
950. J.H. Kastner, M. Richmond, N. Grosso, D.A. Weintraub, T. Simon, A. Frank, K. Hamaguchi, H. Ozawa, A. Henden, Nature **430**, 429 (2004). DOI 10.1038/nature02747
951. J.W. McNeil, B. Reipurth, K. Meech, IAU Circ. **8284**, 1 (2004)
952. C. Briceño, A.K. Vivas, J. Hernández, N. Calvet, L. Hartmann, T. Megeath, P. Berlind, M. Calkins, S. Hoyer, Ap. J. Lettersf **606**, L123 (2004). DOI 10.1086/421395
953. S.M. Andrews, B. Rothberg, T. Simon, Ap. J. Lettersf **610**, L45 (2004). DOI 10.1086/423031
954. W.D. Vacca, M.C. Cushing, T. Simon, Ap. J. Lettersf **609**, L29 (2004). DOI 10.1086/422638
955. B. Reipurth, C. Aspin, Ap. J. Lettersf **606**, L119 (2004). DOI 10.1086/421393

956. J.H. Kastner, M. Richmond, N. Grosso, D.A. Weintraub, T. Simon, A. Henden, K. Hamaguchi, A. Frank, H. Ozawa, Ap. J. Lettersf **648**, L43 (2006). DOI 10.1086/507992
957. M. Audard, G.S. Stringfellow, M. Güdel, S.L. Skinner, F.M. Walter, E.F. Guinan, R.T. Hamilton, K.R. Briggs, C. Baldovin-Saavedra, Astron. Astrophys. **511**, A63+ (2010). DOI 10.1051/0004-6361/200913037
958. N. Grosso, K. Hamaguchi, J.H. Kastner, M.W. Richmond, D.A. Weintraub, Astron. Astrophys. **522**, A56+ (2010). DOI 10.1051/0004-6361/200913850
959. J.M. Girart, S. Curiel, L.F. Rodríguez, M. Honda, J. Cantó, Y.K. Okamoto, S. Sako, A. J. **127**, 2969 (2004). DOI 10.1086/383551
960. A.K. Dupree, N.S. Brickhouse, G.A. Doschek, J.C. Green, J.C. Raymond, Ap. J. Lettersf **418**, L41+ (1993). DOI 10.1086/187111
961. M. Landini, B.C. Monsignori Fossi, Astron. Astrophys. Suppl. **7**, 291 (1972)
962. R. Mewe, Solar Physics **22**, 459 (1972). DOI 10.1007/BF00148711
963. J.C. Raymond, B.W. Smith, Ap. J. Suppl. **35**, 419 (1977). DOI 10.1086/190486
964. A. Bar-Shalom, M. Klapisch, J. Oreg, Phys. Rev. A **38**, 1773 (1988). DOI 10.1103/PhysRevA.38.1773
965. J.S. Kaastra, F.A. Jansen, Astron. Astrophys. Suppl. **97**, 873 (1993)
966. E. Landi, M. Landini, K.P. Dere, P.R. Young, H.E. Mason, Astron. Astrophys. Suppl. **135**, 339 (1999). DOI 10.1051/aas:1999449
967. J.S. Kaastra, R. Mewe, H. Nieuwenhuijzen, in *UV and X-ray Spectroscopy of Astrophysical and Laboratory Plasmas*, ed. by K. Yamashita & T. Watanabe (1996), pp. 411–414
968. R.K. Smith, N.S. Brickhouse, D.A. Liedahl, J.C. Raymond, Ap. J. Lettersf **556**, L91 (2001). DOI 10.1086/322992
969. P. Mazzotta, G. Mazzitelli, S. Colafrancesco, N. Vittorio, Astron. Astrophys. Suppl. **133**, 403 (1998). DOI 10.1051/aas:1998330
970. E. Anders, N. Grevesse, Geochimica et Cosmochimica Acta **53**, 197 (1989). DOI 10.1016/0016-7037(89)90286-X
971. A.H. Gabriel, C. Jordan, M.N.R.A.S. **145**, 241 (1969)
972. R. Rosner, W.H. Tucker, G.S. Vaiana, Ap. J. **220**, 643 (1978). DOI 10.1086/155949
973. G.S. Vaiana, R. Rosner, A.R.A.A. **16**, 393 (1978). DOI 10.1146/annurev.aa.16.090178.002141
974. N.S. Brickhouse, in *IAU Colloq. 152: Astrophysics in the Extreme Ultraviolet*, ed. by S. Bowyer & R. F. Malina (1996), pp. 105–+
975. C.R. Canizares, D.P. Huenemoerder, D.S. Davis, D. Dewey, K.A. Flanagan, J. Houck, T.H. Markert, H.L. Marshall, M.L. Schattenburg, N.S. Schulz, M. Wise, J.J. Drake, N.S. Brickhouse, Ap. J. Lettersf **539**, L41 (2000). DOI 10.1086/312823
976. A.C. Brinkman, C.J.T. Gunsing, J.S. Kaastra, R.L.J. van der Meer, R. Mewe, F. Paerels, A.J.J. Raassen, J.J. van Rooijen, H. Bräuninger, W. Burkert, V. Burwitz, G. Hartner, P. Predehl, J.U. Ness, J.H.M.M. Schmitt, J.J. Drake, O. Johnson, M. Juda, V. Kashyap, S.S. Murray, D. Pease, P. Ratzlaff, B.J. Wargelin, Ap. J. Lettersf **530**, L111 (2000). DOI 10.1086/312504
977. J.T. Schmelz, J.L.R. Saba, K.T. Strong, Ap. J. Lettersf **398**, L115 (1992). DOI 10.1086/186590
978. J.U. Ness, N.S. Brickhouse, J.J. Drake, D.P. Huenemoerder, Ap. J. **598**, 1277 (2003). DOI 10.1086/379059
979. P.S. Wojdowski, D.A. Liedahl, M. Sako, S.M. Kahn, F. Paerels, Ap. J. **582**, 959 (2003). DOI 10.1086/344821
980. N.S. Schulz, C.R. Canizares, J.C. Lee, M. Sako, Ap. J. Lettersf **564**, L21 (2002). DOI 10.1086/338862
981. R. Mewe, J. Schrijver, Astron. Astrophys. **65**, 99 (1978)
982. S.M. Kahn, M.A. Leutenegger, J. Cottam, G. Rauw, J.M. Vreux, A.J.F. den Boggende, R. Mewe, M. Güdel, Astron. Astrophys. **365**, L312 (2001). DOI 10.1051/0004-6361:20000093
983. A.K. Pradhan, J.M. Shull, Ap. J. **249**, 821 (1981). DOI 10.1086/159340

984. D. Porquet, J. Dubau, Astron. Astrophys. Suppl. **143**, 495 (2000). DOI 10.1051/aas:2000192
985. J.U. Ness, J.H.M.M. Schmitt, M. Audard, M. Güdel, R. Mewe, Astron. Astrophys. **407**, 347 (2003). DOI 10.1051/0004-6361:20030880
986. N.S. Brickhouse, S.R. Cranmer, A.K. Dupree, G.J.M. Luna, S. Wolk, Ap. J. **710**, 1835 (2010). DOI 10.1088/0004-637X/710/2/1835
987. M. Güdel, S.L. Skinner, M. Audard, K.R. Briggs, S. Cabrit, Astron. Astrophys. **478**, 797 (2008). DOI 10.1051/0004-6361:20078141
988. M. Güdel, Astronomische Nachrichten **329**, 218 (2008). DOI 10.1002/asna.200710917
989. J. Robrade, J.H.M.M. Schmitt, Astron. Astrophys. **449**, 737 (2006). DOI 10.1051/0004-6361:20054247
990. M. Güdel, Y. Nazé, A.A.P.R. **17**, 309 (2009). DOI 10.1007/s00159-009-0022-4
991. J. Robrade, J.H.M.M. Schmitt, A&A **473**, 229 (2007)
992. J.H.M.M. Schmitt, J. Robrade, J.U. Ness, F. Favata, B. Stelzer, Astron. Astrophys. **432**, L35 (2005). DOI 10.1051/0004-6361:200500014
993. H.M. Günther, C. Liefke, J.H.M.M. Schmitt, J. Robrade, J.U. Ness, Astron. Astrophys. **459**, L29 (2006). DOI 10.1051/0004-6361:20066306
994. M. Güdel, S.L. Skinner, S.Y. Mel'Nikov, M. Audard, A. Telleschi, K.R. Briggs, Astron. Astrophys. **468**, 529 (2007). DOI 10.1051/0004-6361:20066318
995. P. Testa, D.P. Huenemoerder, N.S. Schulz, K. Ishibashi, Ap. J. **687**, 579 (2008). DOI 10.1086/591485
996. H.M. Günther, J.H.M.M. Schmitt, Astron. Astrophys. **494**, 1041 (2009). DOI 10.1051/0004-6361:200811007
997. A. Telleschi, M. Güdel, K.R. Briggs, S.L. Skinner, M. Audard, E. Franciosini, Astron. Astrophys. **468**, 541 (2007). DOI 10.1051/0004-6361:20065422
998. M. Güdel, A. Telleschi, Astron. Astrophys. **474**, L25 (2007). DOI 10.1051/0004-6361:20078143
999. S.L. Skinner, M. Audard, M. Guedel, ArXiv e-prints (2011)
1000. S. Czesla, J.H.H.M. Schmitt, Astron. Astrophys. **470**, L13 (2007). DOI 10.1051/0004-6361:20077741
1001. K. Imanishi, K. Koyama, Y. Tsuboi, Ap. J. **557**, 747 (2001). DOI 10.1086/321691
1002. M. Tsujimoto, E.D. Feigelson, N. Grosso, G. Micela, Y. Tsuboi, F. Favata, H. Shang, J.H. Kastner, Ap. J. Suppl. **160**, 503 (2005). DOI 10.1086/432093
1003. F. Favata, G. Micela, B. Silva, S. Sciortino, M. Tsujimoto, Astron. Astrophys. **433**, 1047 (2005). DOI 10.1051/0004-6361:20042019
1004. J. Rho, M.F. Corcoran, Y.H. Chu, W.T. Reach, Ap. J. **562**, 446 (2001). DOI 10.1086/323053
1005. L.K. Townsley, E.D. Feigelson, T. Montmerle, P.S. Broos, Y.H. Chu, G.P. Garmire, Ap. J. **593**, 874 (2003). DOI 10.1086/376692
1006. J.P. Cassinelli, N.A. Miller, W.L. Waldron, J.J. MacFarlane, D.H. Cohen, Ap. J. Lettersf **554**, L55 (2001). DOI 10.1086/320916
1007. N.S. Schulz, in *Revista Mexicana de Astronomia y Astrofisica Conference Series*, *Revista Mexicana de Astronomia y Astrofisica Conference Series*, vol. 15, ed. by J. Arthur & W. J. Henney (2003), *Revista Mexicana de Astronomia y Astrofisica Conference Series*, vol. 15, pp. 220–222
1008. A. ud-Doula, S.P. Owocki, Ap. J. **576**, 413 (2002). DOI 10.1086/341543
1009. M. Gagné, D. Cohen, S. Owocki, A. Ud-Doula, in *The High Energy Universe at Sharp Focus: Chandra Science*, *Astronomical Society of the Pacific Conference Series*, vol. 262, ed. by E. M. Schlegel & S. D. Vrtilek (2002), *Astronomical Society of the Pacific Conference Series*, vol. 262, pp. 31–+
1010. A.W. Mitschang, N.S. Schulz, D.P. Huenemoerder, J.S. Nichols, P. Testa, Ap. J. **734**, 14 (2011). DOI 10.1088/0004-637X/734/1/14
1011. R.J. Murphy, R. Ramaty, D.V. Reames, B. Kozlovsky, Ap. J. **371**, 793 (1991). DOI 10.1086/169944
1012. J.N. Goswami, K.K. Marhas, S. Sahijpal, Ap. J. **549**, 1151 (2001). DOI 10.1086/319434
1013. R. Diehl, F.X. Timmes, P.A.S.P. **110**, 637 (1998). DOI 10.1086/316169

1014. T. Montmerle, New Astronomy Reviews **46**, 573 (2002). DOI 10.1016/S1387-6473(02)00204-X
1015. P.S. Teixeira, C.J. Lada, E.T. Young, M. Marengo, A. Muench, J. Muzerolle, N. Siegler, G. Rieke, L. Hartmann, S.T. Megeath, G. Fazio, Ap. J. Lettersf **636**, L45 (2006). DOI 10.1086/500009
1016. P.R. Allen, K.L. Luhman, P.C. Myers, S.T. Megeath, L.E. Allen, L. Hartmann, G.G. Fazio, Ap. J. **655**, 1095 (2007). DOI 10.1086/510363
1017. E. Furlan, D.M. Watson, M.K. McClure, P. Manoj, C. Espaillat, P. D'Alessio, N. Calvet, K.H. Kim, B.A. Sargent, W.J. Forrest, L. Hartmann, Ap. J. **703**, 1964 (2009). DOI 10.1088/0004-637X/703/2/1964
1018. R.A. Gutermuth, S.T. Megeath, P.C. Myers, L.E. Allen, J.L. Pipher, G.G. Fazio, Ap. J. Suppl. **184**, 18 (2009). DOI 10.1088/0067-0049/184/1/18
1019. P. Manoj, K.H. Kim, E. Furlan, M.K. McClure, K.L. Luhman, D.M. Watson, C. Espaillat, N. Calvet, J.R. Najita, P. D'Alessio, L. Adame, B.A. Sargent, W.J. Forrest, C. Bohac, J.D. Green, L.A. Arnold, Ap. J. Suppl. **193**, 11 (2011). DOI 10.1088/0067-0049/193/1/11
1020. B. Davies, N. Bastian, M. Gieles, A.C. Seth, S. Mengel, I.S. Konstantopoulos, M.N.R.A.S. **411**, 1386 (2011). DOI 10.1111/j.1365-2966.2010.17777.x
1021. A. Roman-Lopes, Z. Abraham, A. J. **131**, 951 (2006). DOI 10.1086/499307
1022. S. Wang, L.W. Looney, Ap. J. **659**, 1360 (2007). DOI 10.1086/512661
1023. R. Kurtev, J. Borissova, L. Georgiev, S. Ortolani, V.D. Ivanov, Astron. Astrophys. **475**, 209 (2007). DOI 10.1051/0004-6361:20066706
1024. B. Davies, D.F. Figer, C.J. Law, R.P. Kudritzki, F. Najarro, A. Herrero, J.W. MacKenty, Ap. J. **676**, 1016 (2008). DOI 10.1086/527350
1025. L.A. Zapata, S. Leurini, K.M. Menten, P. Schilke, R. Rolffs, C. Hieret, A. J. **136**, 1455 (2008). DOI 10.1088/0004-6256/136/4/1455
1026. M. Messineo, D.F. Figer, B. Davies, R.M. Rich, E. Valenti, R.P. Kudritzki, Ap. J. Lettersf **683**, L155 (2008). DOI 10.1086/591673
1027. M.G. Guarcello, G. Micela, F. Damiani, G. Peres, L. Prisinzano, S. Sciortino, Astron. Astrophys. **496**, 453 (2009). DOI 10.1051/0004-6361/200810671
1028. F.C. Adams, P.C. Myers, Ap. J. **553**, 744 (2001). DOI 10.1086/320941
1029. C.J. Clarke, I.A. Bonnell, L.A. Hillenbrand, Protostars and Planets IV pp. 151–+ (2000)
1030. R.S. Klessen, A. Burkert, Ap. J. **549**, 386 (2001). DOI 10.1086/319053
1031. P.C. Myers, Ap. J. Lettersf **496**, L109+ (1998). DOI 10.1086/311256
1032. M. Tapia, P. Persi, M. Roth, Astron. Astrophys. **316**, 102 (1996)
1033. P. Persi, M. Roth, M. Tapia, M. Ferrari-Toniolo, A.R. Marenzi, Astron. Astrophys. **282**, 474 (1994)
1034. E.A. Lada, J. Bally, A.A. Stark, Ap. J. **368**, 432 (1991). DOI 10.1086/169708
1035. C.J. Lada, J. Alves, E.A. Lada, A. J. **111**, 1964 (1996). DOI 10.1086/117933
1036. E.A. Lada, Ap. J. Lettersf **393**, L25 (1992). DOI 10.1086/186442
1037. A. Schulz, R. Guesten, R. Zylka, E. Serabyn, Astron. Astrophys. **246**, 570 (1991)
1038. D.L. Depoy, E.A. Lada, I. Gatley, R. Probst, Ap. J. Lettersf **356**, L55 (1990). DOI 10.1086/185749
1039. C.J. Lada, A.A. Muench, E.A. Lada, J.F. Alves, A. J. **128**, 1254 (2004). DOI 10.1086/423294
1040. N. Peretto, P. André, A. Belloche, Astron. Astrophys. **445**, 979 (2006). DOI 10.1051/0004-6361:20053324
1041. P.S. Teixeira, L.A. Zapata, C.J. Lada, Ap. J. Lettersf **667**, L179 (2007). DOI 10.1086/522494
1042. A.A. Muench, E.A. Lada, C.J. Lada, J. Alves, Ap. J. **573**, 366 (2002). DOI 10.1086/340554
1043. C.J. Lada, E.A. Lada, in *The Formation and Evolution of Star Clusters*, *Astronomical Society of the Pacific Conference Series*, vol. 13, ed. by K. Janes (1991), *Astronomical Society of the Pacific Conference Series*, vol. 13, pp. 3–22
1044. L.A. Hillenbrand, S.E. Strom, N. Calvet, K.M. Merrill, I. Gatley, R.B. Makidon, M.R. Meyer, M.F. Skrutskie, A. J. **116**, 1816 (1998). DOI 10.1086/300536
1045. D.E.A. Nürnberger, L. Bronfman, H.W. Yorke, H. Zinnecker, Astron. Astrophys. **394**, 253 (2002). DOI 10.1051/0004-6361:20021022

1046. Z. Jiang, Y. Yao, J. Yang, D. Baba, D. Kato, M. Kurita, C. Nagashima, T. Nagata, T. Nagayama, Y. Nakajima, M. Ishii, M. Tamura, K. Sugitani, Ap. J. **596**, 1064 (2003). DOI 10.1086/378150
1047. C.D. Wilson, W.E. Harris, R. Longden, N.Z. Scoville, Ap. J. **641**, 763 (2006). DOI 10.1086/500577
1048. J.B. Lamb, M.S. Oey, J.K. Werk, L.D. Ingleby, Ap. J. **725**, 1886 (2010). DOI 10.1088/0004-637X/725/2/1886
1049. D. Calzetti, R. Chandar, J.C. Lee, B.G. Elmegreen, R.C. Kennicutt, B. Whitmore, Ap. J. Lettersf **719**, L158 (2010). DOI 10.1088/2041-8205/719/2/L158
1050. B.G. Elmegreen, Ap. J. **539**, 342 (2000). DOI 10.1086/309204
1051. F. Palla, S.W. Stahler, Ap. J. **540**, 255 (2000). DOI 10.1086/309312
1052. B.G. Park, H. Sung, M.S. Bessell, Y.H. Kang, A. J. **120**, 894 (2000). DOI 10.1086/301459
1053. G.H. Herbig, Ap. J. **497**, 736 (1998). DOI 10.1086/305500
1054. A. Moitinho, J. Alves, N. Huélamo, C.J. Lada, Ap. J. Lettersf **563**, L73 (2001). DOI 10.1086/338503
1055. L. Hartmann, A. J. **121**, 1030 (2001). DOI 10.1086/318770
1056. G.H. Herbig, Ap. J. **135**, 736 (1962). DOI 10.1086/147316
1057. I. Iben, Jr., R.J. Talbot, Ap. J. **144**, 968 (1966). DOI 10.1086/148695
1058. W. Herbst, D.P. Miller, A. J. **87**, 1478 (1982). DOI 10.1086/113238
1059. B.G. Elmegreen, Ap. J. **515**, 323 (1999). DOI 10.1086/307011
1060. B.G. Elmegreen, Y. Efremov, R.E. Pudritz, H. Zinnecker, Protostars and Planets IV pp. 179–+ (2000)
1061. J.V. Feitzinger, E. Braunsfurth, Astron. Astrophys. **139**, 104 (1984)
1062. P. Battinelli, Y. Efremov, E.A. Magnier, Astron. Astrophys. **314**, 51 (1996)
1063. G.R. Ivanov, M.N.R.A.S. **257**, 119 (1992)
1064. E. Bica, C.M. Dutra, J. Soares, B. Barbuy, Astron. Astrophys. **404**, 223 (2003). DOI 10.1051/0004-6361:20030486
1065. N.V. Kharchenko, L.K. Pakulyak, A.E. Piskunov, VizieR Online Data Catalog **808**, 291 (2003)
1066. E. Bica, C. Bonatto, C.M. Dutra, Astron. Astrophys. **405**, 991 (2003). DOI 10.1051/0004-6361:20030700
1067. S. van den Bergh, R.D. McClure, Astron. Astrophys. **88**, 360 (1980)
1068. Q. Zhang, S.M. Fall, Ap. J. Lettersf **527**, L81 (1999). DOI 10.1086/312412
1069. K.E. Johnson, H.A. Kobulnicky, Ap. J. **597**, 923 (2003). DOI 10.1086/378585
1070. C.D. Wilson, N. Scoville, S.C. Madden, V. Charmandaris, Ap. J. **599**, 1049 (2003). DOI 10.1086/379344
1071. J.L. Turner, S.C. Beck, in *Recycling Intergalactic and Interstellar Matter, IAU Symposium*, vol. 217, ed. by P.-A. Duc, J. Braine, & E. Brinks (2004), *IAU Symposium*, vol. 217, pp. 208–+
1072. D. Leisawitz, F.N. Bash, P. Thaddeus, Ap. J. Suppl. **70**, 731 (1989). DOI 10.1086/191357
1073. H. Zinnecker, M.J. McCaughrean, B.A. Wilking, in *Protostars and Planets III*, ed. by E. H. Levy & J. I. Lunine (1993), pp. 429–495
1074. W.H. Warren, Jr., J.E. Hesser, Ap. J. Suppl. **34**, 115 (1977). DOI 10.1086/190446
1075. W.H. Warren, Jr., J.E. Hesser, Ap. J. Suppl. **34**, 207 (1977)
1076. A.G.A. Brown, D. Hartmann, W.B. Burton, Astron. Astrophys. **300**, 903 (1995)
1077. R. Genzel, J. Stutzki, A.R.A.A. **27**, 41 (1989). DOI 10.1146/annurev.aa.27.090189.000353
1078. B. Zuckerman, Ap. J. **183**, 863 (1973). DOI 10.1086/152273
1079. V. Pankonin, C.M. Walmsley, M. Harwit, Astron. Astrophys. **75**, 34 (1979)
1080. M. Peimbert, Annals of the New York Academy of Sciences **395**, 24 (1982). DOI 10.1111/j.1749-6632.1982.tb43375.x
1081. A. Schulz, C. Henkel, U. Beckmann, C. Kaseman, G. Schneider, L.A. Nyman, G. Persson, L.G. Gunnarsson, G. Delgado, Astron. Astrophys. **295**, 183 (1995)
1082. R.L. Plambeck, M.C.H. Wright, R. Rao, Ap. J. **594**, 911 (2003). DOI 10.1086/377097

1083. L.E. Kristensen, M. Gustafsson, D. Field, G. Callejo, J.L. Lemaire, L. Vannier, G. Pineau des Forêts, Astron. Astrophys. **412**, 727 (2003). DOI 10.1051/0004-6361:20031276
1084. P. Caselli, P.C. Myers, Ap. J. **446**, 665 (1995). DOI 10.1086/175825
1085. D. Li, P.F. Goldsmith, K. Menten, Ap. J. **587**, 262 (2003). DOI 10.1086/368078
1086. J.P. Williams, R.L. Plambeck, M.H. Heyer, Ap. J. **591**, 1025 (2003). DOI 10.1086/375396
1087. Y. Aso, K. Tatematsu, Y. Sekimoto, T. Nakano, T. Umemoto, K. Koyama, S. Yamamoto, Ap. J. Suppl. **131**, 465 (2000). DOI 10.1086/317378
1088. J. Bally, C.R. O'Dell, M.J. McCaughrean, A. J. **119**, 2919 (2000). DOI 10.1086/301385
1089. J. Franco, G. Tenorio-Tagle, P. Bodenheimer, M. Rozyczka, I.F. Mirabel, Ap. J. **333**, 826 (1988). DOI 10.1086/166792
1090. L.A. Hillenbrand, J.M. Carpenter, Ap. J. **540**, 236 (2000). DOI 10.1086/309309
1091. G.H. Herbig, D.M. Terndrup, Ap. J. **307**, 609 (1986). DOI 10.1086/164447
1092. L.A. Hillenbrand, L.W. Hartmann, Ap. J. **492**, 540 (1998). DOI 10.1086/305076
1093. J. Bally, W.D. Langer, W. Liu, Ap. J. **383**, 645 (1991). DOI 10.1086/170822
1094. T. Preibisch, Y. Balega, K.H. Hofmann, G. Weigelt, H. Zinnecker, New A. **4**, 531 (1999). DOI 10.1016/S1384-1076(99)00042-1
1095. T. Preibisch, A.G.A. Brown, T. Bridges, E. Guenther, H. Zinnecker, A. J. **124**, 404 (2002). DOI 10.1086/341174
1096. K. Tachihara, A. Mizuno, Y. Fukui, Ap. J. **528**, 817 (2000). DOI 10.1086/308189
1097. S. Bontemps, P. André, A.A. Kaas, L. Nordh, G. Olofsson, M. Huldtgren, A. Abergel, J. Blommaert, F. Boulanger, M. Burgdorf, C.J. Cesarsky, D. Cesarsky, E. Copet, J. Davies, E. Falgarone, G. Lagache, T. Montmerle, M. Pérault, P. Persi, T. Prusti, J.L. Puget, F. Sibille, Astron. Astrophys. **372**, 173 (2001). DOI 10.1051/0004-6361:20010474
1098. P. Andre, T. Montmerle, E.D. Feigelson, H. Steppe, Astron. Astrophys. **240**, 321 (1990)
1099. P. Andre, J. Martin-Pintado, D. Despois, T. Montmerle, Astron. Astrophys. **236**, 180 (1990)
1100. P. Andre, D. Ward-Thompson, M. Barsony, Protostars and Planets IV pp. 59–+ (2000)
1101. J.R. Houck, T.L. Roellig, J. van Cleve, W.J. Forrest, T. Herter, C.R. Lawrence, K. Matthews, H.J. Reitsema, B.T. Soifer, D.M. Watson, D. Weedman, M. Huisjen, J. Troeltzsch, D.J. Barry, J. Bernard-Salas, C.E. Blacken, B.R. Brandl, V. Charmandaris, D. Devost, G.E. Gull, P. Hall, C.P. Henderson, S.J.U. Higdon, B.E. Pirger, J. Schoenwald, G.C. Sloan, K.I. Uchida, P.N. Appleton, L. Armus, M.J. Burgdorf, S.B. Fajardo-Acosta, C.J. Grillmair, J.G. Ingalls, P.W. Morris, H.I. Teplitz, Ap. J. Suppl. **154**, 18 (2004). DOI 10.1086/423134
1102. S.C. Simonson, III, Ap. J. **154**, 923 (1968). DOI 10.1086/149813
1103. N.A. Patel, P.F. Goldsmith, M.H. Heyer, R.L. Snell, P. Pratap, Ap. J. **507**, 241 (1998). DOI 10.1086/306305
1104. C.D. Garmany, R.E. Stencel, Astron. Astrophys. Suppl. **94**, 211 (1992)
1105. N.R. Walborn, R.J. Panek, Ap. J. **286**, 718 (1984). DOI 10.1086/162647
1106. S.R. Pottasch, B.A.I.N. **13**, 77 (1956)
1107. H.I. Matthews, Astron. Astrophys. **75**, 345 (1979)
1108. M. de Muizon, D. Rouan, P. Lena, C. Nicollier, J. Wijnbergen, Astron. Astrophys. **83**, 140 (1980)
1109. A. Wootten, A. Sargent, G. Knapp, P.J. Huggins, Ap. J. **269**, 147 (1983). DOI 10.1086/161026
1110. M. Nakano, Y. Tomita, H. Ohtani, K. Ogura, Y. Sofue, P.A.S.J. **41**, 1073 (1989)
1111. R.D. Schwartz, B.A. Wilking, A.L. Giulbudagian, Ap. J. **370**, 263 (1991). DOI 10.1086/169812
1112. C. Codella, R. Bachiller, B. Nisini, P. Saraceno, L. Testi, Astron. Astrophys. **376**, 271 (2001). DOI 10.1051/0004-6361:20010963
1113. N.A. Patel, P.F. Goldsmith, R.L. Snell, T. Hezel, T. Xie, Ap. J. **447**, 721 (1995). DOI 10.1086/175912
1114. M. Kun, L.G. Balazs, I. Toth, A.P.S.S. **134**, 211 (1987). DOI 10.1007/BF01094933
1115. K.W. Chan, T. Onaka, Ap. J. Lettersf **533**, L33 (2000). DOI 10.1086/312603
1116. K. Hamaguchi, M.F. Corcoran, Y. Ezoe, L. Townsley, P. Broos, R. Gruendl, K. Vaidya, S.M. White, T. Strohmayer, R. Petre, Y.H. Chu, Ap. J. Lettersf **695**, L4 (2009). DOI 10.1088/0004-637X/695/1/L4

1117. E.D. Feigelson, K.V. Getman, L.K. Townsley, P.S. Broos, M.S. Povich, G.P. Garmire, R.R. King, T. Montmerle, T. Preibisch, N. Smith, K.G. Stassun, J. Wang, S. Wolk, H. Zinnecker, Ap. J. Suppl. **194**, 9 (2011). DOI 10.1088/0067-0049/194/1/9
1118. M. Gagné, G. Fehon, M.R. Savoy, D.H. Cohen, L.K. Townsley, P.S. Broos, M.S. Povich, M.F. Corcoran, N.R. Walborn, N. Remage Evans, A.F.J. Moffat, Y. Nazé, L.M. Oskinova, Ap. J. Suppl. **194**, 5 (2011). DOI 10.1088/0067-0049/194/1/5
1119. Y. Nazé, P.S. Broos, L. Oskinova, L.K. Townsley, D. Cohen, M.F. Corcoran, N.R. Evans, M. Gagné, A.F.J. Moffat, J.M. Pittard, G. Rauw, A. ud-Doula, N.R. Walborn, Ap. J. Suppl. **194**, 7 (2011). DOI 10.1088/0067-0049/194/1/7
1120. L.K. Townsley, P.S. Broos, Y.H. Chu, R.A. Gruendl, M.S. Oey, J.M. Pittard, Ap. J. Suppl. **194**, 16 (2011). DOI 10.1088/0067-0049/194/1/16
1121. H. Ungerechts, P. Thaddeus, Ap. J. Suppl. **63**, 645 (1987). DOI 10.1086/191176
1122. F. Palla, S.W. Stahler, Ap. J. **581**, 1194 (2002). DOI 10.1086/344293
1123. J. Scalo, in *Physical Processes in Fragmentation and Star Formation, Astrophysics and Space Science Library*, vol. 162, ed. by R. Capuzzo-Dolcetta, C. Chiosi, & A. di Fazio (1990), *Astrophysics and Space Science Library*, vol. 162, pp. 151–176
1124. E.N. Parker, Ap. J. **145**, 811 (1966). DOI 10.1086/148828
1125. A. Gomez de Castro, R.E. Pudritz, Ap. J. **395**, 501 (1992). DOI 10.1086/171671
1126. J. Ballesteros-Paredes, L. Hartmann, E. Vázquez-Semadeni, Ap. J. **527**, 285 (1999). DOI 10.1086/308076
1127. M. Gomez, L. Hartmann, S.J. Kenyon, R. Hewett, A. J. **105**, 1927 (1993). DOI 10.1086/116567
1128. C. Briceño, N. Calvet, S. Kenyon, L. Hartmann, A. J. **118**, 1354 (1999). DOI 10.1086/300997
1129. J. Jijina, P.C. Myers, F.C. Adams, Ap. J. Suppl. **125**, 161 (1999). DOI 10.1086/313268
1130. R. Wichmann, G. Torres, C.H.F. Melo, S. Frink, S. Allain, J. Bouvier, J. Krautter, E. Covino, R. Neuhäuser, Astron. Astrophys. **359**, 181 (2000)
1131. C. Briceno, L.W. Hartmann, J.R. Stauffer, M. Gagne, R.A. Stern, J.P. Caillault, A. J. **113**, 740 (1997). DOI 10.1086/118293
1132. G.H. Herbig, *Can Post-T Tauri Stars Be Found?* (1978), pp. 171–+
1133. G.H. Herbig, F.J. Vrba, A.E. Rydgren, A. J. **91**, 575 (1986). DOI 10.1086/114039
1134. L. Hartmann, J.R. Stauffer, S.J. Kenyon, B.F. Jones, A. J. **101**, 1050 (1991). DOI 10.1086/115747
1135. J. Ballesteros-Paredes, A.P.S.S. **292**, 193 (2004). DOI 10.1023/B:ASTR.0000045018.50154.b0
1136. L. Hartmann, Ap. J. **578**, 914 (2002). DOI 10.1086/342657
1137. T. Onishi, A. Mizuno, A. Kawamura, H. Ogawa, Y. Fukui, Ap. J. **502**, 296 (1998). DOI 10.1086/305867
1138. F.M. Walter, J.M. Alcala, R. Neuhauser, M. Sterzik, S.J. Wolk, Protostars and Planets IV pp. 273–+ (2000)
1139. C. Leitherer, in *The Stellar Initial Mass Function (38th Herstmonceux Conference), Astronomical Society of the Pacific Conference Series*, vol. 142, ed. by G. Gilmore & D. Howell (1998), *Astronomical Society of the Pacific Conference Series*, vol. 142, pp. 61–+
1140. P.T. de Zeeuw, R. Hoogerwerf, J.H.J. de Bruijne, A.G.A. Brown, A. Blaauw, A. J. **117**, 354 (1999). DOI 10.1086/300682
1141. J.A. Eisner, J.M. Carpenter, Ap. J. **598**, 1341 (2003). DOI 10.1086/379102
1142. P.J. Armitage, C.J. Clarke, F. Palla, M.N.R.A.S. **342**, 1139 (2003). DOI 10.1046/j.1365-8711.2003.06604.x
1143. S.S. Russell, G. Srinivasan, G.R. Huss, G.J. Wasserburg, G.J. MacPherson, Science **273**, 757 (1996). DOI 10.1126/science.273.5276.757
1144. M.F. Skrutskie, D. Dutkevitch, S.E. Strom, S. Edwards, K.M. Strom, M.A. Shure, A. J. **99**, 1187 (1990). DOI 10.1086/115407
1145. M. Simon, L. Prato, Ap. J. **450**, 824 (1995). DOI 10.1086/176187
1146. S.J. Wolk, F.M. Walter, A. J. **111**, 2066 (1996). DOI 10.1086/117942

1147. J.S. Bary, D.A. Weintraub, J.H. Kastner, Ap. J. Lettersf **576**, L73 (2002). DOI 10.1086/343064
1148. J.S. Bary, D.A. Weintraub, J.H. Kastner, Ap. J. **586**, 1136 (2003). DOI 10.1086/367719
1149. S.E. Strom, S. Edwards, M.F. Skrutskie, in *Protostars and Planets III*, ed. by E. H. Levy & J. I. Lunine (1993), pp. 837–866
1150. J.J. Hester, R. Gilmozzi, C.R. O'dell, S.M. Faber, B. Campbell, A. Code, D.G. Currie, G.E. Danielson, S.P. Ewald, E.J. Groth, J.A. Holtzman, T. Kelsall, T.R. Lauer, R.M. Light, R. Lynds, E.J. O'Neil, Jr., E.J. Shaya, J.A. Westphal, Ap. J. Lettersf **369**, L75 (1991). DOI 10.1086/185962
1151. C.R. O'dell, Science **259**, 33 (1993)
1152. P. Laques, J.L. Vidal, Astron. Astrophys. **73**, 97 (1979)
1153. W.A. Feibelman, P.A.S.P. **101**, 547 (1989). DOI 10.1086/132466
1154. E. Churchwell, M. Felli, D.O.S. Wood, M. Massi, Ap. J. **321**, 516 (1987). DOI 10.1086/165648
1155. C.R. O'dell, Z. Wen, Ap. J. **436**, 194 (1994). DOI 10.1086/174892
1156. C.R. O'dell, K. Wong, A. J. **111**, 846 (1996). DOI 10.1086/117832
1157. M.J. McCaughrean, C.R. O'dell, A. J. **111**, 1977 (1996). DOI 10.1086/117934
1158. D. Johnstone, D. Hollenbach, J. Bally, Ap. J. **499**, 758 (1998). DOI 10.1086/305658
1159. J. Bally, R.S. Sutherland, D. Devine, D. Johnstone, A. J. **116**, 293 (1998). DOI 10.1086/300399
1160. C.R. O'Dell, A. J. **122**, 2662 (2001). DOI 10.1086/323720
1161. B. Lefloch, J. Cernicharo, D. Cesarsky, K. Demyk, L.F. Rodriguez, Astron. Astrophys. **368**, L13 (2001). DOI 10.1051/0004-6361:20010209
1162. M.J. McCaughrean, J.R. Stauffer, A. J. **108**, 1382 (1994). DOI 10.1086/117160
1163. S. Richling, H.W. Yorke, Communications of the Konkoly Observatory Hungary **103**, 103 (2003)
1164. C.P. Dullemond, A. Natta, Astron. Astrophys. **408**, 161 (2003). DOI 10.1051/0004-6361:20030734
1165. S.J. Kenyon, B.C. Bromley, Ap. J. Lettersf **577**, L35 (2002). DOI 10.1086/344084
1166. S.J. Kenyon, B.C. Bromley, A. J. **127**, 513 (2004). DOI 10.1086/379854
1167. R. Jayawardhana, R.S. Fisher, L. Hartmann, C. Telesco, R. Pina, G. Fazio, Ap. J. Lettersf **503**, L79+ (1998). DOI 10.1086/311524
1168. G. Schneider, B.A. Smith, E.E. Becklin, D.W. Koerner, R. Meier, D.C. Hines, P.J. Lowrance, R.J. Terrile, R.I. Thompson, M. Rieke, Ap. J. Lettersf **513**, L127 (1999). DOI 10.1086/311921
1169. R.A. Webb, B. Zuckerman, J.S. Greaves, W.S. Holland, in *American Astronomical Society Meeting Abstracts*, *Bulletin of the American Astronomical Society*, vol. 197 (2000), *Bulletin of the American Astronomical Society*, vol. 197, pp. 827–+
1170. B. Zuckerman, A.R.A.A. **39**, 549 (2001). DOI 10.1146/annurev.astro.39.1.549
1171. W. Herbst, C.M. Hamilton, F.J. Vrba, M.A. Ibrahimov, C.A.L. Bailer-Jones, R. Mundt, M. Lamm, T. Mazeh, Z.T. Webster, K.E. Haisch, E.C. Williams, A.H. Rhodes, T.J. Balonek, A. Scholz, A. Riffeser, P.A.S.P. **114**, 1167 (2002). DOI 10.1086/344205
1172. J.N. Winn, P.M. Garnavich, K.Z. Stanek, D.D. Sasselov, Ap. J. Lettersf **593**, L121 (2003). DOI 10.1086/378314
1173. R.D. Mathieu, A.M. Ghez, E.L.N. Jensen, M. Simon, Protostars and Planets IV pp. 703–+ (2000)
1174. C. Spangler, A.I. Sargent, M.D. Silverstone, E.E. Becklin, B. Zuckerman, Ap. J. **555**, 932 (2001). DOI 10.1086/321490
1175. H.J. Habing, C. Dominik, M. Jourdain de Muizon, R.J. Laureijs, M.F. Kessler, K. Leech, L. Metcalfe, A. Salama, R. Siebenmorgen, N. Trams, P. Bouchet, Astron. Astrophys. **365**, 545 (2001). DOI 10.1051/0004-6361:20000075
1176. M.C. Wyatt, Ap. J. **598**, 1321 (2003). DOI 10.1086/379064
1177. A.J. Weinberger, E.E. Becklin, B. Zuckerman, Ap. J. Lettersf **584**, L33 (2003). DOI 10.1086/368065

1178. H.H. Aumann, C.A. Beichman, F.C. Gillett, T. de Jong, J.R. Houck, F.J. Low, G. Neugebauer, R.G. Walker, P.R. Wesselius, Ap. J. Lettersf **278**, L23 (1984). DOI 10.1086/184214
1179. D.E. Backman, F. Paresce, in *Protostars and Planets III*, ed. by E. H. Levy & J. I. Lunine (1993), pp. 1253–1304
1180. P. Kalas, J. Larwood, B.A. Smith, A. Schultz, Ap. J. Lettersf **530**, L133 (2000). DOI 10.1086/312494
1181. Z. Wahhaj, D.W. Koerner, M.E. Ressler, M.W. Werner, D.E. Backman, A.I. Sargent, Ap. J. Lettersf **584**, L27 (2003). DOI 10.1086/346123
1182. A. Brandeker, R. Liseau, G. Olofsson, M. Fridlund, Astron. Astrophys. **413**, 681 (2004). DOI 10.1051/0004-6361:20034326
1183. D.W. Koerner, A.I. Sargent, N.A. Ostroff, Ap. J. Lettersf **560**, L181 (2001). DOI 10.1086/324226
1184. D.J. Wilner, M.J. Holman, M.J. Kuchner, P.T.P. Ho, Ap. J. Lettersf **569**, L115 (2002). DOI 10.1086/340691
1185. J.S. Greaves, W.S. Holland, G. Moriarty-Schieven, T. Jenness, W.R.F. Dent, B. Zuckerman, C. McCarthy, R.A. Webb, H.M. Butner, W.K. Gear, H.J. Walker, Ap. J. Lettersf **506**, L133 (1998). DOI 10.1086/311652
1186. A. Natta, V. Grinin, V. Mannings, Protostars and Planets IV pp. 559–+ (2000)
1187. R. Millan-Gabet, F.P. Schloerb, W.A. Traub, Ap. J. **546**, 358 (2001). DOI 10.1086/318239
1188. A. Natta, T. Prusti, R. Neri, D. Wooden, V.P. Grinin, V. Mannings, Astron. Astrophys. **371**, 186 (2001). DOI 10.1051/0004-6361:20010334
1189. C.P. Dullemond, C. Dominik, A. Natta, Ap. J. **560**, 957 (2001). DOI 10.1086/323057
1190. A. Natta, F. Palla, H.M. Butner, N.J. Evans, II, P.M. Harvey, Ap. J. **406**, 674 (1993). DOI 10.1086/172478
1191. D.H. Wooden, in *The Nature and Evolutionary Status of Herbig Ae/Be Stars, Astronomical Society of the Pacific Conference Series*, vol. 62, ed. by P. S. The, M. R. Perez, & E. P. J. van den Heuvel (1994), *Astronomical Society of the Pacific Conference Series*, vol. 62, pp. 138–+
1192. A. Fuente, J. Martin-Pintado, R. Bachiller, R. Neri, F. Palla, Astron. Astrophys. **334**, 253 (1998)
1193. F.C. Adams, G. Laughlin, Icarus **150**, 151 (2001). DOI 10.1006/icar.2000.6567
1194. D. Breitschwerdt, M.J. Freyberg, J. Truemper (eds.). *The Local Bubble and Beyond, Lecture Notes in Physics, Berlin Springer Verlag*, vol. 506 (1998). DOI 10.1007/BFb0104684
1195. R. Egger, B. Aschenbach, in *Roentgenstrahlung from the Universe*, ed. by H. U. Zimmermann, J. Trümper, & H. Yorke (1996), pp. 249–250
1196. I.A. Grenier, Astron. Astrophys. **364**, L93 (2000)
1197. L. Spezzi, P. Vernazza, B. Merín, L.E. Allen, N.J. Evans, II, J.K. Jørgensen, T.L. Bourke, L.A. Cieza, M.M. Dunham, P.M. Harvey, T.L. Huard, D. Peterson, N.F.H. Tothill, The Gould's Belt Team, Ap. J. **730**, 65 (2011). DOI 10.1088/0004-637X/730/2/65
1198. P. Artymowicz, Annual Review of Earth and Planetary Sciences **25**, 175 (1997). DOI 10.1146/annurev.earth.25.1.175
1199. A.M. Lagrange, D.E. Backman, P. Artymowicz, Protostars and Planets IV pp. 639–+ (2000)
1200. S.V.W. Beckwith, T. Henning, Y. Nakagawa, Protostars and Planets IV pp. 533–+ (2000)
1201. S.J. Weidenschilling, *Formation processes and time scales for meteorite parent bodies* (1988), pp. 348–371
1202. J.J. Lissauer, G.R. Stewart, in *Protostars and Planets III*, ed. by E. H. Levy & J. I. Lunine (1993), pp. 1061–1088
1203. S.J. Weidenschilling, J.N. Cuzzi, in *Protostars and Planets III*, ed. by E. H. Levy & J. I. Lunine (1993), pp. 1031–1060
1204. J.B. Pollack, D. Hollenbach, S. Beckwith, D.P. Simonelli, T. Roush, W. Fong, Ap. J. **421**, 615 (1994). DOI 10.1086/173677
1205. P. Meakin, B. Donn, Ap. J. Lettersf **329**, L39 (1988). DOI 10.1086/185172
1206. H. Nakano, A. Kouchi, S. Tachibana, A. Tsuchiyama, Ap. J. **592**, 1252 (2003). DOI 10.1086/375856

1207. L. Grossman, J.W. Larimer, Reviews of Geophysics and Space Physics **12**, 71 (1974)
1208. K.R. Bell, P.M. Cassen, J.T. Wasson, D.S. Woolum, Protostars and Planets IV pp. 897–+ (2000)
1209. R.H. Jones, T. Lee, H.C. Connolly, Jr., S.G. Love, H. Shang, Protostars and Planets IV pp. 927–+ (2000)
1210. W.A. Fowler, J.L. Greenstein, F. Hoyle, Geophysical Journal International **6**, 148 (1962). DOI 10.1111/j.1365-246X.1962.tb02977.x
1211. J.N. Goswami, H.A.T. Vanhala, Protostars and Planets IV pp. 963–+ (2000)
1212. A.G.W. Cameron, in *Protostars and Planets III*, ed. by E. H. Levy & J. I. Lunine (1993), pp. 47–73
1213. G.J. Wasserburg, R. Gallino, M. Busso, J.N. Goswami, C.M. Raiteri, Ap. J. Lettersf **440**, L101 (1995). DOI 10.1086/187771
1214. M. Arnould, G. Paulus, G. Meynet, Astron. Astrophys. **321**, 452 (1997)
1215. A.G.W. Cameron, Meteoritics **30**, 133 (1995)
1216. D.D. Clayton, L. Jin, Ap. J. Lettersf **451**, L87+ (1995). DOI 10.1086/309684
1217. T. Lee, F.H. Shu, H. Shang, A.E. Glassgold, K.E. Rehm, Ap. J. **506**, 898 (1998). DOI 10.1086/306284
1218. F.H. Shu, H. Shang, M. Gounelle, A.E. Glassgold, T. Lee, Ap. J. **548**, 1029 (2001). DOI 10.1086/319018
1219. K.D. McKeegan, M. Chaussidon, F. Robert, in *Lunar and Planetary Institute Science Conference Abstracts*, *Lunar and Planetary Inst. Technical Report*, vol. 31 (2000), *Lunar and Planetary Inst. Technical Report*, vol. 31, pp. 1999–+
1220. S.J. Desch, in *Lunar and Planetary Institute Science Conference Abstracts*, *Lunar and Planetary Institute Science Conference Abstracts*, vol. 30 (1999), *Lunar and Planetary Institute Science Conference Abstracts*, vol. 30, pp. 1962–+
1221. G. Morfill, H. Spruit, E.H. Levy, in *Protostars and Planets III*, ed. by E. H. Levy & J. I. Lunine (1993), pp. 939–978
1222. J.A. Wood, Nature **194**, 127 (1962). DOI 10.1038/194127a0
1223. S.J. Weidenschilling, F. Marzari, L.L. Hood, Science **279**, 681 (1998). DOI 10.1126/science.279.5351.681
1224. E.D. Feigelson, G.P. Garmire, S.H. Pravdo, Ap. J. **572**, 335 (2002). DOI 10.1086/340340
1225. P. Cassen, in *Saas-Fee Advanced Course 31: Extrasolar planets*, ed. by D. Queloz, S. Udry, M. Mayor, W. Benz, P. Cassen, T. Guillot, & A. Quirrenbach (2006), pp. 369–448. DOI 10.1007/978-3-540-31470-73
1226. M. Ollivier, F. Roques, F. Casoli, T. Encrenaz, F. Selsis, *Planetary Systems* (2009). DOI 10.1007/978-3-540-75748-1
1227. P.J. Armitage, *Astrophysics of Planet Formation* (2010) Cambridge University Press
1228. S. Seager & D. Deming, A.R.A.A. **48**, 631 (2010). DOI 10.1146/annurev-astro-081309-130837
1229. R.P. Butler, G.W. Marcy, D.A. Fischer, T.M. Brown, A.R. Contos, S.G. Korzennik, P. Nisenson, R.W. Noyes, Ap. J. **526**, 916 (1999). DOI 10.1086/308035
1230. A. Quirrenbach, in *Saas-Fee Advanced Course 31: Extrasolar planets*, ed. by D. Queloz, S. Udry, M. Mayor, W. Benz, P. Cassen, T. Guillot, & A. Quirrenbach (2006), pp. 1–242. DOI 10.1007/978-3-540-31470-71
1231. J.N. Winn, J.A. Johnson, G.W. Marcy, R.P. Butler, S.S. Vogt, G.W. Henry, A. Roussanova, M.J. Holman, K. Enya, N. Narita, Y. Suto, E.L. Turner, Ap. J. Lettersf **653**, L69 (2006). DOI 10.1086/510528
1232. I.A. Bond, A. Udalski, M. Jaroszyński, N.J. Rattenbury, B. Paczyński, I. Soszyński, L. Wyrzykowski, M.K. Szymański, M. Kubiak, O. Szewczyk, K. Żebruń, G. Pietrzyński, F. Abe, D.P. Bennett, S. Eguchi, Y. Furuta, J.B. Hearnshaw, K. Kamiya, P.M. Kilmartin, Y. Kurata, K. Masuda, Y. Matsubara, Y. Muraki, S. Noda, K. Okajima, T. Sako, T. Sekiguchi, D.J. Sullivan, T. Sumi, P.J. Tristram, T. Yanagisawa, P.C.M. Yock, Ap. J. Lettersf **606**, L155 (2004). DOI 10.1086/420928
1233. A. Udalski, Act.A.A. **53**, 291 (2003)

1234. I.A. Bond, F. Abe, R.J. Dodd, J.B. Hearnshaw, M. Honda, J. Jugaku, P.M. Kilmartin, A. Marles, K. Masuda, Y. Matsubara, Y. Muraki, T. Nakamura, G. Nankivell, S. Noda, C. Noguchi, K. Ohnishi, N.J. Rattenbury, M. Reid, T. Saito, H. Sato, M. Sekiguchi, J. Skuljan, D.J. Sullivan, T. Sumi, M. Takeuti, Y. Watase, S. Wilkinson, R. Yamada, T. Yanagisawa, P.C.M. Yock, M.N.R.A.S. **327**, 868 (2001). DOI 10.1046/j.1365-8711.2001.04776.x
1235. J.N. Winn, J.A. Johnson, A.W. Howard, G.W. Marcy, G.Á. Bakos, J. Hartman, G. Torres, S. Albrecht, N. Narita, Ap. J. **718**, 575 (2010). DOI 10.1088/0004-637X/718/1/575
1236. R.A. Rossiter, Ap. J. **60**, 15 (1924). DOI 10.1086/142825
1237. D.B. McLaughlin, Ap. J. **60**, 22 (1924). DOI 10.1086/142826
1238. D. Queloz, A. Eggenberger, M. Mayor, C. Perrier, J.L. Beuzit, D. Naef, J.P. Sivan, S. Udry, Astron. Astrophys. **359**, L13 (2000)
1239. J.N. Winn, J.A. Johnson, K.M.G. Peek, G.W. Marcy, G.Á. Bakos, K. Enya, N. Narita, Y. Suto, E.L. Turner, S.S. Vogt, Ap. J. Lettersf **665**, L167 (2007). DOI 10.1086/521362
1240. A.S. Wolf, G. Laughlin, G.W. Henry, D.A. Fischer, G. Marcy, P. Butler, S. Vogt, Ap. J. **667**, 549 (2007). DOI 10.1086/503354
1241. J.N. Winn, J.A. Johnson, D. Fabrycky, A.W. Howard, G.W. Marcy, N. Narita, I.J. Crossfield, Y. Suto, E.L. Turner, G. Esquerdo, M.J. Holman, Ap. J. **700**, 302 (2009). DOI 10.1088/0004-637X/700/1/302
1242. J.N. Winn, J.A. Johnson, S. Albrecht, A.W. Howard, G.W. Marcy, I.J. Crossfield, M.J. Holman, Ap. J. Lettersf **703**, L99 (2009). DOI 10.1088/0004-637X/703/2/L99
1243. G.W. Henry, Ap. J. Lettersf **536**, L47 (2000). DOI 10.1086/312726
1244. D. Charbonneau, T.M. Brown, D.W. Latham, M. Mayor, Ap. J. Lettersf **529**, L45 (2000). DOI 10.1086/312457
1245. M. Mayor, D. Queloz, Nature **378**, 355 (1995). DOI 10.1038/378355a0
1246. T. Mazeh, S. Zucker, A. dalla Torre, F. van Leeuwen, Ap. J. Lettersf **522**, L149 (1999). DOI 10.1086/312231
1247. A.W. Howard, G.W. Marcy, S.T. Bryson, J.M. Jenkins, J.F. Rowe, N.M. Batalha, W.J. Borucki, D.G. Koch, E.W. Dunham, T.N. Gautier, III, J. Van Cleve, W.D. Cochran, D.W. Latham, J.J. Lissauer, G. Torres, T.M. Brown, R.L. Gilliland, L.A. Buchhave, D.A. Caldwell, J. Christensen-Dalsgaard, D. Ciardi, F. Fressin, M.R. Haas, S.B. Howell, H. Kjeldsen, S. Seager, L. Rogers, D.D. Sasselov, J.H. Steffen, G.S. Basri, D. Charbonneau, J. Christiansen, B. Clarke, A. Dupree, D.C. Fabrycky, D.A. Fischer, E.B. Ford, J.J. Fortney, J. Tarter, F.R. Girouard, M.J. Holman, J.A. Johnson, T.C. Klaus, P. Machalek, A.V. Moorhead, R.C. Morehead, D. Ragozzine, P. Tenenbaum, J.D. Twicken, S.N. Quinn, H. Isaacson, A. Shporer, P.W. Lucas, L.M. Walkowicz, W.F. Welsh, A. Boss, E. Devore, A. Gould, J.C. Smith, R.L. Morris, A. Prsa, T.D. Morton, ArXiv e-prints (2011)
1248. W.J. Borucki, D. Koch, G. Basri, N. Batalha, T. Brown, D. Caldwell, J. Caldwell, J. Christensen-Dalsgaard, W.D. Cochran, E. DeVore, E.W. Dunham, A.K. Dupree, T.N. Gautier, J.C. Geary, R. Gilliland, A. Gould, S.B. Howell, J.M. Jenkins, Y. Kondo, D.W. Latham, G.W. Marcy, S. Meibom, H. Kjeldsen, J.J. Lissauer, D.G. Monet, D. Morrison, D. Sasselov, J. Tarter, A. Boss, D. Brownlee, T. Owen, D. Buzasi, D. Charbonneau, L. Doyle, J. Fortney, E.B. Ford, M.J. Holman, S. Seager, J.H. Steffen, W.F. Welsh, J. Rowe, H. Anderson, L. Buchhave, D. Ciardi, L. Walkowicz, W. Sherry, E. Horch, H. Isaacson, M.E. Everett, D. Fischer, G. Torres, J.A. Johnson, M. Endl, P. MacQueen, S.T. Bryson, J. Dotson, M. Haas, J. Kolodziejczak, J. Van Cleve, H. Chandrasekaran, J.D. Twicken, E.V. Quintana, B.D. Clarke, C. Allen, J. Li, H. Wu, P. Tenenbaum, E. Verner, F. Bruhweiler, J. Barnes, A. Prsa, Science **327**, 977 (2010). DOI 10.1126/science.1185402
1249. J.J. Lissauer, D.C. Fabrycky, E.B. Ford, W.J. Borucki, F. Fressin, G.W. Marcy, J.A. Orosz, J.F. Rowe, G. Torres, W.F. Welsh, N.M. Batalha, S.T. Bryson, L.A. Buchhave, D.A. Caldwell, J.A. Carter, D. Charbonneau, J.L. Christiansen, W.D. Cochran, J.M. Desert, E.W. Dunham, M.N. Fanelli, J.J. Fortney, T.N. Gautier, III, J.C. Geary, R.L. Gilliland, M.R. Haas, J.R. Hall, M.J. Holman, D.G. Koch, D.W. Latham, E. Lopez, S. McCauliff, N. Miller, R.C. Morehead, E.V. Quintana, D. Ragozzine, D. Sasselov, D.R. Short, J.H. Steffen, Nature **470**, 53 (2011). DOI 10.1038/nature09760

1250. W.J. Borucki, D.G. Koch, G. Basri, N. Batalha, T.M. Brown, S.T. Bryson, D. Caldwell, J. Christensen-Dalsgaard, W.D. Cochran, E. DeVore, E.W. Dunham, T.N. Gautier, III, J.C. Geary, R. Gilliland, A. Gould, S.B. Howell, J.M. Jenkins, D.W. Latham, J.J. Lissauer, G.W. Marcy, J. Rowe, D. Sasselov, A. Boss, D. Charbonneau, D. Ciardi, L. Doyle, A.K. Dupree, E.B. Ford, J. Fortney, M.J. Holman, S. Seager, J.H. Steffen, J. Tarter, W.F. Welsh, C. Allen, L.A. Buchhave, J.L. Christiansen, B.D. Clarke, J.M. Désert, M. Endl, D. Fabrycky, F. Fressin, M. Haas, E. Horch, A. Howard, H. Isaacson, H. Kjeldsen, J. Kolodziejczak, C. Kulesa, J. Li, P. Machalek, D. McCarthy, P. MacQueen, S. Meibom, T. Miquel, A. Prsa, S.N. Quinn, E.V. Quintana, D. Ragozzine, W. Sherry, A. Shporer, P. Tenenbaum, G. Torres, J.D. Twicken, J. Van Cleve, L. Walkowicz, ArXiv e-prints (2011)
1251. A.W. Howard, G.W. Marcy, J.A. Johnson, D.A. Fischer, J.T. Wright, H. Isaacson, J.A. Valenti, J. Anderson, D.N.C. Lin, S. Ida, Science **330**, 653 (2010). DOI 10.1126/science.1194854
1252. J.J. Fortney, M.S. Marley, J.W. Barnes, Ap. J. **659**, 1661 (2007). DOI 10.1086/512120
1253. S. Wolf, G. D'Angelo, Ap. J. **619**, 1114 (2005). DOI 10.1086/426662
1254. D. Lafrenière, R. Jayawardhana, M.H. van Kerkwijk, Ap. J. Lettersf **689**, L153 (2008). DOI 10.1086/595870
1255. B. Zuckerman, I. Song, Astron. Astrophys. **493**, 1149 (2009). DOI 10.1051/0004-6361:200810038
1256. T. Preibisch, A.G.A. Brown, T. Bridges, E. Guenther, H. Zinnecker, A. J. **124**, 404 (2002). DOI 10.1086/341174
1257. D. Lafrenière, R. Jayawardhana, M.H. van Kerkwijk, Ap. J. **719**, 497 (2010). DOI 10.1088/0004-637X/719/1/497
1258. J.A. Johnson, A.W. Howard, B.P. Bowler, G.W. Henry, G.W. Marcy, J.T. Wright, D.A. Fischer, H. Isaacson, P.A.S.P. **122**, 701 (2010). DOI 10.1086/653809
1259. C. Marois, B. Macintosh, T. Barman, B. Zuckerman, I. Song, J. Patience, D. Lafrenière, R. Doyon, Science **322**, 1348 (2008). DOI 10.1126/science.1166585
1260. I. Baraffe, G. Chabrier, T.S. Barman, F. Allard, P.H. Hauschildt, Astron. Astrophys. **402**, 701 (2003). DOI 10.1051/0004-6361:20030252
1261. P. Kalas, J.R. Graham, E. Chiang, M.P. Fitzgerald, M. Clampin, E.S. Kite, K. Stapelfeldt, C. Marois, J. Krist, Science **322**, 1345 (2008). DOI 10.1126/science.1166609
1262. C. Dominik, J. Blum, J.N. Cuzzi, G. Wurm, Protostars and Planets V pp. 783–800 (2007)
1263. C.P. Dullemond, C. Dominik, Astron. Astrophys. **421**, 1075 (2004). DOI 10.1051/0004-6361:20040284
1264. S. Fromang, J. Papaloizou, Astron. Astrophys. **452**, 751 (2006). DOI 10.1051/0004-6361:20054612
1265. H.B. Throop, J. Bally, Ap. J. Lettersf **623**, L149 (2005). DOI 10.1086/430272
1266. F.H. Shu, D. Johnstone, D. Hollenbach, Icarus **106**, 92 (1993). DOI 10.1006/icar.1993.1160
1267. F.C. Adams, D. Hollenbach, G. Laughlin, U. Gorti, Ap. J. **611**, 360 (2004). DOI 10.1086/421989
1268. C. Hayashi, Progress of Theoretical Physics Supplement **70**, 35 (1981). DOI 10.1143/PTPS.70.35
1269. G.M. Kennedy, S.J. Kenyon, Ap. J. **695**, 1210 (2009). DOI 10.1088/0004-637X/695/2/1210
1270. K. Lodders, Ap. J. **591**, 1220 (2003). DOI 10.1086/375492
1271. W. Benz, E. Asphaug, Icarus **142**, 5 (1999). DOI 10.1006/icar.1999.6204
1272. Z.M. Leinhardt, S.T. Stewart, Icarus **199**, 542 (2009). DOI 10.1016/j.icarus.2008.09.013
1273. J.M. Petit, M. Henon, Icarus **66**, 536 (1986). DOI 10.1016/0019-1035(86)90089-8
1274. J.J. Lissauer, S.J. Wolk, C.A. Griffith, D.E. Backman, Ap. J. **465**, 371 (1996). DOI 10.1086/177426
1275. P. Goldreich, Y. Lithwick, R. Sari, Ap. J. **614**, 497 (2004). DOI 10.1086/423612
1276. H. Liu, J.L. Zhou, S. Wang, Ap. J. **732**, 66 (2011). DOI 10.1088/0004-637X/732/2/66
1277. S.N. Raymond, T. Quinn, J.I. Lunine, Icarus **183**, 265 (2006). DOI 10.1016/j.icarus.2006.03.011
1278. A.M. Mandell, S.N. Raymond, S. Sigurdsson, Ap. J. **660**, 823 (2007). DOI 10.1086/512759

1279. J.B. Pollack, O. Hubickyj, P. Bodenheimer, J.J. Lissauer, M. Podolak, Y. Greenzweig, Icarus **124**, 62 (1996). DOI 10.1006/icar.1996.0190
1280. W.K.M. Rice, P.J. Armitage, Ap. J. Lettersf **598**, L55 (2003). DOI 10.1086/380390
1281. J.J. Lissauer, A.R.A.A. **31**, 129 (1993). DOI 10.1146/annurev.aa.31.090193.001021
1282. E. Kokubo, S. Ida, Icarus **143**, 15 (2000). DOI 10.1006/icar.1999.6237
1283. M. Ikoma, K. Nakazawa, H. Emori, Ap. J. **537**, 1013 (2000). DOI 10.1086/309050
1284. Y. Alibert, C. Mordasini, W. Benz, C. Winisdoerffer, Astron. Astrophys. **434**, 343 (2005). DOI 10.1051/0004-6361:20042032
1285. H. Tanaka, S. Ida, Icarus **139**, 350 (1999). DOI 10.1006/icar.1999.6107
1286. M.H. Lee, E.W. Thommes, Ap. J. **702**, 1662 (2009). DOI 10.1088/0004-637X/702/2/1662
1287. K. Tsiganis, R. Gomes, A. Morbidelli, H.F. Levison, Nature **435**, 459 (2005). DOI 10.1038/nature03539
1288. S. Ida, R.M. Canup, G.R. Stewart, Nature **389**, 353 (1997). DOI 10.1038/38669
1289. E. Kokubo, S. Ida, J. Makino, Icarus **148**, 419 (2000). DOI 10.1006/icar.2000.6496
1290. A.E. Saal, E.H. Hauri, M.L. Cascio, J.A. van Orman, M.C. Rutherford, R.F. Cooper, Nature **454**, 192 (2008). DOI 10.1038/nature07047
1291. D. Prialnik, *An Introduction to the Theory of Stellar Structure and Evolution* (2000)
1292. H. Bondi, M.N.R.A.S. **112**, 195 (1952)
1293. F.H. Shu, *The physics of astrophysics. Volume II: Gas dynamics.* (1992)
1294. E. Priest, T. Forbes, *Magnetic Reconnection* (2000)
1295. S.I. Braginskii, Reviews of Plasma Physics **1**, 205 (1965)
1296. J.V. Hollweg, Ap. J. **306**, 730 (1986). DOI 10.1086/164382
1297. P.A. Sweet, in *Electromagnetic Phenomena in Cosmical Physics, IAU Symposium*, vol. 6, ed. by B. Lehnert (1958), *IAU Symposium*, vol. 6, pp. 123–+
1298. E.N. Parker, Physical Review **107**, 830 (1957). DOI 10.1103/PhysRev.107.830
1299. E.N. Parker, Ap. J. Suppl. **8**, 177 (1963). DOI 10.1086/190087
1300. L. Spitzer, *Physics of Fully Ionized Gases* (1962)
1301. E.N. Parker, Ap. J. **122**, 293 (1955). DOI 10.1086/146087
1302. R.H. Kraichnan, Physical Review Letters **42**, 1677 (1979). DOI 10.1103/PhysRevLett.42.1677
1303. E.N. Parker, Ap. J. **408**, 707 (1993). DOI 10.1086/172631
1304. P. Charbonneau, K.B. MacGregor, Ap. J. **486**, 502 (1997). DOI 10.1086/304485
1305. J.A. Markiel, J.H. Thomas, Ap. J. **523**, 827 (1999). DOI 10.1086/307774
1306. Y.C. Kim, P. Demarque, Ap. J. **457**, 340 (1996). DOI 10.1086/176733
1307. B.R. Durney, D.S. De Young, I.W. Roxburgh, Solar Physics **145**, 207 (1993). DOI 10.1007/BF00690652
1308. M. Küker, G. Rüdiger, Astron. Astrophys. **346**, 922 (1999)
1309. J. Goodman, H. Ji, Journal of Fluid Mechanics **462**, 365 (2002). DOI 10.1017/S0022112002008704
1310. G.B. Rybicki, A.P. Lightman, *Radiative processes in astrophysics* (1979)
1311. A. Thorne, U. Litzen, S. Johansson, *Spectrophysics* (1999)
1312. D. Mihalas, *Stellar atmospheres /2nd edition/* (1978)
1313. M. Schwarzschild, *Structure and evolution of the stars.* (1958)
1314. D.R. Alexander, R.L. Rypma, H.R. Johnson, Ap. J. **272**, 773 (1983). DOI 10.1086/161339
1315. C. Helling, J.M. Winters, E. Sedlmayr, Astron. Astrophys. **358**, 651 (2000)
1316. P.H. Hauschildt, D.K. Lowenthal, E. Baron, Ap. J. Suppl. **134**, 323 (2001). DOI 10.1086/320855
1317. E. Herbst, in *The Diffuse Interstellar Bands*, *Astrophysics and Space Science Library*, vol. 202, ed. by A. G. G. M. Tielens & T. P. Snow (1995), *Astrophysics and Space Science Library*, vol. 202, pp. 307–+
1318. W.D. Langer, E.F. van Dishoeck, E.A. Bergin, G.A. Blake, A.G.G.M. Tielens, T. Velusamy, D.C.B. Whittet, Protostars and Planets IV pp. 29–+ (2000)
1319. P.S. Julienne, M. Krauss, B. Donn, Ap. J. **170**, 65 (1971). DOI 10.1086/151188
1320. E. Herbst, Ap. J. **241**, 197 (1980). DOI 10.1086/158331

1321. H.C. van de Hulst, *Light Scattering by Small Particles* (1957)
1322. C.F. Bohren, D.R. Huffman, *Absorption and scattering of light by small particles* (1983)
1323. T. Ueta, M. Meixner, Ap. J. **586**, 1338 (2003). DOI 10.1086/367818
1324. H.P. Gail, E. Sedlmayr, Astron. Astrophys. **206**, 153 (1988)
1325. S.H. Kim, P.G. Martin, P.D. Hendry, Ap. J. **422**, 164 (1994). DOI 10.1086/173714
1326. C. Helling, Role of molecular opacities in circumstellar dust shells. Ph.D. thesis, Institut für Astronomie und Astrophysik, TU Berlin, Hardenbergstrasse 36, D-10632 Berlin, Germany (1999)
1327. V. Ossenkopf, Astron. Astrophys. **280**, 617 (1993)
1328. R.L. Kurucz, Rev. Mex. Astron. Astrofis **23**, 181 (1992)
1329. R.L. Kurucz, in *IAU Colloq. 138: Peculiar versus Normal Phenomena in A-type and Related Stars*, *Astronomical Society of the Pacific Conference Series*, vol. 44, ed. by M. M. Dworetsky, F. Castelli, & R. Faraggiana (1993), *Astronomical Society of the Pacific Conference Series*, vol. 44, pp. 87–+
1330. P.H. Hauschildt, F. Allard, E. Baron, Ap. J. **512**, 377 (1999). DOI 10.1086/306745
1331. S.H. Saar, J.L. Linsky, Ap. J. Lettersf **299**, L47 (1985). DOI 10.1086/184578
1332. S.H. Saar, in *Infrared Solar Physics*, *IAU Symposium*, vol. 154, ed. by D. M. Rabin, J. T. Jefferies, & C. Lindsey (1994), *IAU Symposium*, vol. 154, pp. 493–+
1333. J. Charlton, C. Churchill, *Quasistellar Objects: Intervening Absorption Lines* (2000). DOI 10.1888/0333750888/2366
1334. N.W. Griffiths, C. Jordan, Ap. J. **497**, 883 (1998). DOI 10.1086/305502
1335. S.R. Pottasch, Ap. J. **137**, 945 (1963). DOI 10.1086/147569
1336. P.S. Wojdowski, N.S. Schulz, Ap. J. **616**, 630 (2004). DOI 10.1086/424904
1337. P.F. Winkler, G.W. Clark, T.H. Markert, K. Kalata, H.W. Schnopper, C.R. Canizares, Ap. J. Lettersf **246**, L27 (1981). DOI 10.1086/183546
1338. J.G. Doyle, J.L. Schwob, Journal of Physics B Atomic Molecular Physics **15**, 813 (1982). DOI 10.1088/0022-3700/15/5/023
1339. D.A. Liedahl, in *X-Ray Spectroscopy in Astrophysics*, *Lecture Notes in Physics, Berlin Springer Verlag*, vol. 520, ed. by J. van Paradijs & J. A. M. Bleeker (1999), *Lecture Notes in Physics, Berlin Springer Verlag*, vol. 520, pp. 189–+
1340. G.R. Blumenthal, G.W.F. Drake, W.H. Tucker, Ap. J. **172**, 205 (1972). DOI 10.1086/151340
1341. E.H. Hauri, T. Weinreich, A.E. Saal, M.C. Rutherford, J.A. Van Orman, Science Express **26 May 2011** DOI 10.1126/science.1204626
1342. T. Preibisch, S. Hodgkin, M. Irwin, R.J. Lewis, R.R. King, M.J. McCaughrean, H. Zinnecker, L. Townsley, P. Broos, Ap. J. Suppl. **194**, 10
1343. A. Dalgarno, R.A. McCray, A.R.A.A. **10**, 375 (1972)
1344. K.R. Lang, *Astrophysical formulae* (Springer, New York, 1999)

Index

N.S. Schulz, *The Formation and Early Evolution of Stars*, Astronomy and Astrophysics Library, DOI 10.1007/978-3-642-23926-7, © Springer-Verlag Berlin Heidelberg 2012

MIX
Papier aus verantwortungsvollen Quellen
Paper from responsible sources
FSC® C105338

If you have any concerns about our products,
you can contact us on
ProductSafety@springernature.com

In case Publisher is established outside the EU,
the EU authorized representative is:
Springer Nature Customer Service Center GmbH
Europaplatz 3, 69115 Heidelberg, Germany

Printed by Libri Plureos GmbH
in Hamburg, Germany